VOLUME SIX HUNDRED AND TWELVE

METHODS IN ENZYMOLOGY

High-Density Sequencing Applications in Microbial Molecular Genetics

METHODS IN ENZYMOLOGY

VOLUME SIX HUNDRED AND TWELVE

METHODS IN ENZYMOLOGY

High-Density Sequencing Applications in Microbial Molecular Genetics

Edited by

AGAMEMNON J. CARPOUSIS

LMGM, CBI, Université de Toulouse,
CNRS, Toulouse, France

ELSEVIER

ACADEMIC PRESS

An imprint of Elsevier

Academic Press is an imprint of Elsevier
50 Hampshire Street, 5th Floor, Cambridge, MA 02139, United States
525 B Street, Suite 1650, San Diego, CA 92101, United States
The Boulevard, Langford Lane, Kidlington, Oxford OX5 1GB, United Kingdom
125 London Wall, London, EC2Y 5AS, United Kingdom

First edition 2018

ISBN: 978-0-12-815993-4
ISSN: 0076-6879

For information on all Academic Press publications
visit our website at https://www.elsevier.com/books-and-journals

Publisher: Fiona Pattison
Acquisition Editor: Ashlie Jackman
Editorial Project Manager: Shellie Bryant
Production Project Manager: Divya Krishna Kumar
Cover Designer: Mark Rogers

Typeset by SPi Global, India

CONTENTS

CONTRIBUTORS

Nadra Al-Husini
Department of Biological Sciences, Wayne State University, Detroit, MI, United States

Liis Andresen
Department of Cell and Molecular Biology, Biomedical Center, Uppsala University, Uppsala, Sweden

James R. Aretakis
Department of Biological Sciences, Wayne State University, Detroit, MI, United States

Cecília M. Arraiano
Instituto de Tecnologia Química e Biológica António Xavier, Universidade Nova de Lisboa, Oeiras, Portugal

Amna Asif-Laidin
INSERM U944, CNRS UMR 7212, Université Paris Diderot, Sorbonne Paris Cité, Institut Universitaire d'Hématologie, Hôpital St. Louis, Paris Cedex 10, France

Benura Azeroglu
Institute of Cell Biology, School of Biological Sciences, University of Edinburgh, King's Buildings, Edinburgh, United Kingdom

Richard Baker
Department of Microbiology and Physiological Systems, University of Massachusetts Medical School, Worcester, MA, United States

Anastasia Barkova
INSERM U944, CNRS UMR 7212, Université Paris Diderot, Sorbonne Paris Cité, Institut Universitaire d'Hématologie, Hôpital St. Louis, Paris Cedex 10, France

Andreas J. Bäumler
Department of Medical Microbiology and Immunology, School of Medicine, University of California Davis, Davis, CA, United States

Frédéric Boccard
Institut de Biologie Intégrative de la Cellule, CEA, CNRS, Univ. Paris-Sud, Université Paris-Saclay, Gif-sur-Yvette cedex, France

Olivier Bouchez
Plateforme GeT-PlaGe-Genotoul, INRA Auzeville, Castanet-Tolosan, France

Marie Bouvier
LMGM, CBI, Université de Toulouse, CNRS, Toulouse, France

Denise N. Bronner
Department of Medical Microbiology and Immunology, School of Medicine, University of California Davis, Davis, CA, United States

Anne Buddeweg
Institut für allgemeine Mikrobiologie, Christian-Albrechts-Universität zu Kiel, Kiel, Germany

Mariana X. Byndloss
Department of Medical Microbiology and Immunology, School of Medicine, University of California Davis, Davis, CA, United States

Isabelle Caldelari
Architecture et Réactivité de l'ARN, Université de Strasbourg, IBMC-CNRS, Strasbourg, France

Agamemnon J. Carpousis
LMGM, CBI, Université de Toulouse, CNRS, Toulouse, France

Marie-Claude Carrier
Department of Biochemistry, RNA Group, Université de Sherbrooke, Sherbrooke, QC, Canada

Alper Celik
Department of Microbiology and Physiological Systems, University of Massachusetts Medical School, Worcester, MA, United States

Sandrine Chabas
ARNA Laboratory, INSERM U1212, CNRS UMR 5320, University of Bordeaux, Bordeaux, France

Muriel Cocaign-Bousquet
LISBP, Université de Toulouse, CNRS, INRA, INSA, Toulouse, France

Fabien Darfeuille
ARNA Laboratory, INSERM U1212, CNRS UMR 5320, University of Bordeaux, Bordeaux, France

Michael Daume
Max-Planck-Institut für terrestrische Mikrobiologie, Marburg, Germany

Emma Desgranges
Architecture et Réactivité de l'ARN, Université de Strasbourg, IBMC-CNRS, Strasbourg, France

Sandra Duharcourt
Institut Jacques Monod, CNRS, UMR7592, Sorbonne Paris Cité, Paris, France

Théo Foutel-Rodier
Institut Pasteur, Département Génomes et Génétique, Groupe Régulation Spatiale des Génomes; CNRS, UMR 3525, Paris Cedex 15; Sorbonne Université, Collège Doctoral; Institut Pasteur, Center of Bioinformatics, Biostatistics and Integrative Biology (C3BI), Paris, France

Lutz Frönicke
University of California Davis Genome Center, Davis, CA, United States

Christine Gaspin
MIAT, Université de Toulouse, INRA, Castanet-Tolosan, France

Kenn Gerdes
Department of Biology, Centre for Bacterial Stress Response and Persistence, University of Copenhagen, Copenhagen, Denmark

Laurence Girbal
LISBP, Université de Toulouse, CNRS, INRA, INSA, Toulouse, France

Seth R. Goldman
Waksman Institute and Department of Genetics, Rutgers University, Piscataway, NJ, United States

Ignacio González
MIAT, Université de Toulouse, INRA, Castanet-Tolosan, France

A.M. Mahedi Hasan
Institute of Cell Biology, School of Biological Sciences, University of Edinburgh, King's Buildings, Edinburgh, United Kingdom

Feng He
Department of Microbiology and Physiological Systems, University of Massachusetts Medical School, Worcester, MA, United States

Dong-Hyuk Heo
Department of Biochemistry, University of Oxford, Oxford, United Kingdom

Roman Herzog
Faculty of Biology I, Department of Microbiology, Ludwig-Maximilians-University of Munich, Munich, Germany

Erik Holmqvist
Department of Cell and Molecular Biology, Biomedical Center, Uppsala University, Uppsala, Sweden

Allan Jacobson
Department of Microbiology and Physiological Systems, University of Massachusetts Medical School, Worcester, MA, United States

Grace E. Johnson
Department of Biology, Massachusetts Institute of Technology, Cambridge, MA, United States

Tea Kecman
Department of Biochemistry, University of Oxford, Oxford, United Kingdom

Romain Koszul
Institut Pasteur, Département Génomes et Génétique, Groupe Régulation Spatiale des Génomes; CNRS, UMR 3525, Paris Cedex 15; Institut Pasteur, Center of Bioinformatics, Biostatistics and Integrative Biology (C3BI), Paris, France

Soumaya Laalami
CNRS UMR8261—Université Paris Diderot, Sorbonne Paris Cité, Institut de Biologie Physico-Chimique, Paris, France

Sandrine Laguerre
LISBP, Université de Toulouse, CNRS, INRA, INSA, Toulouse, France

David Lalaouna
Architecture et Réactivité de l'ARN, Université de Strasbourg, IBMC-CNRS, Strasbourg, France

David R.F. Leach
Institute of Cell Biology, School of Biological Sciences, University of Edinburgh, King's Buildings, Edinburgh, United Kingdom

Pascale Lesage
INSERM U944, CNRS UMR 7212, Université Paris Diderot, Sorbonne Paris Cité, Institut Universitaire d'Hématologie, Hôpital St. Louis, Paris Cedex 10, France

Gene-Wei Li
Department of Biology, Massachusetts Institute of Technology, Cambridge, MA, United States

Virginia S. Lioy
Institut de Biologie Intégrative de la Cellule, CEA, CNRS, Univ. Paris-Sud, Université Paris-Saclay, Gif-sur-Yvette cedex, France

Martial Marbouty
Institut Pasteur, Département Génomes et Génétique, Groupe Régulation Spatiale des Génomes; CNRS, UMR 3525, Paris Cedex 15; Institut Pasteur, Center of Bioinformatics, Biostatistics and Integrative Biology (C3BI), Paris, France

Stefano Marzi
Architecture et Réactivité de l'ARN, Université de Strasbourg, IBMC-CNRS, Strasbourg, France

Sara Masachis
ARNA Laboratory, INSERM U1212, CNRS UMR 5320, University of Bordeaux, Bordeaux, France

Eric Massé
Department of Biochemistry, RNA Group, Université de Sherbrooke, Sherbrooke, QC, Canada

Bridget McLaughlin
Comprehensive Cancer Center Flow Cytometry Shared Resource, University of California Davis, Davis, CA, United States

Annick Moisan
MIAT, Université de Toulouse, INRA, Castanet-Tolosan, France

Liliana Mora
CNRS UMR8261—Université Paris Diderot, Sorbonne Paris Cité, Institut de Biologie Physico-Chimique, Paris, France

Claire Morin
Department of Biochemistry, RNA Group, Université de Sherbrooke, Sherbrooke, QC, Canada

Saravuth Ngo
CNRS UMR8261—Université Paris Diderot, Sorbonne Paris Cité, Institut de Biologie Physico-Chimique, Paris, France

Bryce E. Nickels
Waksman Institute and Department of Genetics, Rutgers University, Piscataway, NJ, United States

Sébastien Nouaille
LISBP, Université de Toulouse, CNRS, INRA, INSA, Toulouse, France

Kai Papenfort
Faculty of Biology I, Department of Microbiology, Ludwig-Maximilians-University of Munich; Munich Center for Integrated Protein Science (CIPSM), Munich, Germany

Vânia Pobre
Instituto de Tecnologia Química e Biológica António Xavier, Universidade Nova de Lisboa, Oeiras, Portugal

Harald Putzer
CNRS UMR8261—Université Paris Diderot, Sorbonne Paris Cité, Institut de Biologie Physico-Chimique, Paris, France

Lennart Randau
Max-Planck-Institut für terrestrische Mikrobiologie, Marburg, Germany

Peter Redder
Laboratoire de Microbiologie et Génétique Moléculaires, Centre de Biologie Integrative, University of Toulouse, Toulouse, France

Ruth A. Schmitz
Institut für allgemeine Mikrobiologie, Christian-Albrechts-Universität zu Kiel, Kiel, Germany

Jared M. Schrader
Department of Biological Sciences, Wayne State University, Detroit, MI, United States

Linda Sperling
Institute for Integrative Biology of the Cell (I2BC), CNRS, CEA, Univ. Paris-Sud, Université Paris-Saclay, Gif-sur-Yvette CEDEX, France

Agnès Thierry
Institut Pasteur, Département Génomes et Génétique, Groupe Régulation Spatiale des Génomes; CNRS, UMR 3525, Paris Cedex 15; Institut Pasteur, Center of Bioinformatics, Biostatistics and Integrative Biology (C3BI), Paris, France

David Tollervey
Wellcome Centre for Cell Biology, University of Edinburgh, Edinburgh, United Kingdom

Nicolas J. Tourasse
ARNA Laboratory, INSERM U1212, CNRS UMR 5320, University of Bordeaux, Bordeaux, France

Jai J. Tree
School of Biotechnology and Biomolecular Sciences, University of New South Wales, Sydney, NSW, Australia

Lidia Vasiljeva
Department of Biochemistry, University of Oxford, Oxford, United Kingdom

Nathalie Villa-Vialaneix
MIAT, Université de Toulouse, INRA, Castanet-Tolosan, France

Irina O. Vvedenskaya
Waksman Institute and Department of Genetics, Rutgers University, Piscataway, NJ, United States

Alexander J. Westermann
Institute of Molecular Infection Biology (IMIB), University of Würzburg; Helmholtz Institute for RNA-Based Infection Research (HIRI), Würzburg, Germany

PREFACE

The goal of this volume of *Methods in Enzymology* is to describe diverse applications of high-density sequencing to research in microbial molecular genetics. Microorganisms are well suited for multiscale approaches ranging from single molecules to communities of cells. The protocols described here give an overview of how applications in high-density sequencing have been adapted to studies with prokaryotic and eukaryotic microorganisms. Many of the chapters include a brief introduction, suggestions about pitfalls to avoid, a discussion about interpretation of results, and recommendations for follow-up work to validate conclusions. For some applications, a protocol can be used without modification. For other applications, a protocol can be a roadmap to adapt the method to a different microorganism or to a different experimental condition.

Advances permitting sequence depths of tens of millions of reads per library have opened the way to a diversity of comprehensive genome-wide studies including mapping, processing, and stability of transcripts; measurement of protein synthesis rates; determination of the specificity of RNA- and DNA-binding proteins; and analysis of chromosome structure and integrity. In addition to genome-wide studies, high-density sequencing is applicable to any protocol whose output is a library of sequences. As examples, this volume of *Methods in Enzymology* includes chapters on the functional characterization of toxin–antitoxin systems and the characterization of sequence elements that determine levels of transcription.

It is my hope that the chapters in this volume of *Methods in Enzymology* will stimulate imaginative approaches for the application of high-density sequencing to integrative research with microorganisms. As high-density sequencing becomes widespread and relatively inexpensive, genome-wide analyses are rapidly becoming standard practice. This methodology is part of the foundation of synthetic and systems biology.

AGAMEMNON J. CARPOUSIS
Toulouse, October 2018

CHAPTER ONE

Characterizing the Role of Exoribonucleases in the Control of Microbial Gene Expression: Differential RNA-Seq

Vânia Pobre[1], Cecília M. Arraiano[1]
Instituto de Tecnologia Química e Biológica António Xavier, Universidade Nova de Lisboa, Oeiras, Portugal
[1]Corresponding authors: e-mail address: vaniapobre@itqb.unl.pt; cecilia@itqb.unl.pt

Contents

Abstract

Differential RNA-Seq is a next-generation technology method to determine the significant transcriptomic differences between two and more samples. With this method it is possible to analyze the total RNA content of different samples making it the best global analysis method currently available to study the roles of exoribonucleases in the cell. These enzymes are responsible for the RNA processing and degradation in the cells and therefore affect the total RNA pool in ways not yet fully understood. In *Escherichia coli* there are three main degradative exoribonucleases RNase II, RNase R, and PNPase that degrade the RNA from the 3′ to the 5′-end. These enzymes have several roles in the cell and even though they are degradative enzymes RNase II and PNPase can also protect some RNAs from degradation and PNPase can also act as an RNA polymerase under some conditions. The multiplicity of roles of these exoribonucleases leads to a very high

Methods in Enzymology, Volume 612
ISSN 0076-6879
https://doi.org/10.1016/bs.mie.2018.08.010

number of transcripts that are affected by their absence in the cell. With the differential RNA-Seq it is possible to obtain a much deeper understanding of how these enzymes work and regulate the bacterial gene expression. In this chapter we have described a differential RNA-Seq data analysis protocol applied to the study of exoribonucleases. We also included the protocol for experimental validation of the RNA-Seq data using qPCR and motility assays. Although the methods described in this chapter were applied to the study of the exoribonucleases, they can also be used for other differential RNA-Seq studies.

1. OVERVIEW OF EXORIBONUCLEASES

Exoribonucleases have a multiplicity of roles in the cell metabolism because they control RNA processing and degradation and contribute substantially to the recycling of ribonucleotides in the cell. In *Escherichia coli* there are three main exoribonucleases that are responsible for the 3′ to 5′ degradation of RNA: Exoribonuclease II (RNase II), Exoribonuclease R (RNase R), and Polynucleotide phosphorylase (PNPase). RNase II and RNase R are both hydrolytic enzymes from the RNase II family of enzymes, while PNPase is a phosphorolytic enzyme from the PDX family of enzymes (Fig. 1). Even though all three exoribonucleases degrade the RNA from the 3′ to the 5′ end, they have different mechanisms of action and their capacity to degrade different RNA molecules is also quite different (Andrade, Pobre, Silva, Domingues, & Arraiano, 2009).

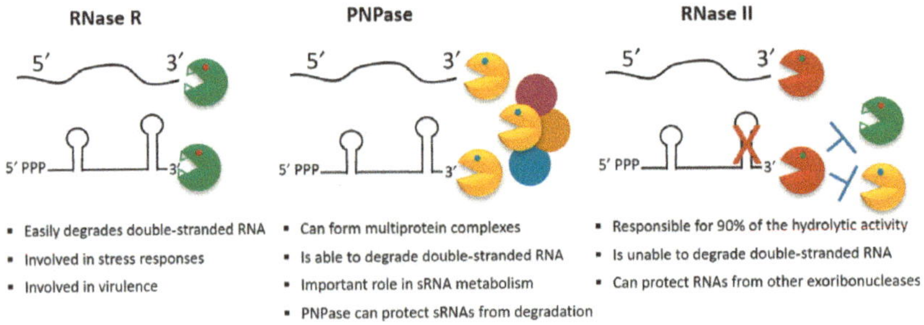

Fig. 1 Schematic representation of the three main *E. coli* exoribonucleases RNase R, PNPase, and RNase II. All three exoribonucleases can degrade single-stranded RNA from the 3′ to the 5′-end. RNase R easily degrades double-stranded RNA by itself. PNPase can degrade double-stranded RNA by itself or when in multiprotein complexes. RNase II action is blocked by RNA secondary structure. However, since this enzyme can degrade the tails of its targets it can protect some RNAs from degradation by the other exoribonucleases.

Since they belong to the same family of enzymes, RNase II and RNase R have some functional and structural similarities. Both have a similar modular organization of domains: an N-terminal region with two CSDs (CSD1 and CSD2), a central catalytic RNB, and a C-terminal S1 domain (Frazão et al., 2006; Reis, Pobre, Silva, Malecki, & Arraiano, 2013). Also they are both processive enzymes although RNase II becomes distributive when the substrate is shorter than 12 nucleotides (Cannistraro & Kennell, 1994). However, RNase R final end product of digestion is a dinucleotide while RNase II final end product is four nucleotides (Cheng & Deutscher, 2002; Frazão et al., 2006). A major difference between these two enzymes is that RNase R can degrade very structured RNA molecules while RNase II cannot. The RNase II structure showed that the access to the catalytic pocket is restricted to single-stranded RNA by steric hindrance at its entrance (Frazão et al., 2006) thus preventing the degradation of double-stranded RNA. On the other hand, RNase R was shown to have a helicase activity that confers the ability to unwind the RNA and therefore the ability to degrade double-stranded RNA (Awano et al., 2010; Hossain, Malhotra, & Deutscher, 2016). Recently the resolution of the RNase R secondary structure revealed that this enzyme has two RNA-binding channels where the 3′ end of the RNA overhang is bound and threaded into the active site thus unwinding the RNA duplex (Chu et al., 2017). This major difference between RNase II and RNase R in the degradation of double-stranded RNA leads to distinct roles in the cell metabolism.

RNase II is responsible for 90% of the hydrolytic activity on poly(A) RNA in *E. coli* crude extracts (Deutscher & Reuven, 1991); however, since the action of the RNase II is blocked by the secondary structures (Spickler & Mackie, 2000), the removal of the poly(A) tails by RNase II can impair the access of the other exoribonucleases to that RNA molecule preventing its degradation (Hajnsdorf, Steier, Coscoy, Teysset, & Régnier, 1994; Marujo et al., 2000; Mohanty & Kushner, 2000a, 2000b; Pepe, Maslesa-Galic, & Simons, 1994) and therefore RNase II can have a significant role in the protection of certain structured RNAs, namely, mRNAs which end by a structured Rho-independent terminators. RNase II is a highly regulated enzyme both at the transcriptional and posttranscriptional levels. RNase II expression is affected by the ribonucleases PNPase, RNase III, and RNase E (Zilhão, Cairrão, Régnier, & Arraiano, 1996; Zilhão, Régnier, & Arraiano, 1995). Moreover, the RNase II protein is also regulated posttranscriptionally by the protein PdeR (Gmr). In a Δ*gmr* mutant, there is a threefold increase in the RNase II levels and activity due to a slower protein turnover (Cairrão, Chora, Zilhão, Carpousis, & Arraiano, 2001). Recently it was also found that RNase II protein can also be acetylated and although this acetylation does not appear to influence the protein stability it does decrease the substrate-binding capacity of RNase II and therefore affects the RNase II activity. During stationary phase the acetylation levels are so high that there is a reduction of approximately 50% of RNase II activity (Song, Wang, Malhotra, Deutscher, & Liang, 2016). Even though the activity of RNase II seems to be considerably lower in stationary phase, this enzyme was found to be essential for the cell survival in stationary phase and during starvation since it regulates RNase PH that is required to initiate ribosome degradation in starving cells (Sulthana, Quesada, & Deutscher, 2017). In exponential phase RNase II was found to be crucial for cell motility and in its absence there is a significant increase in the ability to form biofilms (Pobre & Arraiano, 2015), suggesting that this enzyme has an important role in the control of the planktonic vs sedentary forms of the bacterial life cycle.

RNase R ability to degrade highly structured RNAs was shown to be of great advantage during stress conditions such as cold shock were the structured RNA molecules are more stable (Bárria, Malecki, & Arraiano, 2013; Cairrão, Cruz, Mori, & Arraiano, 2003). In fact RNase R is considered to be a stress induction protein, and it is involved in the response to several stress conditions (Bárria, Pobre, Bravo, & Arraiano, 2016) and like RNase II, deletion of RNase R also significantly impairs cells motility and increases the formation of biofilms (Pobre & Arraiano, 2015). RNase R also has an important role in the regulation of virulence genes in several pathogenic organism such as *Shigella flexneri*, enteroinvasive *E. coli*, *Aeromonas hydrophila*, *Legionella pneumophila* and *Campylobacter jejuni* (Charpentier, Faucher, Kalachikov, & Shuman, 2008; Erova et al., 2008; Haddad et al., 2014; Matos et al., 2014; Matos, Bárria, Pobre, Andrade, & Arraiano, 2012; Tobe, Sasakawa, Okada, Honma, & Yoshikawa, 1992). RNase R is also involved in RNA quality control by degrading defective RNAs (Cheng & Deutscher, 2003; Li, Reimers, Pandit, & Deutscher, 2002) and in protein quality control by affecting *trans*-translation and the rescue of stalled ribosomes (Cairrão et al., 2003; Domingues et al., 2015; Moreira, Domingues, Viegas, Amblar, & Arraiano, 2012). Moreover, RNase R was shown to interact with bacterial ribosomes (Malecki, Bárria, & Arraiano, 2014) and in fact during exponential phase most of the RNase R proteins are bound to ribosomes were they remain stable (Liang & Deutscher, 2013). In stationary phase RNase R no longer binds to ribosomes (Liang & Deutscher, 2013) and its stability is due to the fact that in this growth phase this enzyme is not acetylated (Liang, Malhotra, & Deutscher, 2011).

PNPase is also a processive exoribonuclease that catalyzes the phosphorolytic degradation of RNA releasing nucleoside diphosphates, but if the concentration of inorganic phosphate is very low, this enzyme acts as a polymerase and synthesizes long, highly heteropolymeric tails (Godefroy, 1970; Grunberg-Manago, Oritz, & Ochoa, 1955; Mohanty & Kushner, 2000a, 2000b). PNPase structure showed that this enzyme has a homotrimeric subunit organization with a ring-like architecture (Nurmohamed, Vaidialingam, Callaghan, & Luisi, 2009; Shi, Yang, Lin-Chao, Chak, & Yuan, 2008; Symmons, Jones, & Luisi, 2000). Each monomer has a five-domain arrangement with, two RNase PH domains (PH1 and PH2) that are linked by an α-helical domain at the N-terminus and two RNA-binding domains, KH and S1, in the C-terminal end. Like RNase II, PNPase is also tightly regulated with its levels being affected by RNase III, RNase II, and by PNPase itself (Jarrige, Mathy, & Portier, 2001; Portier, Dondon, Grunberg-Manago, & Régnier, 1987; Robert-Le Meur & Portier, 1992, 1994; Zilhão et al., 1996). Moreover, PNPase levels are also affected by polyadenylation (Mohanty & Kushner, 2002). PNPase degradative activity can also be stalled by secondary structures; however, unlike RNase II, this can be overcome with 3′ polyadenylation of the transcripts (Spickler & Mackie, 2000). Furthermore, PNPase can form multienzyme complexes that can then degrade through structured RNA. PNPase together with RNase E (endoribonuclease), RhlB (a helicase), and enolase (a glycolytic enzyme) forms the degradosome, which is an extremely efficient degradative machine (Miczak, Kaberdin, Wei, & Lin-Chao, 1996; Py, Causton, Mudd, & Higgins, 1994; Py, Higgins, Krisch, & Carpousis, 1996; Vanzo et al., 1998). However, PNPase can form other smaller complexes by binding only to the helicase RhlB or by binding to the poly(A) polymerase (PAP I) both contributing to efficient degradation of secondary structures (Liou, Chang, Lin, & Lin-Chao, 2002; Mohanty, Maples, & Kushner, 2004). Like RNase R, PNPase levels are also known to change in response to several stress conditions (Bárria et al., 2016). In fact, PNPase is essential for the growth of *E. coli* at low temperatures (Awano, Inouye, & Phadtare, 2008; Zangrossi et al., 2000) and under oxidative stress (Wu et al., 2009). Also, PNPase was shown to have a significant role in virulence of several pathogenic bacteria (Matos et al., 2012). For example, in *C. jejuni* PNPase was found to affect several proteins involved in virulence and significantly affect the swimming capacity of this pathogen as well as its ability to invade host cells (Haddad et al., 2012). Unlike RNase II and RNase R, PNPase has a major role in the metabolism of small RNAs (Saramago et al., 2014). PNPase was found to be the main degradative enzyme of sRNAs that control outer membrane proteins (Andrade & Arraiano, 2008). Moreover, free sRNAs, meaning sRNAs that are not bound to their targets nor bound to the RNA chaperone Hfq, are preferably degraded by PNPase (Andrade, Pobre, & Arraiano, 2013; Andrade, Pobre, Matos, & Arraiano, 2012). Surprisingly, it was also shown that PNPase can protect sRNAs from degradation by forming ternary complexes with the sRNA and with Hfq protein (Bandyra, Sinha, Syrjanen, Luisi, & De Lay, 2016). Furthermore, PNPase can also stabilize sRNAs that are bound to Hfq helping to promote the binding to their mRNA targets (Cameron & De Lay, 2016).

Overall these three exoribonucleases have distinct modes of action and although there is a functional overlap each enzyme has specific functions and affect the cells metabolism in different ways.

2. OVERVIEW OF DIFFERENTIAL RNA-SEQ

RNA-Seq is a revolutionary tool, based in next-generation sequencing (NGS) technologies, to obtain the profiling of the full transcriptome of any organism (Wang, Gerstein, & Snyder, 2009). RNA-Seq can have several applications depending on the protocol used for the library preparations and the data analysis. Here we will focus on the differential RNA-Seq method that encompasses the analysis of total RNA from two or more samples with the objective of comparing the transcriptomic differences between the samples. A typical RNA-Seq experiment consists in the following steps: RNA extraction, library preparation, sequencing of the samples, and bioinformatics data analysis (Fig. 2).

The phenol/chloroform is the most common method for extraction of total RNA from microbial samples and is preferred when extracting RNA for RNA-Seq purposes. The main advantage of this protocol is the conservation of all RNA molecules from the sample since most of the other methods/kits designed for total RNA extraction are prone to the loss of specific RNAs, especially smaller RNAs. Moreover, with this method the RNA obtained usually has a very high quality that is essential for a successful RNA-Seq experiment.

There are several commercial platforms available to perform RNA-Seq and the decision of which will be used is very important since each platform has advantages and limitations that significantly affect the results. Although the 454 sequencing from Roche was the first successful NGS system (Margulies et al., 2005), there are currently several others such as Ion torrent, SOLiD from Applied Biosystems, and PacBio that have surpassed the Roche

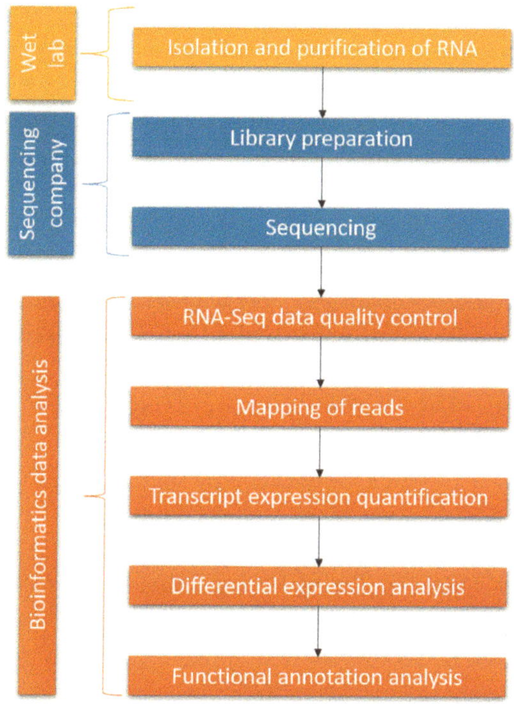

Fig. 2 Schematic of the key steps in a differential RNA-Seq experiment. Purified RNA samples are usually sent to commercial platforms for library preparation and sequencing. Differential RNA-Seq data analysis includes an assessment of the quality of the data, followed by the mapping of the reads, transcript expression quantification, and the differential expression analysis. Additional steps of functional analysis can be done in the case of well-annotated organisms.

systems. However, the Illumina platforms are by far the most used and well known especially since there are a wide range of systems developed by this company. The library preparation protocol depends on which system will be used for the sequencing but in general this step always implies an rRNA depletion followed by the conversion of RNA to cDNA, fragmentation of the cDNA, and ligation of adaptor sequences. It should be highlighted that during this step several important choices must also be made that will greatly influence the results. The first is whether user wants to perform single-end or paired-end RNA-Seq experiments since the library protocol is different for each type of RNA-Seq. Single-end sequencing involves sequencing cDNA from only one end while paired-end sequencing allows users to sequence both ends of a cDNA fragment. Paired-end sequencing facilitates detection of genomic rearrangements and repetitive sequence elements, as well as gene fusions and novel transcripts. However, for differential RNA-Seq experiments, single-end sequencing is sufficient and since normally single-end sequencing is cheaper and the bioinformatic analysis is also easier, this should also be taken into consideration. Another important decision is constructing libraries with or without directionality, i.e., strand-specific libraries. These libraries have the advantage of yielding information about the orientation of transcripts that can be of significant importance for regions with overlapping transcription from opposite directions (David et al., 2006). After the libraries construction the next step is the actual sequencing that is an automated process that can take from several hours to several days depending on the system used.

The raw data obtained from the sequencing is in a fastQ format that includes all the reads (sequences obtained after the sequencing) with a quality score associated. This quality scores can be used to assess the quality of the data and how good the sequencing was. The number of reads obtained will depend on the coverage done during the sequencing of the samples. Usually for a differential RNA-Seq experiment of a bacteria there should be around 20 million reads per sample. The length of the reads can vary from 50 bp to over 1000 bp depending on the system used. Normally for quantification purposes the preferable reads length is 100 bp, if the genome of the organism is already known. The bioinformatics analysis of a differential RNA-Seq experiment includes the data quality control, mapping of reads, transcript expression quantification, and the differential expression analysis. An additional step for functional annotation analysis can be performed if the organism is well annotated, and the differential expression analysis results in a high number of significant transcripts (Fig. 2).

For each of the short reads in the FASTQ file, a corresponding location in the reference genome needs to be determined (mapping of reads). In cases where the organism genome is still unknown, it will be necessary to do a de novo assembly of the transcriptome that can be achieved by using assemblers such as Trinity (Grabherr et al., 2011) or BinPacker (Liu et al., 2016). The mapping of reads to a reference genome can be done using various tools, the most common ones are Bowtie2 (Langmead, Trapnell, Pop, & Salzberg, 2009), TopHat2 (Kim et al., 2013), and BWA (Li & Durbin, 2009). All these tools will try to locate a sequence in the reference genome that matches the read; however, all will give slightly different results. During this step of the data analysis it is important to consider the number of mismatches (number of nucleotides that do not need to have correspondence between the read and the reference sequence) allowed during the mapping as this will significantly impact the overall mapping rate as well as the final results. Generally, it is expected that around 90% of the reads are mapped to the reference genome. The files obtained from the mappers can be in SAM (Sequence Alignment Map) format, that is a plain text file with all the reads aligned to the reference genome, or in BAM format that is the binary version of the SAM file.

The next step in the data analysis is the quantification of the transcripts' expression. This is achieved by counting the number of reads that mapped to each transcript during the mapping of reads. To obtain an accurate quantification for each transcript it is therefore essential to have an annotation of the genome and to attribute a corresponding read to each annotation. For this it is necessary to sort the mapped reads so that the quantification tools can identify correctly which reads belong to each transcript. Samtools can be used to sort, merge, index, and generate alignments in a per-position format (Li et al., 2009). After sorting the mapping files, it is possible to quantify the number of reads using any of the many available tools. The most commonly used are HTSeq (Anders, Pyl, & Huber, 2015), FeatureCounts (Liao, Smyth, & Shi, 2014), and Cuffquant that is actually part of the Cufflinks set of tools (Trapnell et al., 2012). Besides these tools it is also possible to quantify the transcripts' expression using the genome browser Artemis (Carver, Harris, Berriman, Parkhill, & McQuillan, 2012) with the advantage that Artemis can also be used to visualize the mapped reads and discover new genomic features using the mapped reads. It should be noted that the read counts are values that are not normalized and therefore a normalization method should also be done to obtain a more accurate expression value. The most common normalization methods are the RPKM (reads per kilobase of transcript per million mapped reads) or FPKM (fragments per kilobase of transcript per million mapped reads) that take into consideration the library size and the gene length (Mortazavi, Williams, McCue, Schaeffer, & Wold, 2008) and the TMM (Robinson & Oshlack, 2010). Artemis and Cuffquant will calculate the RPKM/FPKM automatically given an annotation file, while HTSeq and FeatureCounts do not have that option integrated.

Differently expressed genes can be identified using the read counts per gene and compare them between samples. Many tools are available for this type of analysis (Soneson & Delorenzi, 2013). However, the most commonly used are the edgeR (Robinson, McCarthy, & Smyth, 2010) and DESeq2 (Love, Huber, & Anders, 2014) packages from Bioconductor. Both packages use nonnormalized read counts to determine differentially expressed genes using a model based on the negative binomial distribution. It is also possible to use Cuffdiff from Cufflinks to calculate the differential expression of the transcripts between two samples.

A differential RNA-Seq experiment can be quite complex, especially the bioinformatics analysis of the data. The final results can be difficult to interpret from a biological point of view so further analysis of the data, such as a functional annotation analysis or a regulatory network construction, might improve the understanding of the data.

3. RNA EXTRACTION

There are several protocols for RNA extraction; however, when preparing total RNA samples for RNA-Seq the phenol/chloroform method should be used to preserve all RNA molecules. This protocol was optimized for *E. coli* sample in exponential phase (OD_{600}~0.5).

1. Collect 20 mL of culture and mix with 20 mL of cold TM stop buffer (Tris–HCl pH 7.2 10 m*M*, $MgCl_2$ 5 m*M*, NaN_3 25 m*M*, and chloramphenicol 0.5 mg/mL). Stop buffer should be kept in the cold!
2. Centrifuge at 4°C, 4000 rpm for 20 min.
3. Resuspend the pellet in 800 μL of lysis buffer (Tris–HCl pH 7.2 10 m*M*, $MgCl_2$ 5 m*M*, Turbo DNase (Ambion) 0.4 U, lysozyme 0.8 mg).
4. Incubate the sample in a water bath at 42°C for 5 min.
5. Flash-freeze the sample with liquid nitrogen and transfer again to the water bath and leave to defrost.
6. Repeat the flash-freeze and defrosting.
7. Flash-freeze a third time but before defrosting add 140 μL of acetic acid 20 m*M* and 90 μL of SDS 10%.
8. Repeat the flash-freeze and defrosting for another three times or until the sample becomes transparent.
9. Add 5 μL of Turbo DNase (Ambion) and incubate at 37°C for 1 h.
10. Transfer the sample to a 2-mL Eppendorf tube, add 1 mL of phenol pH 4.7, and vortex for 2 min.
11. Centrifuge at 4°C, 14,000 rpm for 10 min and collect the supernatant for a new tube.
12. Add 1 mL of phenol pH 4.7 and vortex for 2 min.
13. Centrifuge at 4°C and 14,000 rpm for 10 min and collect the supernatant for a new tube.
14. Add 500 μL of phenol pH 4.7 and 500 μL of chloroform/alcohol isoamyl (24:1) solution.

15. Vortex for 1 min, centrifuge at 4°C and 14,000 rpm for 5 min, and transfer supernatant to a new tube.
16. Add 1 mL of chloroform/alcohol isoamyl (24:1) solution.
17. Vortex for 1 min, centrifuge at 4°C and 14,000 rpm for 5 min, and transfer supernatant to a new tube.
18. Add 0.1 × volume of sodium acetate 3 *M* pH 5.2 and 2.5 × volume of cold ethanol 100%.
19. Precipitate overnight at − 20°C.
20. Centrifuge at 4°C and 14,000 rpm for 45 min.
21. Discard supernatant and add 1 mL of cold ethanol 75% to wash the RNA pellet.
22. Centrifuge at 4°C and 14,000 rpm for 30 min.
23. Discard supernatant and air-dry the RNA pellets in a fume hood.
24. Resuspend pellets in appropriate volume (between 75 and 100 μL) of Milli-Q Water.

4. BIOINFORMATICS ANALYSIS

RNA-Seq data analysis can be very complex and challenging. Although there are several commercial programs that allow the analysis of RNA-Seq data to be performed in a more automatic and less complex way, they are expensive, and the results are not always the best outcome possible. For these reasons we have assembled a protocol for differential RNA-Seq data analysis using license free programs and that we tested and confirmed that the results obtained are reliable. Here, we assume that the reader has a basic knowledge of the Linux operating system and can run programs using simple command lines.

The first step of any RNA-Seq data analysis should always be an assessment of the quality of the data. The FastQC program (https://www.bioinformatics.babraham.ac.uk/projects/fastqc/) is a simple and fast way to determine the quality of the raw data as well as a tool to identify potential problems with the data, such as adaptor contamination or overrepresentation of sequences. With the results from FastQC it is possible to improve the data quality by removing problematic sequences. Seqtk (https://github.com/lh3/seqtk) is a simple tool that can be used to trim the sequences to improve the quality of the data. FastQC program should be rerun after the trimming of the sequences to guaranty that the RNA-Seq data have good quality before proceeding with the analysis.

4.1 Mapping of Reads

There are several available tools for mapping the reads, but we will use Bowtie2 (Langmead & Salzberg, 2012). The first step is to create indexes of the reference genome, in this case we will use *E. coli* genome (NC_000913.3). A fasta file with the full genome (mg1655.fa) and an annotation file (mg1655.gff) were downloaded from the NCBI database. To create the indexes the following command line should be run:

```
bowtie2-build mg1655.fa mg1655_ind
```

This will create several files with the extensions mg1655_ind.1.bt2, mg1655_ind.2.bt2, mg1655_ind.3.bt2, mg1655_ind.4.bt2, mg1655_ind.rev.1.bt2, and mg1655_ind.rev.2.bt2. These files are necessary for the mapping of the reads.

The next step is then mapping the reads to the reference genome. The arguments used during this step will greatly influence the results so careful attention should be given when choosing the arguments to be used. For single-end data the following command line should be used:

```
bowtie2 -x mg1655_ind sample.fastq -un sample_UN -S sample. sam
```

For paired-end data it will be necessary to know the orientation of the two pairs. Here we will assume that the sample_R1.fastq is the forward pair read and that sample_R2.fastq is the reverse pair read and that the library was constructed following the Illumina's Paired-end Sequencing Assay and therefore the pairs are –fr oriented. Then the command line should be:

```
bowtie2 -x mg1655_ind -1 sample_R1.fastq -2 sample_R2.fastq -un
sample_UN -S sample.sam
```

Both commands will create two files: the sample_UN file will contain all reads that were not mapped to the reference genome and the sample.sam file is the mapped reads file that we will be used to continue the data analysis. The unmapped reads can be useful since they represent sequences that were not mapped against the genome and although this might mean that these sequences are trash, it can also mean that there are parts of the genome that are not yet known. For a well-known organism such as *E. coli*, this is unlikely to occur, but for nonmodel organisms this file can be important. Moreover, after Bowtie2 finishes running it will give an alignment summary that will include the mapping statistics. Normally the overall alignment rate should be over 90% and the percentage of reads aligned exactly one time should be clear majority with only a very small percentage

(<1%) aligning more than one time. Problems with the alignment rates should be addressed before proceeding to guaranty that no errors occurred during the mapping of the reads.

4.2 Transcript Expression Quantification

The next step of the RNA-Seq data analysis should be the quantification of the transcripts expression. The Artemis software (Carver et al., 2012) is a convenient tool to use for the quantification of the transcripts expression considering that it is easy to use and since it is a genome browser, it also allows for the visualization of the mapped reads. However, to be able to open the mapped reads in Artemis it is necessary to sort the mapping files. Moreover, Artemis only reads files in BAM format and since the mapping files are in SAM format, the first step is to convert the files from SAM to BAM format using Samtools (Li et al., 2009). The command line should be:

```
samtools view -b -o sample.bam sample.sam
```

This will create a BAM file. To sort this file the following command line should be used:

```
samtools sort —O bam -T tem_bam sample.bam > sample_sort.bam
```

This will sort the mapping file by chromosomal coordinates so that when opening the files in Artemis the program can recognize the position of each read in the genome. The first file to be opened in Artemis should be the genome fasta file (mg1655.fasta). Once this file is loaded in the program the annotation file (mg1655.gff) should be opened using the option "Read an entry…." Only then should the sorted BAM files be loaded to Artemis using the option "Read BAM/VCF…." Once all files are loaded to Artemis it is possible to calculate the number of reads (raw data) and the expression value in RPKM (normalized data) for each transcript. The resulting files should consist of a list of all transcripts and the respective values (either number of reads or RPKM) in each sample.

4.3 Differential Expression Analysis

For the differential expression analysis, we will be using the R packages edgeR (McCarthy, Chen, & Smyth, 2012). Here we will assume that the reader has some knowledge of the R language and is familiar with the usage of R scripts. First, it will be necessary to prepare the files to run the script. Considering that the objective is to identify which transcripts are significantly affected upon deletion of the exoribonucleases we want to compare the expression of each transcript in the different mutants and the WT cells. Assuming that there are three biological replicates for each sample, the table to be prepared must have in the first column the identification of all the transcripts; the second, third, and fourth columns are the number of reads for each transcript for the WT's biological replicates; and the fifth, sixth, and seventh columns are the number of reads for the mutant's biological replicates. For example:

Gene	WT1	WT2	WT3	Mutant1	Mutant2	Mutant3
aacs	4321	4922	5372	4800	5055	5938
aat	741	861	905	618	817	734
accA	7208	8125	7133	8354	10922	7479
accB	5922	7438	5014	5066	8515	4306
accC	24321	27109	23281	27733	38872	28320
...	...	...	...	...	...	...

This table must be saved as a tab delimited file (TableOfCounts.txt) and special attention should be taken so that there are no empty lines in the table as this will cause an error while running the script. The edgeR package has many options but, in this case, we can simply do a pairwise comparison between two groups. The script should be the following:

```
> x <- read.delim("TableOfCounts.txt",row.names="Gene")
> group <- factor(c(1,1,1,2,2,2))
> y <- DGEList(counts=x,group=group)
> y <- calcNormFactors(y)
> design <- model.matrix(~group)
> y <- estimateDisp(y,design)
> et <- exactTest(y)
> topTags(et)
>edgeR_results <- topTags(et, n=4500)
>write.csv2(edgeR_results, "edgeR_results.csv")
>summary(de <- decideTestsDGE(et, p=0.05, adjust="BH"))
>detags <- rownames(et)[as.logical(de)]
>plotSmear(et, de.tags=detags)
>abline(h = c(-2, 2), col = "blue")
>abline(v = c(3), col = "blue")
```

Running this edgeR script will result in a table with the transcripts ranked by FDR values and all transcripts that have an FDR < 0.05 will be considered significant differentially expressed. Since the deletion of an exoribonucleases significantly affects many transcripts another level of filtration of the data should also be considered. The most commonly accepted cutoff is a fold change of 2, so all transcripts that have higher fold changes will be in the final list of significant differentially expressed transcripts; on the contrary, transcripts that have a fold change lower than 2 will be removed from the final list. Here we also introduced another cutoff value where the expression value of the transcripts (LogCPM) is also taken into account. So, all transcripts that have a LogCPM lower than 3 should also be removed and only those with higher expression values are maintained for subsequent analysis. To summarize, the final list of differentially expressed transcripts is comprised by transcripts with FDR < 0.05, fold-change > 2, and LogCPM > 3. Running the above script also results in an MA scatterplot with all the transcripts with an FDR < 0.05 in red, two horizontal lines corresponding to the cutoff of a fold-change of 2 and a vertical line with the cutoff of the LogCPM of 3.

Ideally there should always exist a biological triplicate for all samples to be able to perform a differentially expression analysis; however, since RNA-Seq technologies are still quite expensive, it is not always possible to have the biological replicates. In this situations edgeR provides an alternative analysis of the data where a biological coefficient of variation (BCV) is estimated. For *E. coli* a BCV of 0.2 seems to be the most accurate when comparing the analysis of data with and without biological replicates. In these cases, the edgeR script can be adjusted to the following:

```
> x <- read.delim("TableOfCounts.txt",row.names="Gene")
> group <- factor(c(1,2))
> y <- DGEList(counts=x,group=group)
> y <- calcNormFactors(y)
> bcv <- 0.2
> et <- exactTest(y, dispersion=bcv^2)
> topTags(et)
>edgeR_results <- topTags(et, n=4500)
>write.csv2(edgeR_results, "edgeR_results.csv")
>summary(de <- decideTestsDGE(et, p=0.05, adjust="BH"))
>detags <- rownames(et)[as.logical(de)]
>plotSmear(et, de.tags=detags)
>abline(h = c(-2, 2), col = "blue")
>abline(v = c(3), col = "blue")
```

Running this script will also result in a table and a MA scatterplot like the above with the biological replicates. In these situations, the additional cutoffs for fold-change and LogCPM are of even greater importance.

4.4 Functional Annotation Analysis

When using RNA-Seq technologies to study the roles of the exoribonucleases in the cellular metabolism, it is likely that a very high number of transcripts significantly differentially expressed will be obtained. However, since these enzymes degrade the RNA from the 3′ to the 5′ end, it is not easy to determine only by differential RNA-Seq, exactly the transcripts that are targets of the exoribonucleases. Therefore, a functional annotation approach should be taken, especially since many of the transcripts differentially expressed are probably a consequence of indirect effects of the deletion of these enzymes. Since *E. coli* is a model organism with the vast majority of the genes already annotated and with known function this type of functional analysis is easy to perform. However, for nonmodel organisms this might represent a challenge. A very good and easy tool to perform a functional enrichment analysis is GeneCodis3 (Tabas-Madrid, Nogales-Cadenas, & Pascual-Montano, 2012). There is a website for GeneCodis3 (http://genecodis.cnb.csic.es/) that can be accessed at any time and used freely. The first step is choosing *E. coli* from the list of several model organism available, next the user needs to select the annotations: GO biological process, GO molecular function, and KEGG Pathway. And finally, the user needs to input the transcripts list obtained after the differential expression analysis. After obtaining the results for the functional enrichment the reader should take special care with the transcripts that were not included in the analysis (list given by the program) because this can greatly influence the results. By default, all ncRNAs are not included in the functional enrichment analysis and this can greatly affect the results. The objective of using three different annotations for the enrichment analysis is because it is not always easy to obtain an enrichment using a single annotation. By combining all three annotations (GO biological process, GO molecular function, and KEGG Pathway) it is likely that the user will obtain a good functional enrichment annotation result. Moreover, GeneCodis3 advance options also allows the user to use two different statistical methods, two methods for P-value correction, as well as, an option to change the GO annotation level. All these options should be explored and taken into consideration if running the functional enrichment using the default parameters does not provide any results.

5. EXPERIMENTAL VALIDATION

The amount of information that an RNA-Seq experiment provides greatly surpasses all other experimental methods. The bioinformatics analysis of such data can be complex as described above and, in the end, although a global overview of the data is achieved sometimes it can be difficult to focus on the smaller but most significant results. Moreover, since RNA-Seq results can be significantly affected by experimental and data analysis protocols, it is of great importance to experimentally validate the results obtained by other methods.

5.1 qPCR

Differential RNA-Seq is a quantification of the differences of transcripts expression between two samples. Therefore, the easiest method to validate such data is the real-time qPCR. The $2^{-\Delta\Delta Ct}$ method (Livak & Schmittgen, 2001) was found to have a good correlation with the fold-change calculated using RNA-Seq data (Pobre & Arraiano, 2015) and should therefore be used to experimentally validate this type of data. For any qPCR experiment a housekeeping gene (HK) needs to be selected, this gene expression should not present any variation between the samples to be analyzed. RNA-Seq data can be used to screen for potential HK genes and they should be tested by qPCR to select the best before proceeding. The protocol for qPCR using the Corbett Rotor Gene RG 3000 real-time PCR system and SensiFAST SYBR No-ROX Kit (Bioline) is the following:

1. cDNA synthesis from total RNA using random hexamers.
2. cDNA dilutions (1:5, 1:10, 1:20, 1:50, and 1:75) need to be prepared for the calibration curves.
3. A calibration curve for the housekeeping gene (HK) and for the gene of interest (GI) needs to be run to determine the efficiency of amplification for each pair of primers.
4. All PCR reactions should always be done in triplicate and include:
   a. 1.25 μL cDNA
   b. 0.25 μL Forward primer 10 n*M*
   c. 0.25 μL Reverse primer 10 n*M*
   d. 5 μL SensiFAST SYBR No-ROX Mix
   e. 3.25 μL Milli-Q water
5. A negative control prepared with all the above reagents but without cDNA also needs to be present.

6. The PCR conditions should be the following: 95°C for 2 min, 40 cycles of 95°C for 10 s, 60°C for 15 s, 72°C for 20 s.
7. A melting curve also needs to be obtained from a first step starting from 60°C to 95°C, to control specificities of quantitative PCR reaction for each primer pair.
8. The relative copy number should then be determined using the $2^{-\Delta\Delta Ct}$ method.

Even though the fold-change of transcripts using qPCR and RNA-Seq data have a good correlation, it should be noted that these are distinct methods and it is unlikely that all transcripts can be experimentally validated through this method.

5.2 Motility Assays

Exoribonucleases are known to affect the cells in a number of ways. They influence the cells resistance to stress conditions and even promote changes in their physiology and morphology. Cell motility was found to be greatly affected by the exoribonucleases (Fig. 3) and therefore many transcripts

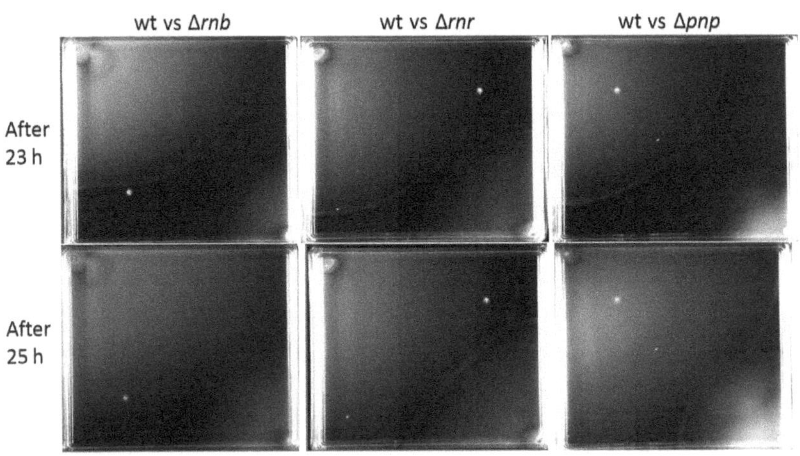

Fig. 3 Deletion of exoribonucleases impairs cell motility. Swimming ability was assessed on LB agar containing 0.25% agar. Strains were inoculated into the swimming media and incubated at 37°C. Pictures were taken regularly to monitor the increase of the swimming halo. Here we show pictures taken 23 and 25 h after inoculation. The *upper left corner* of the plate was inoculated with wild-type cells, and the *bottom right corner* was inoculated with either Δ*rnb*, Δ*rnr*, or Δ*pnp* cells for better comparison between the wild-type and the different mutants. *Reprinted from Pobre, V., & Arraiano, C. M. (2015). Next generation sequencing analysis reveals that the ribonucleases RNase II, RNase R and PNPase affect bacterial motility and biofilm formation in* E. coli. BMC Genomics, 16*, 72, with permission from BMC Publishing.*

related to this pathway are significantly differentially expressed in RNA-Seq data. The motility assays are therefore an easy way to validate the RNA-Seq data. This protocol is quite simple since it only requires the inoculation of swimming media plates (tryptone 10 g/L, NaCl 5 g/L, thymine 50 μg/mL, and agar 0.25%), that are then left at 37°C for several hours. The swimming of the cells can be monitored by taking pictures every 2 or 3 h with a Chemidoc Imaging system (BioRad).

6. CONCLUSION

Exoribonucleases are responsible for the RNA processing and degradation in the cells and therefore greatly affect the cells transcriptome. The differential RNA-Seq method allows for a deeper understanding of the distinct roles of the exoribonucleases. Using this method, it was possible to determine that even though the three exoribonucleases have distinct modes of action, there is functional overlap although each affects cell metabolism in different ways. In this chapter we have described a data analysis method adjusted for the study of the exoribonucleases. However, it is possible to adapt this protocol to other differential RNA-Seq experiments.

ACKNOWLEDGMENTS

This work was financially supported by: Project LISBOA-01-0145-FEDER-007660 (Microbiologia Molecular, Estrutural e Celular) funded by FEDER funds through COMPETE2020—"Programa Operacional Competitividade e Internacionalização (POCI)" and by national funds through FCT—"Fundação para a Ciência e a Tecnologia", including Grant from PTDC/BIA-MIC/1399/2014 and European Union Horizon 2020 Research and Innovation Programme (Grant agreement no. 635536). This work was also partially supported by ONEIDA project (LISBOA-01-0145-FEDER-016417) cofunded by FEEI—"Fundos Europeus Estruturais e de Investimento" from "Programa Operacional Regional Lisboa 2020" and by national funds from FCT—"Fundação para a Ciência e a Tecnologia". V.P. is recipient of an FCT Post-Doctoral fellowship (SFRH/BPD/87188/2012).

REFERENCES

Anders, S., Pyl, P. T., & Huber, W. (2015). HTSeq—A Python framework to work with high-throughput sequencing data. *Bioinformatics*, *31*(2), 166–169.

Andrade, J. M., & Arraiano, C. M. (2008). PNPase is a key player in the regulation of small RNAs that control the expression of outer membrane proteins. *RNA*, *14*(3), 543–551.

Andrade, J. M., Pobre, V., & Arraiano, C. M. (2013). Small RNA modules confer different stabilities and interact differently with multiple targets. *PLoS One*, *8*(1), e52866.

Andrade, J. M., Pobre, V., Matos, A. M., & Arraiano, C. M. (2012). The crucial role of PNPase in the degradation of small RNAs that are not associated with Hfq. *RNA*, *18*(4), 844–855.

Andrade, J. M., Pobre, V., Silva, I. J., Domingues, S., & Arraiano, C. M. (2009). The role of 3′-5′ exoribonucleases in RNA degradation. *Progress in Molecular Biology and Translational Science, 85*, 187–229.

Awano, N., Inouye, M., & Phadtare, S. (2008). RNase activity of polynucleotide phosphorylase is critical at low temperature in *Escherichia coli* and is complemented by RNase II. *Journal of Bacteriology, 190*(17), 5924–5933.

Awano, N., Rajagopal, V., Arbing, M., Patel, S., Hunt, J., Inouye, M., et al. (2010). *Escherichia coli* RNase R has dual activities, helicase and RNase. *Journal of Bacteriology, 192*(5), 1344–1352.

Bandyra, K. J., Sinha, D., Syrjanen, J., Luisi, B. F., & De Lay, N. R. (2016). The ribonuclease polynucleotide phosphorylase can interact with small regulatory RNAs in both protective and degradative modes. *RNA*.

Bárria, C., Malecki, M., & Arraiano, C. M. (2013). Bacterial adaptation to cold. *Microbiology, 159*(Pt. 12), 2437–2443.

Bárria, C., Pobre, V., Bravo, A., & Arraiano, C. (2016). Ribonucleases as modulators of bacterial stress response. In F. de Bruijn (Ed.), *Stress and environmental control of gene expression in bacteria* (pp. 174–184). Wiley-Blackwell.

Cairrão, F., Chora, A., Zilhão, R., Carpousis, A. J., & Arraiano, C. M. (2001). RNase II levels change according to the growth conditions: Characterization of *gmr*, a new *Escherichia coli* gene involved in the modulation of RNase II. *Molecular Microbiology, 39*(6), 1550–1561.

Cairrão, F., Cruz, A., Mori, H., & Arraiano, C. M. (2003). Cold shock induction of RNase R and its role in the maturation of the quality control mediator SsrA/tmRNA. *Molecular Microbiology, 50*(4), 1349–1360.

Cameron, T. A., & De Lay, N. R. (2016). The phosphorolytic exoribonucleases polynucleotide phosphorylase and RNase PH stabilize sRNAs and facilitate regulation of their mRNA targets. *Journal of Bacteriology, 198*(24), 3309–3317.

Cannistraro, V. J., & Kennell, D. (1994). The processive reaction mechanism of ribonuclease II. *Journal of Molecular Biology, 243*, 930–943.

Carver, T., Harris, S. R., Berriman, M., Parkhill, J., & McQuillan, J. A. (2012). Artemis: An integrated platform for visualization and analysis of high-throughput sequence-based experimental data. *Bioinformatics, 28*(4), 464–469.

Charpentier, X., Faucher, S. P., Kalachikov, S., & Shuman, H. A. (2008). Loss of RNase R induces competence development in *Legionella pneumophila*. *Journal of Bacteriology, 190*(24), 8126–8136.

Cheng, Z. F., & Deutscher, M. P. (2002). Purification and characterization of the *Escherichia coli* exoribonuclease RNase R. Comparison with RNase II. *The Journal of Biological Chemistry, 277*(24), 21624–21629.

Cheng, Z. F., & Deutscher, M. P. (2003). Quality control of ribosomal RNA mediated by polynucleotide phosphorylase and RNase R. *Proceedings of the National Academy of Sciences of the United States of America, 100*(11), 6388–6393.

Chu, L. Y., Hsieh, T. J., Golzarroshan, B., Chen, Y. P., Agrawal, S., & Yuan, H. S. (2017). Structural insights into RNA unwinding and degradation by RNase R. *Nucleic Acids Research, 45*(20), 12015–12024.

David, L., Huber, W., Granovskaia, M., Toedling, J., Palm, C. J., Bofkin, L., et al. (2006). A high-resolution map of transcription in the yeast genome. *Proceedings of the National Academy of Sciences of the United States of America, 103*(14), 5320–5325.

Deutscher, M. P., & Reuven, N. B. (1991). Enzymatic basis for hydrolytic versus phosphorolytic mRNA degradation in *Escherichia coli* and *Bacillus subtilis*. *Proceedings of the National Academy of Sciences of the United States of America, 88*(8), 3277–3280.

Domingues, S., Moreira, R. N., Andrade, J. M., Dos Santos, R. F., Barria, C., Viegas, S. C., et al. (2015). The role of RNase R in trans-translation and ribosomal quality control. *Biochimie, 114*, 113–118.

Erova, T. E., Kosykh, V. G., Fadl, A. A., Sha, J., Horneman, A. J., & Chopra, A. K. (2008). Cold shock exoribonuclease R (*VacB*) is involved in *Aeromonas hydrophila* pathogenesis. *Journal of Bacteriology*, *190*(10), 3467–3474.

Frazão, C., McVey, C. E., Amblar, M., Barbas, A., Vonrhein, C., Arraiano, C. M., et al. (2006). Unravelling the dynamics of RNA degradation by ribonuclease II and its RNA-bound complex. *Nature*, *443*, 110–114.

Godefroy, T. (1970). Kinetics of polymerization and phosphorolysis reactions of *Escherichia coli* polynucleotide phosphorylase. Evidence for multiple binding of polynucleotide in phosphorolysis. *European Journal of Biochemistry*, *14*(2), 222–231.

Grabherr, M. G., Haas, B. J., Yassour, M., Levin, J. Z., Thompson, D. A., Amit, I., et al. (2011). Full-length transcriptome assembly from RNA-Seq data without a reference genome. *Nature Biotechnology*, *29*(7), 644–652.

Grunberg-Manago, M., Oritz, P. J., & Ochoa, S. (1955). Enzymatic synthesis of nucleic acidlike polynucleotides. *Science*, *122*(3176), 907–910.

Haddad, N., Matos, R. G., Pinto, T., Rannou, P., Cappelier, J. M., Prevost, H., et al. (2014). The RNase R from *Campylobacter jejuni* has unique features and is involved in the first steps of infection. *The Journal of Biological Chemistry*, *289*(40), 27814–27824.

Haddad, N., Tresse, O., Rivoal, K., Chevret, D., Nonglaton, Q., Burns, C. M., et al. (2012). Polynucleotide phosphorylase has an impact on cell biology of *Campylobacter jejuni*. *Frontiers in Cellular and Infection Microbiology*, *2*, 30. https://doi.org/10.3389/fcimb.2012.00030.

Hajnsdorf, E., Steier, O., Coscoy, L., Teysset, L., & Régnier, P. (1994). Roles of RNase E, RNase II and PNPase in the degradation of the *rpsO* transcripts of *Escherichia coli*: Stabilizing function of RNase II and evidence for efficient degradation in an *ams pnp rnb* mutant. *The EMBO Journal*, *13*(14), 3368–3377.

Hossain, S. T., Malhotra, A., & Deutscher, M. P. (2016). How RNase R degrades structured RNA: Role of the helicase activity and the S1 domain. *The Journal of Biological Chemistry*, *291*(15), 7877–7887.

Jarrige, A. C., Mathy, N., & Portier, C. (2001). PNPase autocontrols its expression by degrading a double-stranded structure in the *pnp* mRNA leader. *The EMBO Journal*, *20*(23), 6845–6855.

Kim, D., Pertea, G., Trapnell, C., Pimentel, H., Kelley, R., & Salzberg, S. L. (2013). TopHat2: Accurate alignment of transcriptomes in the presence of insertions, deletions and gene fusions. *Genome Biology*, *14*(4), R36.

Langmead, B., & Salzberg, S. L. (2012). Fast gapped-read alignment with Bowtie 2. *Nature Methods*, *9*(4), 357–359.

Langmead, B., Trapnell, C., Pop, M., & Salzberg, S. L. (2009). Ultrafast and memory-efficient alignment of short DNA sequences to the human genome. *Genome Biology*, *10*(3), R25.

Li, H., & Durbin, R. (2009). Fast and accurate short read alignment with burrows-wheeler transform. *Bioinformatics*, *25*(14), 1754–1760.

Li, H., Handsaker, B., Wysoker, A., Fennell, T., Ruan, J., Homer, N., et al. (2009). The sequence alignment/map format and SAMtools. *Bioinformatics*, *25*(16), 2078–2079.

Li, Z., Reimers, S., Pandit, S., & Deutscher, M. P. (2002). RNA quality control: Degradation of defective transfer RNA. *The EMBO Journal*, *21*(5), 1132–1138.

Liang, W., & Deutscher, M. P. (2013). Ribosomes regulate the stability and action of RNase R. *The Journal of Biological Chemistry*, *288*, 34791–34798.

Liang, W., Malhotra, A., & Deutscher, M. P. (2011). Acetylation regulates the stability of a bacterial protein: Growth stage-dependent modification of RNase R. *Molecular Cell*, *44*(1), 160–166.

Liao, Y., Smyth, G. K., & Shi, W. (2014). featureCounts: An efficient general purpose program for assigning sequence reads to genomic features. *Bioinformatics*, *30*(7), 923–930.

Liou, G. G., Chang, H. Y., Lin, C. S., & Lin-Chao, S. (2002). DEAD box RhlB RNA helicase physically associates with exoribonuclease PNPase to degrade double-stranded RNA independent of the degradosome-assembling region of RNase E. *The Journal of Biological Chemistry*, *277*(43), 41157–41162.

Liu, J., Li, G., Chang, Z., Yu, T., Liu, B., McMullen, R., et al. (2016). BinPacker: Packing-based De novo transcriptome assembly from RNA-seq data. *PLoS Computational Biology*, *12*(2), e1004772.

Livak, K. J., & Schmittgen, T. D. (2001). Analysis of relative gene expression data using real-time quantitative PCR and the 2(−Delta C(T)) method. *Methods*, *25*(4), 402–408.

Love, M. I., Huber, W., & Anders, S. (2014). Moderated estimation of fold change and dispersion for RNA-seq data with DESeq2. *Genome Biology*, *15*(12), 550.

Malecki, M., Bárria, C., & Arraiano, C. M. (2014). Characterization of the RNase R association with ribosomes. *BMC Microbiology*, *14*, 34.

Margulies, M., Egholm, M., Altman, W. E., Attiya, S., Bader, J. S., Bemben, L. A., et al. (2005). Genome sequencing in microfabricated high-density picolitre reactors. *Nature*, *437*(7057), 376–380.

Marujo, P. E., Hajnsdorf, E., Le Derout, J., Andrade, R., Arraiano, C. M., & Régnier, P. (2000). RNase II removes the oligo(A) tails that destabilize the *rpsO* mRNA of *Escherichia coli*. *RNA*, *6*(8), 1185–1193.

Matos, R. G., Bárria, C., Moreira, R. N., Barahona, S., Domingues, S., & Arraiano, C. M. (2014). The importance of proteins of the RNase II/RNB-family in pathogenic bacteria. *Frontiers in Cellular and Infection Microbiology*, *4*, 68.

Matos, R. G., Bárria, C., Pobre, V., Andrade, J. M., & Arraiano, C. M. (2012). Exoribonucleases as modulators of virulence in pathogenic bacteria. [General Commentary]. *Frontiers in Cellular and Infection Microbiology*, *2*, 65.

McCarthy, D. J., Chen, Y., & Smyth, G. K. (2012). Differential expression analysis of multifactor RNA-Seq experiments with respect to biological variation. *Nucleic Acids Research*, *40*(10), 4288–4297.

Miczak, A., Kaberdin, V. R., Wei, C. L., & Lin-Chao, S. (1996). Proteins associated with RNase E in a multicomponent ribonucleolytic complex. *Proceedings of the National Academy of Sciences of the United States of America*, *93*(9), 3865–3869.

Mohanty, B. K., & Kushner, S. R. (2000a). Polynucleotide phosphorylase, RNase II and RNase E play different roles in the *in vivo* modulation of polyadenylation in *Escherichia coli*. *Molecular Microbiology*, *36*(4), 982–994.

Mohanty, B. K., & Kushner, S. R. (2000b). Polynucleotide phosphorylase functions both as a 3′ right-arrow 5′ exonuclease and a poly(a) polymerase in *Escherichia coli*. *Proceedings of the National Academy of Sciences of the United States of America*, *97*(22), 11966–11971.

Mohanty, B. K., & Kushner, S. R. (2002). Polyadenylation of *Escherichia coli* transcripts plays an integral role in regulating intracellular levels of polynucleotide phosphorylase and RNase E. *Molecular Microbiology*, *45*(5), 1315–1324.

Mohanty, B. K., Maples, V. F., & Kushner, S. R. (2004). The Sm-like protein Hfq regulates polyadenylation dependent mRNA decay in *Escherichia coli*. *Molecular Microbiology*, *54*(4), 905–920.

Moreira, R. N., Domingues, S., Viegas, S. C., Amblar, M., & Arraiano, C. M. (2012). Synergies between RNA degradation and trans-translation in *Streptococcus pneumoniae*: Cross regulation and co-transcription of RNase R and SmpB. *BMC Microbiology*, *12*, 268.

Mortazavi, A., Williams, B. A., McCue, K., Schaeffer, L., & Wold, B. (2008). Mapping and quantifying mammalian transcriptomes by RNA-Seq. *Nature Methods*, *5*(7), 621–628.

Nurmohamed, S., Vaidialingam, B., Callaghan, A. J., & Luisi, B. F. (2009). Crystal structure of *Escherichia coli* polynucleotide phosphorylase core bound to RNase E, RNA and manganese: Implications for catalytic mechanism and RNA degradosome assembly. *Journal of Molecular Biology*, *389*(1), 17–33.

Pepe, C. M., Maslesa-Galic, S., & Simons, R. W. (1994). Decay of the IS10 antisense RNA by 3′ exoribonucleases: Evidence that RNase II stabilizes RNA-OUT against PNPase attack. *Molecular Microbiology, 13*(6), 1133–1142.

Pobre, V., & Arraiano, C. M. (2015). Next generation sequencing analysis reveals that the ribonucleases RNase II, RNase R and PNPase affect bacterial motility and biofilm formation in *E. coli*. *BMC Genomics, 16*, 72.

Portier, C., Dondon, L., Grunberg-Manago, M., & Régnier, P. (1987). The first step in the functional inactivation of the *Escherichia coli* polynucleotide phosphorylase messenger is a ribonuclease III processing at the 5′ end. *The EMBO Journal, 6*(7), 2165–2170.

Py, B., Causton, H., Mudd, E. A., & Higgins, C. F. (1994). A protein complex mediating mRNA degradation in *Escherichia coli*. *Molecular Microbiology, 14*(4), 717–729.

Py, B., Higgins, C. F., Krisch, H. M., & Carpousis, A. J. (1996). A DEAD-box RNA helicase in the *Escherichia coli* RNA degradosome. *Nature, 381*(6578), 169–172.

Reis, F. P., Pobre, V., Silva, I. J., Malecki, M., & Arraiano, C. M. (2013). The RNase II/RNB family of exoribonucleases: Putting the 'Dis' in disease. *Wiley Interdisciplinary Reviews-RNA, 4*(5), 607–615.

Robert-Le Meur, M., & Portier, C. (1992). *E.coli* polynucleotide phosphorylase expression is autoregulated through an RNase III-dependent mechanism. *The EMBO Journal, 11*(7), 2633–2641.

Robert-Le Meur, M., & Portier, C. (1994). Polynucleotide phosphorylase of *Escherichia coli* induces the degradation of its RNase III processed messenger by preventing its translation. *Nucleic Acids Research, 22*(3), 397–403.

Robinson, M. D., McCarthy, D. J., & Smyth, G. K. (2010). edgeR: A bioconductor package for differential expression analysis of digital gene expression data. *Bioinformatics, 26*(1), 139–140.

Robinson, M. D., & Oshlack, A. (2010). A scaling normalization method for differential expression analysis of RNA-seq data. *Genome Biology, 11*(3), R25.

Saramago, M., Bárria, C., Dos Santos, R. F., Silva, I. J., Pobre, V., Domingues, S., et al. (2014). The role of RNases in the regulation of small RNAs. *Current Opinion in Microbiology, 18*, 105–115.

Shi, Z., Yang, W. Z., Lin-Chao, S., Chak, K. F., & Yuan, H. S. (2008). Crystal structure of *Escherichia coli* PNPase: Central channel residues are involved in processive RNA degradation. *RNA, 14*(11), 2361–2371.

Soneson, C., & Delorenzi, M. (2013). A comparison of methods for differential expression analysis of RNA-seq data. *BMC Bioinformatics, 14*, 91.

Song, L., Wang, G., Malhotra, A., Deutscher, M. P., & Liang, W. (2016). Reversible acetylation on Lys501 regulates the activity of RNase II. *Nucleic Acids Research, 44*(5), 1979–1988.

Spickler, C., & Mackie, A. (2000). Action of RNases II and polynucleotide phosphorylase against RNAs containing stem-loops of defined structure. *Journal of Bacteriology, 182*, 2422–2427.

Sulthana, S., Quesada, E., & Deutscher, M. P. (2017). RNase II regulates RNase PH and is essential for cell survival during starvation and stationary phase. *RNA, 23*(9), 1456–1464.

Symmons, M. F., Jones, G. H., & Luisi, B. F. (2000). A duplicated fold is the structural basis for polynucleotide phosphorylase catalytic activity, processivity, and regulation. *Structure, 8*(11), 1215–1226.

Tabas-Madrid, D., Nogales-Cadenas, R., & Pascual-Montano, A. (2012). GeneCodis3: A non-redundant and modular enrichment analysis tool for functional genomics. *Nucleic Acids Research, 40*(Web Server issue), W478–W483.

Tobe, T., Sasakawa, C., Okada, N., Honma, Y., & Yoshikawa, M. (1992). *vacB*, a novel chromosomal gene required for expression of virulence genes on the large plasmid of *Shigella flexneri*. *Journal of Bacteriology, 174*(20), 6359–6367.

Trapnell, C., Roberts, A., Goff, L., Pertea, G., Kim, D., Kelley, D. R., et al. (2012). Differential gene and transcript expression analysis of RNA-seq experiments with TopHat and Cufflinks. *Nature Protocols*, 7(3), 562–578.

Vanzo, N. F., Li, Y. S., Py, B., Blum, E., Higgins, C. F., Raynal, L. C., et al. (1998). Ribonuclease E organizes the protein interactions in the *Escherichia coli* RNA degradosome. *Genes & Development*, *12*(17), 2770–2781.

Wang, Z., Gerstein, M., & Snyder, M. (2009). RNA-Seq: A revolutionary tool for transcriptomics. [Review]. *Nature Reviews. Genetics*, *10*(1), 57–63.

Wu, J., Jiang, Z., Liu, M., Gong, X., Wu, S., Burns, C. M., et al. (2009). Polynucleotide phosphorylase protects *Escherichia coli* against oxidative stress. *Biochemistry*, *48*(9), 2012–2020.

Zangrossi, S., Briani, F., Ghisotti, D., Regonesi, M. E., Tortora, P., & Dehò, G. (2000). Transcriptional and post-transcriptional control of polynucleotide phosphorylase during cold acclimation in *Escherichia coli*. *Molecular Microbiology*, *36*(6), 1470–1480.

Zilhão, R., Cairrão, F., Régnier, P., & Arraiano, C. M. (1996). PNPase modulates RNase II expression in *Escherichia coli*: Implications for mRNA decay and cell metabolism. *Molecular Microbiology*, *20*(5), 1033–1042.

Zilhão, R., Régnier, P., & Arraiano, C. M. (1995). The role of endonucleases in the expression of ribonuclease II in *Escherichia coli*. *FEMS Microbiology Letters*, *130*(2–3), 237–244.

CHAPTER TWO

Conformational Studies of Bacterial Chromosomes by High-Throughput Sequencing Methods

Virginia S. Lioy, Frédéric Boccard[1]
Institut de Biologie Intégrative de la Cellule, CEA, CNRS, Univ. Paris-Sud, Université Paris-Saclay, Gif-sur-Yvette cedex, France
[1]Corresponding author: e-mail address: frederic.boccard@i2bc.paris-saclay.fr

Contents

Methods in Enzymology, Volume 612
ISSN 0076-6879
https://doi.org/10.1016/bs.mie.2018.07.007

Abstract

The development of next-generation sequencing technologies has allowed the application of different methods dedicated to the study of DNA–protein interactions and chromosome conformation to entire bacterial genome. By combining these approaches, the role of various parameters and factors involved in gene expression and chromosome organization can be disclosed at the molecular level over the full genome. Here we describe two methods that profoundly revolutionized our vision of DNA–protein interactions and spatial organization of chromosomes. Chromosome conformation capture (3C) coupled to deep sequencing (3C-seq) enables studies of the genome-wide chromosome folding and its control by different parameters and structural factors. Chromatin immunoprecipitation (ChIP) followed by high-throughput DNA sequencing (ChIP-seq) revealed the extent and regulation of DNA–protein interactions in vivo and highlight the role of structural factors in the control of chromosome organization. In this chapter, we describe a detailed protocol of 3C-seq and ChIP-seq experiments that, when combined, allows the spatial study of the chromosome and the factors that promote specific folding. Data processing and analysis for both experiments are also discussed.

ABBREVIATIONS

AFA adaptive focused acoustics™
BAM compressed binary version of SAM file
BSA bovine serum albumin
CFU colony-forming unit
DAPI 4,6-diamidino-2-phenylindole
EDTA ethylenediaminetetraacetic acid
FA formaldehyde
NGS next-generation sequencing
RE restriction enzyme
SAM sequence alignment map
SDS sodium dodecyl sulfate
TBS Tris-buffered saline
TE Tris–EDTA buffer solution
WIG wiggle file

1. INTRODUCTION

1.1 Chromosome Conformation

The genome of all organisms must be condensed to fit within cells that are typically several thousand-fold smaller than the length of the DNA. In bacteria, DNA is efficiently compacted into a structure called the nucleoid, a dynamic macromolecular complex that simultaneously supports gene regulation and DNA segregation. Nucleoid folding and compaction relies on a multilayer organization that results from a combination of processes (Kleckner et al., 2014; Wang, Llopis, & Rudner, 2013): (i) negative DNA supercoiling, (ii) formation of chromatin-like structures by nucleoid-associated proteins (NAPs), (iii) condensation by structural maintenance of chromosome (SMC) proteins, (iv) macromolecular crowding, and (v) out-of-equilibrium processes such as transcription.

For several decades, numerous studies have focused on the control of chromosome folding and the role played by different DNA-binding proteins on this process. From the analyses of DNA transactions using various site-specific recombination systems, it was proposed that DNA supercoiling in bacteria results in the formation of branched and plectonemic structures, whose precise role(s) and distribution remain poorly understood. Different genetic systems have been used to probe in vivo the structure of *Escherichia coli* or *Salmonella* nucleoid, in particular the γδ Res system (Higgins, Yang, Fu, & Roth, 1996; Manna & Higgins, 1999) and the λ Int system (Garcia-Russell et al., 2004; Valens, Penaud, Rossignol, Cornet, & Boccard, 2004). These genetic tools allowed to infer the spatial distances between genomic loci from the frequencies at which these loci physically interact within a population of cells. Spatially closed regions of the genome contact frequently, while spatially distant regions rarely contact. In *Salmonella*, studies using γδ resolvase revealed the presence of stochastic boundaries between plectonemic supercoil structures, allowing DNA interactions in *cis* between sites located more than 100 kb apart (Higgins et al., 1996). RNA polymerase was subsequently shown to block the movement of plectonemic supercoil in the transcribed track, thus generating supercoil diffusion barriers (Booker, Deng, & Higgins, 2010). In *E. coli*, a genetic assay that measured the frequency of recombination between phage λ *att* sites scattered throughout the chromosome showed that it was organized into a ring composed of four macrodomains and two less-structured regions. The presence of Ori and Ter macrodomains were previously identified by FISH analyses (Niki, Yamaichi, & Hiraga, 2000). Remarkably, by tracking the movement of chromosomal loci labeled by fluorescent markers in time-lapse microscopy during a complete cell cycle, the dynamic behavior of loci belonging to various macrodomains and less constrained regions was shown to be radically different; in macrodomains, constraints on mobility are apparent, whereas in nonstructured regions, markers exhibited a greater motility explaining their ability to interact with flanking macrodomains (Espeli, Mercier, & Boccard, 2008).

1.2 DNA–Protein Interactions

DNA-binding proteins are involved in a number of basic cellular processes including DNA replication, transcription, recombination, repair, and transposition. Similarly, multiple DNA-binding proteins are suspected to contribute to the hierarchical organization of the bacterial chromosome; they include NAPs and SMC complexes. NAPs are the most abundant proteins in bacteria and are proposed to wrap, bend, and bridge the bacterial DNA through binding, influencing the supercoiling level of the DNA. Most of the knowledge on NAPs functions came from in vitro experiments because of the challenge of studying the molecular functions of these proteins in vivo (Gruber, 2014). The same was true for SMC complexes, which were shown to have a main role in chromosome segregation, but their impact on chromosome conformation was uncertain until a few years ago (Nolivos & Sherratt, 2014).

1.3 Combining NGS to 3C and ChIP Approaches

The development of two methods, chromatin immunoprecipitation (ChIP) and chromosome conformation capture (3C), opened new possibilities to identify specific protein–DNA interactions and to study the particular folding of the chromosome induced by DNA-binding proteins. ChIP is based on the use of an antibody against the DNA-binding factor of interest followed by footprinting or PCR quantification of the DNA associated with the immunoprecipitated protein (Kang, Vieira, & Bungert, 2002). On the other hand, the 3C method is based on covalent cross-linking and ligation of DNA fragments that are spatially close; the frequency of selected ligated products is subsequently determined by qPCR (Dekker, Rippe, Dekker, & Kleckner, 2002). One of the most remarkable uses of ChIP was the identification of 10 ParB-binding sites (*parS* sites) in *Bacillus subtilis* (Lin & Grossman, 1998). This discovery was the first step to the understanding of how ParABS systems segregate the bacterial chromosome. In contrast with eukaryotic systems, the 3C method in bacteria was rarely used to uncover specific DNA conformations and became more popular when coupled to next-generation sequencing (NGS) (see later).

The development of NGS technologies dramatically reduced the cost of the procedure, making the sequencing of bacterial genomes more affordable (Van Dijk, Auger, Jaszczyszyn, & Thermes, 2014). NGS boosted the development of many genome-wide techniques, such as microarrays, ChIP on chip, ChIP-seq, RNA sequencing (RNA-seq), and 3C-seq (Fig. 1).

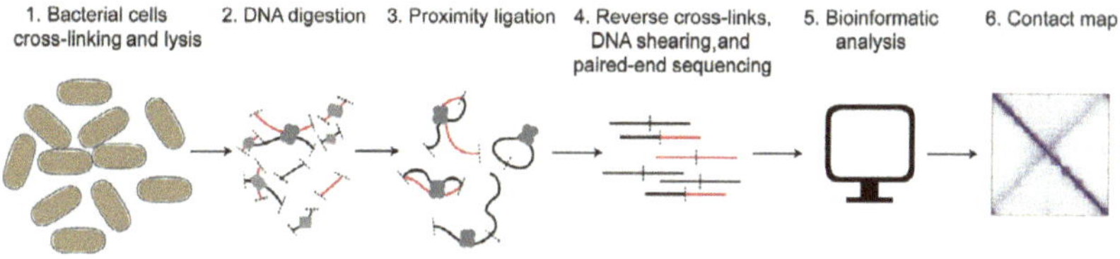

Fig. 1 Outline of 3C-seq protocol. Bacterial nucleoids are cross-linked with formaldehyde, lysed (1) and digested with a restriction enzyme (2). Then DNA is ligated by proximity ligation (3). Finally, the DNA fragments are purified and processed for Illumina sequencing (4). Reads are bioinformatically processed (5) and contact maps are established (6).

Of all these techniques, ChIP-seq and 3C-seq are the most relevant to reveal chromosome architecture: ChIP-seq gives linear information about how proteins bind to DNA and 3C-seq reveals the spatial organization of DNA promoted by the formation of nucleoprotein complexes.

The use of 3C-seq (or HiC) with different bacterial systems revealed important insights about the architecture of chromosome domains, the control of chromosome positioning by the ParABS systems, the activity of condensin complexes, or the effect of NAPs on the nucleoid folding (Le, Imakaev, Mirny, & Laub, 2013; Lioy et al., 2018; Marbouty et al., 2015). Pioneering work using HiC in *C. crescentus* revealed the in vivo functioning of two architectural proteins: HU and SMC. This research showed that HU facilitates local short-range compaction of the genome, whereas the SMC complex aligns the chromosome arms by promoting contacts between distant loci in the chromosome (Le et al., 2013). A few years later, the use of HiC/3C-seq in other bacterial species uncovered the role of SMC complexes in chromosome organization, since they were shown to generate either interarm interactions in bacteria with longitudinal organization (Marbouty et al., 2015; Wang et al., 2015) or promoting intraarm interactions in the Mb range in bacteria with transversal organization (Lioy et al., 2018). Remarkably, 3C contact frequencies in *E. coli* correlated well with collisions between loci, as revealed by the genetic analysis. In addition, the correlation of ChIP-seq data with 3C-seq data allowed the identification of the *modus operandi* of three major NAPs in *E. coli*: HU, Fis, and H-NS (Lioy et al., 2018). Finally, the combination of time course experiments of Hi-C combined with ChIP-seq against the SMC complex in *B. subtilis* disclosed the dynamics of loading and translocation of the condensin complex in vivo (Wang, Brandão, Le, Laub, & Rudner, 2017).

In this chapter, we will present a detailed protocol to perform 3C-seq and ChIP-seq in bacteria. We will discuss the advantages and limitations of these techniques as well as the advantages of combining these protocols to study the dynamics of chromosome folding.

2. 3C-SEQ: THE PROTOCOL

2.1 Buffers and Reagents

- 37% Formaldehyde solution (Sigma-Aldrich).
- 2.5 *M* Glycine solution prepared in sterile water and filtrated using a 0.22-μm unit filter.
- TE buffer: 10 m*M* Tris–HCl pH 7.5, 1 m*M* EDTA. We advise to use 100 × TE buffer (Sigma-Aldrich) as stock solution.
- Ready-Lyse Lysozyme™ 4000 KU.
- Restriction enzyme (RE). We recommend using high concentration stocks.
- RE buffer advised for the RE.
- 10% Sodium dodecyl sulfate (SDS) solution (w/v). Prepared in sterile water. We advise to use 20% SDS (Sigma-Aldrich) as stock solution.
- 10% Triton X-100 solution (v/v). Prepared in sterile water. We advise to use 100% Triton X-100 (Sigma-Aldrich) as stock solution.
- 10 × Ligation buffer: 500 m*M* Tris–HCl pH 7.4, 100 m*M* $MgCl_2$, 100 m*M* DTT. Prepared in sterile water and filtrated using a 0.22 μm membrane. Stock at −20°C in aliquots.
- 10 mg/mL bovine serum albumin (BSA) prepared in sterile water. Store as 1 mL aliquots at −20°C.
- 100 m*M* adenosine-5′-triphosphate (ATP) pH 7.0 prepared in sterile water. We advise to use adenosine-5′-triphosphate from Sigma-Aldrich.
- T4 DNA ligase 5 U/μL (Thermofisher Scientific).
- 20 mg/mL proteinase K in water.
- 500 m*M* EDTA pH 8.0.
- 3 *M* sodium acetate pH 5.2 (CH_3COONa).
- 2-Propanol.
- Phenol:chloroform:isoamyl alcohol (25:24:1—pH 8).
- 100% Ethanol.
- 10 mg/mL DNase-free RNase A.

- 15-mL disposable conical tubes.
- 50-mL disposable conical tubes.
- 1.5-mL microcentrifugue tubes.
- 2-mL microcentrifugue tubes.

2.2 Cross-Linking of Cell Cultures

1. Grow bacteria cells in the desired medium to the desired OD. For *Pseudomonas aeruginosa* 3C-seq experiments, 1×10^8 total CFUs were used for each experiment.
2. Add fresh formaldehyde solution to the bacterial cultures (usually 3% final concentration, see Section 2.4).
3. Incubate 30 min at room temperature (RT) under moderate agitation.
4. Transfer the culture to 4°C and incubate 30 min under moderate agitation.
5. Add glycine to quench the excess of formaldehyde (250 m*M* final concentration).
6. Incubate 20 min at 4°C under moderate agitation.
7. Harvest the cross-linked cells by centrifuging 10 min at 4°C (3500 × *g*). Discard the supernatant.
8. Suspend the cells in 10 mL of growth medium.
9. Harvest the cells by centrifuging 10 min at 4°C (3500 × *g*). Discard the supernatant.
10. Suspend the cells in 1 mL of growth medium and transfer them into a 2-mL microcentrifuge tube.
11. Harvest the cells by centrifuging 10 min at 4°C (3500 × *g*). Discard the supernatant.
12. Keep the pellet at −80°C until use. Storage time longer than 4 weeks is not advised.

2.3 3C Library Preparation

2.3.1 Cell Lysis

13. Thaw the pellet by gently suspending in 550 μL of 1 × TE.
14. Add 4 μL of Ready-Lyse™ Lysozyme solution.
15. Incubate 20 min at RT.
16. Add 30 μL of SDS 10% (0.5% final concentration).
17. Incubate 10 min at RT.

2.3.2 RE Digestion

18. Transfer 50 μL of cells into 1 × 1.5-mL microcentrifugue tube each containing 450 μL of reaction buffer (50 μL of 10 × restriction buffer, 50 μL 10% Triton X-100, and water up to 450 μL).
19. Repeat this step for another 8 × 1.5-mL microcentrifugue tubes.
20. Add 100 U of RE to the 9 × tubes. After incubation, keep one of these tubes as RE digestion control (RE+).
21. Transfer 50 μL of cells into 1 × 1.5-mL microcentrifugue tube containing 450 μL of reaction buffer (50 μL of 10 × restriction buffer, 50 μL 10% Triton X-100, and water up to 450 μL). After incubation, keep this tube as nondigested control (RE−).
22. Incubate 3 h at 37°C under shaking (see Section 2.4).
23. Add 50 μL of proteinase K to the RE+ and RE− controls. Incubate 65°C overnight.
24. Centrifuge the rest of tubes at 16,000 × *g* during 20 min at RT.
25. Discard the supernatants.
26. Suspend each pellet in 500 μL of sterile water.

2.3.3 Ligation

27. Transfer two tubes into 1 × 15-mL falcon tubes, each containing 7 mL of reaction buffer (800 μL of 10 × ligation buffer, 80 μL of 10 mg/mL BSA, 80 μL of 100 m*M* ATP, and sterile water up to 8 mL).
28. Repeat this step for another 3 × 15-mL falcon tubes.
29. Add 125 U of T4 DNA ligase to each tube. Mix by inverting the tubes three times.
30. Incubate 4 h at 16°C in a water bath.
31. Add 100 μL of 500 m*M* EDTA to each tube to stop the reaction.
32. Add 100 μL of proteinase K to each tube to degrade proteins.
33. Mix by gently inverting the tubes three times and incubate overnight at 65°C to reverse the cross-linking.

2.3.4 DNA Purification

34. Cool the 4 × 15-mL falcon tubes, the RE+ and the RE− microfuge tube to RT.
35. Keep the RE+ and the RE− microfuge tubes on ice to be further processed (see step 45).
36. Transfer the content of 2 × 15-mL falcon tubes into 1 × 50-mL falcon tube.
37. Repeat this step with the other 2 × 15-mL tubes.

38. Add to each 50-mL falcon tube 1.6 mL of 3 *M* CH_3COONa (pH 5) and 16 mL of 2-propanol.
39. Mix by gently inverting the tubes three times and incubate at −80°C for 1 h to precipitate the DNA.
    Tubes can be stored at −80°C for several days.
40. Centrifuge the tubes 30 min at 4°C (10,000 × *g*).
41. Carefully, discard the supernatant. Remove traces of supernatant with a pipette.
42. Suspend each pellet in 900 μL of 1 × TE. These tubes contain the 3C libraries.
43. Transfer 450 μL of the 3C library into a 2-mL microcentrifuge tube.
44. Repeat this step with the rest of the 3C library.
45. Add to each tube 450 μL of phenol:chloroform:isoamyl alcohol (25:24:1—pH 8). Include the RE+ and the RE− tubes.
46. Mix by vortexing and then centrifuge the tubes for 5 min at 25°C (16,000 × *g*).
47. Transfer 400 μL of the aqueous phase to a fresh 1.5 mL microfuge.
48. Add 40 μL of 3 *M* CH_3COONa (pH 5.2) and 1 mL of cold ethanol to each tube.
49. Mix by gently inverting the tubes three times and incubate at −80°C for 30 min.
    Tubes can be store at −80°C for several days.
50. Centrifuge the tubes 20 min at 4°C (10,000 × *g*). Discard the supernatants.
51. Wash the DNA pellets with 400 μL of cold 70% ethanol.
52. Centrifuge the tubes 20 min at 4°C (10,000 × *g*). Discard the supernatants.
53. Dry the pellets for 10 min in the SpeedVac™ Concentrator (AL-Q/30°C).
54. Suspend each pellet in 30 μL of 1 × TE supplemented with RNAse A (1 mg/mL final concentration).
55. Incubate the six tubes for 30 min at 37°C.
56. Pool the four tubes of the 3C libraries into a 1.5-mL microfuge tube.
57. Check the quality and the concentration of the 3C library on a 1% agarose gel.

2.4 Notes

1. Culture conditions

Chromosomes of cells grown in either exponential or stationary phase can be analyzed using 3C-seq. When characterizing the impact of mutants involved in chromosome organization, there is a high chance to have anucleated cells or cells with abnormal nucleoids. To limit the defects induced by the mutation, the cells are grown under conditions where growth defects were minimalized. It is interesting to note that these adverse effects will directly influence the ratio DNA:protein:formaldehyde, affecting the cross-linking efficiency and production of valid reads. For these reasons, we recommend to carefully choose the best growth condition to perform 3C-seq. Nucleoid inspection by microscopy of DAPI-stained cells and growth rates followed by CFU determination are good ways to set up the best condition.

2. Formaldehyde cross-linking

Formaldehyde cross-linking is a central step in the success of 3C-seq experiments. Indeed, the probability for a cross-linking event to occur depends on the length of the DNA fragment: the shorter the fragment, the lower the probability to be cross-linked by formaldehyde (Cournac, Marie-Nelly, Marbouty, Koszul, & Mozziconacci, 2012). For this reason, in 3C-seq protocols the advised concentration of FA is 3%, but depending on the strain analyzed and the RE used, this could be further increased (Lioy et al., 2018). Fig. 2 shows an example *P. aeruginosa* contact matrices in the presence of 1%, 3%, or 5% FA. In this particular example, different concentrations of FA will unveil different features. At 1% FA long-range contacts are not favored and CIDs are not well defined. However, folding of the chromosome at very short ranges (10–20 kb) can be observed. When 3% FA is used, CIDs borders begin to be evident and long-range contacts are better defined. Finally, at 5% FA all the features of *P. aeruginosa* chromosome become clear: a local and fine structure of the chromosome is observed, CIDs are well defined and long-range contacts are better outlined.

3. Enzyme selection

One key feature of 3C-seq is the use of REs that cut frequently. This will produce small restriction fragments (ideally around 400 bp) that, when ligated, give a product of around 1 kb in length. 3C-seq libraries are then prepared by purifying and shearing DNA into fragments of 300 bp–1 kb. Within this library, the probability of identifying a valid read (i.e., ligated product) is high, thus purification/enrichment of valid reads by biotinylation followed by streptavidin pull-down is not necessary. We advise the use of 4 bp cutter RE, producing 5′ or 3′ cohesive ends that are available in high concentration (5 U to 10 U/μL). Once the RE is selected, a full screening

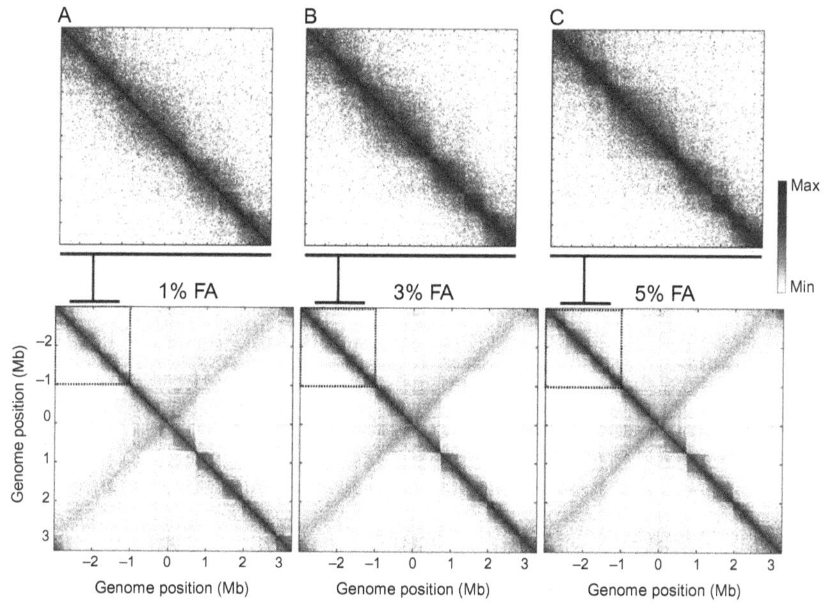

Fig. 2 Normalized contact matrices for *P. aeruginosa* PAO1 cells cross-linked using different concentrations of formaldehyde. Three independent cultures of asynchronous *P. aeruginosa* PAO1 cells were cross-linked using 1% (A), 3% (B), or 5% (C) formaldehyde. The *x* and *y* axes represent the genomic coordinates of the reference genome (*P. aeruginosa PAO1-gif*). *Color scale* reflects the frequency of contacts between two regions of the genome, from *white* (rare contacts) to *dark purple* (frequent contacts). The *inset* shows a zoom of the main diagonal.

of the best restriction buffer should be performed. Usually, commercially advised buffers are the best choice.

Another important parameter to optimize is the RE incubation time. When using optimal RE, 3 h digestion is the recommended time of incubation. However, it could be possible to select RE with high frequency of cutting (e.g., producing restriction fragments of around 100 bp), but digestion times and enzyme concentration should be set up to obtain partial digestion. Fig. 3 shows the profile of cross-linked *P. aeruginosa* cells digested with *Hpa*II during 3 h under increasing concentration of FA. In silico analysis of the 3C-seq data shows that *Hpa*II cuts 1 site every 8 restriction sites at 1% FA, 1 site every 12 restriction sites at 3% FA, and 1 site every 14 restriction sites at 5% FA. This shows that depending on the FA concentration used to cross-link cells, the efficiency of RE digestion will vary. Finally, if 6 bp cutters are selected, we advise performing Hi-C instead of 3C-seq.

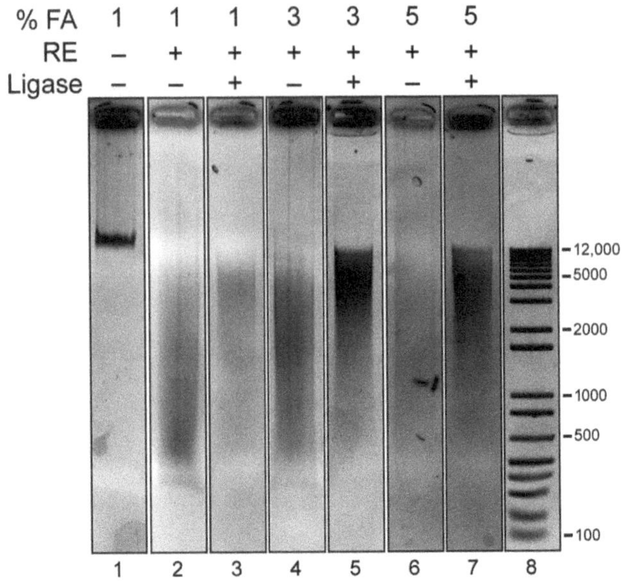

Fig. 3 Profiles of 3C libraries performed using different concentrations of formaldehyde. Genomic DNA is run on a 1% agarose gel after purification. 3C libraries (lane 3, 5, and 7) present a differential and characteristic mobility pattern revealing the ligation of distant DNA fragments brought together by formaldehyde compared to nondigested DNA (lane 1) or digested DNA (lane 2, 4, and 6). Samples were treated as described in the protocol. For simplicity, only nondigested DNA cross-linked with 1% formaldehyde is shown (lane 1). Molecular weight marker is shown in lane 8.

4. Ligation

Ligation is the last step of the 3C protocol. If successful, it will reveal short- and long-range contacts within the chromosome. However, efficient cross-linking and enzyme digestion will directly impact the success of the step. We advise a ligation using high concentration ligase during 4 h at 16°C. Longer incubation times will not affect the final products.

5. Assessing 3C library quality

Once cells are lysed, chromosomes digested, and ligation carried out, the completion of the 3C library needs to be assessed by gel electrophoresis. Although the presence of a shift after ligation does not guarantee the success of the 3C-seq experiment, it is a good indication. Fig. 3 shows the 3C profiles in the presence of different FA concentrations (1%, 3%, and 5%). In all the conditions, a shift in the DNA profile is observed after ligation (lanes +RE/−ligase vs +RE/+ligase).

3. CHIP-SEQ: THE PROTOCOL

3.1 Buffers and Reagents

- 37% Formaldehyde solution (Sigma-Aldrich).
- 2.5 *M* glycine solution prepared in sterile water and filtrated using a 0.22 μm membrane.
- Lysis buffer: 50 m*M* Tris pH 7.4 with 150 m*M* NaCl and 1 m*M* EDTA and cOmplete™, Mini Protease Inhibitor (1 tablet every 10 mL of lysis buffer). To be prepared in sterile water.
- Tris-buffered saline: 50 m*M* Tris–HCl pH 7.4, 150 m*M* NaCl. To be prepared in sterile water.
- Ready-Lyse™ Lysozyme solution (4000 KU).
- 10% Triton X-100 solution (v/v). Prepared in sterile water. We advise to use 100% Triton X-100 (Sigma-Aldrich) as stock solution.
- ANTI-FLAG M2 Affinity Gel (Sigma-Aldrich).
- 3 × FLAG peptide elution solution: Prepared following manufacturer's instructions (Sigma-Aldrich).
- 20 mg/mL proteinase K in water.
- 3 *M* sodium acetate pH 5.2 (CH_3COONa).
- Phenol:chloroform:isoamyl alcohol (25:24:1—pH 8).
- 100% Ethanol.
- 10 mg/mL DNase-free RNase A.
- 15-mL disposable conical tubes.
- 1.5-mL microcentrifuge tubes.
- 2-mL microcentrifuge tubes.

3.2 Cross-Linking of Cell Cultures

1. Grow bacteria cells in the desired medium to the desired OD.
2. Add formaldehyde 1% final concentration to cross-link. Mix carefully.
3. Incubate 30 min with gentle agitation at RT.
4. Add glycine to quench the excess of formaldehyde (250 m*M* final concentration).
5. Incubate 15 min with gentle agitation at RT.
6. Perform two consecutive washes of the cells with 10 mL of 1 × TBS by centrifuging 10 min at 4°C (3500 × *g*). Discard the supernatants and keep the cell pellet.

Pellets can be stored at −80° C for several days.

3.3 Cell Lysis

7. Suspend the cell pellet in 0.5 mL of lysis buffer.
8. Add 4 μL of Ready-Lyse™ Lysozyme solution.
9. Incubate 20 min at RT.
10. Add Triton to 0.5% final concentration.
11. Incubate 10 min at RT.
12. Adjust volume to 1 mL using lysis buffer. Sonicate cells using Focus-ultrasonicator with AFA (Focus-ultrasonicator Covaris S220): *10–30 min, Intensity 4, Peak incident 140 W, Duty cycle 5%, and Cycle per burst 200.*
13. Centrifuge the sonicated cells 30 min at 4°C (18,000 × *g*).
14. Recover the supernatant in a 2-mL falcon tube.
15. Transfer 50 μL into a fresh tube to be used as the input.

3.4 Batch Immunoprecipitation of FLAG Fusions Proteins

16. Incubate the cells extract with the correct amount of ANTI-FLAG M2 Affinity Gel.
17. Incubate ON at 4°C with gentle mixing. Avoid magnetic stirring system since it will destroy the resin beads.
18. Centrifuge 30 s at 4°C (5000–8000 × *g*). Discard the supernatant using a narrow-end pipette tip.
19. Perform consecutive washes of the resin for 10 min at 4°C with gentle mixing. The volume to be added should be equal to 20 times the total packed gel volume.
    a. Two washes with 1 × TBS + 0.05% Tween-20.
    b. Three washes with 1 × TBS.

3.5 Elution of the Immunoprecipitated Protein

20. Prepare the 3 × FLAG Elution according to manufacturer's protocol.
21. Perform two consecutive elution steps of the resin for 30 min at 4°C with gentle mixing. The volume of the 3 × FLAG Elution solution should be 2.5 times the volume of the total packed gel volume.
22. Centrifuge 30 s at 4°C (5000–8000 × *g*). Transfer the supernatants to a fresh test tube.

3.6 Purification of the Coimmunoprecipitated DNA

23. Add RNase to the supernatant and the input sample (30 μg/mL, final concentration).

24. Incubate 1 h at 37°C.
25. Add proteinase K to the supernatant and the input sample (50 μg/mL, final concentration).
26. Incubate at 65°C overnight.
27. Let the tubes cool down at RT for 20 min.
28. Add to each tube 1 volume of phenol:chloroform:isoamyl alcohol (25:24:1—pH 8).
29. Mix by vortexing and then centrifuge the tubes for 5 min at 25°C (16,000 × *g*).
30. Transfer 400 μL of the aqueous phase to a 1.5-mL microfuge tube.
31. Add 40 μL of 3 *M* CH_3COONa (pH 5) and 1 mL of cold ethanol to each tube.
32. Mix by gently inverting the tubes three times and store at −80°C for 30 min.
33. Centrifuge the tubes 20 min at 4°C (10,000 × *g*). Discard the supernatants.
34. Wash the DNA pellets with 400 μL of cold 70% ethanol.
35. Centrifuge the tubes 20 min at 4°C (10,000 × *g*). Discard the supernatants.
36. Dry the pellets for 10 min in the SpeedVac™ Concentrator (AL-Q/30°C).
37. Suspend each pellet in 50 μL of 10 m*M* Tris–HCl pH 8.5.
38. Check the DNA concentration recovered with Qubit fluorometer.

3.7 Notes

1. Growth conditions and cross-linking

Growth conditions and total volume of culture used to ChIP will greatly depend on the protein to be immunoprecipitated. One key point to consider is the binding capacity of the resin. For the ANTI-FLAG M2 affinity gel, 10 μL of packed gel volume (20 μL of gel suspension) binds more than 1 μg of FLAG fusion protein. For this reason, we highly recommend performing a Western blot to determine the best growth conditions and to confirm that the FLAG tag is well expressed. For *P. aeruginosa* ParB-FLAG, ChIP-seq experiments were carried out in 150 mL of cells grown in minimal medium at 30°C and OD = 0.1 (Lagage, Boccard, & Vallet-Gely, 2016).

To cross-link cells, 1% FA should be used since higher concentrations might affect the ChIP process.

2. Cell lysis and sonication
We strongly recommend using a Focus-ultrasonicator with AFA (Covaris S220 system) since we obtained more consistent results than using a Sonicator Tip. The sonication conditions listed in the protocol are adapted for 50 mL of *P. aeruginosa* cells grown in minimal medium and in exponential phase ($\approx 1 \times 10^9$ total cells). Different growth conditions might require adapting the concentration of lysozyme used and the sonication parameters. Fig. 4 shows the size distribution of purified genomic DNA fragments from cross-linked cells after a time course sonication protocol. Long processing times promote a change in the genomic DNA fragment size and distribution. The desired fragment size distribution will depend on the protein to be immunoprecipitated. If specific DNA-binding proteins are immunoprecipitated (e.g., ParB, H-NS, Fis, etc.) then fragments with a mean size of $\approx$250 bp in length are recommended (Fig. 4, lane 4) (Lagage et al., 2016). When nonspecific DNA-binding proteins are studied (e.g., SMC complexes) we recommend performing ChIP on samples with a mean fragment of DNA size of $\approx$800 bp. This will add a second step of DNA shearing to prepare libraries for NGS sequencing (Minnen et al., 2016; Nolivos et al., 2016).

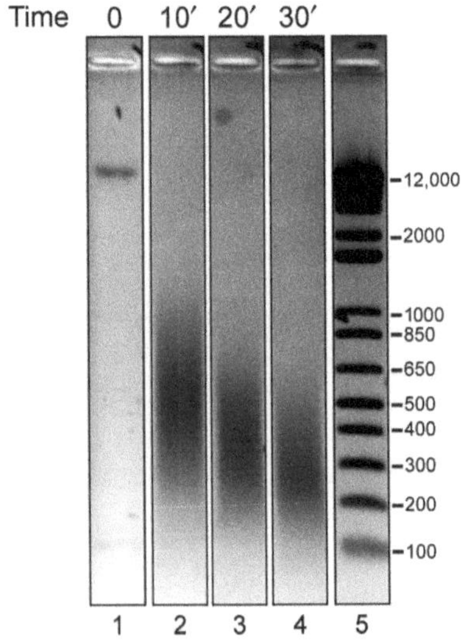

Fig. 4 Size fragment distributions of input DNA after processing cross-linked cells with a Focus-ultrasonicator with AFA. Formaldehyde cross-linked cells were lysed following steps 7–11 of the ChIP-seq protocol and then processed in one miliTUBE 1 mL AFA Fiber using the parameters shown in step 12. A sample of 50 μL was taken for each time point, and 50 μL of lysis buffer was added to keep volume constant. The profile of purified genomic DNA fragments from cross-linked cells after a time course sonication protocol is shown in lanes 1–4. Molecular weight marker is shown in lane 5.

3. Library preparation for NGS sequencing

A detailed protocol for 3C-seq libraries preparation for NGS is already available and will not be further described (Cournac, Marbouty, Mozziconacci, & Koszul, 2016). For ChIP-seq library preparation we recommend choosing a commercial kit, due to the low amount of DNA obtained after ChIP.

Briefly, independent of the technique used (3C-seq or ChIP-seq), the steps used afterward to prepare DNA libraries for NGS are the same:

i. DNA shearing using Focus-ultrasonicators with AFA (omitted step when preparing ChIP-seq DNA libraries with a mean size of $\approx$250 bp in length)
ii. DNA ends repair, 5′ phosphorylation, and dA-tailing
iii. Adapter ligation
iv. Size selection
v. PCR amplification of adaptor ligated DNA
vi. Clean up of amplified library

Our samples are sequenced using Illumina sequencers. Usually one lane of NextSeq 500 (100 million reads) is appropriate to perform 10 independent samples. We do not recommend sequencing 3C-seq and ChIP-seq libraries in the same run, since the output reads for these experiments is different: 3C-seq requires paired-end reads, whereas for ChIP-seq single-end reads are recommended.

4. DATA PROCESSING AND ANALYSIS

4.1 Analysis of 3C-Seq Data

During the last years, several pipelines have been developed to analyze NGS data derived from 3C experiments and to remove experimental biases (Imakaev et al., 2012; Servant et al., 2015; Wingett et al., 2015). Only *HiC-Lib* (based on the iterative correction and eigenvector decomposition; https://bitbucket.org/mirnylab/hiclib) and *SCN* (based on the *sequential component normalization*; https://github.com/koszullab/E_coli_analysis) were successfully used to analyze sequencing data from bacterial 3C experiments (Le et al., 2013; Lioy et al., 2018; Wang et al., 2015). Both methods are appropriate to build normalized contact matrices, and the main difference is the normalization procedure (Cournac et al., 2012; Imakaev et al., 2012). A general description of the main steps in sequencing data processing and filtering is outlined in Fig. 5. The 3C-seq outputs are paired-end FASTQ files, and each read will potentially map to distant sites in the genome. However, it is frequent to find the ligation of two distant DNA fragments in the same read.

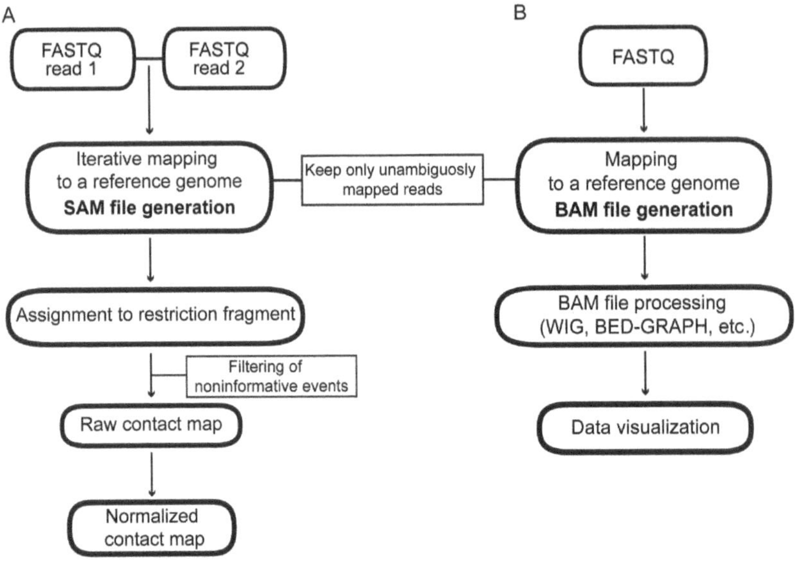

Fig. 5 Outline of a 3C-seq and ChIP-seq pipeline analysis. (A) A 3C-seq pipeline thoroughly maps FASTQ files against the reference genome, generating SAM files. Then SAM files are filtered to retrieve the valid 3C-seq reads which are then used to build a raw contact map. Finally, raw contact maps are normalized to remove experimental bias. (B) FASTQ files are mapped to a reference genome to generate BAM files. BAM files are further processed according to the chosen method of analysis and visualization.

Because of this, individual FASTQ files are mapped by truncating the read to a certain length and aligning this portion to a reference genome and then accumulating alignments over increasing truncation lengths (iterative mapping, Fig. 5A). Afterward, the mapped reads are assigned to the restriction fragment and the data are further filtered to remove noninformative events (loops, weirds, or nondigested fragments) and the efficiency of the RE is determined. Filter reads are then used to establish raw contact maps that will be normalized to remove any experimental bias. As instructions and details of mapping and data filtering were already well described elsewhere (Cournac et al., 2016; Imakaev et al., 2012), they will not be further discussed here.

4.2 Analysis of ChIP-Seq Data

Analysis of ChIP-seq data is simpler than that of 3C-seq (Fig. 5B). Single-end reads are mapped to the reference genome, using a short read aligner (e.g., Bowtie2, http://bowtie-bio.sourceforge.net/bowtie2/index.shtml). Then, mapped files are further processed to convert them into the correct format depending on the visualization tool that will be used. We prefer the wiggle format (WIG) to process ChIP-seq data, since we use a custom made pipeline developed with MATLAB® to smooth, merge, and normalize the data (El Sayyed et al., 2016). Alternatively, a user-friendly interface such as Galaxy (https://usegalaxy.org/) can be used to map the sequencing data to the reference genome. Subsequent data analysis can also be performed using available freeware SeqMonk (https://www.bioinformatics.babraham.ac.uk/projects/seqmonk/) and a spreadsheet.

5. CONCLUSION

The combination of 3C or ChIP methods with NGS has dramatically improved our understanding of chromosome organization in different bacterial models and has allowed us to uncover the activity of condensin-like complexes in shaping bacterial chromosomes. The application of these methods to other bacteria should be straightforward as it does not imply the development of sophisticated tools. The only requirements are knowing the genomic sequence, identifying REs that cut frequently, and determining the proper concentration of formaldehyde to obtain reproducible contacts.

ACKNOWLEDGMENTS

We thank Isabelle Vallet-Gely and Stéphanie Bury-Moné for critical reading of this chapter and to the I2BC genomic facility for high-throughput sequencing.

REFERENCES

Booker, B. M., Deng, S., & Higgins, N. P. (2010). DNA topology of highly transcribed operons in Salmonella enterica serovar Typhimurium. *Molecular Microbiology*, *78*(6), 1348–1364.

Cournac, A., Marbouty, M., Mozziconacci, J., & Koszul, R. (2016). Generation and analysis of chromosomal contact maps of yeast species. *Methods in Molecular Biology*, *1361*, 227–245.

Cournac, A., Marie-Nelly, H., Marbouty, M., Koszul, R., & Mozziconacci, J. (2012). Normalization of a chromosomal contact map. *BMC Genomics*, *13*, 436.

Dekker, J., Rippe, K., Dekker, M., & Kleckner, N. (2002). Capturing chromosome conformation. *Science (New York, N.Y.)*, *295*(5558), 1306–1311.

El Sayyed, H., Le Chat, L., Lebailly, E., Vickridge, E., Pages, C., Cornet, F., et al. (2016). Mapping topoisomerase IV binding and activity sites on the E. coli genome. *PLoS Genetics*, *12*(5), 1–22.

Espeli, O., Mercier, R., & Boccard, F. (2008). DNA dynamics vary according to macrodomain topography in the E. coli chromosome. *Molecular Microbiology*, *68*(6), 1418–1427.

Garcia-Russell, N., Harmon, T., Le, T., Amaladas, N., Mathewson, R., & Segall, A. (2004). Unequal access of chromosomal regions to each other in Salmonella: Probing

chromosome structure with phage lambda integrase-mediated long-range rearrangements. *Molecular Microbiology*, *52*(2), 329–344.
Gruber, S. (2014). Multilayer chromosome organization through DNA bending, bridging and extrusion. *Current Opinion in Microbiology*, *22*, 102–110. Elsevier Ltd.
Higgins, N., Yang, X., Fu, Q., & Roth, J. (1996). Surveying a supercoil domain by using the gamma delta resolution system in Salmonella typhimurium. *Journal of Bacteriology*, *178*(10), 2825–2835.
Imakaev, M., Fudenberg, G., McCord, R. P., Naumova, N., Goloborodko, A., Lajoie, B. R., et al. (2012). Iterative correction of Hi-C data reveals hallmarks of chromosome organization. *Nature Methods*, *9*(10), 999–1003.
Kang, S.-H. L., Vieira, K., & Bungert, J. (2002). Combining chromatin immunoprecipitation and DNA footprinting: A novel method to analyze protein-DNA interactions in vivo. *Nucleic Acids Research*, *30*(10), e44.
Kleckner, N., Fisher, J. K., Stouf, M., White, M. A., Bates, D., & Witz, G. (2014). The bacterial nucleoid: Nature, dynamics and sister segregation. *Current Opinion in Microbiology*, *22*, 127–137.
Lagage, V., Boccard, F., & Vallet-Gely, I. (2016). Regional control of chromosome segregation in Pseudomonas aeruginosa. *PLoS Genetics*, *12*(11), e1006428.
Le, T. B. K., Imakaev, M. V., Mirny, L. A., & Laub, M. T. (2013). High-resolution mapping of the spatial organization of a bacterial chromosome. *Science (New York, N.Y.)*, *342*(6159), 731–734.
Lin, D. C.-H., & Grossman, A. D. (1998). Identification and characterization of a bacterial chromosome partitioning site. *Cell*, *92*(5), 675–685.
Lioy, V. S., Cournac, A., Marbouty, M., Duigou, S., Mozziconacci, J., Espéli, O., et al. (2018). Multiscale structuring of the E. coli chromosome by nucleoid-associated and condensin proteins. *Cell*, *172*(4), 771.e18–783.e18.
Manna, D., & Higgins, N. (1999). Phage Mu transposition immunity reflects supercoil domain structure of the chromosome. *Molecular Microbiology*, *32*(3), 595–606.
Marbouty, M., Le Gall, A., Cattoni, D. I., Cournac, A., Koh, A., Fiche, J.-B., et al. (2015). Condensin- and replication-mediated bacterial chromosome folding and origin condensation revealed by Hi-C and super-resolution imaging. *Molecular Cell*, *59*(4), 588–602. Elsevier Ltd.
Minnen, A., Bürmann, F., Wilhelm, L., Anchimiuk, A., Diebold-Durand, M., & Gruber, S. (2016). Control of Smc coiled coil architecture by the ATPase heads facilitates targeting to chromosomal ParB/parS and release onto flanking DNA. *Cell Reports*, *14*(8), 2003–2016.
Niki, H., Yamaichi, Y., & Hiraga, S. (2000). Dynamic organization of chromosomal DNA in Escherichia coli. *Genes & Development*, *14*(2), 212–223.
Nolivos, S., & Sherratt, D. (2014). The bacterial chromosome: Architecture and action of bacterial SMC and SMC-like complexes. *FEMS Microbiology Reviews*, *38*(3), 380–392.
Nolivos, S., Upton, A. L., Badrinarayanan, A., Muller, J., Zawadzka, K., Wiktor, J., et al. (2016). MatP regulates the coordinated action of topoisomerase IV and MukBEF in chromosome segregation. *Nature Communications*, 7, 10466.
Servant, N., Varoquaux, N., Lajoie, B. R., Viara, E., Chen, C.-J., Vert, J.-P., et al. (2015). HiC-Pro: An optimized and flexible pipeline for Hi-C data processing. *Genome Biology*, *16*(1), 259.
Valens, M., Penaud, S., Rossignol, M., Cornet, F., & Boccard, F. (2004). Macrodomain organization of the Escherichia coli chromosome. *The EMBO Journal*, *23*(21), 4330–4341.
Van Dijk, E., Auger, H., Jaszczyszyn, Y., & Thermes, C. (2014). Ten years of next-generation sequencing technology. *Trends in Genetics: TIG*, *30*(9), 418–426.

Wang, X., Brandão, H. B., Le, T. B. K., Laub, M. T., & Rudner, D. Z. (2017). Bacillus subtilis SMC complexes juxtapose chromosome arms as they travel from origin to terminus. *Science*, *355*(6324), 1–32.

Wang, X., Le, T. B. K., Lajoie, B. R., Dekker, J., Laub, M. T., & Rudner, D. Z. (2015). Condensin promotes the juxtaposition of DNA flanking its loading site in Bacillus subtilis. *Genes & Development*, *29*(15), 1661–1675.

Wang, X., Llopis, P. M., & Rudner, D. Z. (2013). Organization and segregation of bacterial chromosomes. *Nature Reviews Genetics*, *14*, 191–203. (February). Nature Publishing Group. Retrieved February 12, 2013, from http://www.nature.com/doifinder/10.1038/nrg3375.

Wingett, S., Ewels, P., Furlan-Magaril, M., Nagano, T., Schoenfelder, S., Fraser, P., et al. (2015). HiCUP: Pipeline for mapping and processing Hi-C data. *F1000Research*, *1310*, 1–12.

CHAPTER THREE

Large-Scale Measurement of mRNA Degradation in *Escherichia coli*: To Delay or Not to Delay

Sandrine Laguerre*, Ignacio González†, Sébastien Nouaille*, Annick Moisan†, Nathalie Villa-Vialaneix†, Christine Gaspin†, Marie Bouvier‡, Agamemnon J. Carpousis‡, Muriel Cocaign-Bousquet*,[1], Laurence Girbal*

*LISBP, Université de Toulouse, CNRS, INRA, INSA, Toulouse, France
†MIAT, Université de Toulouse, INRA, Castanet-Tolosan, France
‡LMGM, CBI, Université de Toulouse, CNRS, Toulouse, France
[1]Corresponding author: e-mail address: cocaign@insa-toulouse.fr

Contents

Methods in Enzymology, Volume 612
ISSN 0076-6879
https://doi.org/10.1016/bs.mie.2018.07.003

Abstract

In this study, we compared different computational methods used for genome-wide determination of mRNA half-lives in *Escherichia coli* with a special focus on the impact on considering a delay before the onset of mRNA decay after transcription arrest. A wide variety of datasets were analyzed coming from different technical methods for mRNA quantification (microarrays, RNA-seq, and RT-qPCR) and different bacterial growth conditions. The exponential decay of mRNA levels was fitted using both linear and exponential models and with or without a delay. We showed that for all the models, independently of mRNA quantification methods and growth conditions, ignoring the delay resulted in only a modest overestimation of the half-life. For approximately 80% of the mRNAs, differences in mRNA half-life values were less than 34 s. The correlation between half-lives estimated with and without a delay was extremely high. However, the slope of the linear regression between the half-lives with and without a delay tended to decrease with the delay. For the few mRNAs for which taking into account the delay influenced the estimated half-life, the impact was dependent on the model and the growth condition. The smallest impact was obtained for the linear model.

1. INTRODUCTION

Bacterial cells rapidly and efficiently adjust their metabolism in response to changing environments. Adaptation to a newly encountered environment proceeds through the regulation of gene expression. Particularly, the messenger RNA (mRNA) levels adjusted via changes in both rates of mRNA synthesis and degradation.

In *Escherichia coli*, the components of the mRNA degradation machinery, the endoribonuclease RNase E, the 3′ exoribonuclease PNPase, the RNA helicase RhlB, and the enolase, forming the degradosome complex, have been extensively characterized (Andrade, Pobre, Silva, Domingues, & Arraiano, 2009; Carpousis, 2002, 2007). RNA degradation is initiated by an RNase E endonucleolytic cleavage followed by a 3′–5′ exonucleolytic degradation of the mRNA fragments (Bouvier & Carpousis, 2011; Carpousis, Luisi, & McDowall, 2009; Richards, Luciano, & Belasco, 2012). A number of molecular mechanisms regulating the degradation of particular transcripts has been described in *E. coli* (for review, see Mohanty & Kushner, 2016). These mechanisms involve mRNA-binding molecules such as small RNAs (Kavita, de Mets, & Gottesman, 2017; Morita, Maki, & Aiba, 2005; Storz, Opdyke, & Zhang, 2004), proteins (e.g., Hfq (Zhang et al., 2003) and CsrA (Yakhnin et al., 2013)), and/or metabolites in riboswitch (Caron et al., 2012).

In addition to these regulations of specific mRNAs, more general mechanisms regulate bulk mRNA stabilities in the cells. Several studies in *E. coli* addressing the global mechanisms controlling mRNA half-lives at the genome-wide scale have been published over the years (Bernstein, Khodursky, Lin, Lin-Chao, & Cohen, 2002; Bernstein, Lin, Cohen, & Lin-Chao, 2004; Chen, Shiroguchi, Ge, & Xie, 2015; Dar & Sorek, 2018; Esquerre et al., 2014; Moffitt, Pandey, Boettiger, Wang, & Zhuang, 2016; Nouaille et al., 2017; Potts et al., 2017; Selinger, Saxena, Cheung, Church, & Rosenow, 2003). These studies contributed in understanding the role of the degradosome proteins on general mRNA degradation (Bernstein et al., 2004) and of the regulatory protein such as CsrA (Esquerre et al., 2016; Potts et al., 2017). In addition, they showed that genome-wide transcript stability is dependent on gene function (Bernstein et al., 2002; Esquerre et al., 2015; Selinger et al., 2003), operon structure (Dar & Sorek, 2018; Esquerre et al., 2015; Selinger et al., 2003), spatial organization (Moffitt et al., 2016), mRNA abundance (Esquerre et al., 2015; Nouaille et al., 2017), and codon usage (Esquerre et al., 2015). The interplay between mRNA degradation, transcription rate, and growth rate was also investigated at the genome-wide level (Chen et al., 2015; Esquerre et al., 2014).

Determination of mRNA half-lives in *E. coli* requires two key steps after transcription arrest by rifampicin addition: mRNA quantification over the time and then the modeling of degradation kinetics. For large-scale half-life estimations, different well-known techniques allowing medium or high-throughput mRNA quantifications (RT-qPCR, microarrays, and RNA-seq) are routinely used. For the modeling aspects, things are slightly different. The classic exponential model based on linear regression on a semilog mRNA level plot has been used for decades for individual mRNA half-life estimations. This method has then been naturally extended to genome-wide analyses (Bernstein et al., 2002, 2004; Esquerre et al., 2014; Potts et al., 2017), but other models are now emerging. Recently, a new approach introduced more complex exponential models considering a delay before the onset of the decay (Chen et al., 2015; Dar & Sorek, 2018; Moffitt et al., 2016) and a degradation-resistant residual mRNA level at the end of the degradation process (Moffitt et al., 2016). The delay in exponential decay is hypothesized to be the time required for the RNA polymerase to complete the transcription initiated prior rifampicin addition. The duration of the delay should therefore be linearly dependent on the length of the transcript (Chen et al., 2015; Moffitt et al., 2016).

The purpose of this study is to describe and compare the different computational methods used for genome-wide determination of mRNA half-lives in *E. coli* and, more particularly, to focus on the question of the impact of the delay before mRNA decay on the estimated mRNA half-life. Different models were considered in this study: the linear model applied to log transformed data vs exponential models. We also tested various mRNA quantification methods (microarrays, RNA-seq, and RT-qPCR) and growth conditions (exponential and stationary phases in batch culture and slow growth in a continuous culture) of *E. coli* cells. Our results show that for each model, taking a delay into account before the onset of the mRNA decay has a modest impact on mRNA half-lives for most of the mRNAs, whatever the mRNA quantification method and growth conditions. With a delay, the estimated mRNA half-lives are generally slightly lower than those estimated without a delay, but both variables were still highly correlated. The extent of the delay nevertheless affects the slope of the linear regression between mRNA half-lives estimated with and without a delay in a growth condition- and model-dependent manner.

2. MEASURING mRNA DEGRADATION RATES

2.1 Methods

Methods to measure mRNA degradation rates fall into two categories. The first approach involves pulse-chase methods in which the cellular pool of mRNA is labeled, for example, with radioactive uridine, and then the label is chased with cold uridine. Turnover is measured as the rate of decrease of radioactivity in the cellular pool mRNA. In principle, this method has the advantage of not perturbing cell growth or metabolism. Nevertheless, it is generally accepted that pulse-chase methods are not feasible in cultures of bacteria since mRNA lifetimes are comparable to the time required for external nucleosides to equilibrate with cellular pools of nucleosides. The second and commonly used approach involves stopping transcription and then measuring the decay of preexisting mRNA. Although it is possible to heat inactivate transcription in mutant strains of *E. coli* expressing a thermolabile RNA polymerase (Little & Dennis, 1979), the most widespread method uses the drug rifampicin to shutoff transcription. Classical mRNA turnover measurements before the era of genomics were made using radioactively labeled bulk RNA or Northern blotting of a specific transcript. Degradation was treated as a first-order reaction. mRNA levels measured from 2 to 20 min after the addition of rifampicin were fitted to an exponential decay curve. In general, values ranging from 100% to 20% of the maximal level were used to determine a half-life. As will be discussed in the following sections, the advent of genome-wide studies has made it necessary to develop automatic methods to treat large data sets resulting in the determination of degradation rates of thousands of messages.

2.2 Rifampicin

Rifampicin is a wide-spectrum antibiotic in the frontline of the treatment of tuberculosis, leprosy, and other bacterial infections that are difficult to cure. The antibacterial activity of rifampicin is due to its inhibition of transcription (Alifano, Palumbo, Pasanisi, & Tala, 2015). Molecular genetic studies showed that resistance to rifampicin in *E. coli* arises from mutations in the *rpoB* gene resulting in amino acid substitutions or small deletions in a specific region of the β-subunit of RNA polymerase (Jin & Gross, 1988, 1989). Structural and mechanistic studies showed that rifampicin blocks growth of an RNA chain past 2 or 3 nucleotides. Rifampicin binds to the exit channel of the RNA polymerase by making specific contacts that involve the β-subunit of RNA polymerase. Rifampicin blocks the exit channel thereby inhibiting transcription (Campbell et al., 2001). Elongating RNA polymerase is insensitive to rifampicin inhibition because RNA in the exit channel blocks rifampicin binding. Thus, runoff transcription continues after the addition of rifampicin.

2.3 Practical Considerations

Lags in the inhibition of transcription after rifampicin addition are due to two effects. First, rifampicin entry into the cell can take several minutes. For this reason, high concentrations of rifampicin, typically $500\,\mu g\,mL^{-1}$, are used to achieve full inhibition as rapidly as possible. Although this strategy is effective in many laboratory strains of bacteria, the rate of entry of rifampicin is a concern when studying environmental or pathogenic bacteria that have drug-resistant cell capsules. Second, since elongating RNA polymerase is insensitive to rifampicin, initiated transcripts will be transcribed to completion. The average elongation rate in *E. coli* is 42 nucleotides per second under normal growth conditions (Proshkin, Rahmouni, Mironov, & Nudler, 2010). Runoff transcription will generally terminate 1–3 min after the addition of rifampicin except for a few very long transcription units. A question that often arises is whether rifampicin has secondary effects, for example, on translation. The response to this question is twofold.

First, since resistant *E. coli* strains with mutations in the *rpoB* gene appear to grow normally in the presence of rifampicin, secondary effects are likely to be minor. Second, an important consideration in mRNA degradation measurements using rifampicin is to determine initial rates of degradation when the physiological state of the cell is the same as before drug addition. An obvious concern is that depletion of mRNA leads to the progressive arrest of translation and thereby increasing concentrations of free mRNA-binding proteins, ribosomes, and stable RNAs such as tRNAs. These changes could account for seemingly paradoxical results such as an increase in the level of an mRNA shortly after the addition of rifampicin or the apparent stabilization of an mRNA or an mRNA fragment at late times after the addition of rifampicin.

3. GENOME-WIDE mRNA QUANTIFICATION

3.1 mRNA Extraction and Quantification

After transcription arrest, mRNA levels are quantified over time. Usually, a reference time point (T0) is collected just before rifampicin addition to use for setting the initial mRNA levels in the cells (Bernstein et al., 2002; Esquerre et al., 2014). The main techniques used in *E. coli* to follow mRNA level decay are microarrays (Bernstein et al., 2002, 2004; Esquerre et al., 2014; Lin, Singh, Bernstein, & Lin-Chao, 2008; Selinger et al., 2003) or RNA-seq experiments (Chen et al., 2015; Dar & Sorek, 2018; Moffitt et al., 2016; Potts et al., 2017; this study) for large-scale mRNA quantification and RT-qPCR for medium size quantification (Chen et al., 2015; Nouaille et al., 2017; Potts et al., 2017). All methods require total RNA extraction (using RNA purification kit (Nouaille et al., 2017; Potts et al., 2017; Selinger et al., 2003), phenol-based isolation methods (Chen et al., 2015; Dar & Sorek, 2018; Esquerre et al., 2014), or the RNAsnap protocol (Moffitt et al., 2016; Stead et al., 2011)) followed by DNase I treatment to remove genomic DNA contamination and then complementary DNA (cDNA) synthesis by a reverse transcriptase. In RNA-seq experiments, rRNAs are removed (usually using the Ribo-Zero kit for Gram-negative bacteria, Illumina) before retrotranscription (Chen et al., 2015; Moffitt et al., 2016; Potts et al., 2017; this study). For microarrays, labeled cDNA was originally hybridized onto homemade PCR product arrays (Bernstein et al., 2002; Lin et al., 2008) which are now replaced by commercial high-density oligonucleotide arrays (Affymetrix (Selinger et al., 2003), Nimblegen (Esquerre et al., 2014)). cDNA libraries constructed for

RNA-seq experiments are mainly sequenced using Illumina HiSeq 2000 (Chen et al., 2015; Moffitt et al., 2016; Potts et al., 2017) and HiSeq 2500 (this study). In this study, reads from nine samples were aligned to *E. coli* K12 MG1655 genome sequence (ENA, Assembly GCA_000005845.2) using Bowtie2 version 2.1.0. Counting was performed with HTSeq-count version 0.9.1 using the "intersection non empty" mode. Bedtools 2-2.20.1 were used for more general and common genomic questions.

3.2 Normalization

A step of normalization before comparing transcript levels across samples is also necessary to account for technical variations between time points. Exogenous RNA spike-ins are commonly used for normalization (e.g., in house in vitro transcribed RNAs (Chen et al., 2015; Moffitt et al., 2016) or commercial RNAs like the ERCC RNA spike-ins (Ambion) (Dar & Sorek, 2018; Potts et al., 2017; this study)). Spike-ins are either added to the sample before RNA extraction (Chen et al., 2015; Moffitt et al., 2016 #8) or to purified total RNA (Potts et al., 2017; Selinger et al., 2003; this study). A second approach based on internal normalization controls is also used in microarrays and RT-qPCR experiments. Normalization coefficients for microarrays are computed from rRNA spot intensities (Bernstein et al., 2002; Lin et al., 2008) or invariant probeset intensities (Esquerre et al., 2014). In RT-qPCR experiments, housekeeping gene amplifications (e.g., *ihfB*) are used as internal normalization controls (Nouaille et al., 2017). Finally, mRNAs with low spot intensity on microarrays (Bernstein et al., 2002; Selinger et al., 2003) or low read count in RNA-seq experiments (Potts et al., 2017) are excluded in some cases from further analysis.

4. mRNA HALF-LIFE ESTIMATION

mRNA decay over time after transcription arrest is fitted using basic or more complex exponential decay models. The most sophisticated one to date is from Moffitt et al. (2016):

$$\mathrm{mRNA}(t) = \mathrm{mRNA}_0, \quad \text{if } 0 < t < \alpha,$$
$$\mathrm{mRNA}(t) = \mathrm{mRNA}_f + (\mathrm{mRNA}_0 - \mathrm{mRNA}_f)\, e^{-k_{\mathrm{deg}} * (t-\alpha)}, \quad \text{if } t \geq \alpha,$$

where mRNA_0 is the initial level of mRNA, mRNA(t) is the level of subsequent time point, mRNA_f is the residual mRNA level at the end of the decay process, α is the delay before the onset of the exponential decay after transcription arrest by rifampicin, and k_{deg} is the rate of exponential decay. The mRNA half-life $t_{1/2}$ is then inversely proportional to k_{deg}: $t_{1/2}=\ln(2)/k_{deg}$. In this work, these parameters were estimated using three approaches based on a linear model (model I), an exponential model (model II) (Chen et al., 2015), and finally on an exponential model with the residual mRNA level ($mRNA_f$) (model III) (Moffitt et al., 2016).

4.1 Decay Models

Model I: The simplest way to estimate the exponential decay rate is to firstly perform a log transformation of the normalized mRNA levels and then to use a linear model on the log transformed values. Model I allows including the delay α before the onset of the exponential decay using a piecewise linear model. It does not include any residual mRNA level at the end of the decay process.

$$\log(y) = \log(\mathrm{mRNA}_0), \text{ if } 0 < t < \alpha,$$
$$\log(y) = \log(\mathrm{mRNA}_0) - k_{\mathrm{deg}}{}^*(t-\alpha), \text{ if } t \geq \alpha,$$

The parameters log($mRNA_0$), k_{deg}, and α were estimated using the segmented package (Muggeo, 2008) of the R programming language (R 3.4.3 2017-11-30).

If no delay is included (α is set to 0), model I becomes a simple linear model on log transformed values, as classically used in the literature (Bernstein et al., 2002, 2004; Esquerre et al., 2014; Potts et al., 2017). Log transformed mRNA levels thus follow the equation:

$$\log(y) = \log(\mathrm{mRNA}_0) - k_{\mathrm{deg}}{}^*t$$

The parameters log($mRNA_0$) and k_{deg} were estimated using the lm function of the R programming language, and log($mRNA_0$) and k_{deg} were, respectively, retrieved as the intercept and the slope fitted by the function.

Model II: This model corresponds to an exponential model with the delay α before the onset of the exponential decay and without any residual mRNA level $mRNA_f$ according to the following equations (Chen et al., 2015):

$$\mathrm{mRNA}(t) = \mathrm{mRNA}_0, \quad \text{if } 0 < t < \alpha,$$
$$\mathrm{mRNA}(t) = \mathrm{mRNA}_0\, e^{-k_{\mathrm{deg}}*(t-\alpha)}, \quad \text{if } t \geq \alpha.$$

The parameters $mRNA_0$, k_{deg}, and α were estimated using the nls function of the R programming language.

Model III: This model corresponds to model II implemented with the possibility that the mRNA decay is not complete and leads to a residual $mRNA_f$ at the end of the decay process. Model III corresponds to the following equations (Moffitt et al., 2016):

$$mRNA(t) = mRNA_0, \quad \text{if } 0 < t < \alpha,$$

$$mRNA(t) = mRNA_f + (mRNA_0 - mRNA_f)e^{-k_{deg}*(t-\alpha)}, \quad \text{if } t \geq \alpha.$$

The parameters $mRNA_0$, $mRNA_f$, k_{deg}, and α were estimated using the nls function of the R programming language after log transformation of mRNA levels.

Parameters of models II and III were estimated in the presence of a delay, as described in the above equations, and also in the absence of a delay when the α value was set to 0.

4.2 Accuracy Criteria of the Estimation

Selection of reliable mRNA half-life estimations is based on accuracy criteria of these estimations. Accuracy criteria are usually the R^2 for linear model (Bernstein et al., 2002; Potts et al., 2017) or a measure based on the coefficient of variation of the estimates of k_{deg} ($error_{k_{deg}}/k_{deg}$) (Chen et al., 2015; Esquerre et al., 2014; Moffitt et al., 2016). More specifically, the measure used in this study is the RMSE divided by the mean of the observed values (RMSE being the sum of the square roots of the residuals of the model divided by the number of values). After examining the plots of the observed values across time, ordered by this accuracy measure, we defined an empirical threshold of 0.15 below which half-lives were considered as accurate enough.

5. EFFECT OF THE DELAY BEFORE THE ONSET OF EXPONENTIAL DECAY ON mRNA HALF-LIFE ESTIMATION

5.1 Datasets of mRNA Levels Over the Time

To examine the effect on mRNA half-life estimation of considering or not the delay for RNA polymerase to terminate transcription initiated before the addition of rifampicin, we used three datasets of mRNA levels over the time. They corresponded to total RNA samples collected over the time (between 7 and 12 data points) after rifampicin addition ($500\,\mu g\,mL^{-1}$) in *E. coli* MG1655 cells growing at 37°C at different growth rates: at $0.63\,h^{-1}$ in exponential phase in a batch culture on M9 minimum medium supplemented with 3 gL^{-1} glucose (Esquerre et al., 2014), at zero-growth in stationary phase in a batch culture on LB (C. Dressaire et al., unpublished data), and at 0.1 h^{-1} a slow growth imposed in a continuous culture on M9 minimum medium supplemented with 3 gL^{-1} glucose (Esquerre et al., 2014). The genome-wide quantifications of mRNA levels were performed with microarrays for the exponential and stationary phases (C. Dressaire et al., unpublished data; Esquerre et al., 2014), whereas for the slow growth condition RNA-seq (this study) and RT-qPCR (Esquerre et al., 2014) were used.

5.2 Estimations of the Delay Before the Onset of the mRNA Exponential Decay

When a delay is hypothesized before mRNA decay, the three models, model I (linear), model II (exponential), and model III (exponential with residual mRNA level), provide the estimations of the delay α and the mRNA half-life ($t_{1/2}$ with delay). We applied the three models to estimate the delay and the $t_{1/2}$ with delay to the three mRNA level datasets corresponding to different growth conditions (exponential, stationary, and slow growth as described earlier). We only considered half-lives estimated with accuracy (coefficient of variation of the linear regression <0.3 for model I and coefficient of variation of RMSE <0.15 for models II and III). Extremely high half-lives above 100 min were excluded from the analysis. One can notice that the number of selected half-lives was dependent on the model. Independently of the growth conditions, low and high numbers of selected half-lives were obtained for model III and model I, respectively (Table 1).

The estimated delays in mRNA decay were then clustered into five bins for the three models in each growth condition (Table 1). The delay spanned between 0 and 18 min with the maximal value obtained with model II whatever the growth condition. At least 60% of the mRNAs exhibited a delay estimated to 0. The highest numbers of zero-delay were obtained at slow growth condition with a percentage always above 86% whatever the model used. Delays lower or equal to 2 min were obtained for 86% (model II in stationary phase) to 99% (models II and III during slow growth) of the mRNAs. When we summed the three growth conditions, more than 68% of the mRNAs displayed an estimated delay of 0, more than 93% an estimated delay lower or equal to 2 min, and only 2% a delay above 5 min. Together these results showed that a short delay ($\leq$2 min) was estimated for most of the mRNAs for all the models and growth conditions.

Table 1 Repartition of the Delay Values

Delay (min)	Exponential Phase	Stationary Phase	Slow Growth in Continuous Culture	Pool of the Three Growth Conditions
Model I				
0	2074 (64%)	1607 (60%)	3332 (88%)	7013 (72%)
(0, 2]	957 (30%)	782 (29%)	380 (10%)	2119 (22%)
(2, 3]	143 (4%)	88 (3%)	52 (1%)	283 (3%)
(3, 5]	65 (2%)	70 (3%)	25 (1%)	160 (2%)
(5, 14]	0 (0%)	154 (6%)	2 (0%)	156 (2%)
Total	3239 (100%)	2701 (100%)	3791 (100%)	9731 (100%)
Model II				
0	1435 (60%)	1606 (60%)	2045 (86%)	5086 (68%)
(0, 2]	824 (34%)	707 (26%)	319 (13%)	1850 (25%)
(2, 3]	99 (4%)	170 (6%)	0 (0%)	269 (4%)
(3, 5]	33 (1%)	66 (2%)	12 (0%)	111 (1%)
(5, 18]	0 (%)	121 (5%)	5 (0%)	126 (2%)
Total	2391 (100%)	2670 (100%)	2381 (100%)	7442 (100%)
Model III				
0	1493 (67%)	1931 (76%)	1519 (89%)	4943 (76%)
(0, 2]	598 (27%)	371 (15%)	169 (10%)	1138 (18%)
(2, 3]	99 (4%)	99 (4%)	0 (0%)	198 (3%)
(3, 5]	32 (1%)	57 (2%)	21 (1%)	110 (2%)
(5, 15]	0 (0%)	96 (4%)	4 (0%)	100 (2%)
Total	2222 (100%)	2554 (100%)	1713 (100%)	6489 (100%)

The delays before the onset of the mRNA exponential decay were estimated from the mRNA degradation kinetics using the three models I (linear), II (exponential), and III (exponential with a residual mRNA level) in the three growth conditions (exponential, stationary, and slow growth) and broken into five bins. The number of selected mRNAs in each delay bin is given. For each model, the number of mRNAs obtained in a given bin of delay across the three growth conditions were pooled and provided in the column entitled "pool of the three growth conditions." We also summed the number of mRNAs (line "Total") per model and growth condition.

5.3 Comparison of mRNA Half-Lives Estimated With and Without a Delay

Our goal was to study the influence of introducing a delay before the onset of the mRNA decay on the estimated value of the mRNA half-life. We

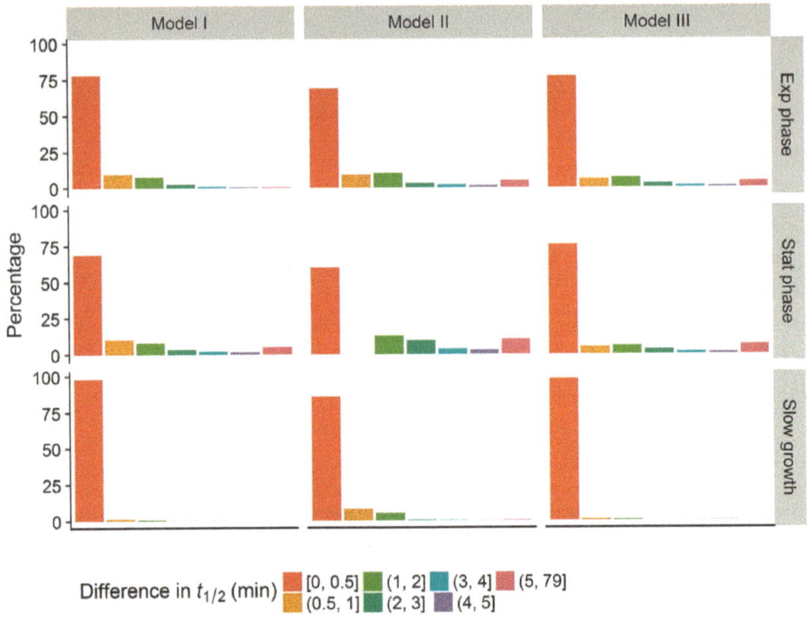

Fig. 1 Distribution of the absolute differences between mRNA half-lives estimated with and without a delay before the onset of the mRNA exponential decay. The differences are broken in six bins. The distribution is shown for each model and growth condition.

therefore estimated mRNA half-life of all mRNAs without taking into account a delay ($t_{1/2}$ without delay) by setting the delay to a zero value ($\alpha = 0$) in models I, II, and III. We first assessed the absolute difference between $t_{1/2}$ with delay and $t_{1/2}$ without delay. Fig. 1 shows the distribution of the values of this difference according to the growth conditions and the used models. When considering all growth conditions and models, the absolute differences of $t_{1/2}$ values spanned from 0 to 79 min, 90% were below 1.8 min, and 80% were below 0.56 min. Analysis of the sign of the differences showed that most (97.5%) of the half-lives estimated with a delay were lower or equal to the half-lives estimated without a delay indicating that the introduction of a delay tended to decrease the estimated half-life value. We conclude that not taking the delay into account in the estimation of the mRNA half-life results, for approximately 80% of the mRNAs, in an overestimation of the half-life by less than 34 s.

5.4 Correlation Between mRNA Half-Lives Estimated With and Without a Delay

To examine the relationship between half-lives estimated with and without a delay, we assessed the correlation between the estimated values of those two parameters. Fig. 2A shows scatter plots of the estimated half-lives without delay vs with delay from the three models in all growth conditions. We can observe that the half-lives estimated with a delay were never meaningfully higher than the half-lives estimated without a delay. When considering all sets of values, the correlation coefficient between $t_{1/2}$ without delay and $t_{1/2}$ with delay was 0.99 showing an extremely high correlation. The detailed values of the correlation coefficient for each model and growth condition are given in Fig. 2B. All the correlation coefficients were above 0.95 except for model II in slow growth condition, where the value was equal to 0.77. To illustrate the high correlation between half-lives estimated with and without a delay, the points were shaded according to their distance to the $y = x$ line (Fig. 2C). The Euclidean distance from each gene represented by its values of $t_{1/2}$ with and without delay and the $y = x$ line spanned between 0 and 56 min, but 88% of the values were strictly lower than 1 min and 94% of the values were lower than 2 min. Altogether these results demonstrate that whatever the growth condition and the model used, half-lives estimated with and without a delay were very consistent.

5.5 Influence of the Delay Extent

To investigate the potential effect of the delay value on the correlation between half-lives estimated with and without a delay, linear regressions of $t_{1/2}$ with delay on $t_{1/2}$ without delay were performed within each bin of delay for each condition and model. Fig. 3A presents the scatter plots of the half-life values estimated with and without a delay colored according to the delay bin. Results of linear regressions (slope estimation, number of estimated $t_{1/2}$ in the delay bin, adjusted R^2) are provided in Fig. 3B. For each condition and model, whatever the delay, the linear relationship between half-lives estimated with and without a delay was of good quality (minimal value of adjusted R^2 was of 0.84). We showed that the values of the slope of the linear regression between the two estimated parameters varied with the delay: the slope decreased when the delay increased. The extent of slope reduction (up to sevenfold) was dependent on the growth conditions. The smallest slope reduction (no more than 25% slope reduction for a delay between 3 and 5 min) was generally obtained in stationary phase, when the mRNAs are significantly stabilized (Chen et al., 2015; C. Dressaire et al., unpublished data).

The boxplots of the differences between half-lives estimated with and without a delay were calculated according to the delay (Fig. 4). For the three growth conditions, the differences were the smallest for model I compared

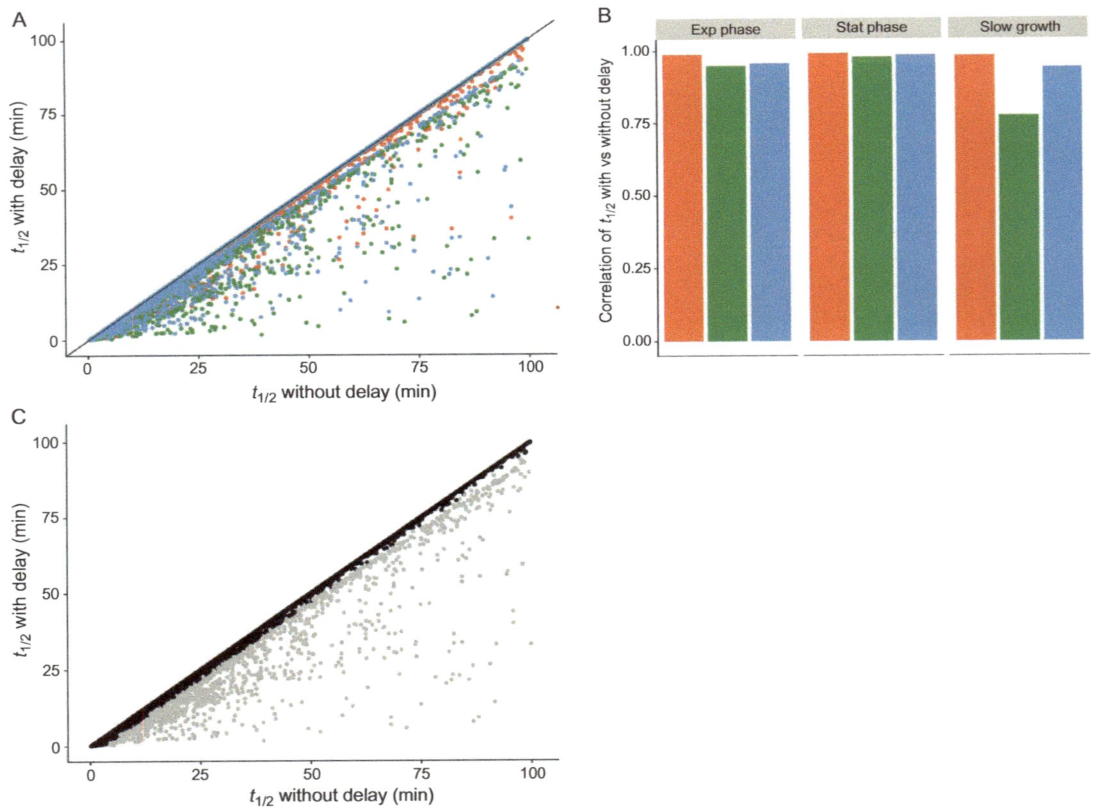

Fig. 2 Correlation between mRNA half-lives estimated with and without a delay before the onset of the mRNA exponential decay. (A) Plot of the $t_{1/2}$ estimated with a delay vs $t_{1/2}$ estimated without a delay using model I (points in *red*), model II (points in *green*), and model III (points in *blue*), in the three growth conditions, exponential, stationary, and slow growth ($N = 23{,}662$ pairs of $t_{1/2}$ values). The line $y = x$ line is depicted in *black*. (B) Coefficients of correlation between $t_{1/2}$ estimated with and without a delay for the three models, model I (bars in *red*), model II (bars in *green*), and model III (bars in *blue*), in the three growth conditions. (C) Plot similar to (A) in which the points are shaded according to their distance to the $y = x$ line. The points at a distance lower or equal to 2 min are in *black*, whereas the points at a distance above 2 min are in *gray*.

A

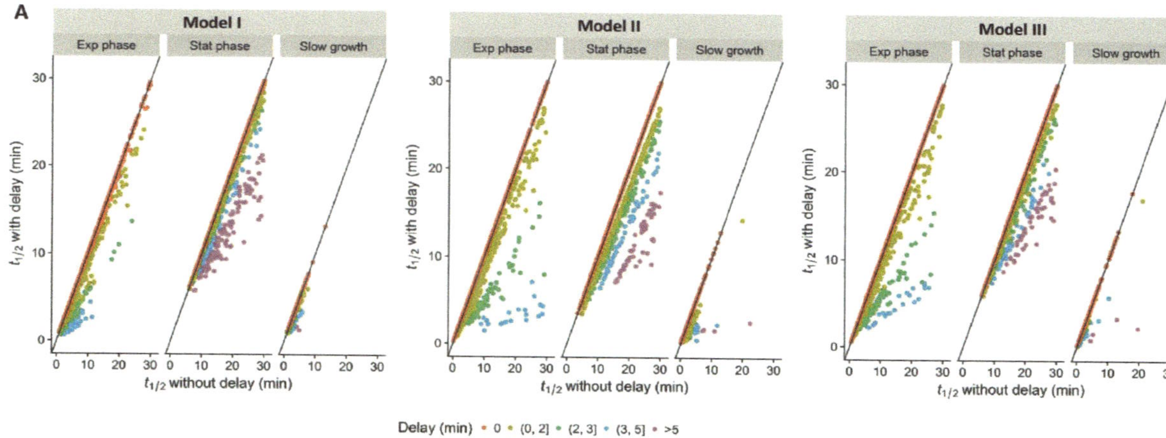

B

	Model I					Model II					Model III				
Delay (min)	0	(0.2]	(2.3]	(3.5]	(5.14]	0	(0.2]	(2.3]	(3.5]	(5.18]	0	(0.2]	(2.3]	(3.5]	(5.15]
Exponential phase															
Slope	0.99	0.84	0.54	0.31	NA	1	0.86	0.45	0.14	NA	1	0.86	0.49	0.21	NA
N	2052	951	142	65	NA	1435	824	99	33	NA	1493	598	99	32	NA
Adj. R^2	1	0.99	0.99	0.96	NA	1	0.99	0.95	0.84	NA	1	0.99	0.95	0.93	NA
Stationary phase															
Slope	1	0.96	0.91	0.85	0.69	1	0.92	0.86	0.74	0.49	1	0.94	0.89	0.76	0.53
N	1030	495	45	56	115	1606	707	170	66	121	1931	371	99	57	96
Adj. R^2	1	1	1	1	0.99	1	1	1	1	0.95	1	1	1	1	0.95
Slow growth															
Slope	0.99	0.92	0.73	0.59	NA	1	0.61	NA	0.27	NA	1	0.84	NA	0.55	NA
N	3332	380	52	25	NA	2045	319	NA	12	NA	1519	169	NA	21	NA
Adj. R^2	1	1	1	0.99	NA	1	0.96	NA	0.88	NA	1	1	NA	0.98	NA

Fig. 3 Influence of the delay value on the correlation between mRNA half-lives estimated with a delay and without a delay. (A) Plots of the $t_{1/2}$ estimated with a delay vs $t_{1/2}$ estimated without a delay for each bin of delay for the three models I (linear), II (exponential), and III (exponential with a residual mRNA level) in the three growth conditions (exponential, stationary, and slow growth). The line $y = x$ line is depicted in *black*. (B) The table provides the slopes, the numbers of $t_{1/2}$ values (N), and the adjusted R^2 of the linear regressions of the $t_{1/2}$ estimated with a delay vs $t_{1/2}$ estimated without a delay for each bin of delay for all the models in the three growth conditions. In these linear regressions the intercept was forced to 0, which makes sense biologically. *NA*, not applicable.

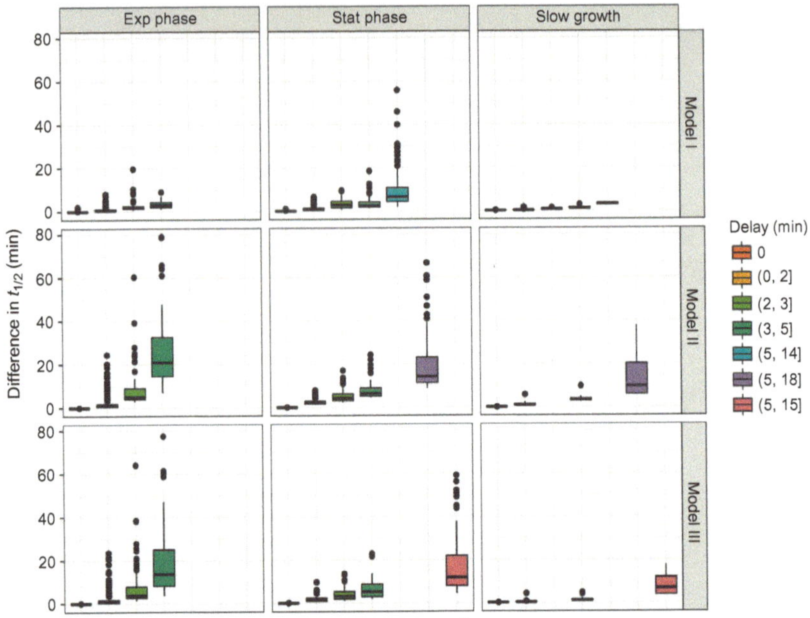

Fig. 4 Boxplots of the difference between half-lives estimated with and without a delay according to the delay bin for each growth condition and model.

to models II and III, even for the highest delay values. This result shows that the impact of the delay was higher in models II and III compared to model I. Therefore, it seems that in the studied growth conditions, the linear model would be the model in which not taking the delay into account would have the smallest impact on mRNA half-life estimation.

5.6 Validation on RT-qPCR mRNA Quantification Datasets

In the slow growth condition (growth rate of $0.1\,h^{-1}$) obtained in a continuous culture, RT-qPCR technique was used to quantify the level of 11 mRNAs over time (Esquerre et al., 2014). For each mRNA, we plotted the C_t values as a function of the time and estimated the k_{deg} slope using the linear model I. In this case, the mRNA half-life was equal to $1/k_{deg}$. Ten mRNA half-lives were estimated with accuracy (coefficient of variation of the slope <0.3). The delays estimated using the piecewise linear model I were below 0.56 min for 9 on 10 mRNAs. For these mRNAs, the difference between half-lives estimated with and without a delay was very small (below 0.11 min). For the remaining gene (*ompC*), the delay was estimated to 7 min leading to a difference of 7 min between half-lives estimated with

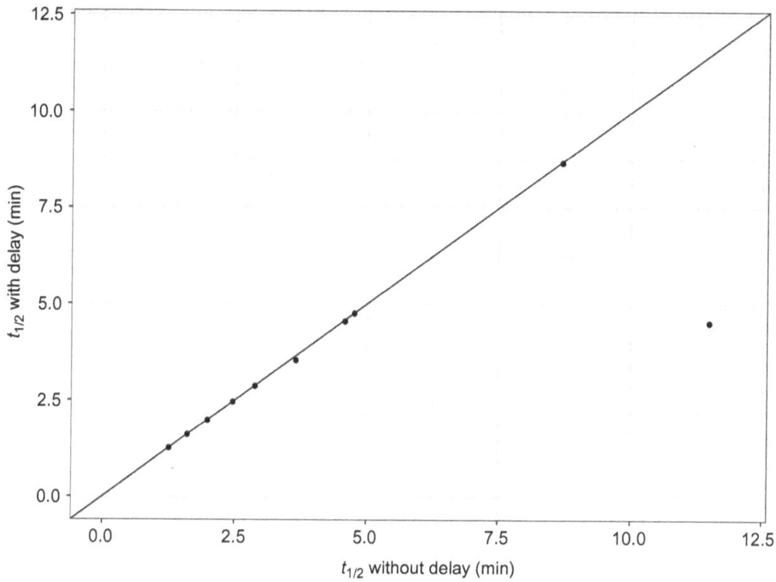

Fig. 5 Plot of the $t_{1/2}$ estimated with a delay vs $t_{1/2}$ estimated without a delay using RT-qPCR for mRNA level quantification. The line $y=x$ line is depicted in *black*.

and without a delay. *ompC* is a monocistronic gene with not an especially long mRNA (1183 nt). Its transcript is one of the most stable and most expressed transcripts under the slow growth condition (Esquerre et al., 2014). These results show that for most of the mRNAs, including a delay in the linear modeling of the mRNA decay quantified by RT-qPCR did not strongly affect their half-life estimation, leading to an extremely high correlation between $t_{1/2}$ with and without a delay (Fig. 5).

6. CONCLUDING REMARKS

In this study, we demonstrated that considering the delay before the onset of mRNA exponential decay in the modeling of the degradation kinetics had a small impact on genome-wide mRNA half-life estimation: for about 80% of the mRNAs the half-life value was reduced at the maximum by 34 s. This conclusion, which was obtained using both linear and nonlinear models, held in *E. coli* cells growing at different growth rates and in stationary phase. The small impact of including a delay in the estimation of mRNA half-life was also independent of the large-scale methods used for mRNA quantification such as microarrays and RNA-seq and was also confirmed for the RT-qPCR method commonly used in studies focusing on a more limited number of mRNAs. mRNA half-lives estimated by the widely used simple linear model were the least sensitive to the introduction of the delay in the modeling approach. Nevertheless it should be noted that delays higher than 5 min were obtained for 2% of the mRNAs using either the linear model or the nonlinear models (Table 1). The intersection constituted a group of 82 mRNAs with a delay above 5 min for all the models. This group did not exhibit any significant bias toward high mRNA length (mean length value of 1128 bp for this group compared to 1002 bp for the mRNAs with a delay lower than 5 min in all the models). For these specific mRNAs, taking into account the delay before mRNA decay may significantly influence their half-life estimation.

Using our mRNA level datasets, estimation of both the delay and the mRNA half-life using nonlinear models reduced the number of selected half-lives passing the accuracy criteria. Our datasets contained between 7 and 12 data points. Accurate estimation of both the delay and the mRNA half-life with nonlinear models may require a higher number of data points of RNA level measurement (Rasch, 1977). Due to the high degree of correlation between half-life estimations with and without a delay before the onset of mRNA exponential decay, to the smaller differences between these estimated $t_{1/2}$ and to the consistently higher number of reliably determined half-lives, the linear model without a delay seems to be the method of choice for work involving comparison of mRNA stability under different growth conditions or between a mutant strain and its wild-type control. However, a formal comparison of the three models using simulations will have to be performed to confirm this result.

ACKNOWLEDGMENTS

This work was funded by the Agence Nationale de la Recherche (ANR-13-BSV6-0005). The authors thank Q. Morin-Ogier for technical help.

REFERENCES

Alifano, P., Palumbo, C., Pasanisi, D., & Tala, A. (2015). Rifampicin-resistance, rpoB polymorphism and RNA polymerase genetic engineering. *Journal of Biotechnology*, *202*, 60–77. https://doi.org/10.1016/j.jbiotec.2014.11.024.

Andrade, J. M., Pobre, V., Silva, I. J., Domingues, S., & Arraiano, C. M. (2009). In C. Condon (Ed.), *Progress in molecular biology and translational science: Vol. 85. The role of 3′–5′ exoribonucleases in RNA degradation* (pp. 187–229). Chapter 5. Elsevier. https://doi.org/10.1016/s0079-6603(08)00805-2.

Bernstein, J. A., Khodursky, A. B., Lin, P. H., Lin-Chao, S., & Cohen, S. N. (2002). Global analysis of mRNA decay and abundance in Escherichia coli at single-gene resolution using two-color fluorescent DNA microarrays. *Proceedings of the National Academy of Sciences of the United States of America*, *99*(15), 9697–9702. https://doi.org/10.1073/pnas.112318199.

Bernstein, J. A., Lin, P. H., Cohen, S. N., & Lin-Chao, S. (2004). Global analysis of Escherichia coli RNA degradosome function using DNA microarrays. *Proceedings of the National Academy of Sciences of the United States of America*, *101*(9), 2758–2763. https://doi.org/10.1073/pnas.0308747101.

Bouvier, M., & Carpousis, A. J. (2011). A tale of two mRNA degradation pathways mediated by RNase E. *Molecular Microbiology*, *82*(6), 1305–1310. https://doi.org/10.1111/j.1365-2958.2011.07894.x.

Campbell, E. A., Korzheva, N., Mustaev, A., Murakami, K., Nair, S., Goldfarb, A., et al. (2001). Structural mechanism for rifampicin inhibition of bacterial RNA polymerase. *Cell*, *104*(6), 901–912.

Caron, M. P., Bastet, L., Lussier, A., Simoneau-Roy, M., Masse, E., & Lafontaine, D. A. (2012). Dual-acting riboswitch control of translation initiation and mRNA decay. *Proceedings of the National Academy of Sciences of the United States of America*, *109*(50), E3444–E3453. https://doi.org/10.1073/pnas.1214024109.

Carpousis, A. J. (2002). The Escherichia coli RNA degradosome: Structure, function and relationship in other ribonucleolytic multienzyme complexes. *Biochemical Society Transactions*, *30*(2), 150–155.

Carpousis, A. J. (2007). The RNA degradosome of Escherichia coli: An mRNA-degrading machine assembled on RNase E. *Annual Review of Microbiology*, *61*, 71–87. https://doi.org/10.1146/annurev.micro.61.080706.093440.

Carpousis, A. J., Luisi, B. F., & McDowall, K. J. (2009). Endonucleolytic initiation of mRNA decay in Escherichia coli. *Progress in Molecular Biology and Translational Science*, *85*, 91–135. https://doi.org/10.1016/s0079-6603(08)00803-9.

Chen, H., Shiroguchi, K., Ge, H., & Xie, X. S. (2015). Genome-wide study of mRNA degradation and transcript elongation in Escherichia coli. *Molecular Systems Biology*, *11*(1), 781. https://doi.org/10.15252/msb.20145794.

Dar, D., & Sorek, R. (2018). Extensive reshaping of bacterial operons by programmed mRNA decay. *PLoS Genetics*, *14*(4), e1007354. https://doi.org/10.1371/journal.pgen.1007354.

Esquerre, T., Bouvier, M., Turlan, C., Carpousis, A. J., Girbal, L., & Cocaign-Bousquet, M. (2016). The Csr system regulates genome-wide mRNA stability and transcription and thus gene expression in Escherichia coli. *Scientific Reports*, *6*, 25057. https://doi.org/10.1038/srep25057.

Esquerre, T., Laguerre, S., Turlan, C., Carpousis, A. J., Girbal, L., & Cocaign-Bousquet, M. (2014). Dual role of transcription and transcript stability in the regulation of gene expression in Escherichia coli cells cultured on glucose at different growth rates. *Nucleic Acids Research*, *42*(4), 2460–2472. https://doi.org/10.1093/nar/gkt1150.

Esquerre, T., Moisan, A., Chiapello, H., Arike, L., Vilu, R., Gaspin, C., et al. (2015). Genome-wide investigation of mRNA lifetime determinants in Escherichia coli cells cultured at different growth rates. *BMC Genomics*, *16*, 275. https://doi.org/10.1186/s12864-015-1482-8.

Jin, D. J., & Gross, C. A. (1988). Mapping and sequencing of mutations in the Escherichia coli rpoB gene that lead to rifampicin resistance. *Journal of Molecular Biology*, *202*(1), 45–58.

Jin, D. J., & Gross, C. A. (1989). Characterization of the pleiotropic phenotypes of rifampin-resistant rpoB mutants of Escherichia coli. *Journal of Bacteriology*, *171*(9), 5229–5231.

Kavita, K., de Mets, F., & Gottesman, S. (2017). New aspects of RNA-based regulation by Hfq and its partner sRNAs. *Current Opinion in Microbiology*, *42*, 53–61. https://doi.org/10.1016/j.mib.2017.10.014.

Lin, P. H., Singh, D., Bernstein, J. A., & Lin-Chao, S. (2008). Genomic analysis of mRNA decay in E. coli with DNA microarrays. *Methods in Enzymology*, *447*, 47–64. https://doi.org/10.1016/S0076-6879(08)02203-9.

Little, R., & Dennis, P. P. (1979). Expression of RNA polymerase and ribosome component genes in Escherichia coli mutants having conditionally defective RNA polymerases. *Journal of Bacteriology*, *137*(1), 115–123.

Moffitt, J. R., Pandey, S., Boettiger, A. N., Wang, S., & Zhuang, X. (2016). Spatial organization shapes the turnover of a bacterial transcriptome. *eLife*, *5*, e13065. https://doi.org/10.7554/eLife.13065.

Mohanty, B. K., & Kushner, S. R. (2016). Regulation of mRNA decay in bacteria. *Annual Review of Microbiology*, *70*, 25–44. https://doi.org/10.1146/annurev-micro-091014-104515.

Morita, T., Maki, K., & Aiba, H. (2005). RNase E-based ribonucleoprotein complexes: Mechanical basis of mRNA destabilization mediated by bacterial noncoding RNAs. *Genes & Development*, *19*(18), 2176–2186. https://doi.org/10.1101/gad.1330405.

Muggeo, V. M. R. (2008). Segmented: An R package to fit regression models with broken-line relationships. *R News*, *8*(1), 20–25.

Nouaille, S., Mondeil, S., Finoux, A. L., Moulis, C., Girbal, L., & Cocaign-Bousquet, M. (2017). The stability of an mRNA is influenced by its concentration: A potential physical mechanism to regulate gene expression. *Nucleic Acids Research*, *45*(20), 11711–11724. https://doi.org/10.1093/nar/gkx781.

Potts, A. H., Vakulskas, C. A., Pannuri, A., Yakhnin, H., Babitzke, P., & Romeo, T. (2017). Global role of the bacterial post-transcriptional regulator CsrA revealed by integrated transcriptomics. *Nature Communications*, *8*(1), 1596. https://doi.org/10.1038/s41467-017-01613-1.

Proshkin, S., Rahmouni, A. R., Mironov, A., & Nudler, E. (2010). Cooperation between translating ribosomes and RNA polymerase in transcription elongation. *Science*, *328*(5977), 504–508. https://doi.org/10.1126/science.1184939.

Rasch, D. (1977). Sample size determination for estimating the parameter of an exponential distribution. *Biometrical Journal*, *19*(7), 521–528.

Richards, J., Luciano, D. J., & Belasco, J. G. (2012). Influence of translation on RppH-dependent mRNA degradation in Escherichia coli. *Molecular Microbiology*, *86*(5), 1063–1072. https://doi.org/10.1111/mmi.12040.

Selinger, D. W., Saxena, R. M., Cheung, K. J., Church, G. M., & Rosenow, C. (2003). Global RNA half-life analysis in Escherichia coli reveals positional patterns of transcript degradation. *Genome Research*, *13*(2), 216–223. https://doi.org/10.1101/gr.912603.

Stead, M. B., Marshburn, S., Mohanty, B. K., Mitra, J., Pena Castillo, L., Ray, D., et al. (2011). Analysis of Escherichia coli RNase E and RNase III activity in vivo using tiling microarrays. *Nucleic Acids Research*, *39*(8), 3188–3203. https://doi.org/10.1093/nar/gkq1242.

Storz, G., Opdyke, J. A., & Zhang, A. (2004). Controlling mRNA stability and translation with small, noncoding RNAs. *Current Opinion in Microbiology*, 7(2), 140–144. https://doi.org/10.1016/j.mib.2004.02.015.

Yakhnin, A. V., Baker, C. S., Vakulskas, C. A., Yakhnin, H., Berezin, I., Romeo, T., et al. (2013). CsrA activates flhDC expression by protecting flhDC mRNA from RNase E-mediated cleavage. *Molecular Microbiology*, *87*(4), 851–866. https://doi.org/10.1111/mmi.12136.

Zhang, A., Wassarman, K. M., Rosenow, C., Tjaden, B. C., Storz, G., & Gottesman, S. (2003). Global analysis of small RNA and mRNA targets of Hfq. *Molecular Microbiology*, *50*(4), 1111–1124. https://doi.org/10.1046/j.1365-2958.2003.03734.x.

CHAPTER FOUR

FASTBAC-Seq: Functional AnalysiS of Toxin–Antitoxin Systems in BACteria by Deep Sequencing

Sara Masachis*, Nicolas J. Tourasse*, Sandrine Chabas*, Olivier Bouchez†, Fabien Darfeuille*,[1]

*ARNA Laboratory, INSERM U1212, CNRS UMR 5320, University of Bordeaux, Bordeaux, France
†Plateforme GeT-PlaGe-Genotoul, INRA Auzeville, Castanet-Tolosan, France
[1]Corresponding author: e-mail address: fabien.darfeuille@inserm.fr

Contents

Methods in Enzymology, Volume 612
ISSN 0076-6879
https://doi.org/10.1016/bs.mie.2018.08.033

Abstract

As the number of bacterial genomes and transcriptomes increases, so does the number of newly identified toxin–antitoxin (TA) systems. However, their functional characterization remains challenging, often requiring the use of overexpression vectors that can lead to misinterpretations of in vivo results. To fill this gap, we developed a systematic approach called FASTBAC-Seq (Functional AnalysiS of Toxin–Antitoxin Systems in BACteria by Deep Sequencing). Combining life/death phenotypic selection with next-generation sequencing, FASTBAC-Seq allows the rapid identification of loss-of-function (toxicity) mutations in toxin-encoding genes belonging to TA loci with nucleotide resolution. Here, we present the setup used on the first-time application of FASBACT-Seq to characterize a member of the *aapA*/IsoA family of type I TA systems hosted on the chromosome of the major human gastric pathogen *Helicobacter pylori*. We propose FASBACT-Seq as a powerful tool for the functional characterization of TA systems that can in addition uncover key elements for the understanding of gene expression regulation in bacteria.

1. INTRODUCTION

Toxin–antitoxin (TA) systems are simple genetic loci composed of a small toxic protein that targets key cellular events and a labile antitoxin counterpart that prevents toxin action. Discovered 30 years ago on plasmids acting as postsegregational killing systems (Gerdes et al., 1986; Weaver, 2012), recent studies have shown their wide distribution on bacterial chromosomes, where understanding their role is less intuitive (Fozo, Hemm, & Storz, 2008; Kawano, 2012; Vogel, Argaman, Wagner, & Altuvia, 2004). Despite the extensive identification of new chromosomally encoded TA systems, tools for their study are scarce. Traditional approaches often use overexpression vectors to address the functionality of putative toxins (Fozo et al., 2008; Jahn & Brantl, 2013; Unoson & Wagner, 2008). However, overexpression can lead to the observation of inaccurate effects of toxin expression. In addition, these approaches cannot identify *cis*-acting elements controlling toxin expression. This represents a major limitation for the study of type I TA systems, where, due to the RNA nature of the antitoxin component, most regulatory events occur at the co- and posttranscriptional levels (for a review, see Masachis & Darfeuille, 2018).

Furthermore, the observation of so-called suppressor mutants upon toxin overexpression is frequently reported (Jahn, Preis, Wiedemann, & Brantl, 2012; Silvaggi, Perkins, & Losick, 2005). However, little focus has been given for their understanding. Some classical techniques such as site-directed mutagenesis (SDM) or more systematic methods such as oligonucleotide-generated random sequence libraries (Mok, Patel, & Li, 2010) or error-prone mutagenesis polymerase chain reaction (PCR) (Guo et al., 2014) have been used to identify functional sequence determinants of toxic peptides belonging to type I TA systems. However, these approaches were strongly focused on toxin-coding sequences, underestimating the potential regulatory role of noncoding regions embedded within toxin-encoding mRNAs.

Here, we developed a high-throughput systematic approach called FASTBAC-Seq (Functional AnalysiS of Toxin–Antitoxin Systems in BACteria by Deep Sequencing) that combines a simple genetic selection of suppressors with next-generation sequencing (NGS) technology. Indeed, with this method, we take advantage of the lethality induced by the expression of the toxin to search for toxicity suppressor mutations. To this end, we need to first introduce point mutations inactivating the antitoxin promoter in the chromosome of the bacterium, then select the surviving clones, isolate their genomic DNA, and map the mutations onto the TA locus by deep sequencing (Fig. 1).

In this chapter, we present a detailed FASTBAC-Seq protocol that was applied to study the *aapA3*/IsoA3 system, belonging to a recently discovered family of type I TA systems hosted in the chromosome of the major human pathogen *Helicobacter pylori* (Arnion et al., 2017). This approach has recently enabled us to identify a myriad of otherwise unpredictable regulatory elements embedded within the *aapA3* toxin-encoding mRNA that act in concert with the IsoA3 antitoxin RNA and that are essential to tightly control toxin expression.

2. MATERIALS

2.1 Molecular Biology Equipment

1. Benchtop centrifuge.
2. Microvolume UV–vis spectrophotometer (Nanodrop 2000, or equivalent).
3. Safe-lock tubes 0.2, 1.5 mL.
4. Thermo cycler.
5. Vortex.

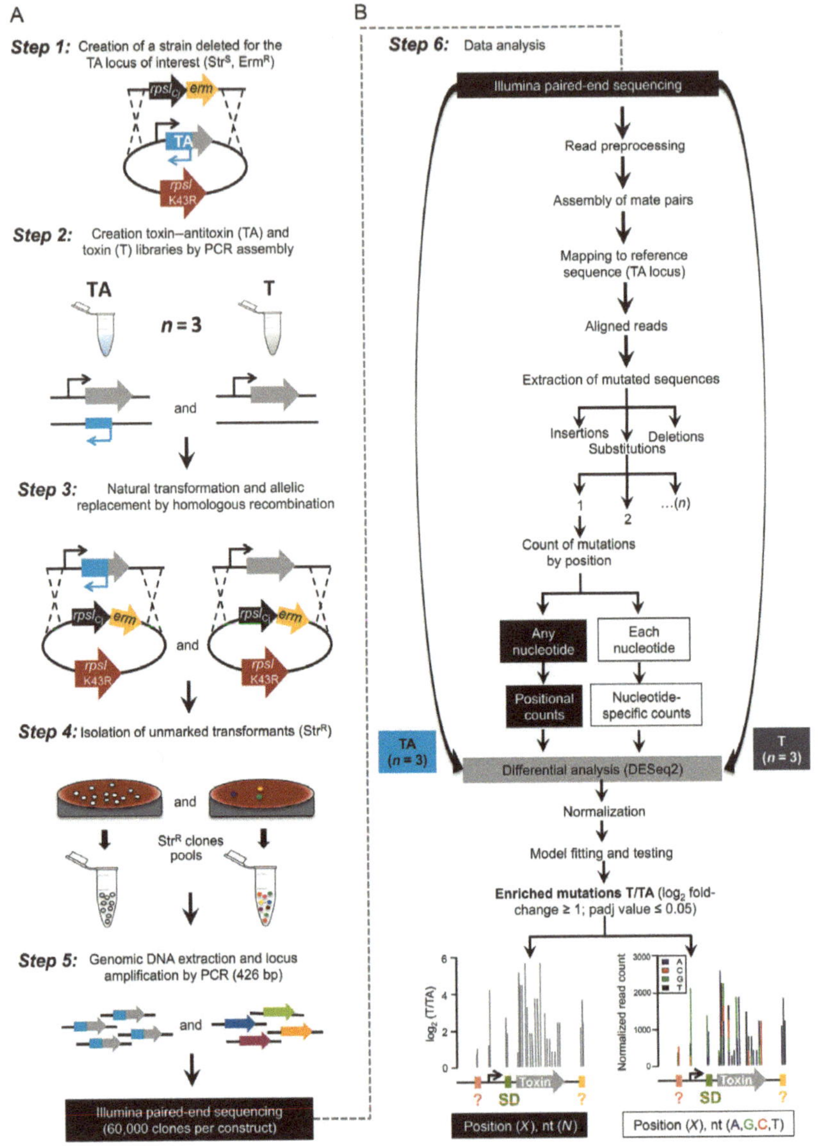

Fig. 1 Experimental workflow for the FASTBAC-Seq approach. (A) Data generation and (B) data analysis.

2.2 General Equipment for Bacterial Cultures

1. Inoculating loops (1, 10 μL, plastic).
2. Incubator for bacterial plates.
3. Petri dishes.
4. Spectrophotometer or cell density meter, and cuvettes.
5. Vortex.

2.3 *E. coli*-Specific Equipment

1. Heat block for 1.5-mL tubes.
2. Swing-out rotor, refrigerated centrifuge.
3. Serological pipets 5 mL (plastic).
4. Shaker.
5. Water bath.
6. 50-mL falcon tubes.

2.4 *H. pylori*-Specific Equipment

1. Anaerobic jar system (Oxoid).
2. Anoxomat® Mark II (Advanced Instruments).
3. Q-tips.

2.5 Molecular Biology Reagents

1. Gene-specific DNA oligonucleotides.
2. DreamTaq DNA polymerase (ThermoFischer Scientific).
3. Phusion® High-Fidelity DNA Polymerase (New England Biolabs).
4. Pfu Ultra High-Fidelity DNA Polymerase (Agilent).
5. PCR and DNA Fragment Purification Kit (Neo Biotech).
6. Quick Bacteria Genomic DNA Extraction Kit (Neo Biotech).
7. High Purity Plasmid Miniprep Kit (Neo Biotech).
8. pGEM®-T Vector System.
9. DpnI (Promega).

2.6 *E. coli* Strains and Cell Culture

1. One shot TOP10 chemically competent *E. coli* cells.
2. Lennox Broth (LB) liquid medium: 1% (w/v) peptone 140, 0.5% (w/v) yeast extract, and 85.6 m*M* sodium chloride (Oxoid).
3. Lennox agar: LB medium (see earlier), 1.2% agar (Oxoid).
4. Ampicillin (Merck) PCR primers.

2.7 *H. pylori* Strains and Cell Culture

1. *H. pylori* 26695 wild-type (WT) strain.
2. *H. pylori* 26695 *rpsl*K43R strain in which the *rpsl* gene contains the K43R mutation providing streptomycin resistance.
3. pSP60 plasmid carrying the double-gene cassette *rpsl-erm* (Dailidiene, Dailide, Kersulyte, & Berg, 2006).
4. Columbia blood agar base (CAB): 0.5% (w/v) enzymatic digest of casein, 0.8% (w/v) enzymatic digest of animal tissue, 1% (w/v) yeast enriched

peptone, 0.1% (w/v) corn starch, 85.6 m*M* sodium chloride, and 1.4% (w/v) agar (Oxoid).

5. Laked horse blood (used at 7% on CAB plates) (ThermoFischer Scientific).
6. *H. pylori* selective supplement (Dent, used as one vial for 500 mL of medium) (Oxoid).
7. Brain heart infusion (BHI) liquid medium: 0.5% (w/v) beef heart, 1.25% (w/v) calf brains, 0.25% (w/v) disodium hydrogen phosphate, 0.2% (w/v) D-(+)-glucose, and 1% (w/v) peptone (Oxoid).
8. Streptomycin sulfate salt (Merck).
9. Erythromycin (Merck).

3. METHODS: A HIGH-THROUGHPUT GENETIC SELECTION OF INTRINSIC TOXICITY SUPPRESSORS IN SIX STEPS

3.1 Step 1: Deletion of the TA Locus Using the *rpsl*$_{Cj}$*-erm* Cassette

The genetic selection presented here requires the previous generation of an *H. pylori* strain in which the TA locus of interest has been replaced by the double-cassette *rpsl*$_{Cj}$*-erm* in a *rpsl* K43R mutant (Str^R) background (ΔTA::*rpsl*$_{Cj}$*-erm*/*rpsl*K43R) as described by Dailidiene et al. (2006) (step 1, Fig. 1). This double-cassette codes for a WT copy of the *rpsl* gene from *Campylobacter jejuni* (*Cj*) that has a dominant streptomycin sensitivity (Str^S) effect over the chromosomal *H. pylori rpsl* copy carrying the K43R mutation (*rpsl* K43R, conferring streptomycin resistance, Str^R). This allows the selection of unmarked transformant strains (Str^R) having replaced the *rpsl*$_{Cj}$*-erm* cassette by the PCR products of interest. The second gene in the cassette is an erythromycin resistance gene (*erm*) that enables the selection of *rpsl*$_{Cj}$*-erm* containing strains during the generation of the ΔTA locus strain.

To generate this TA-deleted strain, we took advantage of the natural competence and highly active recombination machinery of *H. pylori* (Dorer, Cohen, Sessler, Fero, & Salama, 2013). Thus, *H. pylori* chromosomal manipulations in this, and the next steps of the FASTBAC-Seq process, were conveniently performed by homologous recombination of PCR-generated DNA fragments upon natural transformation. The creation of the ΔTA::*rpsl*$_{Cj}$*-erm* deletion cassette was performed by PCR assembly of the *rpsl*$_{Cj}$*-erm* cassette with upstream and downstream fragments corresponding to approximately 400 nucleotide (nt)-long flanking regions around the

TA locus. These flanking fragments allow targeted homologous recombination to occur and can be amplified from a WT *H. pylori* strain.

3.1.1 *Creation of a* rpsl$_{Cj}$-erm-*Based TA-Deletion Cassette by PCR Assembly*

1. Prepare genomic DNA from WT *H. pylori* 26695 strain following a routine protocol (as described in Arnion et al., 2017).
2. Design two gene-specific DNA oligonucleotide pairs to amplify an upstream and a downstream region to the TA locus of approximately 400 nt each. These regions will allow homologous recombination to occur. For each oligonucleotide pair, the internal primer (the *antisense* of the upstream pair and the *sense* of the downstream pair) will determine the amplitude of the deletion, while the external one will determine the region that will be used for homologous recombination. Both internal primers should contain a 5′-overhang tail of approximately 25 nt with homology to the *rpsl$_{Cj}$-erm* cassette to allow PCR assembly (see note 1 in Section 3.1.1.1 for examples used in our case study).
3. Amplify by PCR the desired upstream and downstream fragments designed in point 2 of Section 3.1.1 using the genomic DNA isolated in point 1 of Section 3.1.1 as template. The use of a high-fidelity DNA polymerase is recommended. In our example, the Phusion® High-Fidelity DNA Polymerase (error rate of 4.4×10^{-7} errors per nt per cycle) and 35 amplification cycles (98°C 3 min (98°C 30 s, 54°C 30 s, 72°C 1 min) × 35, 72°C 2 min) were employed to amplify a 415- and 418-nt upstream and downstream fragments, respectively.
4. Purify the PCR products using the PCR and DNA Fragment Purification Kit (Neo Biotech) or equivalent. Quantify them by using a micro-volume UV–vis spectrophotometer (Nanodrop 2000, or equivalent).
5. Using the primer pair described in note 2 of Section 3.1.1.1 and the pSP60 plasmid as DNA template, amplify the *rpsl$_{Cj}$-erm* cassette with a high-fidelity DNA polymerase (i.e., Phusion® High-Fidelity DNA Polymerase or equivalent). An optimal PCR program to amplify the 1533-nt fragment is: 98°C 30 s (54°C 30 s, 72°C 1 min 30 s) × 35, 72°C 2 min.
6. Purify the PCR products as performed in point 4 of Section 3.1.1.
7. Assemble by PCR the three purified DNA fragments from points 4 and 6 of Section 3.1.1. To this aim, use the two external primers described in point 2 of Section 3.1.1 (*sense* from the upstream region primer pair and *antisense* from the downstream region primer pair). The use of a

high-fidelity DNA polymerase is required. In our example, the Phusion® High-Fidelity DNA Polymerase and 35 amplification cycles (98°C 3 min (98°C 30 s, 54°C 45 s, 72°C 2 min 30 s) × 35, 72°C 2 min) were employed to amplify a 2366 nt-long fragment corresponding to the *aapA3*/IsoA3 deletion cassette Δ*aapA3*/IsoA3::*rpsl*$_{Cj}$-*erm*.

8. Purify and quantify the PCR products as in point 4 of Section 3.1.1.

3.1.1.1 Notes

1. Primer pairs used in our example to delete the *aapA3*/IsoA3 TA locus in the *H. pylori* 26695 strain using the *rpsl*$_{Cj}$-*erm* cassette (23-nt sequences homologous to the *rpsl*$_{Cj}$-*erm* cassette are underlined) are:
   - Upstream region:
     Sense primer: 5′ GCATTATAAAATGAAATCC 3′
     Antisense primer: 5′ GTTTAATCCATAGTTATAAAGCACA AAAAGAGGGATTTTAAAAG 3′
   - Downstream region:
     Sense primer: 5′ GCTATAAATTATTTAATAAGTAACCGC TTGCTCTAGCTTTTTG 3′
     Antisense primer: 5′CTAGCCACGCTCTATTAGAG3′
2. Primers used for *rpsl*$_{Cj}$-*erm* cassette amplification from pSP60:
   Sense primer: 5′ TGCTTTATAACTATGGATTAAAC 3′
   Antisense primer: 5′ TTACTTATTAAATAATTTATAGC 3′

3.1.2 H. pylori *Natural Transformation and Isolation of Recombinant Strains*

1. Thaw the *H. pylori rpsl* K43R strain (Str^R) from a −80°C glycerol stock making a small patch (≈1.5 cm^2) on a petri dish containing CAB media supplemented with 7% laked horse blood, Dent and 10 μg/mL streptomycin.
2. Place the plate into an anaerobic jar system and incubate within an incubator for bacterial plates at 37°C under microaerophilic conditions (6% O_2, generated with an Anoxomat® Mark II System or equivalent) for 48 h. These are standard *H. pylori* culture conditions and will be used throughout our protocol.
3. Recover some cells by sliding a Q-tip over the bacterial patch and transfer into a new (prewarmed at 37°C) petri dish made as in point 1 of Section 3.1.2 covering the whole plate surface.
4. Incubate the plate for 16 h as in point 2 of Section 3.1.2.

5. With a Q-tip, recover the whole bacterial layer and suspend in a 1.5-mL tube containing 1 mL BHI medium.
6. Measure the OD_{600} on a spectrophotometer or cell density meter. An optimal OD_{600} value should be in the range 7–10.
7. Inoculate the volume equivalent to 1 OD_{600} in a petri dish containing CAB media supplemented with 7% laked horse blood and Dent. With an inoculating loop (10 μL, plastic) make a bacterial patch of approximately 1 cm^2.
8. Incubate the plate for 4 h.
9. Add 1 μg of purified PCR assembly product from point 8 of Section 3.1.1 to the bacterial patch and mix homogeneously with the help of an inoculating loop.
10. Incubate the plate for 16 h.
11. With a Q-tip, recover the whole bacterial layer and spread in a fresh plate containing 10 μg/mL erythromycin (antibiotic allowing the selection of recombinant strains).
12. Incubate the plate 4–6 days until isolated colonies appear.
13. Once isolated colonies appear, extract genomic DNA as in point 1 of Section 3.1.1 and using gene-specific DNA oligonucleotides verify by Sanger sequencing the proper targeted insertion of the $rpsl_{Cj}$-*erm* cassette on the chromosomal TA locus. Positive strains (ΔTA::$rpsl_{Cj}$-*erm*/*rpsl*K43R, Str^S/Erm^R) are ready to use for the FASTBAC-Seq approach.

3.2 Step 2: Creation of TA and Toxin (T) Locus by PCR Assembly

FASTBAC-Seq is a method based on the lethality induced by the chromosomal expression of the toxin. To inactivate the antitoxin promoter, we transformed *H. pylori* with PCR fragments encompassing the TA locus containing two point mutations in the −10 box of the antitoxin gene. To generate these DNA fragments, we used SDM PCR. However, due to *H. pylori* natural competence and its highly active recombination machinery, we observed that even the smallest trace of WT genomic DNA in the PCR product used for transformation lead to the enrichment of transformants having integrated a WT sequence (nontoxic). This phenomenon strongly limits the emergence of suppressor strains and must be avoided.

To overcome this limitation, we introduced a subcloning step in *E. coli*. However, since our toxin is also lethal in *E. coli*, it was necessary to split the

TA locus into two separate *E. coli* vectors. These vectors were then used to create the antitoxin promoter inactivation by SDM. Finally, the PCR product containing the TA locus was reconstructed by PCR assembly and used for natural transformation of *H. pylori* by homologous recombination.

3.2.1 TA Locus Subcloning in E. coli

1. Using genomic DNA of the *H. pylori* 26695 strain isolated in point 1 of Section 3.1.1 as template split the TA locus into two fragments by PCR. Primer pairs should be designed to allow the amplification of "Upstream-TA" and "Downstream-TA" regions containing around 400 nt of upstream and downstream sequence around the TA locus, respectively (step 1, Fig. 2). This region will enable homologous recombination to occur during the transformation step (Section 3.3). Importantly, as described earlier, primers should have an overlapping region of around 25 nt to allow subsequent PCR assembly of the fragments (Section 3.2.3).
2. Purify and quantify the PCR products as in point 4 of Section 3.1.1.
3. Clone each of the fragments into a separate pGEM-T vector system or equivalent.
4. Transform the ligated plasmids into one shot TOP10 chemically competent *E. coli* cells (or equivalent). Plasmid-containing cells are selected in a petri dish containing LB agar medium supplemented with 100 μg/mL ampicillin (step 2, Fig. 2).
5. Extract the plasmids by using the High Purity Plasmid Miniprep Kit (Neo Biotech) or equivalent. Confirm by Sanger sequencing that the inserts do not contain undesired mutations. At this stage, two plasmids have been generated, containing each one half of the TA locus, a 400-nt homology region, and 25 overlapping nucleotides to allow PCR assembly: pGEM-T-Up-WT and pGEM-T-Down-WT (step 2, Fig. 2).

3.2.1.1 Notes

1. The PCR described in point 1 of Section 3.2.1 must be performed with a DNA polymerase that adds 3′-end dA overhangs in order to allow the subsequent "TA cloning" approach (i.e., pGEM®-T Vector system or equivalent). In our example, the DreamTaq DNA polymerase (error rate of 2.2×10^{-5} errors per nt per cycle) was used following the manufacturer's instructions (step 1, Fig. 2).

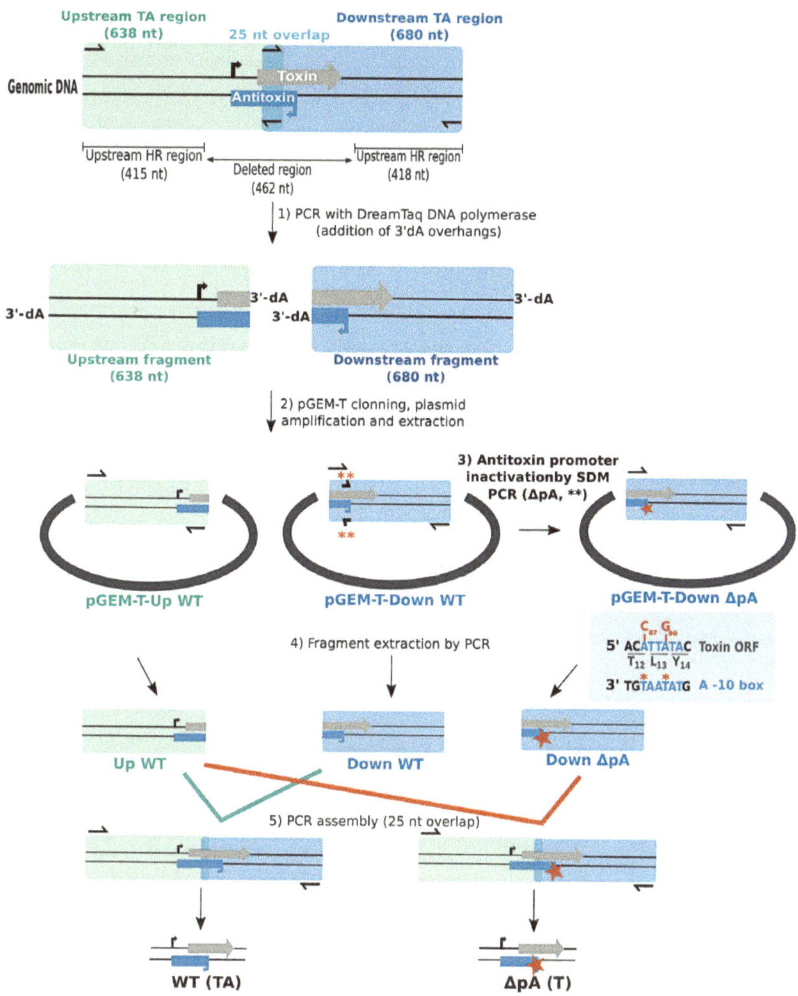

Fig. 2 Experimental workflow for the inactivation of a *cis*-encoded antisense RNA antitoxin promoter.

3.2.2 Inactivation of the Antitoxin Promoter by SDM PCR

In our example, SDM PCR, a technique allowing the creation of specific, targeted changes in double-stranded plasmid DNA, was used to generate the TA locus version carrying the inactivated antitoxin promoter. Custom-designed overlapping oligonucleotides containing the desired mutations were employed. This primer design results in a doubly nicked plasmid (recircularized) that can be directly transformed into *E. coli*.

1. Design an overlapping primer pair to introduce the minimal number of point mutations needed to inactivate the antitoxin promoter. Use a high-fidelity DNA polymerase to perform the SDM PCR. In our example, Pfu Ultra High-Fidelity DNA Polymerase was used and the previously isolated pGEM-T-Down-WT plasmid (point 5 of Section 3.2.1) served as DNA template. PCR protocols were carried out following the manufacturer's instructions.
2. Add 1.5 μL of DpnI (Promega) per 50 μL of PCR product and incubate 30 min at 37°C. DpnI has a $G^{me}A$▼TC cutting site dependent on the N6-methylation of the adenine residue for activity. This property enables the targeted degradation of the template plasmid DNA as PCR products are unmethylated.
3. Purify and quantify the PCR products as in point 4 of Section 3.1.1.
4. Use these purified products to transform one shot TOP10 chemically competent *E. coli* cells (or equivalent).
5. Follow points 4–5 in Section 3.2.1 to isolate the pGEM-T-Down-ΔpA plasmid (containing the downstream fragment with the −10 box of the antitoxin promoter inactivated, ΔpA) (step 3, Fig. 2).

3.2.2.1 Notes

1. In the cases where the antitoxin is a *cis*-encoded antisense RNA overlapping an open reading frame (ORF), you must preserve the toxin active while introducing mutations to inactivate the antitoxin promoter. In our case, two synonymous point mutations were introduced in the −10 box: adenine at position +87 from the toxin transcription start site (TSS) was mutated to cytosine, and adenine at position +90 from the toxin TSS was mutated to guanine (step 3, Fig. 2). With these mutations, we managed to preserve the coding sequence of the toxin while completely inactivating the transcription of the antitoxin (Arnion et al., 2017).
   - Primers used for antitoxin promoter inactivation in our case study (−10 box in bold, mutations underlined):
     Sense primer: 5′ TGGAAAAC**CTTGTA**CTTTGAGT 3′.
     Antisense primer: 5′ ACTCAAAG**TACAAG**GTTTTCCA 3′

3.2.3 Reconstruction of the TA Locus by PCR Assembly

1. Using the previously isolated plasmids pGEM-T-Up-WT and pGEM-T-Down-WT (point 5 in Section 3.2.1) and pGEM-T-Down-ΔpA (point 5 in Section 3.2.2) as DNA templates, extract by PCR amplification the inserted fragments using the same primer pairs as in point 1 of Section 3.2.1. In our example, the Phusion® High-Fidelity DNA

Polymerase and 35 amplification cycles (98°C 3 min (98°C 30 s, 54°C 30 s, 72°C 1 min)°×°35, 72°C 2 min) were employed to amplify a 638-nt long WT "Up" fragment and 680-nt long WT and ΔpA "Down" fragments that contain a 25-nt long overlapping region to allow subsequent PCR assembly (step 4, Fig. 2).

2. Purify and quantify the PCR products as in point 4 of Section 3.1.1.
3. For PCR assembly, a new primer pair is created by using the two former external primers employed in point 1 of Section 3.2.1: forward primer binding to the 5′ end of the "Up" fragment will form a pair with the reverse primer binding to the 3′ end of the "Down" fragment. In our example, the Phusion® High-Fidelity DNA Polymerase and 35 amplification cycles (98°C 3 min (98°C 30 s, 54°C 45 s, 72°C 1 min) × 35, 72°C 2 min) were used to amplify a 1318-nt long DNA fragment. At the end of this step, two different constructs are generated: (1) TA (WT), by assembling both WT "Up" and "Down" fragments and (2) T (ΔpA), by assembling the WT "Up" fragment with the antitoxin −10 box mutated ΔpA "Down" fragment (step 5, Fig. 2; step 2 in Fig. 1).
4. Purify and quantify the PCR products as in point 4 of Section 3.1.1.

3.2.3.1 Notes

1. Note that it is not necessary to use an error-prone DNA polymerase. We have observed that the error rate of a high-fidelity polymerase (e.g., Phusion® High-Fidelity DNA Polymerase) generates a sufficient number of mutations to obtain optimum locus coverage of single-nucleotide mutations. Nevertheless, if aiming at covariation analysis of double and triple mutations, the use of an error-prone DNA polymerase (i.e., saturation mutagenesis strategy) may improve the statistical power.
2. PCR assembly must be performed in triplicates ($n = 3$) to be used next as substrates for three independent biological transformation replicates.

3.3 Step 3: Natural Transformation of *H. pylori* With the PCR-Generated TA Loci

1. Thaw the ΔTA::*rpsl*$_{Cj}$-*erm*/*rpsl*K43R strain (Str^S/Erm^R) generated in point 13 of Section 3.1.2 from a −80°C glycerol stock on a CAB plate supplemented with 10 μg/mL erythromycin.
2. Incubate for 48 h.
3. Recover some cells by sliding a Q-tip over the bacterial patch and transfer into a fresh (prewarmed at 37°C) petri dish made as in point 1 of Section 3.3.
4. Incubate for 16 h.

5. With a Q-tip, recover the whole bacterial layer and suspend in a 1.5 mL tube containing 1 mL BHI medium.
6. Measure the OD_{600} on a spectrophotometer or cell density meter.
7. Inoculate the volume equivalent to 1 OD_{600} in a Petri dish containing CAB media supplemented with 7% laked horse blood and Dent. With an inoculating loop (10 μL, plastic) make a bacterial patch of approximately 1 cm^2. Inoculate as many plates as independent transformations need to be tested. In our example, three transformations were performed: TA, T, and H_2O ("no DNA" control).
8. Incubate the plates for 4 h.
9. Add 1 μg of purified PCR assembly products from point 4 of Section 3.2.2 to the bacterial patches and mix homogeneously with the help of an inoculating loop (step 3, Fig. 1).
10. Incubate for 16 h.
11. With a Q-tip, recover the whole bacterial layer and suspend in a 1.5 mL tube containing 1 mL BHI medium.
12. As an indirect physiological quality control, measure the OD_{600} on a spectrophotometer or cell density meter. An expected optimal OD_{600} should be around 4–5.

3.3.1 *Notes*

1. From point 2 of Section 3.3, three independent biological replicates ($n = 3$) must be generated and maintained throughout the rest of the experiment.
2. Incubation times need to be strictly respected to ensure an optimal and reproducible physiological state of the bacterial cells during transformation.

3.4 Step 4: Selection of Transformants

1. From the 1 mL cell suspension generated in point 11 of Section 3.3, make 10-fold serial dilutions by adding 100 μL of the cell suspension to a new 1.5 mL tube containing 900 μL BHI medium.
2. CAB plates were used to inoculate 100 μL of the 10^{-5}, 10^{-6}, and 10^{-7} dilutions. In parallel, CAB plates containing 10 μg/mL streptomycin were used to plate 100 μL of the 10^{-2}, 10^{-3}, and 10^{-4} dilutions for the water control (H_2O) and T construct, and 10^{-3}, 10^{-4}, and 10^{-5} dilutions for transformation with TA construct. While plating, make a small ($\approx$1 cm^2) patch with the help of an inoculating loop (10 μL, plastic).
3. Incubate the plates 3–4 days until isolated colonies appear.

3.4.1 Determining the Transformation Efficiency

To estimate the optimal serial dilution allowing for the isolation of a maximum number of colonies and the sequencing of a similar number of transformants coming from the TA and T (expected to present a reduced number of colonies due to toxicity) transformations, the determination of the transformation efficiencies with each PCR library is essential.

1. Determine, as shown in Table 1, the number of transformants (Str^R) per total colony forming units (CFU) (Transformation yield) using the serial dilution plates generated in point 3 of Section 3.4. Targeted homologous recombination is expected to induce a replacement of the *rpsl-erm*$_{Cj}$ cassette by the TA or T PCR products. Thus, recombinant strains will lose the resistance to erythromycin (Erm^S) and recover the resistance to streptomycin (*rpsl*K43R, Str^R).

3.4.1.1 Notes

1. We observed a strong reduction in the transformation efficiency when the PCR library carrying the inactivation of the antitoxin promoter (T, ΔpIsoA3) was used (>1.5 log-fold, T, Fig. 3). However, this value was significantly higher (fivefold) than the one obtained when no DNA was used for transformation ($H_2 0$), reflecting the emergence of suppressor mutants.
2. In our example, we used nine plates per biological replicate of the 10^{-1} dilution of the cell suspension transformed with the T construct and three plates per biological replicate of the 10^{-3} dilution for the TA construct transformation, allowing the study of approximately 60,000 clones per construct and per replicate (step 4, Fig. 1).

3.4.2 Differentiating Transformants From Revertants: The StrR/ErmR Screen

A major drawback regarding the use of the *rpsl*$_{cj}$*-erm* cassette is the high rate of emergence of revertants due to mutations in the *rpsl*$_{Cj}$ allele. Thus, the use of an additional step to discriminate transformants (having lost the *rpsl*$_{Cj}$*-erm* cassette, Str^R/Erm^S) from revertants (carrying mutations on the *rpsl*$_{Cj}$ gene, Str^R/Erm^R) is essential for the experimental setup. To that aim, we performed the so-called streptomycin/erythromycin resistance screen (Str^R/Erm^R screen).

1. Divide a CAB containing plate supplemented with 10 μg/mL streptomycin into 15–30 numbered and equal parts (depending on the number of strains to be analyzed). Repeat this step with a 10 μg/mL erythromycin-containing plate.

Table 1 Determining the *H. pylori* Transformation Efficiency

	Inoculated	−Str			+Str			Transformation		
	Volume (μL)	Dilution	No. Colonies	CFU/mL	Dilution	No. Colonies	CFU/mL	Yield	Average	SD
H_2O	100	10^{-5}	7654	7.65E+09	10^{-2}	187	1.87E+05	2.44E−05	1.88E−05	8.74E−06
	100	10^{-6}	820	8.20E+09	10^{-3}	19	1.90E+05	2.32E−05		
	100	10^{-7}	115	1.15E+10	10^{-4}	1	1.00E+05	8.70E−06		
TA	100	10^{-5}	7850	7.85E+09	10^{-3}	Uncountable	—	—	0.0052	0.00027
	100	10^{-6}	894	8.94E+09	10^{-4}	450	4.50E+07	5.03E−03		
	100	10^{-7}	94	9.40E+09	10^{-5}	51	5.10E+07	5.43E−03		
T	100	10^{-5}	7565	7.57E+09	10^{-2}	624	6.24E+05	8.25E−05	7.68E−05	3.30E−06
	100	10^{-6}	846	8.46E+09	10^{-3}	65	6.50E+05	7.68E−05		
	100	10^{-7}	97	9.70E+09	10^{-4}	7	7.00E+05	7.22E−05		

The number of streptomycin-resistant CFU/mL (Str^R CFU/mL) upon transformation with the TA or T PCR assembly products or with H_2O ("no DNA" control) was calculated using serial dilution plates.

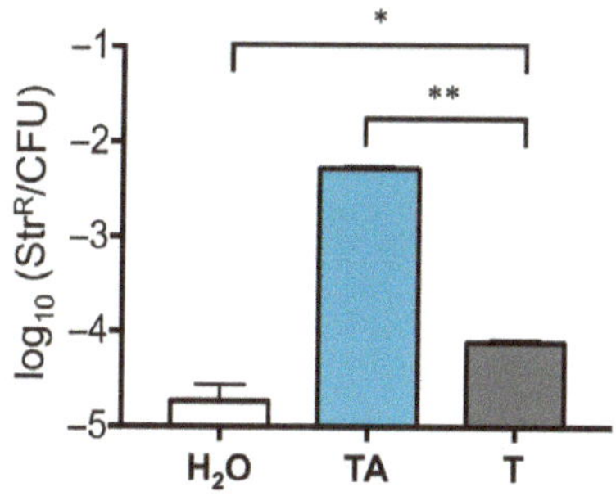

Fig. 3 Efficiency of *H. pylori* transformation during FASTBAC-Seq. Number of Str^R/CFU upon transformation was calculated. Error bars represent s.d.; $n = 3$ replicates $^{**}P < 0.0001$; $^{*}P < 0.0006$; P values according to unpaired t-test.

Table 2 Differentiating Transformants From $rpsl_{Cj}$ Revertants

	No. Colonies Tested	No. Colonies Str^R	No. Colonies Erm^R	% Erm^R ($rpsl_{Cj}$ Suppressors)	% Erm^S (Recombinants)
H_2O	24	24	24	100	0
TA	24	24	2	8.3	91.7
T	48	48	33	68.8	31.3

24 colonies transformed with the TA PCR product or with H_2O ("no DNA" control) and 48 colonies transformed with the toxic T PCR product were subjected to the streptomycin/erythromycin resistance screen. The percentage of $rpsl_{Cj}$ revertants ($rpsl_{Cj}$ suppressors, Str^R, Erm^R) and recombinants (Str^R, Erm^S) was calculated.

2. From one of the previously obtained streptomycin-containing dilution plates generated in point 3 of Section 3.4, select a representative number of colonies (i.e., at least 10% of the total number) to be subjected to the Str^R/Erm^R screen.
3. Systematically pick one colony with the help of an inoculating loop (1 μL, plastic) and streak onto one of the parts in the erythromycin-containing plate (i.e., part no 1).
4. Using the same inoculating loop, make a small patch ($\approx 0.5\,cm^2$) onto the equivalent numbered part in the streptomycin-containing plate (i.e., part no. 1).
5. Keeping the same inoculating loop, expand the initial streak in the erythromycin-containing plate until a small patch ($\approx 0.5\,cm^2$) is generated.
6. Incubate the plates for 24 h.
7. Once the colonies become visible, calculate, as shown in Table 2, the transformation yields obtained with each construct (TA or T) as well as with the "no DNA" control (H_2O).

3.5 Step 5: Sample Preparation and Sequencing by Illumina Paired-End Approach

3.5.1 Pooling of Isolated Strains and Genomic DNA Extraction

1. Once the transformation efficiency has been determined as described in Section 3.4.1, the isolation of recombinant colonies can be performed optimally and in biological triplicates. To this end, follow all the steps described in Sections 3.3 and 3.4 taking care of adapting the serial dilutions to your case of study as described in the scaling process.
2. Pool all the isolated colonies growing in selective media (Str^R) (keeping the independence of the three biological replicates for each independent transformation (TA and T)), in a 1.5-mL tube containing 200 μL of BHI.
3. Centrifuge 5 min at 13,000 rpm in a benchtop centrifuge.
4. Discard the supernatant and use the pellet to extract genomic DNA as explained in point 1 of Section 3.1.1 (step 5, Fig. 1).

3.5.2 The Nested PCR Strategy

As shown in Table 2, the use of the $rpsl_{Cj}$-*erm* counter-selection cassette leads to a high appearance rate of $rpsl_{Cj}$ revertants upon transformation with the toxic T construct (43%–84% in our case, depending on the experiment). Thus, avoidance of PCR amplification from these strains is key to generate NGS substrates of optimum quality (containing exclusively the fragment of interest, the TA locus).

To this aim, we designed a pair of primers that are nested to the ones used to generate the ΔTA::$rpsl_{Cj}$-*erm*/*rpsl*K43R strain. As shown in Fig. 4, if the size of the TA locus deletion is smaller than the size of the region where the primers used for Illumina sequencing bind (Fig. 4, deletion 1), PCR will lead to the generation of two amplicons: (1) the TA locus (426 nt), from the strains having undergone homologous recombination and (2) the ΔTA::$rpsl_{Cj}$-*erm* cassette, from the $rpsl_{Cj}$ revertants. This issue will be predominant for the T-transformed strains and strongly limits the sequencing depth of suppressor strains. This can be easily solved by making a larger deletion while constructing the ΔTA::$rpsl_{Cj}$-*erm*/*rpsl*K43R strain (Fig. 4, deletion 2). This strategy will enable the use of nested primers to generate the amplicon for Illumina sequencing, primers binding within the deleted regions that will be exclusively reintroduced upon homologous recombination.

1. Using the genomic DNA isolated in point 4 of Section 3.5.1 as template, amplify by PCR the TA locus using gene-specific DNA oligonucleotides containing DNA adapters for Illumina paired-end sequencing

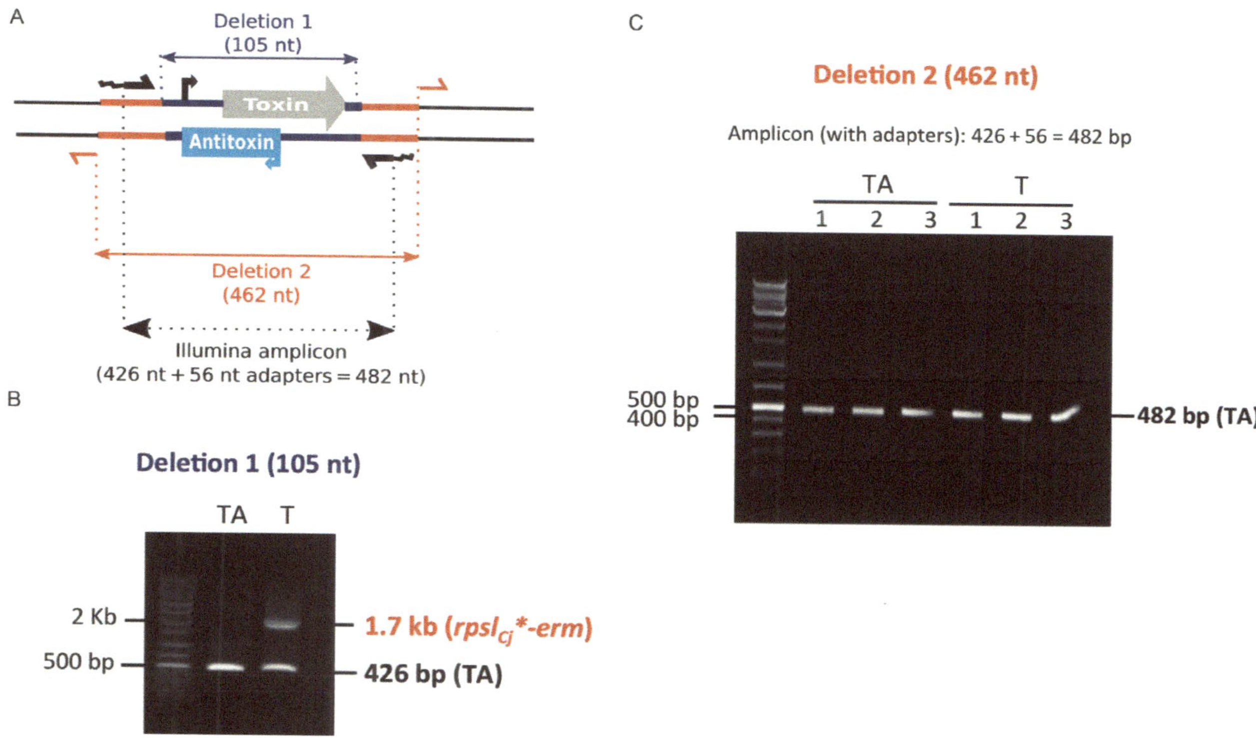

Fig. 4 Nested PCR strategy to generate specific TA locus amplicon for Illumina sequencing. (A) Schematic representation of the two different deletion strategies and sample amplification for Illumina sequencing. (B) Results obtained upon Illumina sequencing of the amplicon amplified in a "deletion 1" background. Upon transformation with T (ΔpA), a band corresponding to the *rpsl*$_{Cj}$-*erm* cassette from the *rpsl*$_{Cj}$ revertants is observed (*rpsl*$_{Cj}$*-*erm*, 1.7 kb) in addition to the 426-nt amplicon corresponding to the TA locus that can be observed in both TA (WT) and T (ΔpA) transformations. (C) Results obtained upon Illumina sequencing of the amplicon amplified in a "deletion 2" background. Only one amplicon observed upon transformation with both TA (WT) and T (ΔpA) constructs. The amplicon corresponds to the TA locus with the DNA Illumina sequencing adapters (482 nt). Numbers 1, 2, and 3 indicate the biological replicates.

approach. In our example, the amplicon was 482-nt long (426-nt TA locus + 28-nt adapters × 2 = 482 nt) (step 5, Fig. 1).

2. Purify and quantify the PCR products as in point 4 of Section 3.1.1.

3.5.3 *Illumina Paired-End Sequencing*

1. Sequencing was performed using an Illumina MiSeq instrument in paired-end mode 2 × 250 nt (overlapping reads) at the Plateforme GeT-PlaGe-Genotoul Centre INRA, Castanet-Tolosan, France.

3.5.3.1 Notes

1. Primer pair used in our example to generate the DNA templates for Illumina paired-end sequencing approach (28-nt long DNA adapters are underlined):

   Sense primer: 5′ <u>CTTTCCCTACACGACGCTCTTCCGATCT</u> CTATCCAATAAAGATAAGC 3′

   Antisense primer: 5′ <u>GGAGTTCAGACGTGTGCTCTTCCGATC</u> <u>T</u>GCACTCTATGAGGGGATTTAG 3′

2. For point 1 of Section 3.5.2, we used Phusion® High-Fidelity Hot Start DNA Polymerase and 35 amplification cycles.

3.6 Step 6: Computational Analysis of the NGS Data

Computational analyses were performed on a computer running under Linux CentOS 6.7 environment. All software and commands described below may be run on any Unix/Linux-like operating system including Mac OS X. Analyses were done with the software versions indicated below. Newer versions may be used but commands may require adjustments in case particular options or tools have changed. The starting NGS input files are the raw sequence files in FASTQ format (Cock, Fields, Goto, Heuer, & Rice, 2010) provided by the sequencing center, with paired-end reads split into two separate files for each replicate (in total six files for TA (WT) samples and six files for T (ΔpA) samples).

3.6.1 *Read Reprocessing and Assembly*

1. Check read quality using FastQC version 0.11.3 (https://www.bioinformatics.babraham.ac.uk/projects/fastqc/).
2. Trim low-quality 3′ ends of reads using cutadapt version 1.1 (Martin, 2011) (https://cutadapt.readthedocs.org/) and a base quality threshold of 28: option "-q 28" (corresponding to a base calling error slightly

above 10^{-3}; threshold based on the "green area" in the "per base sequence quality" section at the top the FastQC report).

3. Discard reads having an average base quality lower than 28 using prinseq-lite version 0.20.4 (Schmieder & Edwards, 2011) (http://prinseq.sourceforge.net/): option "-min_qual_mean 28."
4. Repeat the process described in points 1–3 for each FASTQ file (i.e., 12 times).
5. From the pair of filtered FASTQ files for paired-end reads of a given sample, recover read pairs for which both mates passed the quality filtering steps by means of cmpfastq (http://compbio.brc.iop.kcl.ac.uk/software/cmpfastq.php).
6. Assemble mate reads into a single sequence using PANDAseq version 2.9 (Masella, Bartram, Truszkowski, Brown, & Neufeld, 2012) (https://github.com/neufeld/pandaseq): options "-N -o 30 -O 0 -t 0.6 -A simple_bayesian -C empty." A FASTQ file with assembled reads is created, which will serve as input for further downstream analyses.
7. Repeat points 5–6 for each replicate sample (i.e., six times).

3.6.1.1 Notes

1. Additional read processing (i.e., 5′-end trimming, removal of adapter fragments, and removal of reads containing undetermined nucleotides) may be required depending on the quality of sequencing data, as judged from the FastQC report. Processing options may be different among datasets. "cutadapt" and "prinseq-lite" provide numerous options for customized processing.
2. The "-t" and "-A" parameters in PANDAseq may need to be tuned to ensure satisfactory read pair assembly.
3. In our study, about 5 million read pairs (combining the three biological replicates) could be assembled for the TA (WT) and T (ΔpA) samples.

3.6.2 Read Alignment

1. Align assembled reads onto the 426-nt TA locus reference sequence by the BWA-SW algorithm of BWA version 0.7.12 (Li & Durbin, 2009) (https://sourceforge.net/projects/bio-bwa/): BWA-SW options "-a 1 -b 3 -q 5 -r 2 -z 1," using the FASTQ file of assembled reads generated in point 6 of Section 3.6.1 as input. This produces alignments in SAM format. Convert them to BAM format using samtools version 1.2 (Li et al., 2009) (https://sourceforge.net/projects/samtools/): command

"view," options "-S –b -h" (see http://samtools.github.io/hts-specs/ for details on the SAM/BAM formats).

2. To extract sequences showing only substitutions compared to the reference, filter the BAM file based on the CIGAR string (that describes how the reads align to the reference) and NM tag (that gives the edit distance to the reference) using bamtools version 2.3.0 (Barnett, Garrison, Quinlan, Stromberg, & Marth, 2011) (https://github.com/pezmaster31/bamtools): command "filter," properties "cigar": "426M," and "tag": "NM:1" (for a single substitution) or "tag": "NM:2" (for two substitutions), and so on. This generates a BAM file containing only the selected sequences.
3. To identify sequences harboring a single deletion, extract sequences of 425-nt length that are aligned with a CIGAR string containing a "1D" operation and a NM tag value of 1. This requires the use of regular expressions to parse the SAM file, i.e., with a one-liner command written in AWK: "length($10)==425 && $6~/^[0-9]+M1D[0-9]+M$/ && $0~/NM:i:1(\t|$)/ {print $0}". For sequences harboring two deletions, use a length of 424, replace "1D" by "2D," and "NM:i:1" by "NM:i:2," and so on for sequences with more deletions. Convert the SAM output containing the selected sequences to BAM format using the "view-b" command of samtools.
4. Similarly to point 3, to identify sequences harboring a single insertion, extract sequences of length 427 that are aligned with a CIGAR string containing a "1I" operation and a NM tag value of 1. For sequences with two insertions, use a length of 428, a "2I" CIGAR operation, and a tag of "NM:i:2," and so on for further insertions.
5. Repeat points 1–4 for each replicate sample (i.e., six times).

3.6.2.1 Notes

1. The filtering points 2–4 may be performed using other NGS analysis software, such as NGSUtils (Breese & Liu, 2013). Commands similar to the AWK command shown in point 3 may be done in other languages such as PERL or Python.
2. In our study, a total of $\approx$1.6 and $\approx$2.6 million single-substitution sequences could be extracted for TA (WT) and T (ΔpA) samples, respectively. In contrast, only $\approx$40,000 and $\approx$100,000 single-deletion sequences and $\approx$5000 and $\approx$7000 single-insertion sequences for the TA (WT) and T (ΔpA) samples, respectively.

3.6.3 Statistical Analysis

This section describes the procedures to carry out the statistical analysis of the differential distribution of single-substitution sequences in the TA (WT) and T (ΔpA) samples. Analyses of double- or triple-substitution sequences may be conducted similarly. However, in our case, sequencing depth was not sufficient. As such mutations occur in much lower frequencies than single-nucleotide substitutions, greater sequencing depth would have been required to generate datasets with sufficient statistical power (Holmqvist, Reimegård, & Wagner, 2013). In addition, the possibility that several mutations within the same sequence may be correlated (i.e., nonindependent), may complicate the statistical treatment. For all these reasons, we focused exclusively on single-nucleotide substitutions. To determine whether substitutions were enriched at particular positions in the T (ΔpA) compared to TA (WT) sequences, a "positional" analysis was conducted by summing together the counts of all sequences that showed a substitution at a given position, regardless of the identity of the substituted nucleotide. A "nucleotide-specific" analysis comparing the amount of each individual sequence was also done to determine whether particular nucleotides were enriched at specific positions.

3.6.4 Nucleotide-Specific Analysis

1. Using the BAM file containing all sequences showing a single substitution generated in point 2 of Section 3.6.2, generate the distribution of nucleotides at each position by means of samtools: command "mpileup," options "-d 5000000 -t DP –min-BQ 0."
2. From the mpileup file compute for each position the frequency of each nucleotide that is different from the reference with a AWK, PERL, or Python script. This corresponds to the number of sequences that exhibit a particular mutated nucleotide at a given position.
3. Repeat points 1–2 for each replicate sample (i.e., six times).

3.6.5 Positional Analysis

1. From the mpileup file generated in point 1 of Section 3.6.4 compute for each position the total frequency of nucleotides that are different from the reference with a AWK, PERL, or Python script. This corresponds to the total number of sequences that exhibit a mutated nucleotide at a given position, regardless of nucleotide identity. This is simply the sum of the nucleotide-specific counts at each position calculated in point 2 of Section 3.6.4.
2. Repeat point 1 for each replicate sample (i.e., six times).

3.6.6 *Differential Expression Analysis*

Differential analyses are conducted following the protocol of Haas et al. (2013). The protocol pertains to the quantification of RNA-Seq data and involves a step of mapping the reads to transcripts to estimate transcript and gene abundances. This step does not apply here and the actual sequence counts computed in Sections 3.6.4 and 3.6.5 are used as abundances. DESeq2 (Love, Huber, & Anders, 2014) is employed for the statistical calculations. DESeq2 uses a generalized linear model (GLM) based on the negative binomial distribution to compute statistical significance and applies a powerful normalization method (based on the calculation of size factors) to compare samples with a wide dynamic range. The methods employed take into account variability among the three biological replicates. Positional and nucleotide-specific count data are analyzed in the same manner. Substitutions are considered as significantly over- or underrepresented in the T (ΔpA) compared to TA (WT) samples if the *P*-value adjusted for multiple testing (false discovery rate [FDR] calculated using the Benjamini–Hochberg [BH] method in DESeq2) was equal or lower than 5% (padj $\leq$ 0.05).

1. Combine sequence counts from all TA (WT) and T (ΔpA) replicates computed in Sections 3.6.4 and 3.6.5 into a single matrix using the "merge_RSEM_output_to_matrix.pl" utility of Trinity version 2.2.0 (Haas et al., 2013) (http://trinityrnaseq.github.io) run with option "--mode counts" (because raw counts are used as abundances).
2. Using this matrix as input perform differential analysis between TA (WT) and T (ΔpA) samples with the "run_DE_analysis.pl" utility of Trinity: options "--method DESeq2," "--min_rowSum_counts 10" (the last option discards positions for which the sum of mutations across all samples is lower than 10). This produces, for each position along the reference, normalized counts of the number of substitutions in TA (WT) and T (ΔpA) samples along with $\log_2$ ratios of fold change and associated adjusted *P* values.

3.6.6.1 Notes

1. The four positions at each extremity of the 426-nt amplicon corresponding to the 5′-end of the DNA oligonucleotides used to generate it (see note 1 in Section 3.5.3.1) could not be included in the statistical analyses because there were no substitutions at these positions in any of the TA (WT) and T (ΔpA) replicates.
2. In our study, single substitutions were significantly enriched in the T (ΔpA) samples relative to the TA (WT) samples at 70 and 72 positions in the positional and nucleotide-specific analyses, respectively (65 positions in common).

3.6.7 Generate Plots of Analysis Results

It is useful to visualize graphically the data and the results of the statistical analyses. For example, barplots can be used to view the distribution of significant mutations along the reference sequence. In addition, for the "positional" analysis, a heatmap and hierarchical tree clustering of samples and/or positions according to sequence count patterns may be performed. This type of clustering and representation is commonly done when analyzing large-scale multivariate data and provides a comparative view of the distribution of mutations in all samples, and one can judge the correlation among replicates. Clustering and heatmap generation are done as described in the protocol of Haas et al. (2013), in which raw counts are normalized into FPKM (fragment per kilobase per million reads mapped) values using the TMM (trimmed mean of *M* values) method, then median centered, and $\log_2$ transformed.

1. Barplots of normalized sequence counts and $\log_2$ ratios of fold change, computed in point 2 of Section 3.6.6, were drawn using the "barplot" function in R version 3.2.0 (http://www.R-project.org/).
2. Using the matrix of raw positional substitution counts created in point 1 of Section 3.6.5, normalize counts into TMM normalized, median centered, $\log_2$ transformed FPKM values using the "run_TMM_normalization_write_FPKM_matrix.pl" utility of Trinity.
3. Using the FPKM matrix perform hierarchical clustering of samples and positions and generate a heatmap by means of the "analyze_diff_expr.pl" utility of Trinity: options "-C 0 -P 1 –gene_dist gene_cor –gene_cor pearson –gene_clust average –sample_dist sample_cor –sample_cor pearson –sample_clust average." A $\log_2$ cut-off of 0 and a *P*-value cut-off of 1 (options "-C 0 -P 1") were set in order to include all sequence positions in the map. The Pearson correlation coefficient is used as distance metric, and average linkage is chosen as the clustering method.

3.6.7.1 Notes

1. Hierarchical clustering and heatmap generation can be done using a subset of selected positions (i.e., those with significant mutations) by changing the values of the $\log_2$ and *P*-value cut-offs ("-C" and "-P" options of "analyze_diff_expr.pl," respectively) in point 3 of Section 3.6.7.
2. "analyze_diff_expr.pl" provides alternative distance metrics (such as Spearman correlation coefficient) and clustering methods (such as complete linkage).

4. A CASE STUDY: DECODING AAPA3 PEPTIDE TOXICITY DETERMINANTS WITH NUCLEOTIDE RESOLUTION

4.1 Comprehensive Understanding of Suppressor Mutations

1. To understand the origin of the suppressor mutations revealed by FASTBAC-Seq, it is important to note that we observed a high rate of mutation (more than 50%) in absence of any selection pressure (WT condition) (data not shown). This result clearly shows that the mutations were already present in the pool of sequences generated by the Taq DNA polymerase. Although we cannot rule out the spontaneous origin of the mutations, it is very unlikely.
2. In absence of antitoxin, only cells having recombined a DNA sequence carrying a mutation suppressing the lethal phenotype are viable. Thus, the transformation efficiency obtained with the PCR construct inactivated for the antitoxin (inactivated IsoA3–10 box (T, ΔpIsoA3)) is drastically reduced due to the small number of sequences containing these "viable" mutations (Fig. 3).

4.2 Domain Identification in the Small Toxic Protein AapA3

1. Focusing on the AapA3 ORF region, a "nucleotide-specific" statistical analysis was performed (Fig. 5).
2. As mentioned in Section 3.6.3, our statistical analysis focused on single-nucleotide substitutions. Importantly, their distribution along the protein sequence is not homogeneous. More interestingly, the nature of such substitutions turned out to be highly informative. It is well known that PCR amplification can create mutations because of nucleotide misincorporation by Taq DNA polymerases. In particular, A→T/G→C transitions generated by T–G mispairs are among the most frequent errors introduced by DNA polymerases (Cadwell & Joyce, 1992; Eckert & Kunkel, 1991). Nevertheless, transversion mutations (purine→pyrimidine or pyrimidine→purine, indicated with an asterisk in Fig. 5) can also be, although with lower frequency, generated during PCR amplification. Strong PCR-generated transition and transversion biases have been previously observed in deep sequencing studies (Brodin et al., 2013). Here, the pool of PCR-generated sequences that were transformed into the bacteria during FASTBAC-Seq contained a large excess of single-nucleotide transitions. Thus, when a transversion

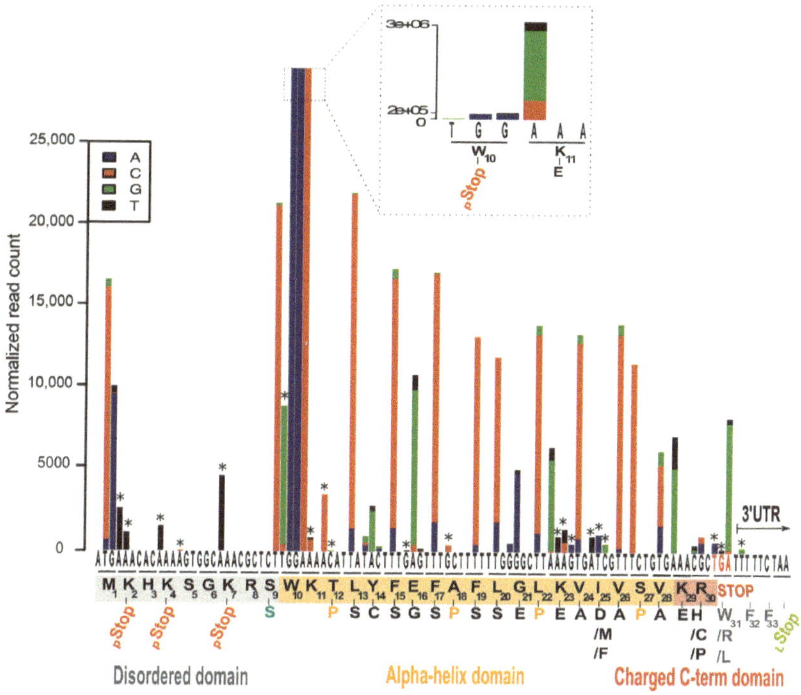

Fig. 5 The AapA3 small toxic protein is organized in three "regions." Statistical analysis of the differential amount of single-nucleotide substitutions in WT (TA; *aapA3*/IsoA3) and ΔpIsoA3 (T; *aapA3*/ΔpIsoA3) samples was carried out using DESeq2 ("nucleotide-specific" analysis; see text for details). Nucleotide substitutions that are significantly enriched (padj $\leq$ 0.05) in the ΔpIsoA3 compared to WT sequences are shown. Codons in the reference sequence are individually underlined and the corresponding encoded amino acid is given (in one-letter code) and numbered below the nucleotide sequence. The alternative amino acid encoded by the most frequent substitution(s) is shown underneath the reference (wild type) amino acid sequence. The substitutions leading to proline residues (potentially disrupting the alpha helix) are highlighted in *yellow*. Data for codons spanned by the three positions (+80, +81, +82, GGA) that have been over-mutated due to PCR amplification biases are shown in the *inset*. *Asterisks* above bars indicate that for the indicated position a transversion mutation was mainly selected (*). The three recognizable domains composing AapA3 peptide: the disordered, the alpha helix, and the positively charged domains are highlighted in *gray*, *yellow*, and *red*, respectively.

mutation is enriched in the absence of antitoxin it is remarkable as the probability of selecting it is very low. Indeed, in our example, most transversions have highly significant *P* values. The selection of a transversion is a strong indicator of biological relevance, i.e., the nature of the nucleotide is crucial for function.

3. Note that, with the exception of the start codon (Methionine in position 1), all suppressor mutations locate at the first and second positions of the codon (Fig. 5). Most of them are transitions, reflecting that toxin inactivation could be easily achieved by the most abundant solutions present in the PCR products. Indeed, single-nucleotide substitutions at the third codon position are rare since they are less prone to be nonsynonymous based on the genetic code. This result helped us validate the cut-off value of padj ≤ 0.05 used for the computational analysis (Section 3.6.6). This value was stringent enough to discard most false positives but permissive enough to identify significant suppressor mutations.
4. A lower mutation rate was observed over the first nine amino acid residues. At positions 2–9, only nonsense mutations (labeled as pStop in Fig. 5) were selected at the first codon position of the lysine codon (AAA to TAA), indicating that the N-terminal region is not crucial for the activity of the AapA3 toxin. A similar result was previously observed during the structural characterization of the AapA1 toxin (belonging to the same family as the AapA3 peptide) where this region was shown to be unfolded (highlighted in gray in Fig. 5).
5. Remarkably, two exceptions can be found in this region: (1) the second most abundant suppressor on the lysine residue at position 2 is a transversion in the second codon position generating a conservative missense mutation from AAA (Lys) to ATA (Ile). It is unlikely that such substitution could have an effect on AapA3 toxicity, especially when no similar mutations have been selected in the rest of the domain. Surprisingly, this new sequence could potentially act as a de novo IsoA3 antitoxin promoter (5′ TATCAT 3′). (2) The second exception is a synonymous mutation identified at the Serine 9 (highlighted in green in Fig. 5) converting the TCT codon into TCC. This suppressor mutation is not acting at the amino acid level but at the translation level.
6. Mutations on the amino acids Trp 10 and Lys 11 were overrepresented in both the TA (WT) and T (ΔpIsoA3)-derived sequences, indicating that they are due to an unanticipated technical artifact linked to the PCR strategy used to generate the DNA substrates for transformation. This led to a strong reduction of the sequencing depth due to the overrepresentation of mutations in the overlapping region used for the PCR assembly (nucleotides 80, 81, and 82 from *aapA3* +1 site). While this artifact impaired the analysis of double and triple mutations due to insufficient sequencing depth, it did not affect the analysis of

single point mutations, which unexpectedly, have been proven sufficient to counteract antitoxin absence.

7. We observed the highest mutation density (58%) in the region Trp 10 to Val 28. Most suppressors on this region are nonconservative mutations: polar residues turned into apolar ones (e.g., Ser 22→Lys 22), or vice versa (e.g., Phe 19→Ser 19). Such mutations modify the peptide hydrophobicity and may potentially alter the interactions with the membrane.
8. Interestingly, at position Thr 12, mainly proline substitutions were selected (transversion in the first codon position: ACA (Thr)→CCA (Pro)). Despite the most abundant solutions in the PCR library would turn Thr (ACA) into Ala (GCA) or Ile (ATA) by transition, or into Ser (TCA), Lys (AAA), or Arg (AGA) by transversion, only the introduction of a proline (CCA) was viable at this particular position. Indeed, in every codon where, following the genetic code, it would be possible to introduce a proline residue by a single-nucleotide substitution, it did happen. These results are in line with previous studies of the AapA1 peptide showing that residues 10–28 form a membrane-interacting alpha-helix domain and demonstrate that similarly to AapA1, this region of the AapA3 peptide may form an alpha helix. Thus, this segment corresponds to the second AapA3 region (residues 10–28), the alpha-helix domain (highlighted in yellow in Fig. 5).
9. Finally, suppressors in the translation stop codon lead to the generation of a longer peptide containing three additional amino acid residues: Trp, Phe, and Phe (the so-called "WFF" motif) until a downstream in-frame stop codon is reached (LStop, Fig. 5). Interestingly, this kind of mutation has already been shown to abolish toxicity of the AapA1 peptide.

5. CONCLUSIONS

5.1 Important Requirements

5.1.1 The Importance of Sequencing "Wild-Type" Transformed Strains

The power of the FASTBAC-Seq method resides in the comparison of strains transformed with WT (TA) and antitoxin-inactivated constructs (T). As exemplified in point 6 of Section 4.2, the parallel sequencing of TA- and T-transformed strains is essential for the identification of technical artifacts deriving from PCR and sequencing biases. Indeed, the mere observation of mutations in samples transformed with the T PCR products is not sufficient to judge their significance, as mutations may be neutral.

Knowledge of the distribution of the same mutations in WT samples is necessary and allows comparative and statistical analysis to identify mutations that are specifically enriched in the absence of antitoxin relative to the WT. The fact that they are statistically enriched shows that they have been selected due to pressure to survive in the presence of a toxin-coding locus in absence of antitoxin.

5.1.2 The Importance of Analyzing (At Least) Three Independent Biological Replicates

The use of biological replicates permits the estimation of intragroup variability and improves statistical power (i.e., the capacity to detect significant differences) and accuracy (i.e., to obtain good estimates of fold changes). Technical variability, on the other hand, is usually small compared to biological variability. Sequencing is highly reproducible and much of the technical variation and biases come from sample and library preparation (Conesa et al., 2016). A minimum of three biological replicates is required, and the more replicates the better. However, generating more replicates increases labor and cost. Statistical power depends on the number of replicates and the sequencing depth. Some studies have shown that increasing the number of biological replicates rather than sequencing depth is a better strategy to improve statistical power (Liu, Zhou, & White, 2014). In contrast, the high reproducibility of only three biological replicates in our case predicts that increasing the number of replicates would not necessarily improve the statistical analysis.

5.2 Advantages

5.2.1 Fast, Systematic, High-Throughput, and Cheap

FASTBAC-Seq can be carried out in less than a month. It allows the rapid identification of suppressor mutations within the whole TA locus. The combination of this classical genetic selection with NGS offers the possibility to explore the complete sequence space with highly informative value upon statistical analysis. Furthermore, if the identification of single-nucleotide mutations is sufficient, the number of sequences required for the analysis can be relatively small and easily affordable.

5.2.2 Study of the Endogenous Expression of TA Systems

FASTBAC-Seq makes possible the study of TA systems at the chromosomal level within its natural host and genetic context. Thus, in contrast to other approaches making use of gene reporters or antibiotic markers,

FASTBAC-Seq allows the in vivo study of toxin expression in endogenous conditions. For this reason, it can accurately identify *cis*-acting regulatory sequences or motifs, not only in the promoter or 5′ untranslated regions but also within the coding and 3′ untranslated regions. In addition, FASTBAC-Seq could be used in different genetic backgrounds, allowing the exploration of secondary suppressor mutations potentially unveiling regulatory pathways.

5.3 Limitations

5.3.1 Exclusive Identification of Loss-of-Function Mutations

One of the main limitations of FASTBAC-Seq is that it relies on the lethal phenotype induced by the chromosomal inactivation of the antitoxin. Thus, it only allows identification of loss-of-function (toxicity) mutations. However, it could in principle be adapted to search for mutations altering other kind of visible phenotype or even gain-of-function mutations. To those ends, one will have to adapt FASTBAC-Seq to an already-known phenotypic sorting (e.g., persistence, biofilm formation, growth rate, cell morphology, etc.) or make use of a reporter system (e.g., green fluorescent protein).

5.3.2 Only Single-Nucleotide Mutations

As mentioned in Section 3.6.3, FASTBAC-Seq currently only allows the identification of single-nucleotide suppressor mutations. The statistical analysis of double or triple mutations as well as other rare events (e.g., long insertions/deletions) would require a greater sequencing depth. To increase sequencing depth and statistical power, saturated mutagenesis (e.g., error-prone PCR) could be used. However, this high mutation rate has the risk of "artificially" reducing recombination efficiency.

5.4 Perspectives

5.4.1 Adapting FASTBAC-Seq to Other Bacterial Species

The simplicity and efficiency of the FASTBAC-Seq method rely on two major features inherent to *H. pylori*. This bacterium is naturally competent and displays a high rate of homologous recombination. Indeed, transformation efficiency has already been shown to be the limiting factor of a similar approach used in *E. coli* (Holmqvist et al., 2013). Nevertheless, FASTBAC-Seq could be adapted to other bacteria provided that a good yield of transformation efficiency can be reached. For instance, it could be used in a modified *E. coli* strain containing the lambda red system that allows the chromosomal inactivation of a single gene in one step (Datsenko & Wanner, 2000). It could also be performed using the CRISPR technology, or a combination of both, the lambda red system and CRISPR (so-called CRMAGE, Ronda, Pedersen, Sommer, & Nielsen, 2016) which would allow the parallel exploration of multiple loci. Alternatively, the method could be adapted to study TA systems on plasmids.

5.4.2 Applying FASTBAC-Seq to Study Other TA Systems

FASTBAC-Seq is a powerful approach that could be used to study any type I TA system. It should help not only to understand the functional determinants of toxicity of some poorly characterized TA systems but also to revisit already-known TA loci for which the precise mechanism of regulation is not well understood (e.g., *ibs*/Sib and *ldr*/Rdl TA systems in *E. coli*). Besides type I TA systems, FASTBAC-Seq could easily be adapted to study other types of TA systems.

ACKNOWLEDGMENTS

We thank all the present and past members of the ARNA laboratory for helpful discussions. We thank D.E. Berg for the generous gift of the *rpsl-erm* cassette. This work was supported by funds from INSERM (U1212), CNRS (UMR5320), Université de Bordeaux, and Agence Nationale de la Recherche (http://www.agence-nationale-recherche.fr/) grants Bactox1 and asSUPYCO. S.M.G. has received funding from the European Union's Horizon 2020 research and innovation programme under the Marie Sklodowska-Curie grant agreement No. 642738. The funders had no role in study design, data collection and analysis, decision to publish, or preparation of the manuscript.

REFERENCES

Arnion, H., Korkut, D. N., Masachis Gelo, S., Chabas, S., Reignier, J., Iost, I., et al. (2017). Mechanistic insights into type I toxin antitoxin systems in Helicobacter pylori: The importance of mRNA folding in controlling toxin expression. *Nucleic Acids Research*, *45*(8), 4782–4795. gkw1343. https://doi.org/10.1093/nar/gkw1343.

Barnett, D. W., Garrison, E. K., Quinlan, A. R., Stromberg, M. P., & Marth, G. T. (2011). BamTools: A C++ API and toolkit for analyzing and managing BAM files. *Bioinformatics*, *27*(12), 1691–1692. https://doi.org/10.1093/bioinformatics/btr174.

Breese, M. R., & Liu, Y. (2013). NGSUtils: A software suite for analyzing and manipulating next-generation sequencing datasets. *Bioinformatics*, *29*(4), 494–496. https://doi.org/10.1093/bioinformatics/bts731.

Brodin, J., Mild, M., Hedskog, C., Sherwood, E., Leitner, T., Andersson, B., et al. (2013). PCR-induced transitions are the major source of error in cleaned ultra-deep pyrosequencing data. *PLoS One*, *8*(7), e70388. https://doi.org/10.1371/journal.pone.0070388.

Cadwell, R. C., & Joyce, G. F. (1992). Randomization of genes by PCR mutagenesis. *PCR Methods and Applications*, *2*(1), 28–33.

Cock, P. J. A., Fields, C. J., Goto, N., Heuer, M. L., & Rice, P. M. (2010). The sanger FASTQ file format for sequences with quality scores, and the Solexa/Illumina FASTQ variants. *Nucleic Acids Research*, *38*(6), 1767–1771. https://doi.org/10.1093/nar/gkp1137.

Conesa, A., Madrigal, P., Tarazona, S., Gomez-Cabrero, D., Cervera, A., McPherson, A., et al. (2016). A survey of best practices for RNA-seq data analysis. *Genome Biology*, *17*(1), 13. https://doi.org/10.1186/s13059-016-0881-8.

Dailidiene, D., Dailide, G., Kersulyte, D., & Berg, D. E. (2006). Contraselectable streptomycin susceptibility determinant for genetic manipulation and analysis of Helicobacter pylori. *Applied and Environmental Microbiology*, *72*(9), 5908–5914. https://doi.org/10.1128/AEM.01135-06.

Datsenko, K. A., & Wanner, B. L. (2000). One-step inactivation of chromosomal genes in Escherichia coli K-12 using PCR products. *Proceedings of the National Academy of Sciences*, *97*(12), 6640–6645. https://doi.org/10.1073/pnas.120163297.

Dorer, M. S., Cohen, I. E., Sessler, T. H., Fero, J., & Salama, N. R. (2013). Natural competence promotes Helicobacter pylori chronic infection. *Infection and Immunity*, *81*(1), 209–215. https://doi.org/10.1128/IAI.01042-12.

Eckert, K. A., & Kunkel, T. A. (1991). DNA polymerase fidelity and the polymerase chain reaction. *PCR Methods and Applications*, *1*(1), 17–24.

Fozo, E. M., Hemm, M. R., & Storz, G. (2008). Small toxic proteins and the antisense RNAs that repress them. *Microbiology and Molecular Biology Reviews*, *72*(4), 579–589. https://doi.org/10.1128/MMBR.00025-08.

Fozo, E. M., Kawano, M., Fontaine, F., Kaya, Y., Kathy, S., Jones, K. L., et al. (2008). Repression of small toxic protein synthesis by the Sib and OhsC small RNAs. *Molecular Microbiology*, *70*(5), 1076–1093. https://doi.org/10.1111/j.1365-2958.2008.06394.x.

Gerdes, K., Bech, F. W., Jorgensen, S. T., Lobner-olesen, A., Rasmussen, P. B., Atlung, T., et al. (1986). Mechanism of postsegregational killing by the hok gene product of the parB system of plasmid Rl and its homology with the relF gene product of the *E. coli* relB operon. *The EMBO Journal*, *5*(8), 2023–2029.

Guo, Y., Quiroga, C., Chen, Q., Mcanulty, M. J., Benedik, M. J., Wood, T. K., et al. (2014). RalR (a DNase) and RalA (a small RNA) form a type I toxin–antitoxin system in *Escherichia coli*. *Nucleic Acids Research*, *42*(10), 6448–6462. https://doi.org/10.1093/nar/gku279.

Haas, B. J., Papanicolaou, A., Yassour, M., Grabherr, M., Blood, P. D., Bowden, J., et al. (2013). De novo transcript sequence reconstruction from RNA-seq using the trinity platform for reference generation and analysis. *Nature Protocols*, *8*(8), 1494–1512. https://doi.org/10.1038/nprot.2013.084.

Holmqvist, E., Reimegård, J., & Wagner, E. G. H. (2013). Massive functional mapping of a 5′-UTR by saturation mutagenesis, phenotypic sorting and deep sequencing. *Nucleic Acids Research*, *41*(12), e122. https://doi.org/10.1093/nar/gkt267.

Jahn, N., & Brantl, S. (2013). One antitoxin—Two functions: SR4 controls toxin mRNA decay and translation. *Nucleic Acids Research*, *41*(21), 9870–9880. https://doi.org/10.1093/nar/gkt735.

Jahn, N., Preis, H., Wiedemann, C., & Brantl, S. (2012). BsrG/SR4 from Bacillus subtilis—The first temperature-dependent type I toxin–antitoxin system. *Molecular Microbiology*, *83*(3), 579–598. https://doi.org/10.1111/j.1365-2958.2011.07952.x.

Kawano, M. (2012). Divergently overlapping *cis*-encoded antisense RNA regulating toxin–antitoxin systems from *E. coli*. *RNA Biology*, *9*(12), 1520–1527. https://doi.org/10.4161/rna.22757.

Li, H., & Durbin, R. (2009). Fast and accurate short read alignment with Burrows–Wheeler transform. *Bioinformatics*, *25*(14), 1754–1760. https://doi.org/10.1093/bioinformatics/btp324.

Li, H., Handsaker, B., Wysoker, A., Fennell, T., Ruan, J., Homer, N., et al. (2009). The sequence alignment/map format and SAMtools. *Bioinformatics*, *25*(16), 2078–2079. https://doi.org/10.1093/bioinformatics/btp352.

Liu, Y., Zhou, J., & White, K. P. (2014). RNA-seq differential expression studies: More sequence or more replication? *Bioinformatics*, *30*(3), 301–304. https://doi.org/10.1093/bioinformatics/btt688.

Love, M. I., Huber, W., & Anders, S. (2014). Moderated estimation of fold change and dispersion for RNA-seq data with DESeq2. *Genome Biology*, *15*(12), 550. https://doi.org/10.1186/s13059-014-0550-8.

Martin, M. (2011). Cutadapt removes adapter sequences from high-throughput sequencing reads. *EMBnet Journal*, *17*(1), 10. http://doi.org/10.14806/ej.17.1.200.

Masachis, S., & Darfeuille, F. (2018). Type I toxin–antitoxin systems: Regulating toxin expression via Shine–Dalgarno sequence sequestration and small RNA binding. *Microbiology Spectrum*, *6*(4). https://doi.org/10.1128/microbiolspec.RWR-0030-2018.

Masella, A. P., Bartram, A. K., Truszkowski, J. M., Brown, D. G., & Neufeld, J. D. (2012). PANDAseq: Paired-end assembler for Illumina sequences. *BMC Bioinformatics*, *13*(1), 31. https://doi.org/10.1186/1471-2105-13-31.

Mok, W. W. K., Patel, N. H., & Li, Y. (2010). Deducing the sequence requirements of, IbsC, a type I toxin in *Escherichia coli*. *The Journal of Biological Chemistry*, *285*(53), 41627–41636. https://doi.org/10.1074/jbc.M110.149179.

Ronda, C., Pedersen, L. E., Sommer, M. O. A., & Nielsen, A. T. (2016). CRMAGE: CRISPR optimized MAGE Recombineering. *Scientific Reports*, *6*(January), 19452. https://doi.org/10.1038/srep19452.

Schmieder, R., & Edwards, R. (2011). Quality control and preprocessing of metagenomic datasets. *Bioinformatics*, *27*(6), 863–864. https://doi.org/10.1093/bioinformatics/btr026.

Silvaggi, J. M., Perkins, J. B., & Losick, R. (2005). Small untranslated RNA antitoxin in Bacillus subtilis. *Journal of Bacteriology*, *187*(19), 6641–6650. https://doi.org/10.1128/JB.187.19.6641-6650.2005.

Unoson, C., & Wagner, E. G. H. (2008). A small SOS-induced toxin is targeted against the inner membrane in *Escherichia coli*. *Molecular Microbiology*, *70*, 258–270. https://doi.org/10.1111/j.1365-2958.2008.06416.x.

Vogel, J., Argaman, L., Wagner, E. G. H., & Altuvia, S. (2004). The small RNA IstR inhibits synthesis of an SOS-induced toxic peptide. *Current Biology*, *14*(24), 2271–2276. https://doi.org/10.1016/j.cub.2004.12.003.

Weaver, K. E. (2012). The par toxin–antitoxin system from Enterococcus faecalis plasmid pAD1 and its chromosomal homologs. *RNA Biology*, *9*(12), 1498–1503. https://doi.org/10.4161/rna.22311.

CHAPTER FIVE

The Challenges of Genome-Wide Studies in a Unicellular Eukaryote With Two Nuclear Genomes

Sandra Duharcourt[*,1], **Linda Sperling**[†]
[*]Institut Jacques Monod, CNRS, UMR7592, Sorbonne Paris Cité, Paris, France
[†]Institute for Integrative Biology of the Cell (I2BC), CNRS, CEA, Univ. Paris-Sud, Université Paris-Saclay, Gif-sur-Yvette CEDEX, France
[1]Corresponding author: e-mail address: sandra.duharcourt@ijm.fr

Contents

Abstract

We present here methods to study a eukaryotic microorganism with two nuclear genomes, both originating from the same zygotic genome. *Paramecium*, like other ciliates, is characterized by nuclear dimorphism, which is the presence of two types of nuclei with distinct organization and functions in the same cytoplasm. The two diploid germline micronuclei (MIC) undergo meiosis and fertilization to transmit the genetic information across sexual generations. The highly polyploid somatic macronucleus (MAC) contains a reduced version of the genome optimized for gene expression. Reproducible programmed DNA elimination of about 30% of the complexity of the 100 Mb MIC genome occurs during development of the MAC along with endoreplication to 800 copies. Large regions that contain transposable elements and other repeats are eliminated, and short single copy remnants of transposable elements, which often interrupt coding sequences, are precisely excised to restore functional open reading frames.

Methods in Enzymology, Volume 612
ISSN 0076-6879
https://doi.org/10.1016/bs.mie.2018.08.012

Genome-wide studies of this process require access to MIC DNA which has long been impossible. The breakthrough with respect to this technical obstacle came with development of a MIC purification protocol involving a critical step of flow cytometry to sort nuclei representing only 0.5% of total genomic DNA. Here, we provide a step-by-step protocol and important tips for purifying nuclei, and present the methods developed for downstream analysis of NGS data.

ABBREVIATIONS

DAPI 4′,6-diamidino-2-phenylindole
DSB double-strand break
FSC forward scatter
GFP green fluorescent protein
IES internal eliminated sequence
IRS IES retention score
MAC macronucleus
MIC micronucleus
NGS next-generation sequencing
NHEJ nonhomologous end joining
ONT Oxford Nanopore Technology
PBS phosphate-buffered saline
PGM PiggyMac
SSC side scatter
TE transposable element
TIR terminal inverted repeat
WGD whole genome duplication

1. INTRODUCTION

Paramecium tetraurelia and other ciliates provide excellent models to study genome dynamics and epigenetic regulation. These unicellular eukaryotes are free-living and belong to the highly diverse ciliate phylum (Lynn, 2008), part of a taxonomic group known as alveolata (Fig. 1A) that also includes apicomplexan parasites and dinoflagellates.

A hallmark of ciliates is their nuclear dualism. In these unicellular organisms, two types of nuclei coexist within the same cytoplasm. In *P. tetraurelia*, each cell contains two identical diploid, transcriptionally silent micronuclei (MICs) and one polyploid transcriptionally active macronucleus (MAC) (Fig. 2). The germline MIC transmits the genetic information to sexual progeny through meiosis, while the lifetime of the somatic MAC is restricted to vegetative growth. At each sexual cycle, the MAC is lost and the functionally distinct MICs and MACs develop from mitotic copies of the zygotic nucleus.

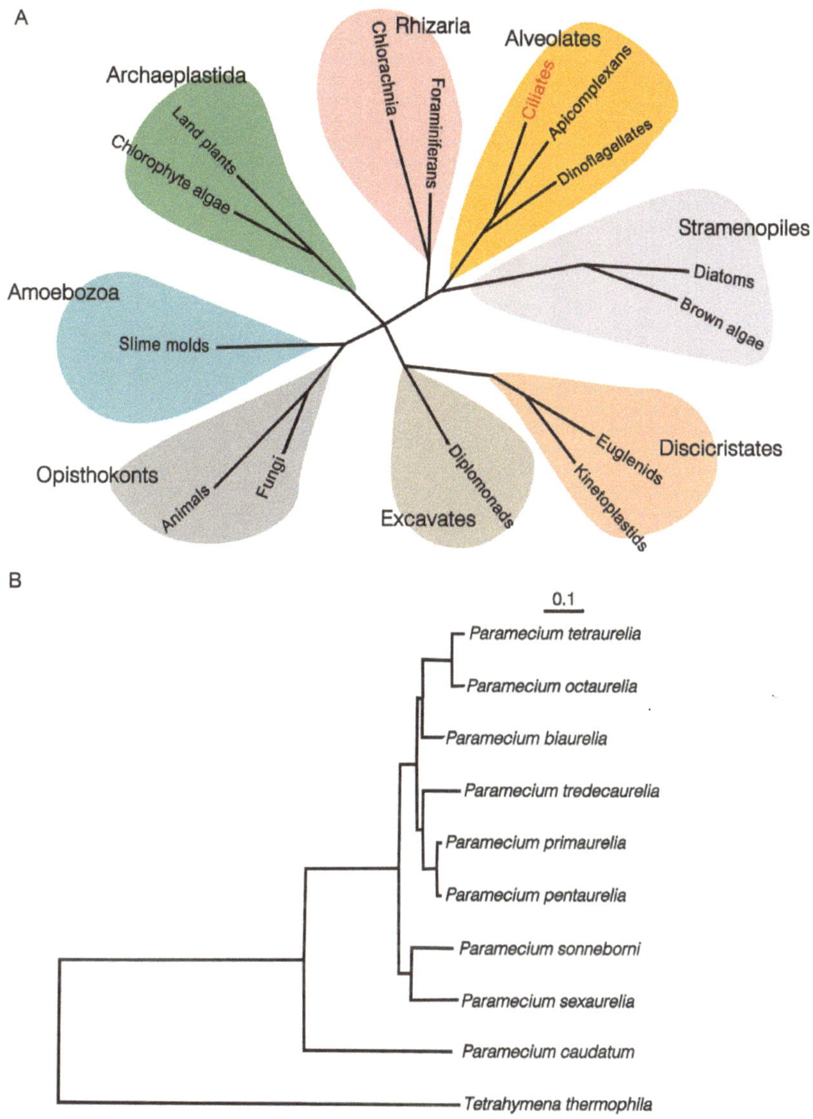

Fig. 1 *Paramecium* phylogeny. (A) Unrooted tree providing a broad picture of the diversity of eukaryotes. *Paramecium* belongs to the ciliate phylum, part of the alveolate clade. Simplified from Baldauf (2008) with adjustment of the alveolate clade according to He et al. (2014). (B). Phylogeny of eight *Paramecium* species from the cryptic *P. aurelia* complex and of *P. caudatum*. *Tetrahymena thermophila* provides the outgroup. Maximum likelihood tree made by concatenating 1061 single copy genes present in all of the *Paramecium* species, kindly provided by Diamantis Sellis and Laurent Duret.

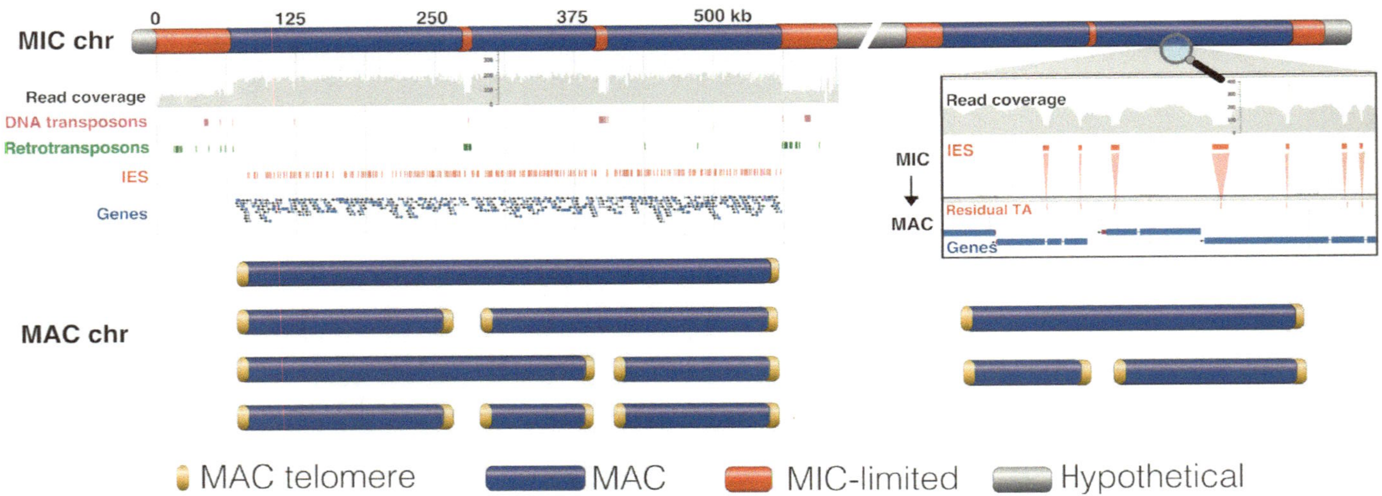

Fig. 2 Genome rearrangements and chromosome organization. An interpretation of MIC chromosome structure is shown at the *top*, with MAC sequences in *blue*, MIC-limited sequences in *red* and hypothetical regions not present in current assemblies of MIC DNA in *gray*. The organization suggested is a monocentric chromosome with two arms; however, the sequence and size of the centromeric region are unknown (*gray region* in *center*). The telomeric sequences that protect MIC chromosome ends are also unknown (*gray* regions at chromosome ends). The organization shown is supported by preliminary *P. tetraurelia* MIC scaffolds, as shown for the *left arm* using a JBrowse image. Tracks, from *top* to *bottom*, show MIC read coverage (200 × for MAC-destined and 100 × for MIC-limited regions), DNA transposons and retrotransposons (Guérin et al., 2017) mapped with the TEannot module of the REPET package, (Flutre, Duprat, Feuillet, & Quesneville, 2011), IESs (Arnaiz et al., 2012), and genes (automated annotations) (Arnaiz et al., 2017). The MAC chromosomes issued from this region are interpreted by drawings in the *bottom half* of the figure. Elimination of MIC-limited regions which contain TEs usually fragments the chromosomes, which are healed by the addition of MAC telomere repeats (mixtures of G_4T_2 and G_3T_3 hexamers) shown in *gold* (often branching from alternative sites, not depicted). Elimination of the two regions of ~10 kb that contain TE insertions (and satellite sequences, not shown) can either fragment the chromosomes or be resolved by rejoining the broken ends (Le Mouël, Butler, Caron, & Meyer, 2003), and the result is a heterogeneous set of MAC chromosomes (Duret et al., 2008 and references therein). On the *right side* of the drawing, a small region of a few kb has been enlarged with a JBrowse image showing IESs, short single copy elements bound by TA dinucleotides, which recombine to leave one TA on the MAC chromosome. Around 80% of the ~45,000 IESs inserted in the MAC-destined regions of the genome interrupt genes, and nearly 50% of protein coding genes contain at least one IES (Arnaiz et al., 2012). IES excision, initiated by a domesticated piggyBac transposase, Pgm, involves a perfectly precise cut and close mechanism (Baudry et al., 2009; Dubois et al., 2017). Note that IESs are eliminated from regions maintained in the MAC but also from MIC-limited regions that contain TEs.

Development of the new MAC involves endoreplication to reach a final ploidy of 800n and extensive programmed genome rearrangements, which reproducibly remove DNA sequences (Fig. 2). Because these events can be experimentally induced by food starvation, which will trigger MIC meiosis, are massive and reproducible, ciliates, and *Paramecium* in particular, are excellent model organisms to study the molecular mechanisms underlying programmed genome rearrangements (Bétermier & Duharcourt, 2014).

In *P. tetraurelia*, the eliminated DNA sequences comprise approximately a third of the MIC genome and include repeated sequences as well as unique internal eliminated sequences (IESs) (Guérin et al., 2017). Molecular and cellular genetics provides a model for this multistep process of DNA elimination. The first "recognition" step allows the specific identification of the target sequences through small RNA-mediated heterochromatin formation, at least for repeated, long sequences (Lepère et al., 2009; Lhuillier-Akakpo et al., 2014; Sandoval, Swart, Arambasic, & Nowacki, 2014). This pathway is very reminiscent of the small RNA guided heterochromatin formation and silencing of repeated sequences in other eukaryotes (Holoch & Moazed, 2015). The second "excision" step consists of tethering the excision complex at the target sites. The excision machinery composed of a catalytically active endonuclease introduces DNA double-strand breaks (DSBs) at the boundaries of germline-limited sequences, leading to the excision of the targeted sequences out of their chromosomal location (Baudry et al., 2009; Dubois et al., 2017). The last "repair" step allows DNA DSB repair at the chromosomal site through classical nonhomologous end joining (NHEJ) pathway (Kapusta et al., 2011; Marmignon et al., 2014), followed by the physical elimination (degradation) of the excised DNA. This multistep process is highly coordinated, as illustrated by the concerted action of the endonuclease and NHEJ repair factors (Marmignon et al., 2014), which likely physically interact within a multicomponent excision complex.

Understanding of the changes that occur during the process of programmed DNA elimination is based on our current knowledge of the germline and somatic genomes (Fig. 2).

The somatic MAC genome of *P. tetraurelia* was sequenced and assembled more than 10 years ago (Aury et al., 2006). The 72 Mb genome is very compact (nearly 80% coding; Arnaiz et al., 2017) and the ~90,000 introns are among the smallest spliceosomal introns ever reported (98% are 20–30 nt, average 25 nt), leading to discovery of translational control of intron splicing (Jaillon et al., 2008). However, the biggest surprise was the discovery that a series of at least three whole genome duplications (WGDs) occurred in the lineage, resulting in a very large number of protein-coding genes for a eukaryotic microbe, nearly 40,000, despite resolution of these dramatic events over evolutionary time by progressive loss of the duplicated genes (Aury et al., 2006; Gout, Kahn, & Duret, 2010; McGrath, Gout, Johri, Doak, & Lynch, 2014). The most recent WGD coincides with the appearance of the *Paramecium aurelia* complex of 15 morphologically identical but reproductively isolated species (Fig. 1B).

Less is known concerning the ~100 Mb germline MIC genome of *P. tetraurelia*. The first genome-wide access to unrearranged DNA involved RNAi knockdown of the *PGM* gene encoding the domesticated transposase required for DNA elimination (Baudry et al., 2009). The "*PGM* DNA" allowed annotation of a genome-wide set of ~45,000 IESs (~3.5 Mb of sequence complexity) (Arnaiz et al., 2012). These elements are small (>99% are 26–500 bp and 93% are smaller than 150 nt), unique copy, and noncoding. Some IESs are copies of DNA transposons, indicating that IESs may be degenerate transposon remnants. The original IES annotation was completely validated when vegetative MIC DNA became available, showing that Pgm is required for all IES excision (Guérin et al., 2017). IES distribution on a MIC region is shown in Fig. 2.

Although the complete transposable element (TE) complement of the genome is still being discovered, the first MIC contigs revealed the presence of five families of non-LTR retrotransposons (LINEs) and of several families of terminal inverted repeat (TIR) DNA transposons (Guérin et al., 2017). Analysis of the purified MIC DNA further revealed that *PGM* DNA does not contain all of the complexity of MIC DNA: a few percent (at least 2.5 Mb) of the total germline DNA complexity is missing from *PGM* DNA. In other words, there are some sequences in the germline that do not require Pgm for their elimination (Guérin et al., 2017 and our unpublished data). The germline-limited DNA also contains highly repeated satellite sequences (2–6 nt microsatellites and minisatellites). Although most of the TE copies in the genome appear to be degenerate and probably do not cause major assembly problems, the satellite repeats do constitute a problem for assembling complete MIC chromosomes, as further discussed in Section 3.4.

2. PROTOCOL: SEPARATING THE MICRONUCLEUS FROM THE MACRONUCLEUS

The purification of *Paramecium* MICs has been a technical challenge for decades. Indeed, the MICs are tiny (3 μm in diameter) and reside in a very large cell (approximately 130 μm long and 30 μm wide). In addition, the gigantic polyploid MAC is fragile and easily breaks down during cell fractionation, especially at the cell lysis step. Thus, separating the MICs from the MACs using cell fractionation does not provide a MIC fraction that is sufficiently pure to exclusively sequence MIC DNA, owing to contamination from the polyploid MAC and also from the bacteria on which *Paramecium* cells feed. In addition to a fractionation step, using a protocol modified from (Preer, Hamilton, & Preer, 1992), it was critical to implement a flow cytometry procedure. The protocol described below is essentially the one published in Guérin et al. (2017). The sorting of the MICs is based on the use of a specific fluorophore that is unambiguously and exclusively associated with the MICs. To achieve this goal, we took advantage of specific fluorescent labeling of the MICs, using a CenH3a-GFP transgene described in Lhuillier-Akakpo, Guérin, Frapporti, and Duharcourt (2016). Transformed clones are selected for their GFP signal/noise ratio, with the precaution that the transgene is not overexpressed and does not affect cell growth or viability.

2.1 MIC Purification

The aim of this protocol is to separate the MICs from the MACs. The first centrifugation step separates most MACs (in the pellet) from most MICs (in the supernatant). The second Percoll centrifugation step eliminates the remaining MACs and some of the large debris (membranes, pieces of cortex, etc.). Most important, the presence of a GFP marker in the MICs is a tremendous help to unambiguously visualize the MICs at every step of the protocol. A noninjected control is mandatory for the downstream steps of flow cytometry.

We found that a better MIC enrichment is obtained with high cell density cultures cultivated at high temperatures (27–30°C; the standard growth temperature for *P. tetraurelia* is 27°C; Beisson et al., 2010). Under these conditions, there are fewer food vacuoles and membrane composition is probably affected (Beisson, Cohen, Lefort-Tran, Pouphile, & Rossignol, 1980), and this appears to facilitate the first steps of cell fractionation. The goal is to obtain 2–3 million cells in 2–3 L at a density >1000 cell/mL that are still in the vegetative phase of the life cycle.

It is important to always add buffers gently to the sample in order to avoid as much as possible osmotic shocks that would result in MIC explosion.

2.1.1 Buffers and Solutions

- Lysis buffer (freshly prepared the day of use or the day before): 10 m*M* Tris–HCl pH 6.8; 10 m*M* $MgCl_2$; 250 m*M* sucrose; 0.2% NP40.
- Washing buffer (freshly prepared the day of use or the day before): 10 m*M* Tris–HCl pH 7.4; 15 m*M* NaCl; 60 m*M* KCl; 5 m*M* $MgCl_2$; 250 m*M* sucrose; 0.5 m*M* EGTA.
- Percoll 50%: 50% Percoll pH 7.5; 250 m*M* sucrose; 10 m*M* $MgCl_2$. Percoll: GE Healthcare (reference 17-0891).

Use Percoll 100% to make Percoll 50% and check pH before use. Percoll 100% has a pH of ~8.4. It is important to carefully equilibrate the pH for 100 mL to pH 7.5 with a small amount of HCl. If necessary, correct with NaOH.

- Phosphate-buffered saline (PBS): 137 m*M* NaCl; 2.7 m*M* KCl; 6.7 m*M* Na_2HPO_4; 1.5 m*M* KH_2PO_4.
- Proteinase K solution (freshly prepared the day of use): 0.5 *M* EDTA pH 9; 1% *N*-lauroylsarcosine; 1% SDS; 1 mg/mL proteinase K (Merck Millipore).
- Tris–EDTA: 10 m*M* Tris–HCl; 1 m*M* EDTA, pH 8.

2.1.2 Cell Culture

Transformed and control *Paramecium* cells are grown in standard *Klebsiella pneumoniae* medium (Beisson et al., 2010) to a final volume of 3 L at a cell density of 1000–1500 cells/mL. The vegetative state of the cells is assessed by nuclear staining with a 33:1 (vol/vol) mix of carmine red (0.5% in 45% acetic acid) and fast green (1% in ethanol). Detection of GFP signal in the MICs is monitored in the transformed cells. Cultures are filtered on eight layers of sterile gauze.

2.1.3 Cell Centrifugation

Cells are centrifuged at 600 × *g* for 1 min in an oil-testing centrifuge (Sigma 6–16, rotor 13116). After each 4 × 100 mL centrifugation step, centrifuged cells are recovered and not centrifuged multiple times. Once the whole culture is centrifuged, the pooled cells are washed in 100 mL of Tris–HCl 10 m*M* pH 7.4 and centrifuged again to obtain a compact pellet. Note the exact volume of the pellet (V).

2.1.4 Cell Lysis

The cell pellet is suspended in two volumes of lysis buffer and kept on ice for 5 min. All subsequent steps are performed at 4°C. Cells are lysed with a

Dounce homogenizer (Dutscher; 045013) until approximately 90% of the cells are broken as observed under a microscope (×20). At this step, it is critical to avoid breaking the MACs into pieces, even if this results in less than 90% of lysed cells.

2.1.5 Centrifugation Step 1

Three volumes of washing buffer are added. The sample is then dispatched in 2-mL Eppendorf tube(s), gently mixed by inversion five times, and centrifuged for 1 min at 200 × *g* at 4°C. In this step, the pellet is not very well attached to the bottom of the tube, so the tube(s) should be handled carefully. The supernatant that contains most of the MICs is recovered. Presence of the MICs (in an aliquot of 1–2 μL of supernatant) is checked under the microscope (×20).

2.1.6 Centrifugation Step 2

The supernatant is transferred into Ultra-clear centrifuge tubes (Beckman Coulter 344059; 2 mL per tube). First, 1 mL of 50% Percoll solution is added drop by drop with a 1000-μL Pipetman, the tube is gently tapped to mix, then 9 mL of 50% Percoll solution are added drop by drop using a 10-mL pipette. The supernatant and the Percoll solution are mixed by gentle pipetting and centrifuged at 24,000 × *g* for 15 min in a SW41Ti swinging rotor (Optima L-80 XP ultracentrifuge, Beckman Coulter). During centrifugation, the Percoll gradient is formed and MICs accumulate at the top of the gradient and MACs at the bottom (Fig. 3). After centrifugation, MICs are carefully recovered in a white-to-brown powdery band with a 200-μL Pipetman into a 2-mL Eppendorf tube with as little as possible additional liquid. The MIC-enriched sample is gently mixed by pipetting to homogenize it. Presence of the MICs (in a 1–2 μL aliquot) is checked under the microscope (×20). Only the MICs should be visible, with no MACs or other cellular debris.

2.1.7 Freezing

The MIC-enriched preparation is diluted 1/1/1 with washing buffer and glycerol 40% (13% glycerol final concentration). After dilution, several hundred MICs per microliter could be counted under a microscope. The MIC-enriched sample is aliquoted (300–500 μL) in 1.5-mL Eppendorf tubes which are frozen at −80°C (do not flash-freeze!) at this step for later use.

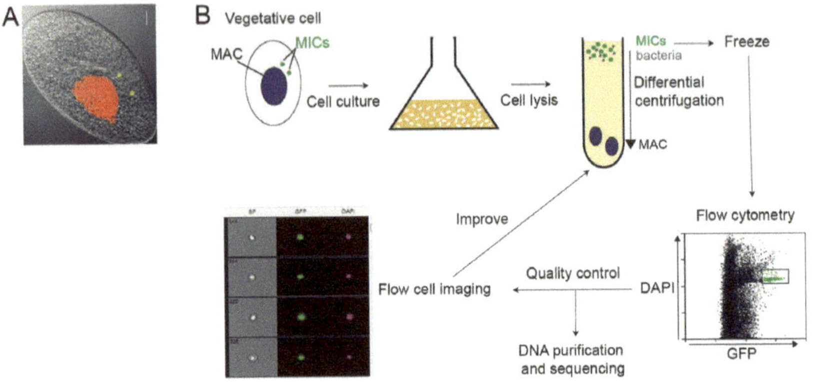

Fig. 3 MIC purification workflow. (A) One vegetative *Paramecium tetraurelia* cell transformed with a MIC-specific GFP transgene. The nuclei are stained with Hoechst (in *red*). The MICs, but not the MAC, are GFP-positive (in *green*). Scale bar is 10 μm. (B) Vegetatively growing GFP transformed *Paramecium* cells are cultivated then lysed. MICs are separated from the MACs by cell fractionation and can be frozen at this step. Sorting of the MICs by flow cytometry is based on GFP and DAPI fluorescence. Quality controls of the MIC-enriched preparation are performed by cell flow imaging and NGS sequencing.

2.2 Sorting by Flow Cytometry

Samples of MICs are thawed on ice, diluted 1/5–1/10 in washing buffer, and stained with DAPI (3 μ*M* final, Invitrogen #D3571). All steps are performed at 4°C. The samples are filtered (30-μm Sysmex filters, 04-004-2326) and sorted on an Influx 500 cell sorter (BD Biosciences) with a 488-nm laser for scatter measurements (forward scatter (FSC) and side scatter (SSC)) and for GFP excitation, and with a 355-nm laser for DAPI excitation. GFP and DAPI staining signals are detected using a 528–38 and 460–50 nm band pass filters, respectively. Phosphate Buffered Saline (Isoflow TM Sheath Fluid, Beckman Coulter) was used as sheath and run at a constant pressure of 15 psi. Frequency of drop formation was approximately 27 kHz. The instrument used a 100-μm nozzle. A threshold on the GFP signal was optimized to increase collecting speed (2500 events per second). Events in GFP and DAPI gates were backgated onto FSC vs SSC to optimize the sorting. Sorting rates typically ranged from 10,000 to 100,000 MICs per hour depending on the preparation. Data were collected using Spigot software. Nuclei were recovered in washing buffer into a 1.5-mL Eppendorf tube.

2.3 Quality Control: Flow Imaging and High-Throughput Sequencing

2.3.1 Flow Cell Imaging

After sorting, ImageStream analysis provides a quality control of the purity and enrichment of the MIC preparation. This analysis was essential for setting up the protocol and improving it. Samples are imaged before and after sorting on a 2 camera, 12 channel ImageStreamX (Amnis/Merck Millipore) imaging flow cytometer at 60 × magnification, using 405, 488, and 785 nm lasers, at respectively, 125, 100, and 0.05 mW. PBS was used as sheath. Brightfield is collected in channels 1 and 9, SSC in channel 6 (745–800 nm bandwidth), GFP in channel 2 (480–560 nm bandwidth), and DAPI in channel 7 (430–505 nm bandwidth). At least 5000 elements should be analyzed for each sample (before and after sorting) in order to detect enough MICs, given their rarity in the sample. Cell classifiers are set for channel 1 area lower limit of 10 to allow the instrument to focus despite low concentration of the sample after sorting. The calibrating beads used for flow cytometry are excluded from the analysis based on their low DAPI and GFP fluorescence signals. Acquisition and analysis are performed using Inspire and IDEAS software, respectively. In general, ~0.2%–3% of all events detected by the Influx cell sorter before sorting correspond to MICs. Over 80% correspond to MICs after sorting. It is worth emphasizing that other elements are not necessarily real contaminants. Only those which contain DNA are real contaminants that are important for downstream steps. The real purity should be estimated by counting the DAPI-positive elements and usually approaches 99%. However, if 2% of DAPI-positive elements are pieces of MACs, this can result in 50% contamination at the DNA level, given MAC ploidy (800n).

2.3.2 Genomic DNA Extraction and Sequencing

After sorting, MICs are treated with three volumes of proteinase K solution at 55°C overnight. Genomic DNA is extracted with the addition of one volume of Tris–HCl-phenol pH 8 with gentle agitation at room temperature for 1 h (no vortex). After centrifugation at 300 × *g* for 15 min, the aqueous phase is recovered, dialyzed twice against 10 m*M* Tris–HCL, 1 m*M* EDTA pH8, 25% ethanol for 2 h, against 10 m*M* Tris–HCL, 1 m*M* EDTA pH8 overnight, then against 1 m*M* Tris pH 8 (no EDTA) for 2 h. DNA is concentrated with a Concentrator plus (Eppendorf) to 50–100 μL. DNA concentration is quantified using QuBit High sensibility kit (Invitrogen) and stored at 4°C. From around 10^6 vegetative *P. tetraurelia* cells, approximately 100 ng of MIC DNA is recovered. This is enough to sequence DNA by a paired-end strategy using Illumina Hi-Seq next-generation sequencing, but is not yet compatible with mate-pair Illumina sequencing or long-read sequencing.

3. ANALYSIS OF SEQUENCE DATA

The workflow discussed in this section assumes proficiency with the Unix/Linux command-line and access to a Linux workstation or equivalent with standard bioinformatics software. In theory, Galaxy (Giardine et al., 2005), a platform designed to allow biologists to carry out NGS analysis without command-line skills is a viable alternative. However, standard Galaxy implementations are not intended to offer the flexibility required for problems that require nonstandard procedures. Implementing a customized Galaxy installation that can run nonstandard procedures is time-consuming and requires advanced informatics training.

We began developing methods for analysis of *Paramecium* germline DNA when we realized that variability in somatic DNA, a consequence of the programmed genome rearrangements, allowed discovery of IESs (or parts of IESs) that were sometimes retained in—or erroneously excised from—the somatic genome (Duret et al., 2008). The analysis of IES retention and IES excision errors led to development of the ParTIES software package (Denby Wilkes, Arnaiz, & Sperling, 2016), which is now routinely used for functional studies, to quantitate the effects of candidate gene knock-down on IES excision genome-wide (Gruchota, Denby Wilkes, Arnaiz, Sperling, & Nowak, 2017; Lhuillier-Akakpo et al., 2014; Maliszewska-Olejniczak et al., 2015). However, our ultimate objective is to assemble and annotate high quality germline genomes for the *P. tetraurelia* model (for studies of rearrangement mechanisms) and for other *Paramecium* strains and species (for comparative genomics). We are particularly interested in discovering the organization of MIC chromosomes. At the moment, we do not even know how many MIC chromosomes there are in *P. tetraurelia* (at least 50 small chromosomes and maybe more; Duret et al., 2008) or have definitive proof that they are indeed monocentric (Lhuillier-Akakpo et al., 2016).

Paramecium germline genomes are relatively small eukaryotic genomes (~100 Mb for *P. tetraurelia*), making it cost-effective to sequence them to high coverage (at least 50–100 ×) with current Illumina sequencing technology. The analysis of the resulting datasets is greatly facilitated by the ease of working with 100% homozygous cultures, thanks to the sexual process of autogamy (autofertilization) found in *P. tetraurelia* and many other species. However, a whole genome duplication (WGD) in the common ancestor of the 15 sibling *P. aurelia* species (Fig. 1B) is still being resolved and about 50% of the preduplication genes remain present in two copies (Aury et al., 2006; McGrath et al., 2014). This large number of duplicate sequences, many sharing >90% nucleotide identity, could lead to assembly collapse. To optimize mapping accuracy and minimize assembly collapse, genome-wide studies of *P. aurelia* species involve paired-end Illumina sequencing, the longer the read size the better (100–250nt for currently available Illumina platforms). The WGD paralogs are also a potential problem for correcting long error-prone reads (PacBio SMRT sequencing or Oxford Nanopore Technology (ONT)) using Illumina short reads, since the long-read error rates are comparable to the rate of divergence between WGD paralogs.

Nonetheless, the biggest challenge for data analysis is the presence of DNA from both MIC and MAC genomes (rearranged and unrearranged DNA) in sequenced samples. Can metagenomics approaches help meet this challenge? Enormous progress has indeed been made in metagenomics, allowing genomes to be characterized and even assembled from environmental samples that contain communities of microorganisms (for an overview, see Ghurye, Cepeda-Espinoza, & Pop, 2016). Total DNA from a *Paramecium* culture resembles an environmental sample as it is expected to contain *Paramecium* germline (MIC) DNA, somatic (MAC) DNA, mitochondrial DNA, as well as bacterial DNA from symbionts (if present), contaminating food bacteria and a tightly associated nonsymbiotic bacterial consortium. Aside from the cost of sequencing such a sample if only the germline DNA is of interest, why would it be a problem today to obtain a meta-assembly of the germline genome? First, it remains very difficult to assemble genomes present in vastly different amounts, which is why rDNA amplification is still widely used for environmental genomics. In *Paramecium*, the germline DNA is not only the most complex genome in the sample, but it is also the least abundant, accounting for only ~0.5% of *Paramecium* genomic DNA (four copies per cell). Even with a sample greatly enriched in germline DNA, thanks to the sorting approach to nuclear purification outlined in Section 2, a second hurdle for a meta-assembly is that the germline and the somatic DNA are largely (~75%) colinear, with frequent small insertions in the germline DNA, making it hard to avoid chimeric contigs that represent neither MIC nor MAC loci, and/or highly fragmented assemblies.

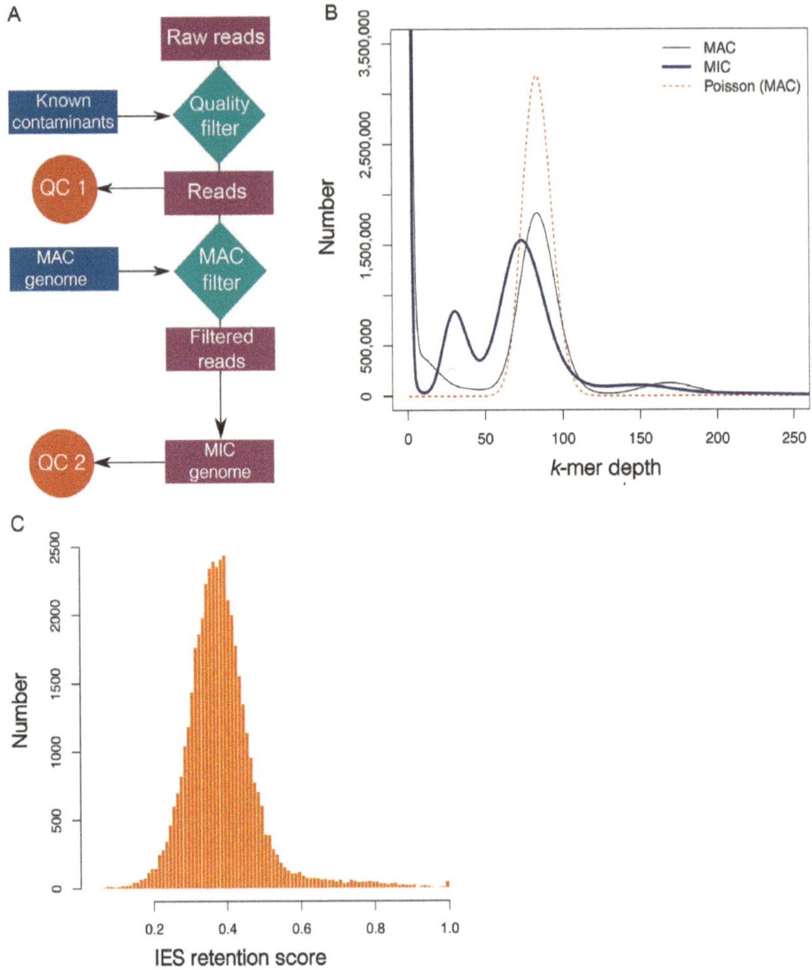

Fig. 4 NGS workflow and methods used to control read quality. (A) Diagram of the workflow: *rectangles* represent datasets, *diamonds* represent bioinformatics calculations that operate on the datasets, and *circles* represent procedures used to control quality. Details are given in the text. (B) Histograms of 17-mer counts of the reads from MIC (*thick blue curve*) and MAC (*thin black curve*) datasets. A Poisson distribution that models the MAC dataset is also shown (*dotted red curve*). (C) Histogram of IES retention scores (IRSs) calculated using the reads (see Fig. 5 for an explanation of IRS).

In this section, we describe a workflow for NGS data processing and assembly (depicted in Fig. 4A) that places a lot of emphasis on filtering and controlling the quality of reads before any analysis is carried out, precisely to deal with the fact that the samples contain mixtures of germline and somatic DNA. Throughout, we favor discussion of the underlying principles, which can be adapted for other genomic analyses in *Paramecium* or applied to other organisms, over step-by-step protocols.

3.1 Preprocessing of Sequencing Reads

Raw Illumina reads must be preprocessed to remover adapters, trim for quality, and remove known contaminants. It is likely that the first two steps have been carried out by the sequencing service. If not, we use Cutadapt (Martin, 2011) to remove the adaptors and Sickle (https://github.com/najoshi/sickle) to trim for quality. In both cases, the minimum quality is usually set at 20 (but can be increased if sequencing depth is high) and the minimum read length is usually set at 75 (but can be increased if the reads are longer than 100 nt). FastQC software (http://www.bioinformatics.babraham.ac.uk/projects/fastqc/) can be used to check read quality at every step of the pipeline.

We also remove reads that correspond to *Paramecium* rDNA, the mitochondrial genome and known contaminants of *Paramecium* monoxenic cultures grown without antibiotics. Namely, *Paramecium* L4440 double-strand RNA feeding vector and 109 complete bacterial genomes that include strains of *Klebsiella* (feeding bacteria), *Pseudomonas* (a frequent contaminant), and *Escherichia coli* (feeding bacteria for RNAi experiments), the major bacterial contaminants we have encountered. The reads are mapped to these contaminant reference sequences using Bowtie 2 (Langmead & Salzberg, 2012), with default parameters.

3.2 Quality Control of Sequencing Reads Using *k*-mer Analysis (QC1)

We have found that analysis of *k*-mers is a very powerful way to evaluate the quality of Illumina datasets. Indeed, analysis of *k*-mers, i.e., all substrings of a given length *k* found in reads, is powerful because it does not require a reference genome assembly. Use of *k*-mers is being developed for more and more applications ranging from metagenomics to genome assembly and transcriptome analysis. We use Jellyfish software (Marçais & Kingsford, 2011) for quality control of datasets, typically by counting all 17-mer substrings that appear in the reads. The fact that samples were obtained from completely homozygous cells simplifies the analysis. In a perfect homozygous dataset with no sequencing errors and no repeated sequences, we can expect the histogram of *k*-mer depth to obey a Poisson distribution with a peak position corresponding to the average sequencing depth. As shown in Fig. 4B, for *P. tetraurelia* MAC reads, there is one large peak, corresponding to 83 × average sequencing depth, in addition to the very large peak at the origin, of *k*-mers that contain sequencing error(s) and are found only 1 or a few times. The small peak at 170 × arises from identical *k*-mers in paralogs produced by the most recent WGD, explaining the doubled *k*-mer depth with respect to the main MAC *k*-mer peak. The superimposed Poisson distribution differs little from the MAC distribution. Excluding the peak at the origin, we can estimate (Li et al., 2010) that the MAC genome complexity is ~76 Mb, in good agreement with the size of *P. tetraurelia* MAC genome assemblies (Aury et al., 2006 and unpublished).

The MIC *k*-mer distribution shown in Fig. 4B is for *P. tetraurelia* sequencing data obtained from sorted MIC DNA (Guérin et al., 2017) and shows two major peaks (excluding the error peak at the origin). The smaller peak is centered at 31 × and corresponds to *k*-mers present only in MIC DNA. A larger peak is centered at 73 × and corresponds to MAC-destined sequences found in both MIC DNA and contaminating MAC DNA. From the ratio of the two peaks (31/73) we can estimate that the MIC DNA corresponds to ~40% of the sample. This confirms an 80-fold MIC DNA enrichment (see Section 2.3 above). We can also roughly estimate from the MIC *k*-mer distribution that there is at least 20 Mb of MIC-limited sequence complexity, by considering only the first peak. This is an underestimate, since highly repeated *k*-mers (depth > 250), that could be of MAC or more often of MIC origin, are not taken into consideration.

The MIC curve in Fig. 4B is for a sample containing both MIC and MAC DNA, presenting two peaks. However, if a sample contains only MIC DNA, there will be only one peak. Therefore, it is important to have independent information to evaluate the purity of the sample. As will be further developed in Section 3.3, MIC DNA enrichment can be independently evaluated, using the ParTIES software (cf. Fig. 5), by determining the extent to which IESs are retained in the reads of a sequencing sample. Fig. 4C shows the histogram of IES retention scores (IRSs) determined by mapping the reads in the MIC sample analyzed in Fig. 4B, to an IES-containing reference and calculating IRS for all of the ~45,000 IESs in the genome. The distribution is close to a Gaussian distribution with a peak at ~0.4 indicating that ~40% of the reads come from the IES-containing MIC DNA, confirming the estimate of MIC purity obtained from the analysis of *k*-mers.

In conclusion, although read quality can (and should) be evaluated with generic software such as FastQC, we find that the wealth of qualitative and

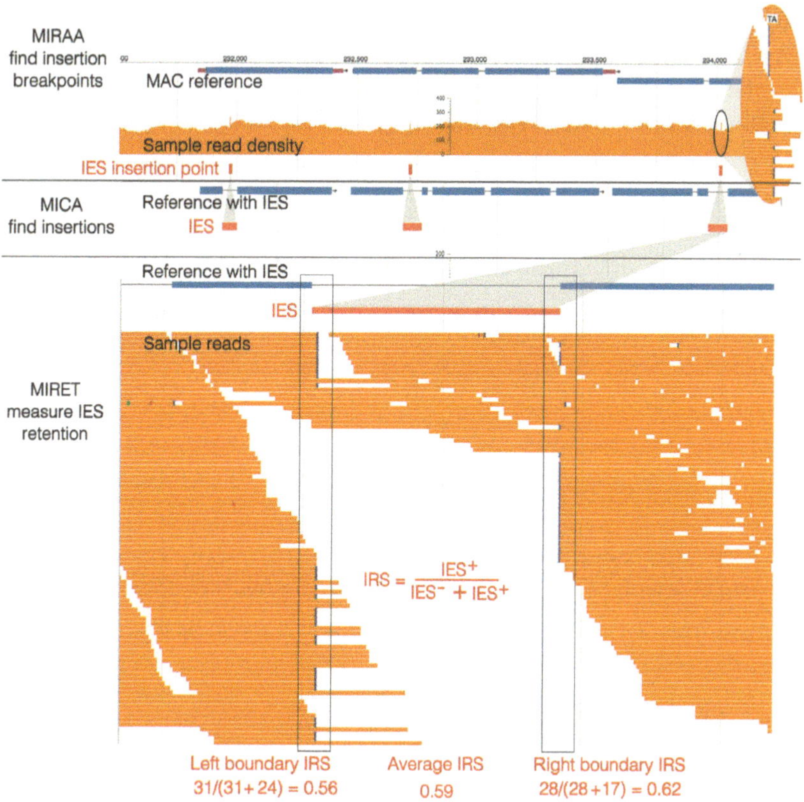

Fig. 5 The ParTIES toolbox for IES studies. Major methods in the ParTIES toolbox are illustrated with JBrowse screenshots (ParameciumDB, https://paramecium.i2bc.paris-saclay.fr). *MIRAA* maps sample reads to a MAC reference genome and looks for breakpoints in the coverage that indicate insertion points. Partially mapped reads contain flanking sequence and (part of) an insertion so always cross the residual TA dinucleotide, leading to an excess of read ends at the insertion site, visible in a density plot as a narrow (usually 2 bp) spike. In the *inset*, partially mapped reads that end at the TA insertion site have a *dark blue edge*. *MICA* performs local assemblies (or optionally uses an assembly with the insertions provided by the user) to determine the sequence of the insertions through comparison to a MAC reference. *MIRET* maps a sample to a reference that contains IESs and determines an IRS for each IES boundary, by counting the IES+ and IES− reads that cross the boundary (the partially mapping IES− reads have a *dark blue edge*). The IRS is the ratio of IES+ reads to all reads crossing the boundary. The average of the left and right boundary scores gives a single IRS for the IES (as in the histogram in Fig. 4C). ParTIES software, distributed under a GPLv3 license, is described in Denby Wilkes et al. (2016) and is available from https://github.com/oarnaiz/ParTIES.

quantitative information that can be obtained by counting *k*-mers in reads provides the most powerful way to appreciate the quality of our *Paramecium* genomic datasets. Indeed, if we do not find a histogram similar to those in Fig. 4B with peak(s) well separated from the big error peak at the origin, we can usually trace back to a problem during library preparation or sequencing. In such cases, we have found that assembly is either impossible or poor, and we consider it more effective to obtain better data than to attempt ad hoc filtering or sampling of the reads.

3.3 Removing MAC Reads That Could Compromise Assembly (MAC Filter)

Contaminating MAC sequences can compromise assembly of the MIC genome, leading to contigs that represent neither MAC nor MIC DNA and/or are highly fragmented. The ParTIES software, developed to identify IESs and study their excision, can be used to identify reads of MAC origin. Fig. 5 illustrates the major methods of ParTIES: MIRAA maps sample reads to a MAC reference assembly to identify insertion breakpoints; MICA carries out local assembly of the reads (or takes a user-provided assembly) and compares the contigs with a MAC reference assembly to identify the IESs; MIRET maps sample reads to a reference that contains IESs to measure the IES retention in the sample by counting IES+ and IES− reads at each IES boundary.

MIRAA breakpoints can be used to identify and remove reads that map perfectly across the MAC junction and therefore come from MAC DNA. This approach does not require prior knowledge of the IESs, and can be applied to any strain or species. The method can be improved if an IES reference set is available for the strain or species, since sometimes MIRAA breakpoints are not found for an IES, owing to the choice of parameters (in particular, the minimum number of excess reads required to identify a breakpoint).

It is in addition possible to identify and remove reads that contain MAC telomere repeats, which are a random mixture of G_4T_2 and G_3T_3 hexanucleotides (Baroin, Prat, & Caron, 1987; McCormick-Graham & Romero, 1996). This is not usually necessary since modern assembly software sets aside low-complexity sequences and highly repeated sequences. For the moment, we do not know whether the MIC chromosome ends are protected by the same sequences as the MAC chromosomes, by other telomere repeats as is the case for the centromere-proximal repeats in MIC telomeres of the ciliate *Tetrahymena* (Kirk & Blackburn, 1995) or by some other mechanism as in *Drosophila*, where retroelement insertions protect chromosome ends (reviewed by Louis, 2002).

For the *P. tetraurelia* MIC sample used for the *k*-mer analysis shown in Fig. 4B, which contains ~40% MIC DNA, the MAC filter removes 5% of the reads. For the *Paramecium biaurelia* MIC sample that contains >90% MIC DNA (used to illustrate assembly quality in Fig. 6; Section 3.5), the MAC filter removes only 1.2% of the reads.

3.4 Assembly Strategies

The MAC-filtered reads can be assembled using software designed for short read assembly. We have used Velvet (Zerbino & Birney, 2008) to assemble contigs from MIC-enriched *P. tetraurelia* DNA (Guérin et al., 2017) and Velvet is the default assembler in the ParTIES software package. Alternatively, 250 nt Illumina paired-end reads sequenced from the ends of short DNA inserts (typically ~400 nt mean size) overlap and can be merged to make libraries of ~400 nt reads with software such as FLASH (Magoč & Salzberg, 2011). The merged reads can be assembled with software optimized for longer reads, or for mixtures of long and short reads, such as the Celera assembler and its descendants (http://wgs-assembler.sourceforge.net/). An overview of assembly strategy and algorithms can be found in Simpson and Pop (2015).

In order to obtain a draft genome assembly, a further step of scaffolding is required. Scaffolding uses additional information (typically from PacBio or ONT long reads, optical mapping or Illumina mate-pair sequencing of large-insert libraries) to join the contigs. Unfortunately, none of these options is currently feasible for DNA produced by sorting MICs, because much larger amounts of starting material are required (10 μg or more as opposed to the 100 ng scale of the method outlined in Section 2).

3.5 Quality Control of the Assembly (QC2)

QUAST software (Gurevich, Saveliev, Vyahhi, & Tesler, 2013) can provide basic assembly metrics along with a wealth of additional information. Assembly metrics provide a first assessment of assembly quality and can also easily be obtained using custom Python, Perl, or R scripts that output the number of contigs, largest contig, smallest contig, complexity (for all contigs, contigs >500 nt, contigs >2 kb), N50, …, N90, G+C content, and the number of Ns, an important metric after scaffolding and gap-filling. Although totally

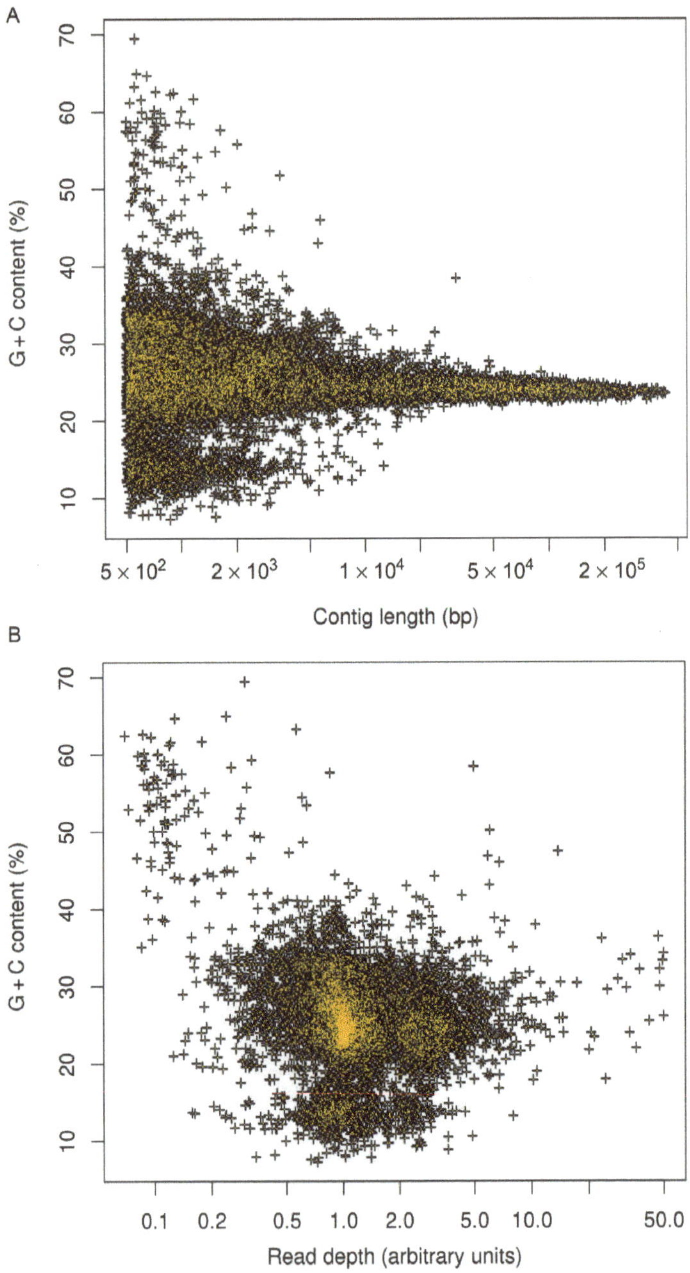

Fig. 6 Quality control of contigs using G+C content, size, and read depth. In the scatter plots, each point represents a contig from a preliminary assembly of *P. biaurelia* MIC DNA prepared after sorting MICs to greater than 90% purity. (A) G+C content as a function of contig length. (B) G+C content as a function of normalized read depth.

standard, assembly metrics are important to communicate a well-accepted and easily understood indication of assembly quality.

We take advantage of the fact that *Paramecium* genomes are AT-rich to use an approach similar to one used to study bacterial communities and symbionts associated with unicellular eukaryotes (Kumar, Jones, Koutsovoulos, Clarke, & Blaxter, 2013). This involves representing the assembly with scatter plots of G + C content, contig length and the number of reads used to build each contig, a proxy for coverage. Note that coverage can be rigorously determined using SAMtools or deepTools (Ramírez et al., 2016) after remapping the reads on the assembly. Bacterial DNA is easily recognized by high G + C content with respect to the *Paramecium* genomic DNA (average ~28% for *P. tetraurelia* and around 23% for most other species; Arnaiz et al., 2017 and unpublished). Scatter plots of a preliminary assembly of *P. biaurelia* MIC DNA are shown in Fig. 6. The observation in Fig. 6A of a narrow horizontal "peak" of contigs (maximum size 430 kb) with similar G + C content (around 23%) and in Fig. 6B of a dense cloud of contigs with this G + C content indicates that it is a good assembly: these contigs represent most of the genome of *P. biaurelia*. The high quality of the assembly is corroborated by an N50 metric of 92 kb (half of the assembly is in contigs larger than 92 kb). The small contigs with high (>40%) G + C content generally have very low read coverage and represent bacterial contaminants. The small contigs with very low G + C content (<18%) have read depth similar to that of the major part of the assembly and represent *Paramecium* sequences, probably including minisatellite and simple-repeats from the MIC-limited portion of the genome. Software designed for metagenomics, such as the Kraken taxonomic sequence classification system (Wood & Salzberg, 2014), can help identify the origin of contigs (or scaffolds) and clean *Paramecium* assemblies by removing contaminant contigs (or scaffolds), usually of bacterial or bacteriophage origin.

Collapse refers to the fact that reads from distinct genomic regions that share high sequence identity are sometimes assembled together. In our experience, given the precautions outlined earlier in this section, the problem does not occur or is extremely rare between WGD paralogs, but it is more frequent for disperse and tandem repeats. Collapse has been observed in both MAC and MIC assembles for some specific genes that contain internal repeats, such as *Paramecium* surface antigens (Prat, Katinka, Caron, & Meyer, 1986) and the highly repeated CenBP, a Sfi1p-like centrin-binding protein (Gogendeau et al., 2007). Collapse may be observed between genes (the case for surface antigens) or within the gene between repeats (the case for CenBP). Collapse can be assessed by mapping the reads to the assembly and calculating depth of coverage, which will be higher in regions where the assembly is collapsed. Potentially collapsed regions identified by depth of coverage can be validated by looking for SNPs, since there will usually be a pattern of alternative nucleotides at some positions in a collapsed region. We use Bowtie 2 with the local option for read mapping to contigs or scaffolds, and deepTools or SAMtools with custom Perl scripts to calculate depth and call SNPs from the resulting bam file.

The completeness of the assembly can be evaluated by remapping genes and IESs to the assembly, if they have been annotated for the species under consideration. Finally, the assembly itself can be used with the ParTIES software and a reference MAC genome to identify IESs, which can then be compared with a reference set if available, and/or provide a new IES reference set for the strain or species if the new assembly is the best available.

4. PERSPECTIVES AND LIMITS

Using the protocols we have described here for sorting nuclei and for high-throughput sequencing analysis, we are able to perform functional studies of rearrangements, in particular, genome-wide identification of IESs and quantification of IES excision errors in different genetic contexts. We can apply these methods to other strains and species in order to use comparative genomics to study the impact of TEs on genome evolution. This involves adaptation of the MIC purification protocol to the different sizes of nuclei and cells in different species followed by sequencing and identification of IESs and TEs.

Sorting nuclei to obtain pure populations, not only of MICs as discussed here, but also of MACs at the end of development (Guérin et al., 2017), opens several exciting possibilities. We can now envisage time course experiments and proteomics studies. The challenge will be to distinguish developing MACs from the maternal MAC fragments at early stages of development and to obtain sufficient quantities for downstream analyses: DNAseq, RNAseq, ChIPseq, mass spectrometry, etc. The motivation for such studies is to decipher the molecular mechanisms of developmental genome rearrangements and the many interacting pathways involved in their epigenetic control.

As mentioned in Section 3, we are still struggling to assemble high quality germline genomes that will reveal the structure of *Paramecium* chromosomes and their variability among *Paramecium* species. We still do not know how many MIC chromosomes there are in *P. tetraurelia* or have definitive proof that the chromosomes are indeed monocentric. Improving the assembly to obtain complete chromosomes necessitates long range information to link contigs or alternatively, long-read sequencing of pure MIC DNA. We are very hopeful that improvement in long-read library preparation and sequencing will allow us to overcome the bottleneck of not having enough MIC DNA. In any event, it may soon become cost-effective to bioinformatically sort MIC long reads after sequencing samples from cells that have not been submitted to flow cytometry.

ACKNOWLEDGMENTS

We are grateful to Olivier Arnaiz for useful discussions and criticism of the manuscript. We thank Laurent Duret for permission to use an unpublished phylogenetic tree of *Paramecium* species and for advice. Our work was supported by the Agence Nationale de la Recherche (ANR-14-CE10–0005–04 "PIGGYPACK"), intramural funding from the CNRS, the Fondation de la Recherche Médicale (Equipe FRM DEQ20160334868), LABEX "Who am I?" (ANR-11-LABX-0071_WHOAMI). We acknowledge the ImagoSeine facility, member of the France BioImaging infrastructure supported by the ANR-10-INSB-04. The sequencing benefited from the facilities and expertise of the I2BC high-throughput sequencing platform.

REFERENCES

Arnaiz, O., Mathy, N., Baudry, C., Malinsky, S., Aury, J.-M., Wilkes, C. D., et al. (2012). The paramecium germline genome provides a niche for intragenic parasitic DNA: Evolutionary dynamics of internal eliminated sequences. *PLoS Genetics*, *8*(10), e1002984. https://doi.org/10.1371/journal.pgen.1002984.

Arnaiz, O., Van Dijk, E., Bétermier, M., Lhuillier-Akakpo, M., de Vanssay, A., Duharcourt, S., et al. (2017). Improved methods and resources for paramecium genomics: Transcription units, gene annotation and gene expression. *BMC Genomics*, *18*(1), 483. https://doi.org/10.1186/s12864-017-3887-z.

Aury, J.-M., Jaillon, O., Duret, L., Noel, B., Jubin, C., Porcel, B. M., et al. (2006). Global trends of whole-genome duplications revealed by the ciliate Paramecium tetraurelia. *Nature*, *444*(7116), 171–178. https://doi.org/nature05230.

Baldauf, S. (2008). An overview of the phylogeny and diversity of eukaryotes. *Journal of Systematics and Evolution*, *46*(3), 263–273.

Baroin, A., Prat, A., & Caron, F. (1987). Telomeric site position heterogeneity in macronuclear DNA of Paramecium primaurelia. *Nucleic Acids Research*, *15*(4), 1717–1728.

Baudry, C., Malinsky, S., Restituito, M., Kapusta, A., Rosa, S., Meyer, E., et al. (2009). PiggyMac, a domesticated piggyBac transposase involved in programmed genome rearrangements in the ciliate Paramecium tetraurelia. *Genes & Development*, *23*(21), 2478–2483. https://doi.org/10.1101/gad.547309.

Beisson, J., Bétermier, M., Bré, M.-H., Cohen, J., Duharcourt, S., Duret, L., et al. (2010). Paramecium tetraurelia: The renaissance of an early unicellular model. *Cold Spring Harbor Protocols*. https://doi.org/10.1101/pdb.emo140.

Beisson, J., Cohen, J., Lefort-Tran, M., Pouphile, M., & Rossignol, M. (1980). Control of membrane fusion in exocytosis. Physiological studies on a Paramecium mutant blocked in the final step of the trichocyst extrusion process. *The Journal of Cell Biology*, *85*(2), 213–227.

Bétermier, M., & Duharcourt, S. (2014). Programmed rearrangement in ciliates: Paramecium. *Microbiology Spectrum*, *2*(6), MDNA3-0035-2014. https://doi.org/10.1128/microbiolspec.MDNA3-0035-2014.

Denby Wilkes, C., Arnaiz, O., & Sperling, L. (2016). ParTIES: A toolbox for Paramecium interspersed DNA elimination studies. *Bioinformatics*, *32*(4), 599–601. https://doi.org/10.1093/bioinformatics/btv691.

Dubois, E., Mathy, N., Régnier, V., Bischerour, J., Baudry, C., Trouslard, R., et al. (2017). Multimerization properties of PiggyMac, a domesticated piggyBac transposase involved in programmed genome rearrangements. *Nucleic Acids Research*, *45*(6), 3204–3216. https://doi.org/10.1093/nar/gkw1359.

Duret, L., Cohen, J., Jubin, C., Dessen, P., Goût, J.-F., Mousset, S., et al. (2008). Analysis of sequence variability in the macronuclear DNA of Paramecium tetraurelia: A somatic view of the germline. *Genome Research*, *18*(4), 585–596. https://doi.org/gr.074534.107.

Flutre, T., Duprat, E., Feuillet, C., & Quesneville, H. (2011). Considering transposable element diversification in de novo annotation approaches. *PLoS One*, *6*(1), e16526. https://doi.org/10.1371/journal.pone.0016526.

Ghurye, J. S., Cepeda-Espinoza, V., & Pop, M. (2016). Metagenomic assembly: Overview, challenges and applications. *The Yale Journal of Biology and Medicine*, *89*(3), 353–362.

Giardine, B., Riemer, C., Hardison, R. C., Burhans, R., Elnitski, L., Shah, P., et al. (2005). Galaxy: A platform for interactive large-scale genome analysis. *Genome Research*, *15*(10), 1451–1455. https://doi.org/10.1101/gr.4086505.

Gogendeau, D., Beisson, J., de Loubresse, N. G., Le Caer, J.-P., Ruiz, F., Cohen, J., et al. (2007). An Sfi1p-like centrin-binding protein mediates centrin-based Ca2+ −dependent contractility in Paramecium tetraurelia. *Eukaryotic Cell*, *6*(11), 1992–2000. https://doi.org/10.1128/EC.00197-07.

Gout, J.-F., Kahn, D., & Duret, L. (2010). The relationship among gene expression, the evolution of gene dosage, and the rate of protein evolution. *PLoS Genetics*, *6*(5), e1000944. https://doi.org/10.1371/journal.pgen.1000944.

Gruchota, J., Denby Wilkes, C., Arnaiz, O., Sperling, L., & Nowak, J. K. (2017). A meiosis-specific Spt5 homolog involved in non-coding transcription. *Nucleic Acids Research*, *45*(8), 4722–4732. https://doi.org/10.1093/nar/gkw1318.

Guérin, F., Arnaiz, O., Boggetto, N., Denby Wilkes, C., Meyer, E., Sperling, L., et al. (2017). Flow cytometry sorting of nuclei enables the first global characterization of Paramecium germline DNA and transposable elements. *BMC Genomics*, *18*(1), 327. https://doi.org/10.1186/s12864-017-3713-7.

Gurevich, A., Saveliev, V., Vyahhi, N., & Tesler, G. (2013). QUAST: Quality assessment tool for genome assemblies. *Bioinformatics*, *29*(8), 1072–1075. https://doi.org/10.1093/bioinformatics/btt086.

He, D., Fiz-Palacios, O., Fu, C.-J., Fehling, J., Tsai, C.-C., & Baldauf, S. L. (2014). An alternative root for the eukaryote tree of life. *Current Biology*, *24*(4), 465–470. https://doi.org/10.1016/j.cub.2014.01.036.

Holoch, D., & Moazed, D. (2015). RNA-mediated epigenetic regulation of gene expression. *Nature Reviews Genetics*, *16*(2), 71–84. https://doi.org/10.1038/nrg3863.

Jaillon, O., Bouhouche, K., Gout, J.-F., Aury, J.-M., Noel, B., Saudemont, B., et al. (2008). Translational control of intron splicing in eukaryotes. *Nature*, *451*(7176), 359–362. https://doi.org/nature06495.

Kapusta, A., Matsuda, A., Marmignon, A., Ku, M., Silve, A., Meyer, E., et al. (2011). Highly precise and developmentally programmed genome assembly in Paramecium requires ligase IV-dependent end joining. *PLoS Genetics*, 7(4), e1002049. https://doi.org/10.1371/journal.pgen.1002049.

Kirk, K. E., & Blackburn, E. H. (1995). An unusual sequence arrangement in the telomeres of the germ-line micronucleus in Tetrahymena thermophila. *Genes & Development*, *9*(1), 59–71.

Kumar, S., Jones, M., Koutsovoulos, G., Clarke, M., & Blaxter, M. (2013). Blobology: Exploring raw genome data for contaminants, symbionts and parasites using taxon-annotated GC-coverage plots. *Frontiers in Genetics*, *4*, 237. https://doi.org/10.3389/fgene.2013.00237.

Langmead, B., & Salzberg, S. L. (2012). Fast gapped-read alignment with Bowtie 2. *Nature Methods*, *9*(4), 357–359. https://doi.org/10.1038/nmeth.1923.

Le Mouël, A., Butler, A., Caron, F., & Meyer, E. (2003). Developmentally regulated chromosome fragmentation linked to imprecise elimination of repeated sequences in paramecia. *Eukaryotic Cell*, *2*(5), 1076–1090.

Lepère, G., Nowacki, M., Serrano, V., Gout, J.-F., Guglielmi, G., Duharcourt, S., et al. (2009). Silencing-associated and meiosis-specific small RNA pathways in Paramecium tetraurelia. *Nucleic Acids Research*, *37*(3), 903–915. https://doi.org/10.1093/nar/gkn1018.

Lhuillier-Akakpo, M., Frapporti, A., Denby Wilkes, C., Matelot, M., Vervoort, M., Sperling, L., et al. (2014). Local effect of enhancer of zeste-like reveals cooperation of epigenetic and cis-acting determinants for zygotic genome rearrangements. *PLoS Genetics*, *10*(9), e1004665. https://doi.org/10.1371/journal.pgen.1004665.

Lhuillier-Akakpo, M., Guérin, F., Frapporti, A., & Duharcourt, S. (2016). DNA deletion as a mechanism for developmentally programmed centromere loss. *Nucleic Acids Research*, *44*(4), 1553–1565. https://doi.org/10.1093/nar/gkv1110.

Li, R., Fan, W., Tian, G., Zhu, H., He, L., Cai, J., et al. (2010). The sequence and de novo assembly of the giant panda genome. *Nature*, *463*(7279), 311–317. https://doi.org/10.1038/nature08696.

Louis, E. J. (2002). Are Drosophila telomeres an exception or the rule? *Genome Biology*, *3*(10), 0007.1–0007.6. https://doi.org/10.1186/gb-2002-3-10-reviews0007.

Lynn, D. (2008). The ciliated protozoa: Characterization, classification, and guide to the literature (3rd ed.). Netherlands: Springer. Retrieved from www.springer.com/fr/book/9781402082382.

Magoč, T., & Salzberg, S. L. (2011). FLASH: Fast length adjustment of short reads to improve genome assemblies. *Bioinformatics (Oxford, England)*, *27*(21), 2957–2963. https://doi.org/10.1093/bioinformatics/btr507.

Maliszewska-Olejniczak, K., Gruchota, J., Gromadka, R., Denby Wilkes, C., Arnaiz, O., Mathy, N., et al. (2015). TFIIS-dependent non-coding transcription regulates developmental genome rearrangements. *PLoS Genetics*, *11*(7), e1005383. https://doi.org/10.1371/journal.pgen.1005383.

Marçais, G., & Kingsford, C. (2011). A fast, lock-free approach for efficient parallel counting of occurrences of k-mers. *Bioinformatics*, *27*(6), 764–770. https://doi.org/10.1093/bioinformatics/btr011.

Marmignon, A., Bischerour, J., Silve, A., Fojcik, C., Dubois, E., Arnaiz, O., et al. (2014). Ku-mediated coupling of DNA cleavage and repair during programmed genome rearrangements in the ciliate Paramecium tetraurelia. *PLoS Genetics*, *10*(8), e1004552. https://doi.org/10.1371/journal.pgen.1004552.

Martin, M. (2011). Cutadapt removes adapter sequences from high-throughput sequencing reads. *EMBnet.Journal*, *17*(1), 10–12. https://doi.org/10.14806/ej.17.1.200.

McCormick-Graham, M., & Romero, D. P. (1996). A single telomerase RNA is sufficient for the synthesis of variable telomeric DNA repeats in ciliates of the genus Paramecium. *Molecular and Cellular Biology*, *16*(4), 1871–1879.

McGrath, C. L., Gout, J.-F., Johri, P., Doak, T. G., & Lynch, M. (2014). Differential retention and divergent resolution of duplicate genes following whole-genome duplication. *Genome Research*, *24*(10), 1665–1675. https://doi.org/10.1101/gr.173740.114.

Prat, A., Katinka, M., Caron, F., & Meyer, E. (1986). Nucleotide sequence of the Paramecium primaurelia G surface protein. A huge protein with a highly periodic structure. *Journal of Molecular Biology*, *189*(1), 47–60.

Preer, L. B., Hamilton, G., & Preer, J. R., Jr. (1992). Micronuclear DNA from Paramecium tetraurelia: Serotype 51 A gene has internally eliminated sequences. *The Journal of Protozoology*, *39*(6), 678–682.

Ramírez, F., Ryan, D. P., Grüning, B., Bhardwaj, V., Kilpert, F., Richter, A. S., et al. (2016). DeepTools2: A next generation web server for deep-sequencing data analysis. *Nucleic Acids Research*, *44*(W1), W160–W165. https://doi.org/10.1093/nar/gkw257.

Sandoval, P. Y., Swart, E. C., Arambasic, M., & Nowacki, M. (2014). Functional diversification of dicer-like proteins and small RNAs required for genome sculpting. *Developmental Cell*, *28*(2), 174–188. https://doi.org/10.1016/j.devcel.2013.12.010.

Simpson, J. T., & Pop, M. (2015). The theory and practice of genome sequence assembly. *Annual Review of Genomics and Human Genetics*, *16*(1), 153–172. https://doi.org/10.1146/annurev-genom-090314-050032.

Wood, D. E., & Salzberg, S. L. (2014). Kraken: Ultrafast metagenomic sequence classification using exact alignments. *Genome Biology*, *15*, R46. https://doi.org/10.1186/gb-2014-15-3-r46.

Zerbino, D., & Birney, E. (2008). Velvet: Algorithms for de novo short read assembly using de Bruijn graphs. *Genome Research*, *18*, 821–829. https://doi.org/10.1101/gr.074492.107.

CHAPTER SIX

CLIP-Seq in Bacteria: Global Recognition Patterns of Bacterial RNA-Binding Proteins

Liis Andresen, Erik Holmqvist[1]
Department of Cell and Molecular Biology, Biomedical Center, Uppsala University, Uppsala, Sweden
[1]Corresponding author: e-mail address: erik.holmqvist@icm.uu.se

Contents

Abstract

RNA–protein interactions are at the heart of many central cellular processes, and RNA-binding proteins (RBPs) associate with virtually all RNA molecules in a cell. In bacteria, global RBPs, often in conjunction with small regulatory RNAs, affect physiology and virulence by controlling transcription, translation, and RNA decay. To understand how these regulatory proteins orchestrate global gene expression, detailed maps of their cellular RNA binding sites are required. To this end, cross-linking and immunoprecipitation followed by deep sequencing (CLIP-seq) has revolutionized RBP studies by providing knowledge about global recognition patterns of RBPs in both eukaryotic and bacterial cells. In this chapter, we provide a step-by-step protocol for global mapping of *bona fide* RBP binding sites using CLIP-seq in bacteria. This protocol has been successfully applied for charting the binding sites of Hfq, CsrA, and ProQ, three global regulatory RBPs in *Salmonella enterica* and *Escherichia coli*, and should be readily applicable to other RBPs and bacterial species.

Methods in Enzymology, Volume 612
ISSN 0076-6879
https://doi.org/10.1016/bs.mie.2018.08.008

1. INTRODUCTION

Bacterial genomes encode hundreds of RNA-binding proteins (RBPs). The most abundant and well-characterized bacterial RBPs are ribosomal proteins (r-proteins) that constitute the structural complex of the translational machinery. Together with other major classes of RBPs, including RNA helicases, tRNA synthases, and ribonucleases (RNases), r-proteins are essential for central cellular processes. In addition to these "housekeeping" activities, RBPs orchestrate gene expression by regulating RNA synthesis and decay, and translation of mRNA into proteins. In bacteria, global regulatory RBPs are key players in posttranscriptional control of many physiological processes including biofilm formation, carbon metabolism, stress adaptation, motility, and virulence, as manifested by proteins Hfq, CsrA, and ProQ (reviewed in Chao & Vogel, 2010; Holmqvist & Vogel, 2018; Kusmierek & Dersch, 2018; Vakulskas, Potts, Babitzke, Ahmer, & Romeo, 2015).

Although interactions between RNA and protein are mainly looked at from an RNA perspective—i. e., the effects that an RBP has on the fate of its RNA target—the interaction can serve the opposite purpose. For example, the activity of posttranscriptional regulator CsrA is inhibited by direct interaction with sRNAs that outcompete the protein from its target mRNAs (Liu et al., 1997; Weilbacher et al., 2003). Another interesting example is the *Enterococcus faecalis* protein RelQ, a (p)ppGpp synthetase whose enzymatic activity is inhibited by RNA binding (Beljantseva et al., 2017; Hauryliuk & Atkinson, 2017).

Notwithstanding the thorough characterization of some RBPs, the biological role of many RBPs is incompletely defined or unknown, much owing to the lack of knowledge about their cellular RNA targets. Recently developed RBP identification protocols that rely on cross-linking and phase separation suggest the existence of many more bacterial RBPs than previously known (L. Andresen et al., unpublished). Characterization of these putative RBPs requires high-throughput methods for mapping their cellular ligands and binding sites.

1.1 Studying RNA–Protein Interactions

Much of our knowledge about global RBPs such as Hfq and CsrA was generated before the era of high-throughput sequencing. The cellular functions of these proteins were initially discovered through phenotypes of mutant strains, while mechanistic insights were obtained through structural biology and biochemistry. With microarray technology and subsequently high-throughput sequencing at hand, it has become possible to monitor RNA–protein interactions, as well as effects on gene expression, in a transcriptome-wide manner.

The target specificity of an RBP can be globally studied from in vitro binding reactions. Methods, such as RNA Bind-n-Seq (Lambert et al., 2014), RNAcompete (Cook et al., 2017; Ray et al., 2009), HiTS-RAP (Ozer et al., 2015; Tome et al., 2014), and RNA-MaP (Buenrostro et al., 2014), permit quantification of the diversity of motifs recognized by RBPs and of the binding strength of RBPs to individual binding sites. Although these methods are powerful in describing the diversity of the RBP binding motifs, and provide information on interaction affinities, they suffer from in vitro limitations. First, randomized RNA pools that are used to cover diversity of RNA species do not quantitatively or qualitatively match cellular RNA pools. Second, in vitro systems cannot recapitulate the complexity of the cell. Thus, they do not necessarily reflect the RBP binding patterns under physiological conditions. Third, in vitro binding reactions require purification of the protein of interest in its active state, a process that can often be both complex and variable.

To study more physiologically relevant interactions, RNA molecules associated with a specific RBP in cells can be detected by RBP immunoprecipitation and identification of coimmunoprecipitated RNA, often referred to as RIP (RNA immunoprecipitation; Köster & Staiger, 2014; Sahr & Buchrieser, 2013). This approach is relatively simple to perform but has several important limitations. First, since RIP is performed under nondenaturing conditions, contaminating protein and RNA may hamper the possibility to differentiate between direct and indirect interactions. Second, the identified RBP-associated full-length RNA molecules do not specify RBP binding sites. Third, RIP does not capture transient RNA–protein interactions, which biases the method toward high-affinity RNA targets. These limitations stimulated the development of cross-linking and immunoprecipitation (CLIP), a stringent method to faithfully identify direct RNA ligands and reveal RBP binding sites at high resolution.

2. METHOD OVERVIEW

What makes CLIP principally different from RIP is the first step of the protocol—ultraviolet (UV) irradiation of living cells (Fig. 1). UV light promotes covalent bond formation between an amino acid of the RBP

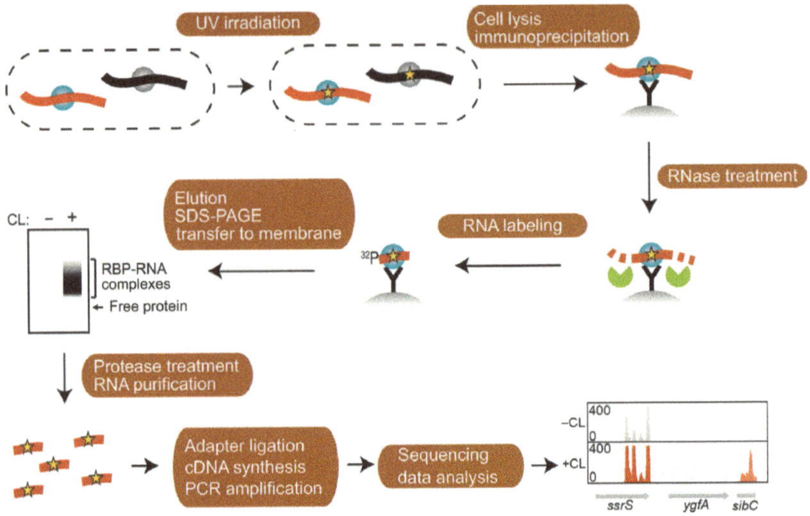

Fig. 1 Schematic representation of the CLIP procedure for identification of RNA–protein interactions in living cells. Cells are irradiated with UV light, leading to formation of a covalent bond between protein and RNA. This is followed by lysis and immunoprecipitation using antibodies against the epitope tag of the RBP or with protein-specific antibodies. Partial on-bead RNase digestion results in smaller RBP–RNA complexes, which are then subjected to radioactive labeling. Labeled complexes are separated on SDS-PAGE and further purified by transfer to a nitrocellulose membrane. Protein bound RNA is released from the membrane with proteinase K treatment and further purified using phenol:chloroform:isoamyl alcohol extraction. For the library preparation, adapters are ligated to the RNA fragments followed by reverse transcription into cDNA. After PCR amplification, the library is ready for high-throughput sequencing followed by data analysis. As an example of a ProQ target identified by CLIP-seq, an IGV (integrative genomics viewer) view of read coverage from a CLIP-seq experiment is shown. Note that the background signal from highly abundant RNAs (*ssrS*) can be present in the libraries from both cross-linked (+CL) and noncross-linked samples (–CL), whereas peaks specific to the RBP of interest (*sibC*) will be enriched only in libraries prepared from UV cross-linked cultures.

placed at "zero distance" (≤2 Å) from an RNA nucleotide base. The introduction of a covalent bond between RBP and RNA ensures RBP–RNA complex stability throughout the protocol and allows for stringent purification procedures. The RBP of interest, together with its covalently bound RNA ligands, is captured from a cell lysate by immunoprecipitation. During immunoprecipitation, covalently bound RNA ligands are trimmed by RNases to leave only the RBP-protected part of the RNA—the RBP "footprint"—intact. After radioactive RNA labeling to allow visualization of the RBP–RNA complex, the eluted material is separated on a denaturing protein gel and transferred to a membrane, to release noncovalent interactions (Fig. 2). The part of the membrane containing the radioactively labeled RBP–RNA complexes is excised and the RNA fragments are released by protease treatment. Finally, the RNA fragments are converted into a DNA library amenable for high-throughput sequencing (Fig. 3).

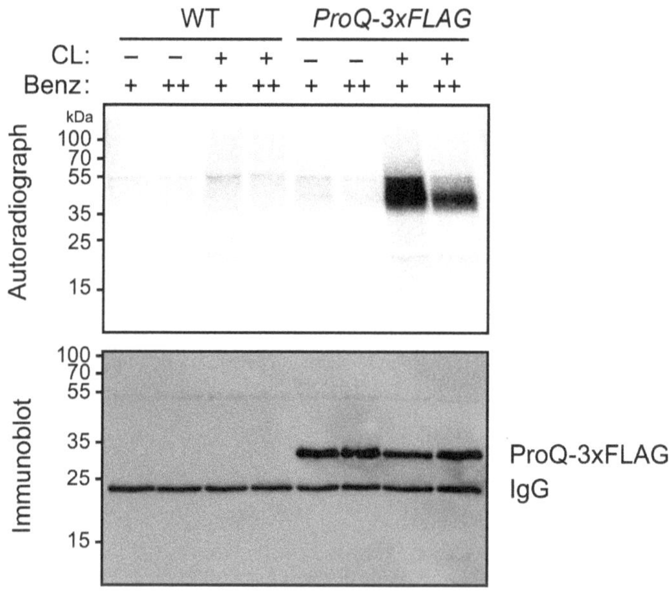

Fig. 2 Example autoradiograph and respective Western blot image for ProQ-3xFLAG immunoprecipitated samples. Radioactive signal indicates the presence of RNA–protein complexes in the cross-linked samples (CL). Size distribution of the complexes depend on the amount of nuclease (Benz) used. Equal presence of ProQ-3xFLAG in cross-linked and noncross-linked samples is confirmed with Western blotting using anti-FLAG antibodies produced in mouse.

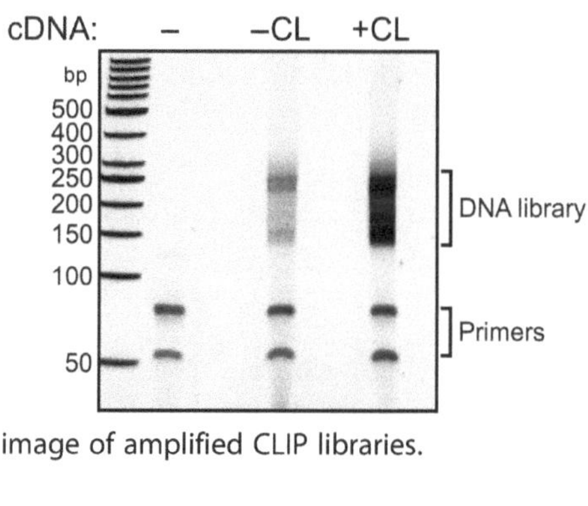

Fig. 3 Example gel image of amplified CLIP libraries.

2.1 CLIP: History and Development

CLIP was first described in a pioneering study from Robert Darnell's group (Ule et al., 2003). While this study used Sanger sequencing to identify around 300 RNA fragments cross-linked to Nova proteins, the advent of high-throughput sequencing soon enabled identification of hundreds of thousands of CLIP-generated RNA fragments (Licatalosi et al., 2008). Since then, a plethora of variants of the original method has been established, each providing an alternative and/or improved version of distinct steps in the protocol, including condition and equipment for UV irradiation, RNA fragmentation, protein purification, adapter ligation, cDNA synthesis, and library preparation (recently reviewed in Lee & Ule, 2018).

2.2 CLIP: Strengths and Limitations

The strengths of the CLIP approach are: (i) due to covalently coupled RBP and its target RNA, stringent washing steps can be applied minimizing scoring of secondary interactions, (ii) CLIP provides an RBP target profile as a snapshot from live cells, enabling use of the method for studying condition-dependent changes in the protein binding pattern, (iii) analysis of sequencing reads for single nucleotide mutations introduced in the cross-linking site enables the mapping of protein binding sites at single-nucleotide resolution.

As every method CLIP has its limitations: (i) it is subject to cross-linking biases, (ii) due to low UV cross-linking efficiency (magnitude of 1%–5% when tested with purified protein and labeled RNA; Darnell, 2010) CLIP-seq rarely achieves saturation thus limiting the identification of binding sites, (iii) it does not enable quantitative analysis of binding strength.

2.3 CLIP in Bacteria

The first study applying a CLIP-related strategy in bacteria was the use of UV cross-linking and analysis of cDNA (CRAC) to map Hfq binding sites in pathogenic *Escherichia coli* (Tree, Granneman, McAteer, Tollervey, & Gally, 2014). Since then, CLIP or CRAC has been applied to Hfq, CsrA, and ProQ in *Salmonella enterica*, Hfq, CsrA, ProQ, YbeY, RelA, and RNase E in *E. coli*, and VapC proteins in *Mycobacterium tuberculosis* (Holmqvist et al., 2016; Holmqvist, Li, Bischler, Barquist, & Vogel, 2018; McAteer et al., 2018; Potts et al., 2017; Sy et al., 2018; Tree et al., 2014; Wang et al., 2018; Waters et al., 2017; Winther, Roghanian, & Gerdes, 2018; Winther, Tree, Tollervey, & Gerdes, 2016). These studies have provided transcriptome-wide RBP–RNA interaction maps, determined cellular

RBP consensus binding sequences and/or structure motifs, revealed unexpected modes of regulation, and expanded the regulons of several global RBPs significantly.

This book chapter provides a step-by-step protocol for performing CLIP-seq in bacteria, including cross-linking, immunoprecipitation, RNA isolation, and conversion of the purified RNA into a DNA library amenable for high-throughput sequencing. This protocol has successfully been used to determine the cellular binding sites of Hfq and CsrA in *S. enterica*, and of ProQ in *S. enterica* and *E. coli*. Downstream bioinformatic analyses of sequence reads have been excellently reviewed elsewhere (Li, Förstner, & Chao, 2018) and will not be described here.

3. EXPERIMENTAL METHOD

3.1 General Considerations

When preparing for a CLIP experiment it is important to consider protein-to-protein differences that may influence the success of the experiment. It is prerequisite that the protein of interest is soluble in the lysis buffer and expressed at detectable levels. If protein levels are too low to yield enough RNA ligands for sequencing, slight overexpression from a titratable promoter construct may be considered. The use of a highly specific antibody toward the native protein, or a commercial antibody toward an epitope tag introduced into the protein, ensures specific immunoprecipitation. The protocol described here is designed for immunoprecipitation of proteins carrying a 3xFLAG epitope tag, but can be modified if the use of a protein-specific antibody is desired (see Note 1). The growth conditions and harvesting time can be freely selected depending on the expression levels of the protein of interest and for studying condition-dependent RNA target profiles. Note that higher cell densities, as well as a growth medium that absorbs less UV light, might be beneficial because of the inherently low UV cross-linking efficiency that may result in low yield of cross-linked complexes.

3.2 Equipment

- Temperature-controlled shakers for growing bacteria
- Refrigerated centrifuges for 50 and 2 mL tubes
- Mixer Mill MM 400 (Retsch) or similar for bacterial lysis (including adapters for 2.0 mL tubes)

- UV chamber with UV-C lamps (e.g., UVC 500 UV Cross-linker, GE Healthcare)
- Rotator for 2 and 15 mL tubes
- Magnetic stand for 2 mL tubes
- Thermomixer (Eppendorf) or similar device
- Equipment for SDS-PAGE
- Equipment for semidry protein transfer
- Phosphorimager with phosphor screens
- PCR cycler
- UV table
- Qubit fluorometer or a Bioanalyzer

3.3 Materials

- Sample tubes (2, 15, and 50 mL)
- *E*-flasks (2 L)
- BioAssay dishes, 245 × 245 mm (Corning)
- Glass beads 0.1 mm diameter (BioSpec)
- Anti-FLAG® M2 Magnetic Beads (Sigma-Aldrich)
- Benzonase® Nuclease (Sigma-Aldrich)
- Calf intestinal alkaline phosphatase (10 U/μL, New England Biolabs)
- T4 polynucleotide kinase (10 U/μL, ThermoFisher)
- γ-^{32}P-ATP (3000 Ci/mmol, PerkinElmer)
- PageRuler™ Plus Prestained Protein Ladder (ThermoFisher)
- Mini-PROTEAN® TGX™ Precast Gels (Bio-Rad)
- Nitrocellulose membrane (Protran BA85, 0.45 μm)
- Plastic filer
- Radioactively labeled marker pen (prepare by dipping the tip of a regular marker pen in a dilute solution of γ-^{32}P-ATP).
- Transparent film
- Razor blades (sterile)
- DNA LoBind 2 mL tubes (Eppendorf)
- Proteinase K (20 mg/mL, ThermoFisher)
- SUPERase In RNase inhibitor (20 U/μL, ThermoFisher)
- Phase-Lock Heavy 2 mL tubes (Quantabio)
- GlycoBlue™ Coprecipitant (15 mg/mL, ThermoFisher)
- NEBNext Multiplex Small RNA Library Prep Set (New England Biolabs)

- PCR tubes
- SYBR™ Gold Nucleic Acid Gel Stain (ThermoFisher)
- GeneRuler 50 bp DNA ladder (ThermoFisher)
- LongAmp® *Taq* 2x Master Mix (New England Biolabs)
- MinElute PCR Purification Kit (Qiagen)
- Spin-X® Centrifuge Tube Filter (Corning)
- Forward primer (5′ AATGATACGGCGACCACCG 3′)
- Reverse primer (5′ CAAGCAGAAGACGGCATACG 3′)

3.4 Solutions and Buffers

- LB medium
- NP-T buffer (50 m*M* NaH_2PO_4, 300 m*M* NaCl, 0.05% Tween, pH 8.0)
- High-salt buffer (50 m*M* NaH_2PO_4, 1 *M* NaCl, 0.05% Tween, pH 8.0)
- CIP buffer (100 m*M* NaCl, 50 m*M* Tris–HCl pH 7.4, 10 m*M* $MgCl_2$)
- PNK buffer (50 m*M* Tris–HCl pH 7.4, 10 m*M* $MgCl_2$, 0.1 m*M* spermidine)
- Protein loading buffer (60 m*M* Tris–HCl pH 6.8, 0.01% bromophenol blue, 2% SDS, 10% glycerol, 1.5% DTT)
- SDS-PAGE running buffer (25 m*M* Tris, 192 m*M* glycine, 0.1% SDS)
- Western blot transfer buffer (25 m*M* Tris, 192 m*M* glycine, 20% methanol)
- 2 × Proteinase K buffer (100 m*M* Tris–HCl pH 7.9, 10 m*M* EDTA, 1% SDS).
- 9 *M* Urea
- Phenol:chloroform:isoamyl alcohol (25:24:1)
- Ethanol 99.5% and 80%
- 3 *M* NaOAc (pH 5.2)
- Sterile water
- Loading buffer GLII (formamide containing 0.025% bromophenol blue, 0.025% xylene cyanol, 0.025% SDS, and 18 m*M* EDTA)
- 1 × TBE buffer (100 m*M* Tris base, 100 m*M* boric acid, 2 m*M* EDTA)

3.5 Procedure

The following procedure has been optimized for RBPs expressed in Gram-negative bacteria such as *E. coli* and *S. enterica*. Application to Gram-positive bacteria may require some changes to the protocol, in particular regarding growth conditions and cell lysis.

3.5.1 Growth of Bacteria, UV Cross-linking, and Cell Lysis

1. Grow a single colony in 10mL LB medium overnight at 37°C with 220rpm shaking.
2. Dilute 2mL overnight culture in 250mL fresh LB medium prewarmed to 37°C in a 2-L *E*-flask. Grow at 37°C with 220rpm shaking.
3. When the optical cell density reaches OD_{600} of 2.0, immediately transfer 100mL of culture to a BioAssay dish and place in UV chamber with UV-C lamps.
4. Apply 800mJ/cm^2 UV-C light to irradiate the culture and cross-link RNA–protein complexes. Meanwhile, transfer 100mL of nonirradiated culture into two 50-mL Falcon tubes on ice (see Note 2).
5. Decant the UV-irradiated culture into two 50-mL Falcon tubes on ice (see Note 3).
6. Pellet cells in a centrifuge precooled to 4°C at 6000 × *g* for 15min.
7. Remove supernatant by decanting. Remove the remaining supernatant using a 1-mL pipette (see Note 4).
8. Suspend pellets in 800 μL NP-T buffer by pipetting up and down.
9. For each pellet, prepare a 2-mL microcentrifuge tube filled with 1mL glass beads. Cool on ice.
10. Transfer the cell suspension from each pellet to a glass bead-containing tube on ice.
11. Place tubes in a precooled grinding mill adapter, put the adapter in the grinding mill, and grind at 30Hz for 10min. Cool on ice. For alternative lysis methods see Note 5.
12. Centrifuge 15min at 16,000 × *g* and 4°C.
13. Transfer supernatant to clean 2.0 mL tube, centrifuge 15min at 16,000 × *g* and 4°C.
14. Pool the cleared lysates from same condition (cross-linked or noncross-linked) into one cooled 15-mL Falcon tube each (see Note 6). Dilute 10 × with cold NP-T buffer on ice.

3.5.2 Immunoprecipitation

15. For each sample, transfer 30 μL anti-FLAG magnetic bead suspension to a new 2.0-mL microcentrifuge tube (see Note 7). Wash three times for 1min in 500 μL NP-T buffer on a rotating wheel (see Note 8).
16. After the last wash, suspend beads in 500 μL NP-T buffer and transfer to the cleared and diluted lysate. Rotate the lysate-bead suspension for 1h at 4°C.

17. Collect beads by centrifugation at 800 × *g*, suspend in 1.0 mL NP-T buffer, and transfer to a clean 2.0-mL microcentrifuge tube.
18. Wash two times for 1 min in 500 μL High-salt buffer and two times for 1 min in 500 μL NP-T buffer using a rotating wheel. Keep on ice.

3.5.3 RNase Treatment, Dephosphorylation, and Radioactive Labeling

19. Prepare the Benzonase solution: For each sample, dilute 2.5 units of Benzonase nuclease in 100 μL NP-T buffer with 1 m*M* $MgCl_2$ (see Note 9).
20. Magnetize beads, remove supernatant, and add 100 μL Benzonase solution. Incubate in a Thermomixer for 10 min at 37°C with 800 rpm shaking. Cool 2 min on ice.
21. Wash once with 500 μL High-salt buffer and two times with 500 μL CIP buffer. Keep on ice.
22. Prepare the CIP solution: For each sample, dilute 10 units of calf intestinal alkaline phosphatase in 100 μL CIP buffer.
23. Magnetize beads, remove supernatant, and add 100 μL CIP solution. Incubate in a Thermomixer for 30 min at 37°C with 800 rpm shaking. Cool 2 min on ice.
24. Wash one time with 500 μL High-salt buffer and two times with 500 μL PNK buffer. Keep on ice.
25. Prepare IP verification sample: Transfer one-tenth of the bead suspension to a fresh microcentrifuge tube, magnetize, and remove supernatant. Add 20 μL Protein loading buffer to the beads, incubate for 5 min at 90°C, and cool on ice for 2 min. Magnetize and transfer the supernatant to a clean microcentrifuge tube. Store at −20°C (see Note 10).
26. Prepare the PNK solution: For each sample, dilute 10 units of T4 polynucleotide kinase and 10 μCi γ-^{32}P-ATP in 100 μL PNK buffer.
27. Magnetize beads, remove supernatant, and add 100 μL PNK solution. Incubate for 30 min at 37°C. Cool 2 min on ice (see Note 11).
28. Wash three times with 500 μL NP-T buffer. After the final wash, suspend the beads in 20 μL Protein loading buffer, incubate 3 min at 95°C, magnetize, and transfer the supernatant to a new microcentrifuge tube. Repeat this procedure once. Keep on ice.

3.5.4 SDS-PAGE and Transfer to Membrane

29. Assemble SDS-PAGE gel system according to instructions.
30. Load the eluted RNA–protein complexes in Protein loading dye on the gel keeping an empty well between each sample to avoid

cross-contamination. Also load the prestained protein size marker on each side of the samples.

31. Run the gel until the front is 5 mm from the bottom of the glass plates (see Note 12).
32. Remove one of the glass plates, cut off the gel front and discard as radioactive waste.
33. Transfer the RBP–RNA complexes to a nitrocellulose membrane using a semidry blotting system at 4°C (see Note 13).
34. Transfer the membrane to one side of a cut-open plastic filer. Cut off the upper left corner of the membrane. Mark each corner of the membrane, as well as the bands of the protein size marker, with a radioactively labeled marker pen. Close the filer and place a phosphor screen on top.
35. After 30–60 min exposure, scan the phosphor screen (see Fig. 2 upper gel for typical autoradiograph).
36. Print the image in full size on a transparent film. Draw a rectangle around the radioactive signal from the RBP–RNA complex, and at the equivalent position in the noncross-linked sample. Use a sterile razor blade to cut out the marked rectangles.
37. Place the transparent film on top of the membrane and align the radioactive marks on the film with the color marks from the marker pen on the membrane. When aligned, fix the membrane and the transparent film with scotch tape.
38. Use a sterile razor blade to cut out the membrane pieces within the empty rectangles of the transparent film. For each sample use a new razor blade.
39. Cut each membrane piece into eight smaller pieces and transfer to a 2.0-mL low-bind microcentrifuge tube (see Note 14).

3.5.5 RBP Digestion and RNA Extraction

40. Prepare the PK solution: For each sample, dilute 0.4 mg Proteinase K and 10 units of SUPERase In RNase inhibitor in 400 μL Proteinase K buffer.
41. Transfer 400 μL PK solution to the tube containing the membrane pieces. Incubate in a Thermomixer for 30 min at 37°C with shaking at 1000 rpm.
42. Add 100 μL Proteinase K buffer with 9 *M* Urea and incubate another 30 min with the same conditions.
43. Centrifuge Phase-lock tubes 1 min at 16,000 × *g* before use.

44. Centrifuge Proteinase K treated sample 2 min at 16,000 × *g* and transfer 450 μL of the supernatant to a Phase-lock tube.
45. Add 450 μL phenol:chloroform:isoamyl alcohol and mix by incubating for 5 min at 30°C in a Thermomixer with shaking at 1000 rpm.
46. Centrifuge for 12 min at 16,000 × *g* and 4°C.
47. Transfer 400 μL of the aqueous phase to a clean low-bind microcentrifuge tube. Add 1.2 mL ice-cold 99.5% ethanol, 40 μL 3 *M* NaOAc pH 5.2, and 1 μL GlycoBlue, and then precipitate for 30 min at −80°C.
48. Centrifuge for 30 min at 16,000 × *g* and 4°C. Remove the supernatant and add 300 μL ice-cold 80% ethanol.
49. Centrifuge for 15 min at 16,000 × *g* and 4°C. Remove the supernatant with a 1-mL pipette.
50. Centrifuge for 1 min at 16,000 × *g* and 4°C. Remove residual ethanol with a 10-μL pipette (see Note 15).
51. Dry the pellets with the lid of the tube open for 2 min at room temperature.
52. Add 10 μL nuclease-free water to the pellet, incubate 3 min in a Thermomixer at 65°C with shaking at 900 rpm. Cool on ice for 2 min. Store the purified RNA at −20°C.

3.5.6 Adaptor Ligation and cDNA Synthesis

To convert the RNA into a cDNA library compatible with sequencing on Illumina instruments, we routinely use the NEBNext Multiplex Small RNA Library Prep Set. All steps can be done in a PCR machine.

53. Dilute the 3′ SR Adaptor 1:10 in 10 μL sterile water.
54. Mix 2.5 μL RNA with 0.5 μL diluted 3′ SR Adaptor and 0.5 μL sterile water in a 0.2-mL PCR tube. As a negative control, mix 3 μL sterile water with 0.5 μL diluted 3′ SR Adaptor.
55. Incubate samples 2 min at 70°C followed by 2 min incubation on ice.
56. Add 5 μL 3′ Ligation Reaction Buffer and 1.5 μL 3′ Ligation Enzyme mix.
57. Incubate 60 min at 25°C.
58. Add 0.25 μL SR RT Primer and 2.5 μL sterile water.
59. Incubate 5 min at 75°C, 15 min at 37°C, and 15 min at 25°C.
60. During the last 15 min incubation in step 59, prepare for the 5′ Adaptor ligation:
    a. Mix 1 μL of the 5′ SR Adaptor with 9 μL of sterile water in a 0.2-mL PCR tube.
    b. Incubate 2 min at 70°C and then place on ice.

61. When step 59 is completed, add to the tubes 0.5 μL diluted and denatured 5′ SR Adaptor, 0.5 μL 5′ Ligation Reaction Buffer, and 1.25 μL 5′ Ligation Enzyme mix.
62. Incubate 60 min at 25°C.
63. To synthesize cDNA, add 4 μL First Strand Synthesis Reaction Buffer, 0.5 μL Murine RNase Inhibitor, and 0.5 μL ProtoScript II Reverse Transcriptase.
64. Incubate 60 min at 50°C followed by 15 min at 70°C. The cDNA can be stored at −20°C.

3.5.7 Initial cDNA Library Amplification to Optimize PCR Cycle Number

The amount of cDNA may differ substantially depending on the RBP studied, which will affect the number of PCR cycles required for generating the amount of DNA necessary for sequencing. At the same time, excessive amplification that increases the risk of introducing amplification biases should be avoided. It is therefore important to first find the optimal PCR cycle number before generating the final library.

65. For each cDNA sample, mix 25 μL LongAmp® *Taq* 2x Master Mix, 1.25 μL SR Primer, 1.25 μL Index 1 Primer, 12.5 μL sterile water, and 10 μL cDNA in a 0.2-mL PCR tube.
66. Split each sample into three 0.2-mL PCR tubes with 16.7 μL each to allow testing different PCR cycle numbers.
67. Run the following PCR program: Initial denaturation for 30 s at 94°C followed by 20 rounds of (15 s at 94°C, 30 s at 62°C, 15 s at 70°C).
68. During the PCR cycling, remove one of the three aliquots from each sample after 16, 18, and 20 cycles, respectively. Put samples on ice.
69. To each sample, add 16 μL of Loading buffer GLII. These samples can be stored at −20°C.
70. Prepare 6% polyacrylamide, 7 *M* urea gel in 1 × TBE buffer. Load 16 μL of each PCR sample, along with the GeneRuler 50 bp DNA ladder (dilute four times in GLII buffer before loading). Run the gel until the lower dye is close to the bottom.
71. Stain the gel for 20 min in 1 × TBE buffer with a 1:10,000 dilution of SYBR Gold gel stain.
72. Visualize the PCR products using a UV table. At the optimal cycle number, bands in the range of 130–300 bp should be clearly visible. Smears of substantially longer fragments indicate overamplification. At higher cycle numbers, a band of 120 bp may be visible, which represents ligated adaptors lacking an insert.

3.5.8 cDNA Library Amplification With Optimal PCR Cycle Number

When the optimal number of PCR cycles has been determined, the remaining cDNA is amplified.

73. For each cDNA sample, mix 25 μL LongAmp® *Taq* 2 × Master Mix, 1.25 μL SR Primer, 12.5 μL sterile water, and 10 μL cDNA in a 0.2-mL PCR tube.
74. Add 1.25 μL Index primer so that each sample has a unique index.
75. Run the following PCR program: Initial denaturation for 30 s at 94°C followed by X (optimal cycle number) rounds of (15 s at 94°C, 30 s at 62°C, 15 s at 70°C).
76. Purify reactions using the MinElute PCR Purification Kit. Elute in 10 μL sterile water.
77. Add 10 μL of Loading buffer GLII and separate on a 6% polyacrylamide, 7 *M* urea gel in 1 × TBE buffer, along with the GeneRuler 50 bp DNA ladder (dilute four times in GLII buffer before loading). Leave one empty lane between the samples to avoid cross-contamination. Run the gel until the lower dye is close to the bottom.
78. Stain the gel for 20 min in 1 × TBE buffer with a 1:10,000 dilution of SYBR Gold gel stain.
79. Visualize the PCR products using a UV table and excise bands between 140 and 300 bp using a sterile razor blade (see Fig. 3, for example, gel image). Cut each gel piece into eight small pieces and transfer to a 2.0-mL LoBind tube.
80. Add 500 μL DNA Gel Elution buffer and incubate in a thermomixer at 16°C with 1200 rpm shaking overnight.
81. Briefly centrifuge samples to collect gel pieces at the bottom of the tube.
82. Transfer supernatant to a Spin-X® Centrifuge Tube Filter placed in a 2.0-mL LoBind tube.
83. Centrifuge for 1 min at 7000 × *g* and collect supernatant.
84. Precipitate at −80°C with 1.5 mL 99.5% ethanol (ice-cold), 1 μL linear acrylamide (10 mg/mL), and 50 μL 3 *M* sodium acetate (pH 5.2).
85. Centrifuge for 30 min at 16,000 × *g* and 4°C. Remove the supernatant and add 300 μL ice-cold 80% ethanol.
86. Centrifuge for 15 min at 16,000 × *g* and 4°C. Remove the supernatant with a 1-mL pipette.
87. Centrifuge for 1 min at 16,000 × *g* and 4°C. Remove residual ethanol with a 10 μL pipette.
88. Dry pellets with open lid for 2 min at room temperature.

89. Add 10 μL sterile water to the pellet, incubate 3 min in a Thermomixer at 65°C with shaking at 900 rpm. Cool on ice for 2 min. Store the purified DNA at −20°C.
90. Measure the DNA concentration using a Qubit fluorometer or a Bioanalyzer. If concentrations are too low to allow sequencing, the DNA library can be reamplified.

3.5.9 Optional: Reamplification of DNA Libraries

91. Mix 25 μL LongAmp® *Taq* 2 × Master Mix, 2 μL forward primer (10 μ*M*), 2 μL reverse primer (μ*M*), 19 μL sterile water, and 2 μL amplified DNA library in a 0.2-mL PCR tube.
92. Run the following PCR program: Initial denaturation for 30 s at 94°C followed by 4–6 cycles of (15 s at 94°C, 30 s at 60°C, 15 s at 65°C).
93. Purify reactions using the MinElute PCR Purification Kit. Elute in 10 μL sterile water.
94. To check the quality of the PCR products, analyze 2 μL on a 6% polyacrylamide, 7 *M* urea gel as described earlier.

3.5.10 Sequencing

The final DNA libraries can be sequenced on Illumina sequencers. Paired-end sequencing is preferential in that it offers a means to distinguish sequencing errors from mutations introduced in earlier steps, including mutations that occur during cDNA synthesis at the cross-link site in the RNA fragments.

3.6 Notes

1. Highly specific commercial antibodies are recommended. If antibodies are provided in solution they should be coupled to magnetic beads prior use (e.g., using Dynabeads magnetic beads). If a protein-specific antibody is used, a no-antibody control should be included to the CLIP experiment.
2. Required cross-linking energy should be optimized for different instruments and growth conditions. Cells can be washed and suspended in PBS for more efficient cross-linking.
3. All the subsequent protocol steps until SDS-PAGE should be performed at 4°C or on ice, unless stated otherwise, to avoid RNA degradation by contaminating nucleases.
4. Cell pellets can be snap-frozen and stored at −80°C for at least 6 months.

5. Grind milling can be replaced with alternative lysis methods. We have successfully used the FastPrep24™ benchtop homogenizer in combination with 0.1 mm glass beads for the same purpose.
6. Keep aliquots of the different stages and test them by Western blot to monitor the efficiency of lysis, binding, and washing steps.
7. The amount of anti-FLAG magnetic beads required depends on the expression level of the RBP and should be optimized. Too little of bead suspension will result in sample loss due to incomplete immunoprecipitation of the RBP, too much will increase unspecific background.
8. When working with magnetic beads it is important to be quick when taking off/adding liquid. Do not let them dry out.
9. The Benzonase concentration has to be optimized specifically for each RBP to yield a partially digested RNA pool (see Fig. 2). Overdigestion results in short unmappable sequencing reads.
10. IP verification samples will be analyzed separately with Western blotting (see Fig. 2 lower panel).
11. In some cases the PNK labeling reaction can result in phosphorylation of the immunopurified protein which is visible as a protein-sized band in both cross-linked and noncross-linked sample on the autoradiograph (Tawk, Sharan, Eulalio, & Vogel, 2017).
12. Free radioactive ATP will be present at the front of the gel. To avoid radioactive contamination of the lower running buffer, make sure not to the let the front run out of the gel.
13. Available Western blotting transfer system can be used following manufacturers' instructions.
14. Tip: Membrane pieces can be collected using a sterile syringe needle.
15. Due to the addition of GlycoBlue, the pellet is clearly visible. However, be aware that the pellet may not stick very well to the surface of the low-bind tubes.

ACKNOWLEDGMENTS

We thank Cedric Romilly, Christian Müller, Steffi Jimmy, and Gerhart Wagner for constructive feedback on the chapter. E.H. is supported by the Wenner-Gren Foundations, the Swedish Research Council (2016–03656) and the Swedish Foundation for Strategic Research (ICA 16–0021).

REFERENCES

Beljantseva, J., Kudrin, P., Andresen, L., Shingler, V., Atkinson, G. C., Tenson, T., et al. (2017). Negative allosteric regulation of Enterococcus faecalis small alarmone synthetase RelQ by single-stranded RNA. *Proceedings of the National Academy of Sciences*, *114*, 3726–3731.

Buenrostro, J. D., Araya, C. L., Chircus, L. M., Layton, C. J., Chang, H. Y., Snyder, M. P., et al. (2014). Quantitative analysis of RNA-protein interactions on a massively parallel array reveals biophysical and evolutionary landscapes. *Nature Biotechnology*, *32*, 562–568.

Chao, Y., & Vogel, J. (2010). The role of Hfq in bacterial pathogens. *Current Opinion in Microbiology*, *13*, 24–33.

Cook, K. B., Vembu, S., Ha, K. C. H., Zheng, H., Laverty, K. U., Hughes, T. R., et al. (2017). RNAcompete-S: Combined RNA sequence/structure preferences for RNA binding proteins derived from a single-step in vitro selection. *Methods*, *126*, 18–28.

Darnell, R. B. (2010). HITS-CLIP: Panoramic views of protein-RNA regulation in living cells. *Wiley Interdisciplinary Reviews. RNA*, *1*, 266–286.

Hauryliuk, V., & Atkinson, G. C. (2017). Small Alarmone Synthetases as novel bacterial RNA-binding proteins. *RNA Biology*, *14*, 1695–1699.

Holmqvist, E., Li, L., Bischler, T., Barquist, L., & Vogel, J. (2018). Global maps of ProQ binding in vivo reveal target recognition via RNA structure and stability control at mRNA 3′ ends. *Molecular Cell*, *70*, 971–982.

Holmqvist, E., & Vogel, J. (2018). RNA-binding proteins in bacteria. *Nature Reviews Microbiology*. https://doi.org/10.1038/s41579-018-0049-5.

Holmqvist, E., Wright, P. R., Li, L., Bischler, T., Barquist, L., Reinhardt, R., et al. (2016). Global RNA recognition patterns of post-transcriptional regulators Hfq and CsrA revealed by UV crosslinking *in vivo*. *The EMBO Journal*, *35*, 991–1011.

Köster, T., & Staiger, D. (2014). Arabidopsis protocols. In J. J. Sanchez-Serrano & J. Salinas (Eds.), *Methods in molecular biology: Vol. 1062*. (pp. 679–695). Springer.

Kusmierek, M., & Dersch, P. (2018). Regulation of host–pathogen interactions via the post-transcriptional Csr/Rsm system. *Current Opinion in Microbiology*, *41*, 58–67.

Lambert, N., Robertson, A., Jangi, M., McGeary, S., Sharp, P. A., & Burge, C. B. (2014). RNA bind-n-Seq: Quantitative assessment of the sequence and structural binding specificity of RNA binding proteins. *Molecular Cell*, *54*, 887–900.

Lee, F. C. Y., & Ule, J. (2018). Advances in CLIP technologies for studies of protein-RNA interactions. *Molecular Cell*, *69*, 354–369.

Li, L., Förstner, K. U., & Chao, Y. (2018). Computational analysis of RNA-protein interactions via deep sequencing. In Y. Wang & M. Sun (Eds.), *Transcriptome data analysis: Methods and protocols* (pp. 171–182). New York, NY: Springer New York.

Licatalosi, D. D., Mele, A., Fak, J. J., Ule, J., Kayikci, M., Chi, S. W., et al. (2008). HITS-CLIP yields genome-wide insights into brain alternative RNA processing. *Nature*, *456*, 464–469.

Liu, M. Y., Gui, G., Wei, B., Preston, J. F., Oakford, L., Yüksel, Ü., et al. (1997). The RNA molecule CsrB binds to the global regulatory protein CsrA and antagonizes its activity in Escherichia coli. *Journal of Biological Chemistry*, *272*, 17502–17510.

McAteer, S. P., Sy, B. M., Wong, J. L., Tollervey, D., Gally, D. L., & Tree, J. J. (2018). Ribosome maturation by the endoribonuclease YbeY stabilises a type III secretion system transcript required for virulence of enterohemorrhagic Escherichia coli. *The Journal of Biological Chemistry*, *293*, 9006–9016.

Ozer, A., Tome, J. M., Friedman, R. C., Gheba, D., Schroth, G. P., & Lis, J. T. (2015). Quantitative assessment of RNA-protein interactions with high-throughput sequencing-RNA affinity profiling. *Nature Protocols*, *10*, 1212–1233.

Potts, A. H., Vakulskas, C. A., Pannuri, A., Yakhnin, H., Babitzke, P., & Romeo, T. (2017). Global role of the bacterial post-transcriptional regulator CsrA revealed by integrated transcriptomics. *Nature Communications*, *8*, 1596.

Ray, D., Kazan, H., Chan, E. T., Castillo, L. P., Chaudhry, S., Talukder, S., et al. (2009). Rapid and systematic analysis of the RNA recognition specificities of RNA-binding proteins. *Nature Biotechnology*, *27*, 667–670.

Sahr, T., & Buchrieser, C. (2013). Legionella: Methods and protocols, *954*, 583–593. In C. Buchrieser & H. Hilbi (Eds.), *Part of the methods in molecular biology book series. Vol. 954*. Springer.

Sy, B., Wong, J., Granneman, S., Tollervey, D., Gally, D., & Tree, J. J. (2018). High-resolution, high-throughput analysis of Hfq-binding sites using UV crosslinking and analysis of cDNA (CRAC). Bacterial regulatory RNA: Methods and protocols. In V. Arluison & C. Valverde (Eds.), *Methods in molecular biology, Vol. 1737*. (pp. 251–272). Springer.

Tawk, C., Sharan, M., Eulalio, A., & Vogel, J. (2017). A systematic analysis of the RNA-targeting potential of secreted bacterial effector proteins. *Scientific Reports*, 7, 9328.

Tome, J. M., Ozer, A., Pagano, J. M., Gheba, D., Schroth, G. P., & Lis, J. T. (2014). Comprehensive analysis of RNA-protein interactions by high-throughput sequencing-RNA affinity profiling. *Nature Methods*, *11*, 683–688.

Tree, J. J., Granneman, S., McAteer, S. P., Tollervey, D., & Gally, D. L. (2014). Identification of bacteriophage-encoded anti-sRNAs in pathogenic Escherichia coli. *Molecular Cell*, *55*, 199–213.

Ule, J., Jensen, K. B., Ruggiu, M., Mele, A., Ule, A., & Darnell, R. B. (2003). CLIP identifies nova-regulated RNA networks in the brain. *Science*, *302*, 1212–1215.

Vakulskas, C. A., Potts, A. H., Babitzke, P., Ahmer, B. M. M., & Romeo, T. (2015). Regulation of bacterial virulence by Csr (Rsm) systems. *Microbiology and Molecular Biology Reviews*, *79*, 193–224.

Wang, D., McAteer, S. P., Wawszczyk, A. B., Russell, C. D., Tahoun, A., Elmi, A., et al. (2018). An RNA-dependent mechanism for transient expression of bacterial translocation filaments. *Nucleic Acids Research*, *46*, 3366–3381.

Waters, S. A., McAteer, S. P., Kudla, G., Pang, I., Deshpande, N. P., Amos, T. G., et al. (2017). Small RNA interactome of pathogenic *E. coli* revealed through crosslinking of RNase E. *The EMBO Journal*, *36*, 374–387.

Weilbacher, T., Suzuki, K., Dubey, A. K., Wang, X., Gudapaty, S., Morozov, I., et al. (2003). A novel sRNA component of the carbon storage regulatory system of Escherichia coli. *Molecular Microbiology*, *48*, 657–670.

Winther, K. S., Roghanian, M., & Gerdes, K. (2018). Activation of the stringent response by loading of RelA-tRNA complexes at the ribosomal A-site. *Molecular Cell*, *70*, 95–105.

Winther, K., Tree, J. J., Tollervey, D., & Gerdes, K. (2016). VapCs of Mycobacterium tuberculosis cleave RNAs essential for translation. *Nucleic Acids Research*, *44*, 9860–9871.

CHAPTER SEVEN

High-Resolution Profiling of NMD Targets in Yeast

Feng He[1], Alper Celik[2], Richard Baker, Allan Jacobson[1]
Department of Microbiology and Physiological Systems, University of Massachusetts Medical School, Worcester, MA, United States
[1]Corresponding authors: e-mail address: feng.he@umassmed.edu; allan.jacobson@umassmed.edu

Contents

Abstract

Contemporary high-throughput sequencing methods, notably RNA-Seq, permit the systematic identification and characterization of transcripts whose levels change significantly in response to altered biological states. We have described methods for the application of this methodology to a definition of the transcripts regulated by the NMD pathway in the yeast *Saccharomyces cerevisiae*. In short, we outline methods for growing cells of wild-type or NMD-deficient yeast, isolating RNA from the different strains, depleting rRNA from each sample, preparing and sequencing the respective RNA-Seq libraries, and employing multiple software packages to characterize the

[2] Present address: The Centre for Computational Medicine, Peter Gilgan Centre for Research and Learning, 686 Bay Street, Toronto, Ontario, Canada M5G 0A4.

Methods in Enzymology, Volume 612
ISSN 0076-6879
https://doi.org/10.1016/bs.mie.2018.09.005

resulting sequence reads meaningfully. Our experimental approach has identified approximately 900 transcripts that are commonly upregulated when yeast NMD is inactivated (Celik, Baker, He, & Jacobson, 2017).

1. INTRODUCTION

Nonsense-mediated mRNA decay (NMD) is a eukaryotic surveillance mechanism that targets mRNAs undergoing premature translation termination for rapid degradation (He & Jacobson, 2015b; Kervestin & Jacobson, 2012; Lykke-Andersen & Bennett, 2014). The pathway was initially uncovered in *Saccharomyces cerevisiae* and *Caenorhabditis elegans* (Leeds, Peltz, Jacobson, & Culbertson, 1991; Peltz, Brown, & Jacobson, 1993; Pulak & Anderson, 1993) and later shown to be conserved from yeast to humans (Behm-Ansmant et al., 2007; Schoenberg & Maquat, 2012). NMD's function was originally thought to be limited to quality control, i.e., the elimination of mRNAs derived from genes harboring nonsense mutations to prevent the accumulation of potentially deleterious truncated polypeptides (He, Peltz, Donahue, Rosbash, & Jacobson, 1993; Pulak & Anderson, 1993). However, NMD also targets a significant fraction of apparently normal and physiologically functional wild-type mRNAs (Celik, Baker, He, & Jacobson, 2017; Schweingruber, Rufener, Zund, Yamashita, & Muhlemann, 2013), indicating that it also serves as a fundamental posttranscriptional regulatory mechanism for eukaryotic gene expression.

In all organisms examined the activation of NMD requires a set of conserved core regulatory factors, Upf1, Upf2, and Upf3 (He & Jacobson, 2015b; Kervestin & Jacobson, 2012). These three proteins interact with each other, the ribosome, and multiple translation and mRNA decay factors (Kervestin & Jacobson, 2012). Based on these molecular interactions, several potential functions have been proposed for the Upf factors, including remodeling terminating mRNPs (Franks, Singh, & Lykke-Andersen, 2010), releasing and recycling ribosomal subunits (Ghosh, Ganesan, Amrani, & Jacobson, 2010), and recruiting mRNA decay factors (He & Jacobson, 2015a; Nicholson, Josi, Kurosawa, Yamashita, & Muhlemann, 2014; Okada-Katsuhata et al., 2012). However, the exact roles for the Upfs, and their modes of action in NMD, remain largely unknown.

Transcripts targeted by NMD have been investigated using genome-wide high-density DNA microarrays over the last two decades. Depending on the organism or cell type, NMD usually targets about 5%–20% of the transcripts in a typical transcriptome (He et al., 2003; Lelivelt & Culbertson, 1999; Mendell, Sharifi, Meyers, Martinez-Murillo, & Dietz, 2004; Ramani et al., 2009; Rehwinkel, Letunic, Raes, Bork, & Izaurralde, 2005; Weischenfeldt et al., 2008) and these transcripts can be classified into several general categories. One category, exemplifying typical NMD substrates, includes mRNAs with a destabilizing premature termination codon (PTC) in their coding region. These transcripts are generated from endogenous genes harboring nonsense or frameshift mutations (He et al., 2003), pseudogenes (He et al., 2003; McGlincy & Smith, 2008), nonproductively rearranged genetic loci (Li & Wilkinson, 1998), or from alternative splicing events that lead to intron retention or inclusion of a PTC-containing exon (Jaillon et al., 2008; Lareau, Inada, Green, Wengrod, & Brenner, 2007; Lykke-Andersen et al., 2014; Ni et al., 2007). A second category contains mRNA-like transcripts with limited or no apparent coding potential, such as long noncoding RNAs (Kurihara et al., 2009; Lykke-Andersen et al., 2014; Tani, Torimura, & Akimitsu, 2013), small RNAs derived from intragenic regions (Smith et al., 2014; Thompson & Parker, 2007), or transcripts of inactivated transposable elements (He et al., 2003). A third category contains a subset of physiologically relevant transcripts that appear to be "normal," such as mRNAs with upstream open reading frames (uORFs) (Arribere & Gilbert, 2013; Gaba, Jacobson, & Sachs, 2005; He et al., 2003), or with atypically long 3′-UTRs (Kebaara & Atkin, 2009; Singh, Rebbapragada, & Lykke-Andersen, 2008), or normal-looking wild-type mRNAs with no atypical features (He et al., 2003).

High-density DNA microarrays have limited dynamic ranges and are also unable to distinguish different transcript isoforms originated from the same genetic locus or homologous loci. To generate a comprehensive and high-resolution catalog of NMD-regulated transcripts, and to delineate the defining features of these transcripts in NMD targeting, we recently utilized RNA-Seq to reevaluate the effects of deleting the *UPF1*, *UPF2*, or *UPF3* genes on the genome-wide expression of annotated yeast genes. Our new analyses confirm the previous results of microarray studies, but also uncover hundreds of new NMD-regulated transcripts that had escaped previous detection, including many intron-containing pre-mRNAs. Here, we describe the detailed experimental methods and the bioinformatics pipeline of our RNA-Seq experiments for analyzing the endogenous NMD substrates in yeast cells. We present detailed procedures for RNA isolation and construction of RNA-Seq libraries from wild-type and *upf1*Δ, *upf2*Δ, or *upf3*Δ cells, and also outline the software used in our differential gene expression analyses.

2. CONSTRUCTION OF STRAND-SPECIFIC RNA-Seq LIBRARIES FROM WILD-TYPE AND NMD-DEFICIENT YEAST CELLS

Construction of RNA-Seq libraries is a multistep process starting with cell cultures and ending with a pool of complex DNA fragments of specific lengths that are representative of the cellular RNA population. In our experiments, a yeast wild-type strain and its isogenic derivatives harboring deletions of *UPF1*, *UPF2*, or *UPF3* are grown in rich liquid medium to mid-log phase. Yeast cells are harvested by centrifugation and total RNA from these cells is isolated by hot phenol extraction (Herrick, Parker, & Jacobson, 1990). To increase the strand-specific DNA fragments in our RNA-Seq libraries, total RNA used for the library construction is first treated with DNase to eliminate contaminating genomic DNA. To enrich the sequencing reads from the mRNA population in our RNA-Seq libraries, the DNA-free total RNA is then treated with Ribo-Zero components to deplete the majority of ribosomal RNAs. The high-performance dUTP second-strand marking protocol is used to generate all our stand-specific RNA-Seq libraries (Levin et al., 2010), and three independent RNA-Seq libraries are constructed for each of the analyzed yeast strains.

2.1 Yeast Strains and Growth Conditions

The wild-type yeast strain (HFY114) and its isogenic derivatives harboring deletions of *UPF1* (HFY871), *UPF2* (HFY116), or *UPF3* (HFY861) in the W303 background are used in our RNA-Seq experiments. The genotype of HFY114 is *MAT***a** *ade2-1 his3-11,15 leu2-3, 112 trp1-1 ura3-1 can1-100 UPF1 UPF2 UPF3*. The genotypes of HFY871, HFY116, and HFY861 are identical to HFY114 except for the respective *upf1::HIS3*, *upf2::HIS3*, and *upf3::HIS3* knock-out alleles. For all experiments, YEPD plates are used for the starter cultures and YEPD liquid medium is used for the final cultures for total RNA preparation.

2.1.1 Equipment

- VWR 30°C incubator (model 2020)
- Thermo Scientific 30°C shaking incubator (model 493)
- Thermo Fisher tabletop refrigerated centrifuge
- Eppendorf Microcentrifuge
- Beckman DU7400 spectrophotometer

2.1.2 Buffers and Reagents

- YEDP plates (1% yeast extract, 2% bacto-peptone, 2% glucose, and 2% agar)
- YEDP liquid medium (1% yeast extract, 2% bacto-peptone, and 2% glucose)

2.1.3 Procedure

1. Streak out yeast strains from −80°C stocks onto a YEPD plate and incubate the plate in a 30°C incubator for 2 days.
2. Inoculate a 3 mL culture in a 15-mL Falcon tube with the starter cells from the plate and incubate the culture to saturation in a 30°C shaking incubator (200 rpm) overnight.
3. Measure the OD_{600} of the overnight culture and inoculate a new 10 mL culture in a 50-mL Falcon tube with 1.5 OD_{600} of the overnight culture.
4. Incubate the new culture in a 30°C shaking incubator (200 rpm) for about two generations until the OD_{600} of the culture reaches about 0.7.
5. Harvest the yeast cells in the 50-mL tube by centrifugation in a bench-top centrifuge at 5000 rpm for 5 min.
6. Discard the liquid medium from the 50-mL tube and then add 0.5 mL new YEPD liquid medium to the cell pellet.
7. Suspend the cell pellet and transfer the cells to a 2-mL microcentrifuge tube.
8. Pellet the cells by centrifugation in a microcentrifuge at 12,000 rpm for 1 min.
9. Remove the liquid medium and freeze the cell pellet on dry ice immediately
10. Store the cell pellet at −80°C until total RNA isolation.

2.1.4 Notes

1. We remove the liquid medium completely from the cell pellet before freezing on dry ice. This can avoid RNA degradation during total RNA isolation.

2.2 Total RNA Isolation From Yeast Cells

Total cellular RNA is isolated from yeast cells by the hot phenol method (Herrick et al., 1990).

2.2.1 Equipment

- Precision 65°C water bath
- VWR Vortex-Genie 2

- Eppendorf Microcentrifuge
- Beckman DU7400 spectrophotometer

2.2.2 Buffers and Reagents

- RNA buffer A: 50 m*M* NaOAc (pH 5.2), 10 m*M* EDTA.
- Complete RNA buffer A: 45 m*M* NaOAc, 9 m*M* EDTA, 1% SDS, 0.1% DEPC.
- Phenol (Fisher, Cat# A92-500) saturated with RNA Buffer A: prepared by adding 500 mL RNA buffer A and 0.5 g quinoinol to 500 g liquid phenol, equilibrating overnight, followed by removal of most of the aqueous layer, stored at 4°C.
- Phenol/chloroform/isoamyl alcohol (25:24:1) (Ambion, Cat# AM9732)

2.2.3 Procedure

1. Suspend cell pellet in 500 µL of complete RNA buffer A.
2. Add 500 µL of prewarmed (65°C) buffer-saturated phenol.
3. Vortex for 10 s and then place tubes at 65°C for 50 s.
4. Repeat step 3 four more times.
5. Centrifuge for 3 min at 13,000 rpm.
6. Remove the phenol (bottom) phase with a Pasteur pipette and add 500 µL of prewarmed (65°C) buffer-saturated phenol to each tube.
7. Repeat steps 3 and 4.
8. Centrifuge for 8 min at 13,000 rpm
9. Transfer the aqueous (upper) phase to a new tube and add 500 µL of phenol/chloroform/isoamyl alcohol (25:24:1) to each tube.
10. Vortex for 1 min and then centrifuge for 10 min at 13,000 rpm.
11. Repeat steps 9 and 10.
12. Remove the aqueous (upper) phase to a new tube, add 1 mL of ice-cold 100% ethanol, mix by inverting the tubes, and precipitate at −70°C ≥ 1 h.
13. Centrifuge for 15 min and remove the ethanol with a drawn-out Pasteur pipette, taking care to not dislodge the white RNA pellet on the bottom of the tube.
14. Add 1 mL of 70% ethanol to each sample and centrifuge for 5 min.
15. Remove the ethanol with a drawn-out Pasteur pipette and air-dry the RNA pellets at room temperature for about 10 min
16. Suspend the RNA pellets in 100 µL 0.1% DEPC-treated double distilled H_2O.
17. Dilute 5 µL of each sample into 495 µL of H_2O and determine A_{260} and A_{280}.

2.2.4 Notes

1. To avoid carryover of the protein interphase to the next step, we always leave some RNA-containing aqueous phase behind during each phenol/chloroform extraction step.
2. Total RNA generally has an A_{260} to A_{280} ratio between 1.8 and 2.0, and a concentration from 2.0 to 4.0 μg/μL.
3. To assess the integrity of the total RNA samples to be used for the library construction, we usually take an aliquot of each total RNA sample, dilute it to an appropriate concentration, and analyze the profiles of the samples on an Agilent 2100 Bioanalyzer using the Eukaryote Total RNA Pico assay. Our RNA samples usually have an RNA integrity number between 5.5 and 7.5 (Fig. 1A).

2.3 Removal of Genomic DNA Contamination by DNase Treatment

Construction of RNA-Seq libraries involves reverse transcription, adaptor ligation, and PCR amplification. Genomic DNA contamination in the total RNA samples can potentially participate in one or more of these reactions and thus can lead to artifacts or distortion of strand-specific information of the libraries. We treat our total RNA samples with Baseline-ZERO DNase to remove any genomic DNA contamination.

2.3.1 Equipment

- Lab-line 37°C multiblock heater
- Eppendorf Microcentrifuge
- Beckman DU7400 spectrophotometer

2.3.2 Buffers and Reagents

- Baseline-ZERO DNase packaged with 10× Reaction Buffer and Stop Solution (Epicentre, Cat # DB0711K)
- Phenol/chloroform/isoamyl alcohol (25:24:1) (Ambion)
- 3 *M* NaOAc (pH 5.2)

2.3.3 Procedure

1. Take 30 μg total RNA from each sample and bring to a total volume of 17 μL with RNase-free water.
2. Add 2 μL of 10× Baseline-ZERO DNase Reaction Buffer to the sample.
3. Add 1 μL (1 MBU) of Baseline-ZERO DNase to the sample [1 MBU digests 1 μg of linear pUC19 DNA to dNMPs in 10 min at 37°C].

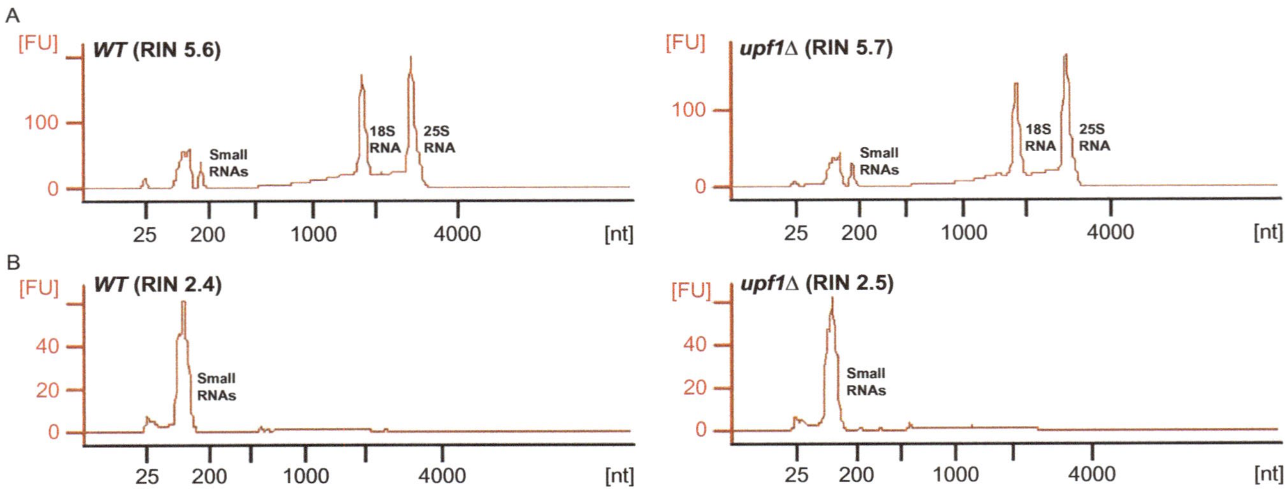

Fig. 1 Representative Bioanalyzer profiles of total RNA samples from wild-type and *upf1Δ* cells used for construction of RNA-Seq libraries. (A) Profiles of total RNA samples before Ribo-Zero treatment. Total RNA samples were diluted to a concentration about 5 ng/μL, and 1.0 μL of each diluted RNA sample was analyzed on an Agilent 2100 Bioanalyzer using the Eukaryote Total RNA Pico assay. (B) Profiles of total RNA samples after Ribo-Zero treatment. After Ribo-Zero treatment, 1.0 μL from each of the total RNA samples in the Fragment, Prime, Finish Mix buffer was diluted with 3.0 μL RNase-free water and 1.0 μL of the resulting diluted RNA sample was analyzed as described in (A). In both (A) and (B), the RNA integrity number and the positions of small RNAs or 18S and 25S rRNAs on the electropherograms are indicated. Notice that after Ribo-Zero treatment, both 18S and 25S rRNAs were essentially eliminated, but the small RNAs were still present in the samples. FU stands for Fluorescence Units.

4. Incubate the reaction at 37°C on a heating block for 30 min.
5. Add 2 μL of Stop Solution and then incubate at 65°C for 5 min.
6. Add 180 μL RNase-free water to the reaction and extract the RNA with 200 μL phenol/chloroform/isoamyl alcohol.
7. Remove about 200 μL of upper (aqueous) phase to a new tube and add 20 μL 3 *M* NaOAc, pH 5.2.
8. Precipitate the RNA with 600 μL ethanol.
9. Suspend the RNA in 50 μL RNase-free water and measure the RNA concentration using a UV spectrophotometer.

2.3.4 Notes

1. At this stage, our RNA samples usually have a concentration about 0.5 μg/μL.

2.4 Removal of rRNAs From the Total RNA Samples by Ribo-Zero Treatment

Yeast rRNAs comprise most of the mass of the total RNA population and their presence in a total RNA sample can lead to low complexity of RNA-Seq libraries and diminished sequencing reads from the mRNA population. Hence, removal of rRNAs from total RNA samples can greatly increase the complexity of RNA-Seq libraries and enhance the sequencing reads from mRNAs, and thus increase the effectiveness and reduce the cost of RNA-Seq experiments. We use the Yeast Ribo-Zero Gold rRNA Removal Kit from Illumina to deplete the rRNAs from the total RNA samples. The procedures outlined below mostly follow the instructions from the manufacturer.

2.4.1 Equipment

- Eppendorf Mastercycler Pro PCR machine
- Lab-line 50°C multiblock heater
- VWR Vortex-Genie 2
- Magnetic rack for 1.5-mL tubes (Life Technologies, Catalog # 123310)
- Eppendorf Microcentrifuge
- Agilent 2100 Bioanalyzer

2.4.2 Buffers and Reagents

Yeast Ribo-Zero Gold rRNA Removal Kit (Illumina, Catalog ID MRZY1306) with the following two parts:

- Part #1, the Magnetic Core Kit, contains Magnetic Beads, Magnetic Bead Resuspension Solution, and RNase-Free Water; stored at 4°C.

- Part #2, the Ribo-Zero rRNA Removal Reagents, contains RiboGuard RNase Inhibitor, Ribo-Zero rRNA Removal Solution, Ribo-Zero rRNA Reaction Buffer, Glycogen (10 mg/mL), 3 *M* Sodium Acetate, and RNase-Free Water; stored at −80°C.

2.4.3 Procedure

2.4.3.1 Prepare the Magnetic Beads

The magnetic beads bind the rRNA removal probes hybridized to yeast rRNAs. Steps 1–11 prepare the magnetic beads for the rRNA removal reaction.

1. Remove the Magnetic Core Kit components from storage at 4°C and allow them to equilibrate to room temperature.
2. Vortex vigorously the Magnetic Beads to homogeneity.
3. Pipette 225 μL of the Magnetic Beads into an RNase-free tube for each reaction.
4. Place each tube on a magnetic stand with cap open for at least 1 min (until the solution becomes clear).
5. With the tube still on the magnetic stand, remove and discard the supernatant.
6. Remove each tube from the magnetic stand, add 225 μL of RNase-free water, and suspend fully the Magnetic Beads by vortexing.
7. Repeat steps 4 through 6.
8. Place each tube on a magnetic stand with the cap open for 1 min. Remove and discard the remaining water.
9. Remove each tube from the magnetic stand, add 65 μL of Magnetic Bead Resuspension Solution, and resuspend the beads by vortexing.
10. Add 1 μL of RiboGuard RNase Inhibitor to each tube of resuspended Magnetic Beads and mix by pipetting 10–15 times.
11. Store the washed Magnetic Beads at room temperature until used in Section 2.4.3.3.

2.4.3.2 Hybridize the rRNA Removal Probes to the rRNAs in the Total RNA Sample

The Ribo-Zero rRNA Removal Solution contains specific probes that can hybridize to the ribosomal RNAs in the total RNA sample. Steps 12 and 13 carry out the hybridization reaction in solution between rRNA removal probes and rRNAs in the total RNA sample.

12. Remove Ribo-Zero rRNA Kit components from storage at −80°C and allow them to equilibrate to room temperature.

13. Assemble the hybridization reaction in a 0.2-mL RNase-free PCR tube.
    - Take 5.0 μg DNA-free total RNA from each sample
    - Bring the sample to a total volume of 26 μL with RNase-free water
    - Add 4 μL Ribo-Zero rRNA Reaction Buffer
    - Add 10 μL Ribo-Zero Removal Solution
14. Mix the reaction fully by pipetting 10–15 times.
15. Incubate the reaction tube in the Mastercycler PCR machine at 68°C for 10 min.
16. Remove the tubes from heat and centrifuge briefly to collect any condensation.
17. Incubate the tubes at room temperature for 5 min.

2.4.3.3 Remove rRNAs From the Total RNA Sample

This part combines probe-hybridized sample solution with the washed, room temperature Magnetic Beads. Steps 18–25 remove yeast rRNAs from the total RNA and yield an rRNA-depleted sample solution.

18. Transfer the probe-hybridized RNA sample (40 μL) to the washed, room temperature Magnetic Beads (65 μL).
19. Without changing the pipette tip, mix immediately and thoroughly the contents of the tube by pipetting 10–15 times.
20. Cap the tubes and vortex at high speed for a minimum of 10 s. Avoid vortexing the beads/sample into the tube cap.
21. Incubate the tubes at room temperature for 5 min.
22. Incubate the tubes on a heating block at 50°C for 5 min.
23. Remove the tubes from the heating block and place them immediately on a magnetic stand for at least 1 min until the solution appears clear.
24. Carefully remove the supernatant (85–90 μL) and transfer to an RNase-free microcentrifuge tube.
25. Place the supernatant (sample depleted of rRNA) on ice and proceed to the next purification step. Alternatively, keep the supernatant at −20°C overnight or at −80°C for long-term storage.

2.4.3.4 Purify the rRNA-Depleted Total RNA Sample by Ethanol Precipitation

rRNA-depleted samples require purification before downstream library preparation. This purification step removes any remaining salts/buffers and also concentrates the depleted samples. Compared to other RNA purification methods, such as using Agencourt RNAClean XP Beads (Beckman Coulter Genomics) or Modified RNeasy MinElute Cleanup Kit (QIAGEN), ethanol precipitation provides optimal recovery of both large and small RNAs. Steps 26–36 carry out RNA purification of rRNA-depleted samples through ethanol precipitation.

26. Bring the volume of each sample to 180 μL using RNase-free water.
27. Add 18 μL of 3 *M* sodium acetate to each sample.
28. Add 2 μL of glycogen (10 mg/mL) to each tube and mix by gentle vortexing.
29. Add 3 volumes (600 μL) of ice-cold 100% ethanol to each tube and mix by gentle vortexing.
30. Place the tubes at −20°C for at least 1 h.
31. Centrifuge the tubes at >10,000 × *g* in a microcentrifuge for 30 min. Carefully remove and discard the supernatant.
32. Wash the pellet with ice-cold 70% ethanol and centrifuge at >10,000 × *g* for 5 min. Carefully remove and discard the supernatant.
33. Repeat step 32 for a total of two 70% ethanol washes.
34. Centrifuge briefly to collect any residual supernatant. Carefully remove and discard the supernatant. Allow the pellet to air-dry at room temperature for 5 min.
35. Dissolve the pellet in 20 μL of the Fragment, Prime, Finish Mix buffer.
36. Place on ice for immediate use, or keep at −20°C overnight or −80°C for long-term storage.

2.4.3.5 Determine Yield and Quality of the rRNA-Depleted Total RNA Samples

In the final step, we take an aliquot of the Ribo-Zero treated and purified total RNA from each sample and analyze the RNA profiles of the samples on an Agilent 2100 Bioanalyzer using the Eukaryote Total RNA Pico assay to assess both rRNA depletion and the yield of the RNA samples.

37. Take 1 μL from each of the rRNA-depleted total RNA samples and analyze on an Agilent RNA6000 PicoChip.

2.4.4 Notes

1. Do not freeze or place the Magnetic Beads on ice, as chilling can damage the beads and affect the kit's performance.
2. The volume of the Ribo-Zero rRNA Removal Solution used per reaction depends on the amount of total RNA in the sample. In our case, we use 10 μL Ribo-Zero Removal Solution for 5 μg total RNA.
3. The supernatant from step 24 contains the rRNA-depleted total RNA sample and this supernatant should not have any visible Magnetic Beads in it before the final ethanol precipitation step.

4. Ribo-Zero treatment almost completely eliminates the rRNAs from the total RNA sample (Fig. 1B). Starting with 5.0 μg total RNA as input, we usually obtain about 100 ng total RNA after rRNA depletion.

2.5 Construction of RNA-Seq Libraries

Construction of RNA-Seq libraries is a multistep process involving RNA fragmentation, first-strand cDNA synthesis, second-strand cDNA synthesis, double-strand cDNA end repair and 3′-end single adenine nucleotide addition, ligation of multiple indexing adapters, and PCR amplification of cDNA fragments. To increase the complexity of our RNA-Seq libraries, we always use rRNA-depleted total RNA as starting material for library construction. We elect to utilize the dUTP second-strand marking method from Illumina to construct our strand-specific RNA-Seq libraries, as this method appears to out-perform all other available protocols (Levin et al., 2010). To increase consistency, we use the components from the Illumina TruSeq Stranded mRNA LT Sample Prep Kit for our RNA-seq library construction. The methods described here mostly follow the instructions recommended by Illumina.

2.5.1 Equipment

- Eppendorf Mastercycler Pro PCR machine
- Magnetic stand-96 (Life Technologies, part # AM10027)
- Agilent 2100 Bioanalyzer

2.5.2 Buffers and Reagents

- TruSeq Stranded mRNA LT Sample Prep Kit—Set A (Illumina, Cat # RS-122-2101) with the following components:
  - Fragment, Prime, Finish Mix (Part # 15032067)
  - First Strand Synthesis Act D Mix (Part # 15031094)
  - Second Strand Marking Master Mix (Part # 15031098)
  - Resuspension Buffer (Part # 15026770)
  - A-Tailing Mix (Part # 15012495)
  - RNA Adapter Index 2 (Part # 15026634)
  - RNA Adapter Index 4 (Part # 15026636)
  - RNA Adapter Index 5 (Part # 15026637)
  - RNA Adapter Index 6 (Part # 15026638)
  - RNA Adapter Index 7 (Part # 15026640)
  - RNA Adapter Index 12 (Part # 15026645)
  - RNA Adapter Index 16 (Part # 15024658)

RNA Adapter Index 19 (Part # 15024661)
Ligation Mix (Part # 15026773)
Stop Ligation Buffer (Part # 15012546)
PCR Master Mix (Part # 15026785)
PCR Primer Cocktail (Part # 15031748)

- SuperScript II Reverse Transcriptase (Invitrogen, cat # 18064-022)
- Agencourt RNAClean XP Kit (Beckman Coulter Genomics, cat# A63987)

2.5.3 Procedure

2.5.3.1 RNA Fragmentation

This part carries out RNA fragmentation and primes the RNA fragments for cDNA synthesis.

1. Preprogram the thermal cycler with the following conditions and save as **Frag–Prime**.
   - choose the preheat lid option and set to 100°C
   - 94°C for 8 min
   - 4°C hold
2. Remove the Fragment, Prime, Finish Mix tube from −20°C storage and thaw the tube on ice.
3. Pipet 18 μL of rRNA-depleted total RNA sample in the Fragment, Prime, Finish Mix buffer into a 0.2-mL PCR tube.
4. Place the tube on the preprogrammed thermal cycler and run the **Frag–Prime** program.
5. Remove the tube from the thermal cycler when it reaches 4°C and centrifuge briefly.
6. Proceed immediately to Section 2.5.3.2.

2.5.3.2 First-Strand cDNA Synthesis

In this section, RNA fragments primed with random hexamers are reverse transcribed using reverse transcriptase to generate the first strand of cDNA. Inclusion of Actinomycin D during the first stand synthesis reaction allows RNA-dependent synthesis, but prevents spurious DNA-dependent synthesis by the reverse transcriptase, and improves the strand specificity of the RNA-Seq libraries.

7. Preprogram the thermal cycler with the following conditions and save as **Synthesize 1st Strand**.
   - Choose the preheat lid option and set to 100°C
   - 25°C for 10 min

- 42°C for 15 min
- 70°C for 15 min
- Hold at 4°C

8. Remove the First Strand Synthesis Act D Mix tube from −20°C storage and thaw it at room temperature.
9. Prepare the required volume of 1:9 solution mix of the SuperScript II and the First Strand Synthesis Act D Mix.
10. Transfer 17 μL of the primed reaction mix to a new 0.2-mL PCR tube and add 8 μL of solution mix of the SuperScript II and the First Strand Synthesis Act D Mix.
11. Pipette the entire reaction mix up and down gently 5–6 times to mix thoroughly and centrifuge briefly the tube.
12. Place the tube on the preprogrammed thermal cycler, and select and run the Synthesize 1st Strand program.
13. Remove the tube from the thermal cycler when the thermal cycler reaches 4°C and proceed immediately to Section 2.5.3.3.

2.5.3.3 Second-Strand cDNA Synthesis

The process here, through the actions of RNase H and *E. coli* DNA polymerase, removes the RNA template and synthesizes a replacement strand to generate double-stranded cDNA. Second-strand synthesis incorporates dUTP in place of dTTP. The presence of dUTP quenches the second strand during PCR amplification, because the polymerase is unable to synthesize a DNA fragment past the incorporated dUTP nucleotide. DNA polymerase used in this reaction also possesses end-repair activity. At the end of this process, we have blunt-ended cDNA fragments.

14. Preheat the thermal cycler to 16°C. Choose the thermal cycler preheat lid option and set to 30°C.
15. Remove the Second Strand Marking Master Mix tube from −20°C storage and thaw it at room temperature.
16. Remove the Resuspension Buffer from 4°C storage and bring it to room temperature.
17. Add 5 μL of Resuspension Buffer to the completed first-strand cDNA tube.
18. Add 20 μL of thawed Second Strand Marking Master Mix to the tube.
19. Pipette gently the entire reaction mix up and down 5–6 times to mix thoroughly.
20. Place the tube on the preheated thermal cycler and incubate the reaction at 16°C for 1 h.

21. Remove the tube from the thermal cycler and put it on the bench.
22. Let the reaction equilibrate to room temperature and then proceed to Section 2.5.3.4.

2.5.3.4 Purification of Double-Stranded cDNA Fragments

1.8 × volume AMPure XP beads are used to purify the double-stranded cDNA fragments from the second-strand reaction mix. The cDNA synthesis reactions are transferred from 0.2-mL PCR tubes to the wells of a PCR plate for this double-stranded cDNA purification step.

23. Remove the AMPure XP beads from 4°C storage and let stand at room temperature for at least 30 min before use.
24. Remove the Resuspension Buffer from 4°C storage and bring it to room temperature.
25. Vortex the AMPure XP beads until they are well dispersed.
26. Transfer the entire 50 μL cDNA synthesis reaction from the 0.2-mL PCR tube to a well of the PCR plate.
27. Add 90 μL of well-mixed AMPure XP beads to each well and pipette gently the entire volume up and down 10 times to mix thoroughly.
28. Incubate the plate at room temperature for 15 min.
29. Place the plate on the magnetic stand at room temperature for 5 min to make sure that all of the beads are bound to the side of the wells.
30. Remove and discard 135 μL supernatant from each well of the plate.
31. With the plate still on the magnetic stand, add 200 μL freshly prepared 80% ethanol to each well without disturbing the beads.
32. Incubate the plate at room temperature for 30 s, and then remove and discard all of the supernatant from each well.
33. Repeat steps 31 and 32 one time for a total of two 80% ethanol washes.
34. Let the plate stand at room temperature for 15 min to air-dry the bead pellets, and then remove the plate from the magnetic stand.
35. Add 17.5 μL Resuspension Buffer to each well of the plate and pipette gently the entire volume up and down 10 times to mix thoroughly.
36. Incubate the plate at room temperature for 2 min.
37. Place the plate on the magnetic stand at room temperature for 5 min.
38. Transfer 15.0 μL supernatant (double-stranded cDNA) from the plate to a new 0.2-mL PCR tube.
39. Store the cDNA tube at −20°C or proceed immediately to Section 2.5.3.5.

2.5.3.5 3′ Adenine Addition to the cDNA Fragments

Using a DNA polymerase lacking 3′ to 5′ proofreading activity, this reaction adds a single "A" nucleotide to the 3′ ends of the blunt cDNA fragments to create common sticky ends. The sticky ends prevent the double-strand cDNA from ligating to one another, and at the same time, enhance the ligation of the cDNA fragments to the indexing adapters with a single "T" nucleotide at their 3′ ends.

40. Preprogram the thermal cycler with the following conditions and save as **ATAIL70**.
    - choose the preheat lid option and set to 100°C
    - 37°C for 30 min
    - 70°C for 5 min
    - Hold at 4°C
41. Remove the Resuspension Buffer from 4°C storage and let it equilibrate to room temperature.
42. Remove the A-Tailing Mix tube from −20°C storage and thaw it at room temperature.
43. Remove the double-stranded cDNA tube from −20°C storage (if it is stored) and thaw it at room temperature.
44. Add 2.5 μL of Resuspension Buffer to the 15.0 μL double-stranded cDNA in the 0.2-mL PCR tube.
45. Add 12.5 μL of thawed A-Tailing Mix to the tube and pipette the entire volume up and down gently 10 times to mix thoroughly.
46. Place the tube on the preprogrammed thermal cycler, and select and run the ATAIL70 program.
47. Remove the tube from the thermal cycler when the thermal cycler temperature is 4°C and then proceed immediately to Section 2.5.3.6.

2.5.3.6 Ligation of Indexing Adapters

This process ligates multiple indexing adapters to the sticky ends of the double-stranded cDNA fragments and prepares the cDNA fragments for PCR amplification and hybridization to a flow cell of the sequencing machine.

48. Review Illumina Sample Pooling Guide and choose the appropriate combination of indexing adapters for samples to be pooled in a single sequencing lane.
49. Choose the thermal cycler preheat lid option and set to 100°C. Preheat the thermal cycler to 30°C.
50. Remove the appropriate RNA Adapter tubes and the Stop Ligation Buffer tube from −20°C storage and thaw them at room temperature.
51. Remove the Resuspension Buffer from 4°C storage and let it equilibrate to room temperature.
52. Add 2.5 μL of Resuspension Buffer to the 30 μL A-tailed cDNA mix in each of 0.2-mL PCR tubes.

53. Remove the Ligation Mix tube from −20°C storage and add 2.5 μL of Ligation Mix to each tube.
54. Add 2.5 μL of the thawed RNA Adapter Index to each tube. Pipette the entire volume up and down gently 10 times to mix thoroughly.
55. Place the tubes on the preheated thermal cycler and incubate at 30°C for 10 min.
56. Remove the tubes from the thermal cycler.
57. Add 5.0 μL of Stop Ligation Buffer to each tube to terminate the ligation reaction. Gently pipette the entire volume up and down 10 times to mix thoroughly.
58. Proceed immediately to the free adapter removal step.

2.5.3.7 Removal of Free Adapters From the Ligation Products

This part carries out two consecutive rounds of 1.0 × volume AMPure XP Bead selections to remove the free indexing adapters and to enrich for the cDNA fragments from the ligation reactions. Elimination of the free adapters minimizes the production of primer dimers during the PCR amplification of library cDNA fragments. The ligation reactions are transferred from each 0.2-mL PCR tube to wells of a PCR plate for this purification step.

59. Remove the AMPure XP Beads from 4°C storage and let stand at room temperature for at least 30 min before use.
60. Vortex the AMPure XP Beads for at least 1 min until they are well dispersed.
61. Transfer the 42.0 μL ligation products from the 0.2-mL tubes to wells of a PCR plate.
62. Add 42.0 μL of mixed AMPure XP Beads to each well of the PCR plate. Pipette the entire volume up and down gently 10 times to mix thoroughly.
63. Incubate the plate at room temperature for 15 min.
64. Place the plate on the magnetic stand at room temperature for 5 min or until the liquid is clear.
65. Remove and discard 79.5 μL supernatant from each well of the plate. Take care not to disturb the beads.
66. With the plate on the magnetic stand, add 200 μL freshly prepared 80% ethanol to each well without disturbing the beads.
67. Incubate the plate at room temperature for 30 s, and then remove and discard all of the supernatant from each well. Take care not to disturb the beads.

68. Repeat steps 66 and 67 one time for a total of two 80% ethanol washes.
69. With the plate on the magnetic stand, let the samples air-dry at room temperature for 15 min.
70. Remove the plate from the magnetic stand.
71. Add 52.5 μL Resuspension Buffer to each well of the plate. Pipette the entire volume up and down gently 10 times to mix thoroughly.
72. Incubate the plate at room temperature for 2 min.
73. Place the plate on the magnetic stand at room temperature for 5 min or until the liquid is clear.
74. Transfer 50 μL supernatant from each well of the plate to the corresponding well of a new PCR plate. Take care not to disturb the beads.
75. Vortex the AMPure XP Beads until they are well dispersed.
76. Add 50 μL of mixed AMPure XP Beads to each well of the plate for a second cleanup. Pipette the entire volume up and down gently 10 times to mix thoroughly.
77. Incubate the plate at room temperature for 15 min.
78. Place the plate on the magnetic stand at room temperature for 5 min or until the liquid is clear.
79. Remove and discard 95 μL supernatant from each well of the plate. Take care not to disturb the beads.
80. With the plate on the magnetic stand, add 200 μL freshly prepared 80% ethanol to each well. Take care not to disturb the beads.
81. Incubate the plate at room temperature for 30 s, and then remove and discard all of the supernatant from each well. Take care not to disturb the beads.
82. Repeat steps 80 and 81 one time for a total of two 80% ethanol washes.
83. With the plate on the magnetic stand, let the samples air-dry at room temperature for 15 min, and then remove the plate from the magnetic stand.
84. Add 22.5 μL Resuspension Buffer to each well of the plate. Pipette the entire volume up and down gently 10 times to mix thoroughly.
85. Incubate the plate at room temperature for 2 min.
86. Place the plate on the magnetic stand at room temperature for 5 min or until the liquid is clear.
87. Transfer 20.0 μL supernatant from each well of the plate to a 0.2-mL PCR tube. Take care not to disturb the beads.
88. Store the purified ligation products at −20°C or proceed immediately to Section 2.5.3.8.

2.5.3.8 PCR Amplification of Library cDNA Fragments

An inefficient ligation reaction can result in cDNA fragments with an adapter attached to just one end or with no adapter attached to both ends. These DNA fragments cannot form clusters in the flow cell. Fragments with an adapter on only one end can attach to the surface of the flow cell, but cannot be bridge-amplified. Fragments with no adapter on both ends cannot even attach to the surface of the flow cell. The PCR reaction uses a pair of primers specific for the end sequences of the adapters to enrich for DNA fragments with adapters attached to both ends of the molecules in the cDNA library.

89. Preprogram the thermal cycler with the following conditions and save as **PCR**.
- choose the preheat lid option and set to 100°C
- 98°C for 30 s
- 15 cycles of
  98°C for 10 s
  60°C for 30 s
  72°C for 30 s
- 72°C for 5 min
- Hold at 4°C

90. Remove the PCR Master Mix and PCR Primer Cocktail from −20°C storage and thaw them at room temperature.

91. Add 5 μL of thawed PCR Primer Cocktail to 20 μL purified ligation mix in the 0.2-mL PCR tube.

92. Add 25 μL of thawed PCR Master Mix to the tube. Pipette the entire volume up and down gently 10 times to mix thoroughly.

93. Place the tube on the preprogrammed thermal cycler, and select and run **PCR.**

2.5.3.9 Purification of PCR-amplified Library DNA Fragments

This final part employs a 1.0× volume AMPure XP Bead selection to remove the PCR primers and to purify the PCR-amplified DNA fragments from the reaction mix. The PCR reactions are transferred from each of 0.2-mL PCR tubes to wells of a PCR plate for this purification step.

94. Remove the AMPure XP Beads from 4°C storage and let stand at room temperature for at least 30 min before use.

95. Remove the Resuspension Buffer from 4°C storage and let it equilibrate to room temperature.

96. Remove the PCR tube from the thermal cycler and transfer the final 50 μL PCR products to a well of the PCR plate.

97. Vortex the AMPure XP Beads until they are well dispersed.
98. Add 50 μL of the mixed AMPure XP Beads to the 50 μL PCR products in each well. Pipette the entire volume up and down gently 10 times to mix thoroughly.
99. Incubate the plate at room temperature for 15 min.
100. Place the PCR plate on the magnetic stand at room temperature for 5 min or until the liquid is clear.
101. Remove and discard 95 μL supernatant from each well of the PCR plate.
102. With the PCR plate on the magnetic stand, add 200 μL freshly prepared 80% ethanol to each well without disturbing the beads.
103. Incubate the PCR plate at room temperature for 30 s, and then remove and discard all of the supernatant from each well.
104. Repeat steps 102 and 103 one time for a total of two 80% ethanol washes.
105. With the PCR plate on the magnetic stand, let the samples air-dry at room temperature for 15 min, and then remove the plate from the magnetic stand.
106. Add 32.5 μL Resuspension Buffer to each well of the PCR plate. Pipette gently the entire volume up and down 10 times to mix thoroughly.
107. Incubate the PCR plate at room temperature for 2 min.
108. Place the PCR plate on the magnetic stand at room temperature for 5 min or until the liquid is clear.
109. Transfer 30 μL supernatant from each well of the PCR plate to a 0.2-mL PCR tube.
110. Take 1.0 μL of the library for quantification and quality control and store the rest of the library at −20°C.

2.5.3.10 Analyzing the Quality and the Yield of RNA-Seq Libraries

As a final step, we take an aliquot from each of the RNA-Seq libraries and analyze the library profiles on an Agilent 2100 Bioanalyzer using the High Sensitivity DNA Assay to assess the size distribution of the library DNA fragments and to determine the DNA concentration of the libraries.

111. Take 1 μL of DNA from each library and combine with 3 or 19 μL of DNase- and RNase-free water (Ambion) depending on library DNA concentration.
112. Load 1 μL of each diluted library sample on an Agilent DNA 1000 chip and run the chip on Agilent 2100 Bioanalyzer.

2.5.4 Notes

1. We usually use ~100 ng of rRNA-depleted total RNA as starting material for each library construction.
2. Depending on the desired size ranges of library DNA fragments, the time of RNA fragmentation at 94°C can be adjusted.
3. Inclusion of Actinomycin D to the first-strand synthesis reaction prevents spurious DNA-dependent synthesis from potential contaminating genomic DNA by the reverse transcriptase.
4. The dUTP second-strand marking method is, in our experience, the most effective protocol to generate strand-specific RNA-Seq libraries.
5. We follow the Illumina Sample Pooling Guide and choose A/C and G/T balanced combinations of indexing adapters for samples to be pooled together for sequencing.
6. We perform two consecutive rounds of 1.0 × volume AMPure XP Bead selection to remove the free indexing adapters from the cDNA ligation reactions. Elimination of the free adapters is important because it minimizes the production of primer dimers during the subsequent PCR amplification of library cDNA fragments.
7. We usually minimize the number of PCR cycles during cDNA fragment amplification to avoid skewing the representation of the library. Our RNA-Seq libraries contain two prominent peaks of DNA fragments, one around 250 bp and another (most likely PCR dimers) around 520 bp (Fig. 2), and usually have a DNA concentration of 10–20 ng/μL and a total yield 300–600 ng per library.

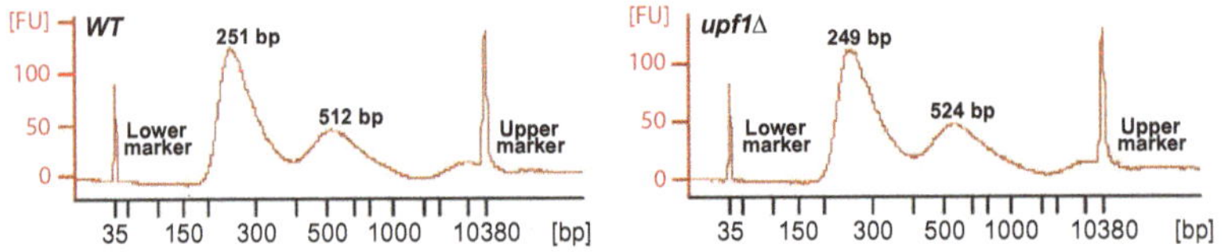

Fig. 2 Representative Bioanalyzer profiles of RNA-Seq libraries from wild-type and *upf1Δ* cells. Final PCR-amplified DNA fragments from each cDNA library were purified using 1.0 × volume AMPure XP. 1.0 μL of purified DNA fragments taken from each sample was diluted with 19.0 μL water, and 1.0 μL of each diluted DNA sample was then analyzed on an Agilent 2100 Bioanalyzer using the High Sensitivity DNA Assay. The positions of the lower and upper markers, and the two predominant peaks of DNA fragments in the libraries, are indicated. FU stands for Fluorescence Units.

3. SEQUENCING OF TOTAL RNA cDNA LIBRARIES FROM WILD-TYPE AND NMD-DEFICIENT YEAST CELLS

Total RNA cDNA libraries were sequenced on the Illumina HiSeq4000 platform at Beijing Genomics Institute. In our experiments, independently constructed libraries from wild-type, *upf1*Δ, *upf2*Δ, and *upf3*Δ strains were pooled into a single lane and sequenced from single ends in a 50-cycle reaction. After sequencing and demultiplexing, about 100 million raw sequence reads of 49 nt in length in Fastq format were generated for each of the pooled libraries. Three independent libraries were sequenced for each strain and our RNA-Seq experiment yielded about 1.6 billion raw reads in total.

4. IDENTIFICATION OF TRANSCRIPTS TARGETED BY NMD

Transcripts targeted by NMD are defined as those whose abundance changes significantly in response to inactivation of the NMD pathway and are identified through bioinformatic analysis of the sequencing data from the wild-type, *upf1*Δ, *upf2*Δ, and *upf3*Δ strains. Our bioinformatic pipeline starts with construction of a comprehensive reference transcriptome and selects a set of specific analytic modules to accomplish the following three major tasks: (1) evaluating the quality of sequence reads in each library and filtering the potentially uninformative reads from the library, (2) mapping the quality reads in each library to the respective transcripts of origin and calculating the total reads mapped to each of the individual transcripts in the transcriptome, and (3) normalizing the sequence reads in the libraries and testing individual transcripts for differential expression between the different libraries.

Several factors led us to use a reference transcriptome instead of a reference genome to carry out our bioinformatic analyses. First, the yeast genome contains hundreds of intron-containing genes and each of these loci has the potential to produce both spliced and unspliced transcript isoforms with different sensitivities to NMD (He et al., 1993). Second, mapping the sequence reads originating from spliced exon–exon junctions to the genomic loci still presents a significant challenge for most sequence read mapping tools

(Engstrom et al., 2013). In contrast, mapping these sequence reads to spliced mRNAs is relatively easy and also possibly error-free. Finally, as a model organism, the yeast *S. cerevisiae* has advanced genome annotations, thus making the generation of a comprehensive transcriptome with all potential transcript isoforms an easy task.

Bioinformatic tools for quality control are somewhat standardized and they all perform essentially the same functions, including assessing the quality and composition of sequence reads in the library and filtering potential biased or contaminating sequence reads from the library. As an initial step in our bioinformatics pipeline, we use the FastQC software package to assess the overall quality of all our sequence libraries. In contrast to tools used for quality control, bioinformatics tools for sequence read mapping, transcript-based read counting, and individual transcript differential expression testing are diverse and appear to lack a common consensus (Seyednasrollah, Laiho, & Elo, 2015; Tang, Sun, Shimizu, & Kadota, 2015). We use the RSEM software package (Li & Dewey, 2011) for sequence read mapping and individual transcript quantification because this program uses a probabilistic model based on an expectation maximization (EM) algorithm and can effectively handle multireads in the library that map to either different transcript isoforms from the same genetic locus or homologous transcripts from different genetic loci. Effective mapping of multireads is important for accurate transcript quantification because the yeast genome contains both duplicated genes and repetitive genetic elements. Software packages that are unable to effectively handle multireads can lead to either overestimation or underestimation of the levels of expression of a significant number of yeast transcripts. We used the DESeq software package (Anders & Huber, 2010) for testing differential expression of individual transcripts because this program, at the time, was a better choice than most alternatives. DESeq models the RNA-Seq transcript counts based on a negative binomial distribution, a more suitable a priori assumption for overdispersed data than the Poisson distribution, which might underestimate the variance in the counting data (Anders & Huber, 2010; Seyednasrollah et al., 2015; Tang et al., 2015).

4.1 Software Packages

Reference transcriptome construction: In-house Perl scripts

Library quality control: FastQC (http://www.bioinformatics.babraham.ac.uk/projects/fastqc/) and Bowtie 2 (http://bowtie-bio.sourceforge.net/bowtie2/index.shtml)

Sequence read mapping and transcript quantification: RSEM v. 1.2.11 (http://deweylab.biostat.wisc.edu/rsem/src/rsem-1.2.11.tar.gz) in conjunction with Bowtie 1.0.0 (https://sourceforge.net/projects/bowtie-bio/files/bowtie/1.0.0/)

Differential expression testing: DESeq (https://bioconductor.org/packages/release/bioc/html/DESeq.html)

Other data analysis and visualization: R v. 3.2.4 with ggplot2 and gplots packages

4.2 Procedure

4.2.1 Constructing the Reference Transcriptome

The R64-2-1 S288C reference genome assembly (sacCer3) from the *Saccharomyces* Genome Database (SGD; https://www.yeastgenome.org/) is used to construct a comprehensive list of transcripts of the yeast transcriptome in FASTA format. We took advantage of the different FASTA files stored in the database and downloaded two separate FASTA files from the database. One file contains the "coding" sequences and the other contains the "gene" sequences of all annotated loci. A custom Perl program (available on request) is used to compare each "coding" entry in the first FASTA file to its "gene" entry in the second FASTA file. If these entries differ in sequence, a new unspliced transcript isoform entry is generated using the "gene" coordinates and appended to the first "coding" FASTA file. The same procedure is used for annotated functional RNA genes, many of which contain introns. By these procedures, we generated a reference transcriptome comprised of 6726 transcripts that includes all annotated protein-coding sequences, functional and noncoding RNAs, and the unspliced isoforms of all intron-containing genes. Each transcript in our transcriptome FASTA file also has a unique gene identifier in the sequence header from which transcript to gene mapping can be derived for gene level expression analysis. The FASTA and transcript-to-gene mapping files are provided in Supplemental Materials available online at https://doi.org/10.1016/bs.mie.2018.09.005.

4.2.2 Assessing the Quality and the General Mapping Statistics of RNA-Seq Libraries

Before mapping the sequence reads in each RNA-Seq library to the reference transcriptome, we typically run two levels of quality control. First, FASTQC with default parameters is used to assess the overall quality of sequence reads in the library and to identify potential contaminants such as reads from the common adapters. The second level of quality control uses Bowtie 2 to map the sequence reads in each library to the R64-2-1 S288C reference yeast genome to obtain general mapping statistics, such as the overall mapping rate and the percentages of unique and multimapping reads. All our RNA-Seq libraries have high quality sequence reads, do not have significant sequence reads biased at a particular nucleotide position, and are virtually free of adapter contamination. All our RNA-Seq libraries yield greater than 90% overall mapping rates, less than 10% of sequence reads for functional RNAs, ~70% of uniquely mapped reads, and 10%–15% multiple-mapped reads. Because our libraries all are of high sequence quality and high mapping rates, our bioinformatic pipeline does not include data filtering after the quality control step.

4.2.3 Mapping Sequence Reads to the Reference Transcriptome and Estimating the Reads Mapped to Individual Transcripts

We use the RSEM software package to map the library sequence reads to the reference transcriptome and to quantify the reads mapped to individual transcripts. Analysis by RSEM is carried out in two major steps: indexing the reference transcriptome and calculating expression. At the end of these two steps, we also use analytical tools in R to assess the overall read count distribution of the libraries and the correlation between replicate libraries (Fig. 3A).

We use the following command lines to generate a Bowtie-indexed transcriptome for sequence read mapping and this step is required only once per transcriptome:

```
rsem-prepare-reference --transcript-to-gene-map mapping.txt \
                       --bowtie \
                       --bowtie-path path to Bowtie executables \
                       transcriptome.fasta \
                       transcriptome_location/transcriptome
```

We carry out sequence read mapping using the loop below, assuming that all the library files in fastq format are in the same folder and the folder does not contain any other files:

```
libs=$(ls ./fastq_location)
for i in ${libs[*]}
        do rsem-calculate-expression \
                -p 16
                --bowtie-m 30 \
                --no-bam-output \
                --forward-prob 0 \
                $i transcriptome_location/trascriptome \
                $i-aligned
        done
```

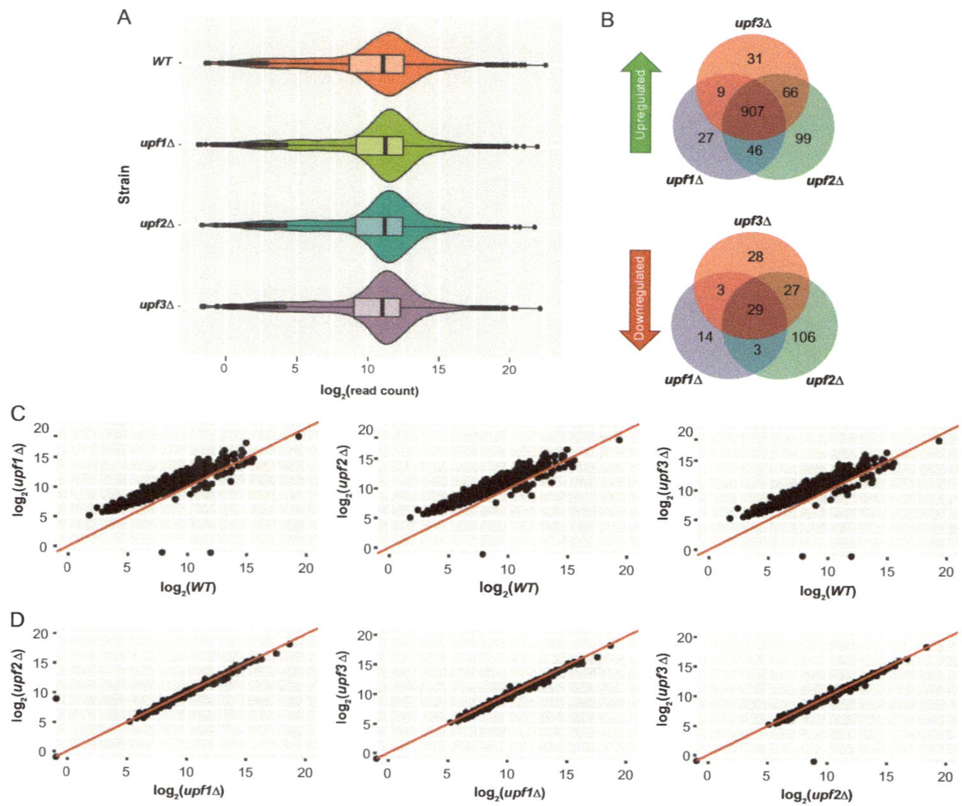

Fig. 3 See legend on next page.

At the end of analysis, RSEM generates two major results output files. One is labeled as library-genes.results and the other is labeled as library-isoforms.results. The former file contains the estimated reads mapped to each gene in the transcriptome and the latter file contains the estimated reads mapped to each transcript isoform. We use the library-isoforms.results as input for differential expression analysis. If we are interested in visualizing the reads mapped to individual transcripts, the command `--no-bam-output` can be omitted to obtain a sorted .bam file for visualization in a genome browser such as IGV.

As an additional the quality control, we also import the library-isoforms.results from RSEM into R, and use geom_violin, geom_boxplot, and geom_point plotting functions from the ggplot2 package to assess the distribution of reads mapped to the transcriptome for each library and to examine the correlation of the mapped reads between different replicate libraries. As shown in Fig. 3A, our RNA-Seq libraries from the wild-type, *upf1Δ*, *upf2Δ*, and *upf3Δ* strains have very similar read distributions over the entire transcriptome.

Fig. 3 Identification of transcripts targeted by the yeast NMD factors Upf1, Upf2, and Upf3. (A) RNA-Seq libraries from *WT*, *upf1Δ*, *upf2Δ*, and *upf3Δ* strains exhibit similar overall read count distributions for 7473 transcripts in the yeast transcriptome. The RSEM program (Li & Dewey, 2011) was used to map the sequence reads of each library to the 7473-transcript transcriptome and to generate estimated read counts of individual transcripts. *Violin* and *box-plots* were used to visualize the average read count distribution of the transcripts in the transcriptomes of the indicated strains from three independent experiments. (B) Transcripts differentially expressed (either up- or downregulated) in *upf1Δ*, *upf2Δ*, and *upf3Δ* strains show significant overlap. The estimated read counts of individual transcripts in each library from RSEM were imported into the Bioconductor DESeq package (Anders & Huber, 2010) and transcripts differentially expressed in the *upf1Δ*, *upf2Δ*, or *upf3Δ* strains were identified by comparison to the data obtained from the wild-type strain. Venn diagrams were used to display the relationships among the sets of transcripts that are either up- or downregulated in *upf1Δ*, *upf2Δ*, and *upf3Δ* strains. (C) All three *UPF* deletion strains exhibit similar genome-wide expression patterns. Scatterplots were used to compare the estimated read counts for the 936 commonly differentially expressed transcripts identified in (B) between WT and *upf1Δ*, *upf2Δ*, or *upf3Δ* strains. The vast majority of differentially expressed transcripts in *UPF* deletion strains showed upregulation and only a small number of transcripts showed downregulation. The $y=x$ line is shown in *red*. (D) Transcripts commonly regulated by Upf1, Upf2, and Upf3 each have virtually identical expression levels in *upf1Δ*, *upf2Δ*, or *upf3Δ* strains. As in (C), scatterplots were used to compare the estimated read counts of the 936 NMD-regulated transcripts between *upf1Δ and upf2Δ*, *upf1 and upf3Δ*, *and upf2Δ* or *upf3Δ* strains. In (A), (C), and (D), $\log_2$ transformed values are used for the analyses.

4.2.4 Identification of Transcripts Differentially Expressed in NMD-Deficient Cells

We use the Bioconductor DESeq package (Anders & Huber, 2010) to test the differential expression of individual transcripts between the wild-type and the *upf1*Δ, *upf2*Δ, or *upf3*Δ strains. In this analysis, the estimated read counts mapped to individual transcripts from the RSEM library-isoforms. results for each library are first imported into R and rounded to the nearest integer using the R `round()` command to make them compatible with DESeq. The resulting expression data are saved as `counts_combined` and then read into DESeq for differential expression analysis. We use the built-in functions of DESeq for all steps of the analysis: estimating normalization factors; fitting the dispersions for individual transcripts; and, testing differential expression. Based on the observed replicate variability of our libraries, we apply a false discovery threshold of 0.01 (1%) instead of an arbitrary fold change cutoff as the criterion for differential expression.

The code snippet below is used for the identification of transcripts differentially expressed between the wild-type and *upf1*Δ libraries with three independent replicates, assuming that rounded count values from RSEM are stored in a `data.frame` named `counts_combined` with column names indicated as `WT_1`, `WT_2`, and so on. The same analysis can be also carried out between the wild-type and *upf2*Δ or *upf3*Δ libraries. This code snippet can also be converted into a function or a newer S4 R object method to automate differential expression analysis of many libraries. The results from these analyses can be stored in memory, written to files, or imported to a small database.

```
design<- data.frame(condition=c(rep("WT",3),rep(UPF1,3)),
                    libType=c("single-end"),
                    row.names=c("WT_1","WT_2","WT_3",
                                UPF1_1,UPF1_2,UPF1_3))
cds<-newCountDataset(counts_combined, design[,1])
cds<-estimateSizeFactors(cds)
cds<-estimateDispersions(cds)
de<-nbiomTest(cds, WT, UPF1)
# remove transcripts that are not detected in any of the libraries
wt_upf1_de<-de[de$baseMean>0,]
wt_upf1_de$sig<-FALSE
wt_upf1_de$sig[wt_upf1_de$padj<0.01]<-TRUE
```

At the end of analysis, DESeq generates a new data.frame of the transcriptome for each strain with the data columns showing the average input values, fold changes, initial statistical test *P* values, adjusted *P* values after Benjamini–Hochberg multiple testing correction, and Boolean identifiers of differential expression. The up- and downregulation of individual transcripts in NMD-deficient cells is inferred from the fold-change values by comparison to wild-type cells.

4.2.5 Example of Downstream Analysis and Validation of Bioinformatics Results

The above bioinformatic pipeline identifies a list of transcripts differentially expressed in each of the *upf*Δ strains. Our earlier results (Celik et al., 2017) illustrate the types of downstream analyses typically undertaken to interrogate and validate these results. To examine functional relationships of Upf1, Upf2, and Upf3 in NMD, we imported the differentially expressed transcript lists and their associated expression data into R. Using Venn diagrams, we examined the relationships of the transcripts differentially expressed in the *upf1*Δ, *upf2*Δ, or *upf3*Δ strains (Fig. 3B). We found that transcripts up- or downregulated in these strains largely overlap. Based on this analysis, we identified 907 commonly upregulated and 29 commonly downregulated transcripts in the *upf1*Δ, *upf2*Δ, and *upf3*Δ strains. Using scatter plots, we compared the levels of expression of the transcripts commonly regulated by NMD in different strains (Fig. 3C and D). We found that these NMD-regulated transcripts have almost identical expression patterns and levels in each of the *upf1*Δ, *upf2*Δ, or *upf3*Δ strains, indicating that Upf1, Upf2, and Upf3 function together and are commonly required for yeast NMD.

Compared to the results of previous microarray analyses (He et al., 2003), the RNA-Seq experiments summarized here identified more than 400 new transcripts targeted by yeast NMD. A significant fraction of these newly identified transcripts are intron-containing pre-mRNAs. We found that about 16% of yeast intron-containing genes produce NMD-sensitive unspliced isoforms (Celik et al., 2017). To validate our bioinformatics results, we assessed the levels of expression of six newly identified NMD-regulated transcripts in wild-type, *upf1*Δ, *upf2*Δ, and *upf3*Δ strains by northern blotting (Fig. 4). As expected, the four intron-containing pre-mRNAs (*HRB1*, *RPL22B*, *NHP6B*, and *MTR2*) and two noncoding RNAs (*IRT1* and *ICR1*) all showed increased accumulation in *upf1*Δ, *upf2*Δ, or *upf3*Δ cells.

4.3 Notes

1. The straightforward method for constructing the transcriptome FASTA file described here is just one possible approach. An alternative would be

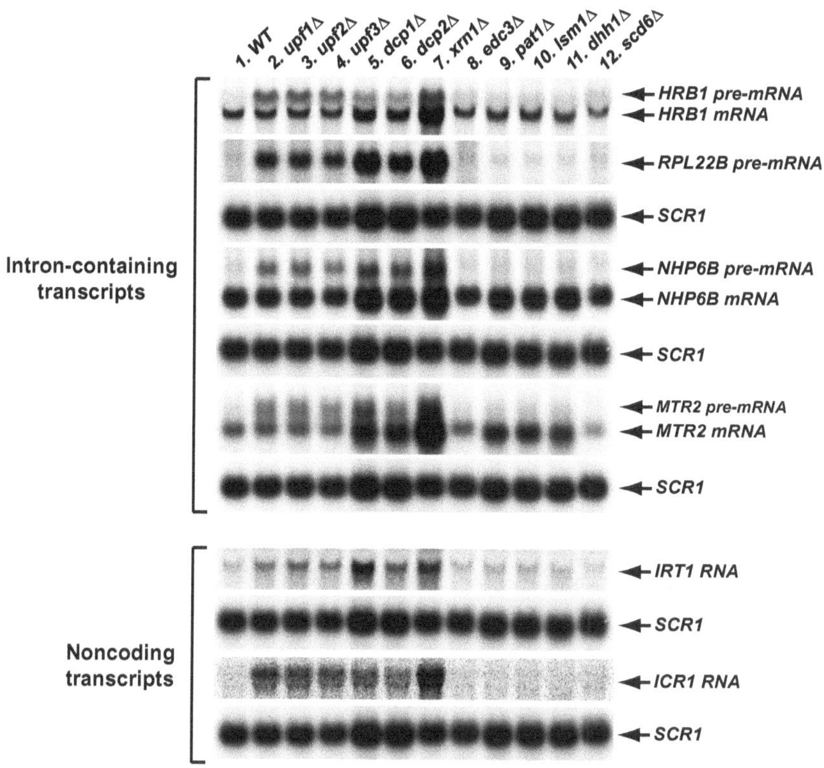

Fig. 4 Northern blotting validation of novel NMD substrates identified by RNA-Seq analyses. RNA-Seq experiments identified more than four hundred transcripts regulated by NMD that had not been identified as such previously. These transcripts include many intron-containing pre-mRNAs and several noncoding RNAs. To validate the bioinformatics analyses, four intron-containing transcripts (*HRB1*, *RPL22B*, *NHP6B*, and *MTR2*) and two "noncoding" RNAs (*ICR1* and *IRT1*) were analyzed by northern blotting for their respective expression levels in wild-type, *upf1Δ*, *upf2Δ*, or *upf3Δ* strains. To assess the specificity and the potential decay mechanisms for these transcripts, RNA from *dcp1Δ*, *dcp2Δ*, *xrn1Δ*, *edc3Δ*, *pat1Δ*, *lsm1Δ*, *dhh1Δ*, and *scd6Δ* strains was also included in this experiment. In each case, a random-primed mRNA-specific probe was hybridized to the blot and *SCR1* served as the loading control.

to filter and edit the R64-2-1 S288C reference genome feature file (GFF3) to include the desired transcripts along with user-generated entries for the unprocessed isoforms. The gffread utility of the cufflinks package (http://cole-trapnell-lab.github.io/cufflinks/) can be used to convert between GFF3 and GTF2 formats and to generate FASTA sequence files from either. Additionally, RSEM accepts GTF2-formatted annotation files as an alternative to FASTA to generate its indexes.

2. Bowtie is the default mapping program called by RSEM. Alternately, Bowtie 2 or STAR could be used instead. Since our reference transcriptome contains both spliced and unspliced transcript isoforms, we use bowtie mapping since the ability of the aligner to handle gapped reads is not needed. Also, using bowtie ensures that reads originating from splice junctions are aligned uniquely to the spliced isoforms.
3. DESeq models count data based on a negative binomial distribution and assumes integer values as input. RSEM uses a probabilistic model to estimate transcript read counts, and the resulting estimates are noninteger. Preprocessing of the RSEM output before importing into DESeq is an important and necessary step of this analysis.
4. For the analysis described in Section 4.2.5, the transcriptome contained an additional 747 transcripts consisting of mRNA's of "dubious" ORF's, noncoding RNA's, 2-μm plasmid transcripts, mitochondrially encoded mRNA's, and unspliced versions of these.

5. SUMMARY AND CONCLUSIONS

Here, we describe methods for the assembly and characterization of RNA-Seq libraries that specifically facilitate identification of the substrates of the NMD pathway in yeast, i.e., the transcripts whose abundance is altered significantly by mutations in any of the three *UPF* genes. We first present methods for growing cells of wild-type, *upf1Δ*, *upf2Δ*, or *upf3Δ* yeast, isolating total cellular RNA from these strains, depleting DNA and rRNA from each total RNA sample, and then preparing strand-specific RNA-Seq libraries. Further, we describe the application of multiple software packages to the sequence data obtained from the libraries. These bioinformatics analyses first facilitate quality control of each set of sequencing reads, then allow mapping of reads to their appropriate origins in a newly defined transcriptome, and ultimately provide quantitative measurements of the extent of differential expression in the NMD-deficient libraries when compared to those from wild-type cells. Collectively, these methods provide a high-resolution profile of approximately 900 transcripts that are commonly upregulated when NMD is inactivated and serve as a basis for further understanding of the mechanism for NMD substrate selection (Celik et al., 2017).

ACKNOWLEDGMENTS

This work was supported by Grants to A.J. (5R01 GM27757-37 and 1R35GM122468-02) from the U.S. National Institutes of Health.

REFERENCES

Anders, S., & Huber, W. (2010). Differential expression analysis for sequence count data. *Genome Biology, 11*(10), R106. https://doi.org/10.1186/gb-2010-11-10-r106.

Arribere, J. A., & Gilbert, W. V. (2013). Roles for transcript leaders in translation and mRNA decay revealed by transcript leader sequencing. *Genome Research, 23*(6), 977–987. https://doi.org/10.1101/gr.150342.112.

Behm-Ansmant, I., Kashima, I., Rehwinkel, J., Sauliere, J., Wittkopp, N., & Izaurralde, E. (2007). mRNA quality control: An ancient machinery recognizes and degrades mRNAs with nonsense codons. *FEBS Letters, 581*(15), 2845–2853. https://doi.org/10.1016/j.febslet.2007.05.027.

Celik, A., Baker, R., He, F., & Jacobson, A. (2017). High-resolution profiling of NMD targets in yeast reveals translational fidelity as a basis for substrate selection. *RNA, 23*(5), 735–748. https://doi.org/10.1261/rna.060541.116.

Engstrom, P. G., Steijger, T., Sipos, B., Grant, G. R., Kahles, A., Ratsch, G., et al. (2013). Systematic evaluation of spliced alignment programs for RNA-seq data. *Nature Methods, 10*(12), 1185–1191. https://doi.org/10.1038/nmeth.2722.

Franks, T. M., Singh, G., & Lykke-Andersen, J. (2010). Upf1 ATPase-dependent mRNP disassembly is required for completion of nonsense-mediated mRNA decay. *Cell, 143*(6), 938–950. https://doi.org/10.1016/j.cell.2010.11.043.

Gaba, A., Jacobson, A., & Sachs, M. S. (2005). Ribosome occupancy of the yeast *CPA1* upstream open reading frame termination codon modulates nonsense-mediated mRNA decay. *Molecular Cell, 20*(3), 449–460. https://doi.org/10.1016/j.molcel.2005.09.019.

Ghosh, S., Ganesan, R., Amrani, N., & Jacobson, A. (2010). Translational competence of ribosomes released from a premature termination codon is modulated by NMD factors. *RNA, 16*(9), 1832–1847. https://doi.org/10.1261/rna.1987710.

He, F., & Jacobson, A. (2015a). Control of mRNA decapping by positive and negative regulatory elements in the Dcp2 C-terminal domain. *RNA, 21*(9), 1633–1647. https://doi.org/10.1261/rna.052449.115.

He, F., & Jacobson, A. (2015b). Nonsense-mediated mRNA decay: Degradation of defective transcripts is only part of the story. *Annual Review of Genetics, 49*, 339–366. https://doi.org/10.1146/annurev-genet-112414-054639.

He, F., Li, X., Spatrick, P., Casillo, R., Dong, S., & Jacobson, A. (2003). Genome-wide analysis of mRNAs regulated by the nonsense-mediated and 5' to 3′ mRNA decay pathways in yeast. *Molecular Cell, 12*(6), 1439–1452.

He, F., Peltz, S. W., Donahue, J. L., Rosbash, M., & Jacobson, A. (1993). Stabilization and ribosome association of unspliced pre-mRNAs in a yeast *upf1*-mutant. *Proceedings of the National Academy of Sciences of the United States of America, 90*(15), 7034–7038.

Herrick, D., Parker, R., & Jacobson, A. (1990). Identification and comparison of stable and unstable mRNAs in *Saccharomyces cerevisiae*. *Molecular and Cellular Biology, 10*(5), 2269–2284.

Jaillon, O., Bouhouche, K., Gout, J. F., Aury, J. M., Noel, B., Saudemont, B., et al. (2008). Translational control of intron splicing in eukaryotes. *Nature, 451*(7176), 359–362. https://doi.org/10.1038/nature06495.

Kebaara, B. W., & Atkin, A. L. (2009). Long 3′-UTRs target wild-type mRNAs for nonsense-mediated mRNA decay in *Saccharomyces cerevisiae*. *Nucleic Acids Research, 37*(9), 2771–2778. https://doi.org/10.1093/nar/gkp146.

Kervestin, S., & Jacobson, A. (2012). NMD: A multifaceted response to premature translational termination. *Nature Reviews. Molecular Cell Biology, 13*(11), 700–712. https://doi.org/10.1038/nrm3454.

Kurihara, Y., Matsui, A., Hanada, K., Kawashima, M., Ishida, J., Morosawa, T., et al. (2009). Genome-wide suppression of aberrant mRNA-like noncoding RNAs by NMD in Arabidopsis. *Proceedings of the National Academy of Sciences of the United States of America, 106*(7), 2453–2458. https://doi.org/10.1073/pnas.0808902106.

Lareau, L. F., Inada, M., Green, R. E., Wengrod, J. C., & Brenner, S. E. (2007). Unproductive splicing of SR genes associated with highly conserved and ultraconserved DNA elements. *Nature*, *446*(7138), 926–929. https://doi.org/10.1038/nature05676.

Leeds, P., Peltz, S. W., Jacobson, A., & Culbertson, M. R. (1991). The product of the yeast *UPF1* gene is required for rapid turnover of mRNAs containing a premature translational termination codon. *Genes & Development*, *5*(12A), 2303–2314.

Lelivelt, M. J., & Culbertson, M. R. (1999). Yeast Upf proteins required for RNA surveillance affect global expression of the yeast transcriptome. *Molecular and Cellular Biology*, *19*(10), 6710–6719.

Levin, J. Z., Yassour, M., Adiconis, X., Nusbaum, C., Thompson, D. A., Friedman, N., et al. (2010). Comprehensive comparative analysis of strand-specific RNA sequencing methods. *Nature Methods*, 7(9), 709–715. https://doi.org/10.1038/nmeth.1491.

Li, B., & Dewey, C. N. (2011). RSEM: Accurate transcript quantification from RNA-Seq data with or without a reference genome. *BMC Bioinformatics*, *12*, 323. https://doi.org/10.1186/1471-2105-12-323.

Li, S., & Wilkinson, M. F. (1998). Nonsense surveillance in lymphocytes? *Immunity*, *8*(2), 135–141.

Lykke-Andersen, J., & Bennett, E. J. (2014). Protecting the proteome: Eukaryotic cotranslational quality control pathways. *The Journal of Cell Biology*, *204*(4), 467–476. https://doi.org/10.1083/jcb.201311103.

Lykke-Andersen, S., Chen, Y., Ardal, B. R., Lilje, B., Waage, J., Sandelin, A., et al. (2014). Human nonsense-mediated RNA decay initiates widely by endonucleolysis and targets snoRNA host genes. *Genes & Development*, *28*(22), 2498–2517. https://doi.org/10.1101/gad.246538.114.

McGlincy, N. J., & Smith, C. W. (2008). Alternative splicing resulting in nonsense-mediated mRNA decay: What is the meaning of nonsense? *Trends in Biochemical Sciences*, *33*(8), 385–393. https://doi.org/10.1016/j.tibs.2008.06.001.

Mendell, J. T., Sharifi, N. A., Meyers, J. L., Martinez-Murillo, F., & Dietz, H. C. (2004). Nonsense surveillance regulates expression of diverse classes of mammalian transcripts and mutes genomic noise. *Nature Genetics*, *36*(10), 1073–1078. https://doi.org/10.1038/ng1429.

Ni, J. Z., Grate, L., Donohue, J. P., Preston, C., Nobida, N., O'Brien, G., et al. (2007). Ultraconserved elements are associated with homeostatic control of splicing regulators by alternative splicing and nonsense-mediated decay. *Genes & Development*, *21*(6), 708–718. https://doi.org/10.1101/gad.1525507.

Nicholson, P., Josi, C., Kurosawa, H., Yamashita, A., & Muhlemann, O. (2014). A novel phosphorylation-independent interaction between SMG6 and UPF1 is essential for human NMD. *Nucleic Acids Research*, *42*(14), 9217–9235. https://doi.org/10.1093/nar/gku645.

Okada-Katsuhata, Y., Yamashita, A., Kutsuzawa, K., Izumi, N., Hirahara, F., & Ohno, S. (2012). N- and C-terminal Upf1 phosphorylations create binding platforms for SMG-6 and SMG-5:SMG-7 during NMD. *Nucleic Acids Research*, *40*(3), 1251–1266. https://doi.org/10.1093/nar/gkr791.

Peltz, S. W., Brown, A. H., & Jacobson, A. (1993). mRNA destabilization triggered by premature translational termination depends on at least three *cis*-acting sequence elements and one *trans*-acting factor. *Genes & Development*, 7(9), 1737–1754.

Pulak, R., & Anderson, P. (1993). mRNA surveillance by the *Caenorhabditis elegans smg* genes. *Genes & Development*, 7(10), 1885–1897.

Ramani, A. K., Nelson, A. C., Kapranov, P., Bell, I., Gingeras, T. R., & Fraser, A. G. (2009). High resolution transcriptome maps for wild-type and nonsense-mediated decay-defective *Caenorhabditis elegans*. *Genome Biology*, *10*(9), R101. https://doi.org/10.1186/gb-2009-10-9-r101.

Rehwinkel, J., Letunic, I., Raes, J., Bork, P., & Izaurralde, E. (2005). Nonsense-mediated mRNA decay factors act in concert to regulate common mRNA targets. *RNA*, *11*(10), 1530–1544. https://doi.org/10.1261/rna.2160905.

Schoenberg, D. R., & Maquat, L. E. (2012). Regulation of cytoplasmic mRNA decay. *Nature Reviews. Genetics*, *13*(4), 246–259. https://doi.org/10.1038/nrg3160.

Schweingruber, C., Rufener, S. C., Zund, D., Yamashita, A., & Muhlemann, O. (2013). Nonsense-mediated mRNA decay—Mechanisms of substrate mRNA recognition and degradation in mammalian cells. *Biochimica et Biophysica Acta*, *1829*(6–7), 612–623. https://doi.org/10.1016/j.bbagrm.2013.02.005.

Seyednasrollah, F., Laiho, A., & Elo, L. L. (2015). Comparison of software packages for detecting differential expression in RNA-seq studies. *Briefings in Bioinformatics*, *16*(1), 59–70. https://doi.org/10.1093/bib/bbt086.

Singh, G., Rebbapragada, I., & Lykke-Andersen, J. (2008). A competition between stimulators and antagonists of Upf complex recruitment governs human nonsense-mediated mRNA decay. *PLoS Biology*, *6*(4), e111. https://doi.org/10.1371/journal.pbio.0060111.

Smith, J. E., Alvarez-Dominguez, J. R., Kline, N., Huynh, N. J., Geisler, S., Hu, W., et al. (2014). Translation of small open reading frames within unannotated RNA transcripts in *Saccharomyces cerevisiae*. *Cell Reports*, 7(6), 1858–1866. https://doi.org/10.1016/j.celrep.2014.05.023.

Tang, M., Sun, J., Shimizu, K., & Kadota, K. (2015). Evaluation of methods for differential expression analysis on multi-group RNA-seq count data. *BMC Bioinformatics16*, 361. https://doi.org/10.1186/s12859-015-0794-7.

Tani, H., Torimura, M., & Akimitsu, N. (2013). The RNA degradation pathway regulates the function of GAS5 a non-coding RNA in mammalian cells. *PLoS One*, *8*(1), e55684. https://doi.org/10.1371/journal.pone.0055684.

Thompson, D. M., & Parker, R. (2007). Cytoplasmic decay of intergenic transcripts in *Saccharomyces cerevisiae*. *Molecular and Cellular Biology*, *27*(1), 92–101. https://doi.org/10.1128/MCB.01023-06.

Weischenfeldt, J., Damgaard, I., Bryder, D., Theilgaard-Monch, K., Thoren, L. A., Nielsen, F. C., et al. (2008). NMD is essential for hematopoietic stem and progenitor cells and for eliminating by-products of programmed DNA rearrangements. *Genes & Development*, *22*(10), 1381–1396. https://doi.org/10.1101/gad.468808.

CHAPTER EIGHT

Generation of a Metagenomics Proximity Ligation 3C Library of a Mammalian Gut Microbiota

Théo Foutel-Rodier[*,†,‡,§], Agnès Thierry[*,†,§], Romain Koszul[*,†,§,1], Martial Marbouty[*,†,§,1]

[*]Institut Pasteur, Département Génomes et Génétique, Groupe Régulation Spatiale des Génomes, Paris Cedex 15, France

[†]CNRS, UMR 3525, Paris Cedex 15, France

[‡]Sorbonne Université, Collège Doctoral, Paris, France

[§]Institut Pasteur, Center of Bioinformatics, Biostatistics and Integrative Biology (C3BI), Paris, France

[1]Corresponding authors: e-mail address: romain.koszul@pasteur.fr; martial.marbouty@pasteur.fr

Contents

Abstract

Microbial species thrive in very diverse environments and play fundamental roles in their equilibrium and dynamics. Metagenomics consists in extracting, sequencing, and studying the DNA present in ecosystems to better understand their regulation. Ideally, the maximal amount of information would be gathered from the full sequences of the genomes, episomes, and phages present in the microbial communities. Current high-throughput DNA sequencing produces reads ranging in size from a few dozen base pairs for the most commonly used technologies to several kb for emerging single-molecule real-time sequencing techniques. Although valuable information can be extracted from processing these DNA sequences into contigs, reconstructing full genomes remains a difficult task. Clustering contigs according to their similarities or read coverage covariations gives some insights on these genomes, but remains limited as viral sequences, or recent horizontal gene transfers, often differ from their host genomes. We recently developed meta3C, a proximity ligation approach that bins contigs in a sequence-independent way by quantifying and exploiting their tridimensional collisions frequencies in vivo. This technique has demonstrated a great potential to reconstruct genomes as well as to assign plasmids and phages to their hosts. It nevertheless requires a specific processing of the microbial samples before sequencing, which has to be carefully planned.

Methods in Enzymology, Volume 612
ISSN 0076-6879
https://doi.org/10.1016/bs.mie.2018.08.001

ABBREVIATIONS

ATP	adenosine triphosphate
bp	base pair
BSA	bovine serum albumin
FA	formaldehyde
PBS	phosphate buffer saline
RT	room temperature
TE (buffer)	Tris–EDTA (buffer)

1. INTRODUCTION

A tacit and ideal objective of metagenomics studies is to characterize the genomes of all the species present in given microenvironments. Today, the most popular high-throughput DNA-sequencing technologies, such as Illumina, result in large number of small reads that are then assembled into larger DNA segments (contigs). Ideally, this assembly step would result in full genomes. However, due to the small read size (a few hundred bp), the presence of repeated sequences longer than the read size remains challenging to characterize and include in the final assembly. New sequencing technologies such as MinION (Oxford Nanopore; Mikheyev & Tin, 2014) or single-molecule real-time sequencing (Pacific Bioscience; Eid et al., 2009) can produce longer reads up to dozens of kb but whose accuracy remains significantly lower than Illumina sequencing, despite continuous improvements (Ardui, Ameur, Vermeesch, & Hestand, 2018; Jain et al., 2018). Recently, hybrid methods combining next-generation sequencing and long-read sequencing to scaffold contigs have been used to improve assembly quality (Mahmoud, Zywicki, Twardowski, & Karlowski, 2017). Yet, DNA assembly alone still cannot fully reconstruct most genomes from a metagenomic sample, and binning steps consisting in pooling together contigs belonging to the same genome remain necessary to reach at a better picture of the genomic structure of the community.

Metagenomic binning algorithms such as CONCOCT (Alneberg et al., 2014) or MetaBat (Kang, Froula, Egan, & Wang, 2015) use nucleotide frequency and sequencing read coverage covariation to pool contigs together. The sum of all contigs present in such cluster eventually covers a full genome. However, these algorithms fail to reconstruct the complete metagenome not only because repeated elements shared between species remain problematic to bin but also because they are unable to assign mobile DNA, such as viruses and mobile genetic elements, to their microbial host(s) (Dick et al., 2009). Proximity ligation is a recent genomic approach that takes advantage of the frequency of collision between DNA segments of a cell population to infer (1) whether they occupy the same cellular compartment (hence belong to the genome of the same species) and (2) their genomic distance along the DNA molecule (Flot, Marie-Nelly, & Koszul, 2015). The quantification of DNA collision frequencies is achieved thanks to alternative usages of the chromosome conformation captures approach (3C/Hi-C) (Dekker, Rippe, Dekker, & Kleckner, 2002; Lieberman-Aiden et al., 2009), a technique aimed at studying the 3D organization of the genome of a given species in a variety of conditions. Briefly, 3C/Hi-C requires first "freezing" the 3D higher-order architecture of the genome of each cell by the addition of a chemical fixative agent such as formaldehyde. This cross-linking molecule will generate covalent bridges between DNA and proteins, resulting in the tethering together of DNA segments physically close in 3D, but not necessary in 1D. After restriction of the DNA molecules trapped in these large complexes and their religation, DNA segments that were frequently trapped together will then be adjacent along the religated molecules. Sequencing the extremities of these molecules provides a count of the number of times each event occurs over the population of cells, resulting in the generation of a contact matrix where each vector reflects the contact frequency of a fragment with all the other fragments of the genome. This information reflects the average 3D configuration of the genome being studied. We and others showed that when applied to a controlled mix of species, ligation events involving DNA segments belonging to different species were sufficiently rare so that each genome was clearly distinguishable from the others based on its internal 3D contacts (Beitel et al., 2014; Burton, Liachko, Dunham, & Shendure, 2014; Marbouty et al., 2014). We further showed that this technique, dubbed meta3C, could indeed bin together genomes of species of complex, natural microbial ecosystems (Marbouty, Baudry, Cournac, & Koszul, 2017; Marbouty et al., 2014). Not only full genomes could be binned to high levels of completeness (as shown by the scaffolds obtained by applying the proximity ligation scaffolding program GRAAL; Marie-Nelly et al., 2014), but the bacterial hosts of unknown plasmids and phages could also be inferred from the 3D contacts between these molecules and the genome, as they all shared the same cellular compartment (Marbouty et al., 2017, 2014).

Meta3C follows the steps indicated in Fig. 1. Briefly, DNA is first cross-linked to preserve its higher-order structure before cells are broken down to release their lysate. There are many ways to break down cells but as

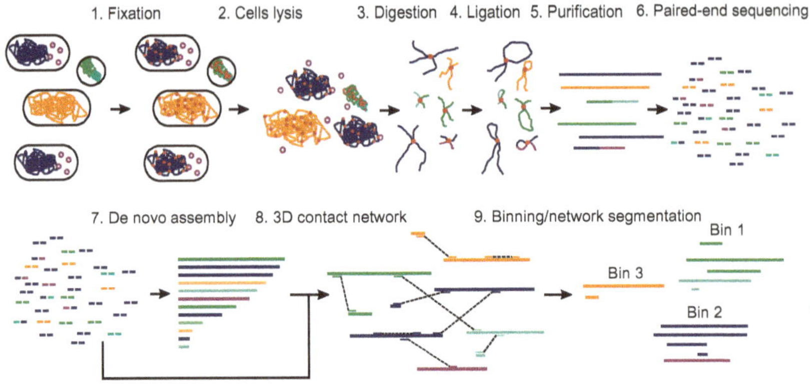

Fig. 1 Principles of a meta3C experiment. (1) A microbial community is chemically fixed using formaldehyde, resulting in covalent bridges between proteins and DNA, and freezing the macromolecular complex in the mix, including chromosomes. (2) The cells present in this community are mechanically disrupted using beads. (3) DNA molecules present in this sample are digested by a restriction enzyme. (4) DNA fragments trapped in cross-linked complexes are ligated together using T4 DNA ligase. (5) All DNA molecules from the mix are then purified and (6) processed through pair-end sequencing. (7) All individual reads can be used to generate a de novo assembly of the gut microbiome. (8) Each read of a pair is then aligned against the contigs generated by this assembly, occasionally resulting in a link between two contigs. (9) Using the resulting contact network, where contigs consist of nodes linked by the pair-end information, the genome content can be inferred through partitioning algorithms.

metagenomes can have a wide range of organisms, mechanical disruption (via bead beating) is a good choice as it can efficiently disrupt many types of organisms (Lim, Song, Kim, Lee, & Nam, 2018). After release, cells lysate are digested using restriction enzymes to cut genomic DNA to a smaller size. Mammalian gut microbiota feature organisms with varying DNA composition, with both AT-rich and GC-rich organisms potentially present. Hence, cutting genomic DNA using a single restriction enzyme can lead to poor recovery of contacts between DNA belonging to organisms with high GC and/or AT content as they would have a lower number of cutting sites in their genomes (Marbouty et al., 2017). To reduce this bias, for each sample/condition, we generate two libraries using different but complementary restriction enzymes: MluCI (AATT) and HpaII (CCGG). Both libraries are pulled together before processing and sequencing. After digestion, free DNA and small molecular weight complexes are discarded by centrifugation (most of the 3C signal is contained within the heavy complexes) and remaining DNA is religated in dilute condition to generate linear 3C hybrid fragments. Finally, DNA is purified before being processed for sequencing.

This method article describes step by step the protocol used to generate a mammalian gut microbiota meta3C library ready to be processed for sequencing. The analysis of the sequencing data is not described here but can be found in Marbouty et al. (2017) (see also the computational pipeline provided on Github; https://github.com/koszullab/metaTOR).

2. META3C LIBRARY PREPARATION

See Note 1.

2.1 Sample Collection and Cross-linking

2.1.1 Buffers and Reagents

- 36%–37% liquid formaldehyde (v/v) (Sigma-Aldrich, Saint-Louis, MO, USA). Do not keep stock solution for more than 2 weeks once opened. Keep away from light.
- 2.5 *M* glycine solution. Weight 75.07 g of glycine and add to 200 mL water. Add water to 400 mL and dissolve glycine using a magnetic stirrer and then sterilize on a 0.22 μm filtering unit and store at room temperature (RT). Dissolution can take several hours.
- Sterile water.
- Phosphate buffer saline (PBS).

2.1.2 Equipment

- Disposable 15 and 50 mL conical tubes.
- Disposable 5 mL tubes.
- 0.22 μm filtration unit.
- Refrigerated tabletop centrifuge with 50 mL tube rotor (10,000 × *g* at 4°C).
- Refrigerated tabletop centrifuge with 5 mL tube rotor (16,000 × *g* at 4°C).
- Sterile blade (scalpel or razor blade).
- Cut P1000 tips. Cut wide enough so feces do not block tip. Use a sterile scalpel or a sterile razor blade to cut several tips at once from a new sterile box of tips. It is possible to autoclave cut tips afterward for maximum sterility.

2.1.3 Cross-linking

- Dilute formaldehyde to ~5% in PBS 1 ×.
- Put ~250 μg human feces in a 50 mL tube (see Note 2).
- Add 25 mL 5% formaldehyde. If feces were to already be in solution before this step, add 37% FA instead so that final FA concentration is 5%.
- Vortex at high speed for 30 s.

- Incubate samples 30 min at RT on tabletop shaker with rapid shaking (200–300 rpm).
- Incubate samples 30 min at 4°C on tabletop shaker with moderate shaking (120 rpm).
- Add 5 mL of 2.5 *M* glycine solution to quench formaldehyde.
- Incubate 5 min at RT on shaker with moderate shaking (120 rpm).
- Incubate 15 min at 4°C on shaker with moderate shaking (120 rpm).
- Centrifuge 10 min at 4°C, 4000 × *g*.
- Carefully remove the supernatant. The pellet could be unstable. Leave a bit of liquid on top of it.
- Add 3 mL sterile water and vortex briefly to suspend.
- Transfer the pellet and liquid to a 5 mL tube using cut P1000 pipet tips.
- Centrifuge 5 min at 4°C, 16,000 × *g*.
- Carefully remove as much of the supernatant as possible.
- Flash freeze the pellet on dry ice for 5 min.
- Store the pellet at −80°C.

2.2 Library Preparation

The meta3C library preparation can be separated in two steps, the making of the 3C library and the processing to prepare the library for sequencing, by storing the meta3C library at −20°C. In this part we will describe the 3C library preparation steps.

2.2.1 *Buffers and Reagents*

- Tris–EDTA (TE) buffer (10 m*M* Tris–HCl pH 7.5 + 1 m*M* EDTA).
- 10% Sodium dodecyl sulfate (SDS) (w/v) (Sigma-Aldrich, Saint-Louis, MO, USA).
- 10% Triton X-100 (Sigma-Aldrich, Saint-Louis, MO, USA).
- Complete Mini EDTA-free (protease inhibitor cocktail) (Roche Diagnostics, Mannheim, Germany). One tablet for 10 mL, make fresh.
- 1000 U MluCI (restriction enzyme, New England Biolabs (NEB), Ipswich, MA, USA) (see Note 3).
- 1000 U HpaII (restriction enzyme, NEB, Ipswich, MA, USA) (see Note 3).
- 10 × NEB 1 digestion buffer or corresponding NEB buffer if enzymes other than MluCI or HpaII are used. Do not use the new NEB buffers (NEBuffer™ 1.1, NEBuffer™ 2.2, …) as they lack dithiothreitol (DTT).
- Cold water (keep at 4°C).
- 1 *M* DTT (Sigma-Aldrich, Saint-Louis, MO, USA).

- 10 × ligation buffer (without ATP) (500 m*M* Tris–HCl pH 7.4, 100 m*M* $MgCl_2$, 100 m*M* DTT). Add 100 mL of Tris–HCl pH 7.5, 20 mL of 1 *M* $MgCl_2$, and 20 mL of 1 *M* DTT in a 500 mL cylinder. Add water to 200 mL, mix, and sterilize using a 0.22 μm filtering unit. Divide into 10 mL aliquots and store at 4°C.
- 10 mg/mL bovine serum albumin (BSA). BSA (Sigma-Aldrich, Saint-Louis, MO, USA) dissolved in water and stored in 1 mL aliquot at −20°C.
- 100 m*M* ATP in water, pH 6–7 (Roche Diagnostics, Mannheim, Germany).
- 250 U T4 DNA ligase (5 U/μL, Thermo Fisher Scientific, Waltham, MA, USA).
- Isopropanol.
- 0.5 *M* EDTA in water, pH 8.
- Proteinase K (20 mg/mL).
- 3 *M* sodium acetate pH 5.2. Weight 204.12 g of sodium acetate and add to water to 400 mL in a graduated cylinder. Adjust pH to 5.2 with glacial acid acetic. Adjust to 500 mL final volume with water. Sterilize with 0.22 μm filtering unit and store at RT.
- 10:9:1 phenol–chloroform–isoamyl alcohol mix pH 8.2.
- Ice-cold 100% ethanol (keep at −20°C).
- 50 μg/mL DNAse-free RNAse A (Euromedex, Souffelweyersheim, France).
- 1 × TAE buffer (40 m*M* Tris–acetate, 1 m*M* EDTA in water).
- Agarose.

2.2.2 Equipment

- Precellys™: device (Precellys™ 24 Tissue homogenizer, Bertin Technologies, Montigny le Bretonneux, France) and corresponding tubes (soil grinding lysing kit SK38). Other devices used to mechanically lyse cells could also be used.
- Refrigerated tabletop centrifuge with 50 mL tube rotor (10,000 × *g* at 4°C).
- Refrigerated tabletop centrifuge with 1.5 mL and 2 mL tube rotor (16,000 × *g* at 4°C).
- 16°C water bath.
- 37°C dry heater or incubator.
- LoBind 1.5 mL Eppendorf tubes (Eppendorf, Hamburg, Germany).
- Agarose gel electrophoresis tank and generator.

2.2.3 Procedure

Mechanical cell disruption

- Thaw pellet on ice.
- Suspend pellet in 4 mL TE buffer supplemented with protease inhibitors and transfer to four Precellys™ SK38 lysing tubes.
- Fill the tubes with TE buffer supplemented with protease inhibitors if necessary.
- Run the Precellys™ with following parameters: V6700, 20 s on–30 s off ×3.
- Do a total of three such Precellys™ runs. Wait 5 min between each run. Put sample on ice to cool them down between runs to prevent heat degradation. Some Precellys™ machines have an automatic cooling using dry ice allowing to perform the three runs without waiting time between them.
- Transfer solution to a 5 mL Eppendorf tube (see Note 4).
- Add 200 μL TE buffer to Precellys™ tube to wash beads then add it to the lysate.
- Complete the recovered lysate to 4.5 mL with TE buffer supplemented with protease inhibitors.
- Add 225 μL 10% SDS (0.5% final) to solubilize proteins. Mix by inverting and incubate 20 min at 65°C, then put on ice.

DNA digestion

- Prepare four digestion mix in 5 mL Eppendorf tubes at RT:
  - 3 mL water.
  - 500 μL 10 × NEB1 digestion buffer.
  - 500 μL 10% Triton.
- Transfer 1 mL of lysate to each tube and mix by inverting.
- Put 250 μL of digestion reaction from each tube in a tube labeled ND (control for nondigested DNA).
- Add 1000 U restriction enzyme to each tube (MluCI for two tubes and HpaII for the two others) (see Note 3).
- Incubate 3 h at 37°C with moderate shaking.
- Take 500 μL of the digestion reaction from each tube and put them in a tube labeled D-MluCI and D-HpaII for each respective enzyme (control for digested DNA). The main tubes will be labeled 3C-MluCI and 3C-HpaII throughout the protocol below.

Free DNA elimination (only for 3C-MluCI and 3C-HpaII)

- Centrifuge the reaction for 20 min at 16,000 × *g* at 4°C.
- Take out supernatant to only keep heavy DNA–protein cross-linked complex.
- Suspend each pellet in 500 μL cold water (see Note 5).

Ligation (only for 3C-MluCI and 3C-HpaII)

- In two 50 mL tubes prepare two ligation mixes with:
  - 13 mL cold water.
  - 1.6 mL 10 × ligation buffer.
  - 160 μL 100 × BSA.
- Transfer solutions of HpaII restriction reaction to one tube containing a ligation mix and solutions of MluCI restriction reaction to the other tube containing a ligation mix.
- Add 160 μL ATP (1 m*M* final) and 250 U T4 DNA ligase to each ligation reaction.
- Mix by inverting and incubate 4 h in a 16°C water bath.

Protein denaturation

- To the control tubes (D-MluCI, D-HpaII, and ND) add:
  - 20 μL 0.5 *M* EDTA.
  - 20 μL 10% SDS.
  - 10 μL proteinase K (20 mg/mL).
- To the ligation tubes (3C-MluCI and 3C-HpaII) add:
  - 200 μL 0.5 *M* EDTA.
  - 200 μL 10% SDS.
  - 100 μL proteinase K (20 mg/mL).
- Incubate overnight at 65°C.

First DNA precipitation (only for 3C-MluCI and 3C-HpaII)

- Cool tubes on ice.
- Precipitate DNA by adding 1.6 mL 3 *M* sodium acetate (0.3 *M* final) to each tube.
- Transfer ligation mixes to new 50 mL conical tubes as the old ones are likely to break in the centrifuge due to the overnight storage at 65°C.
- Mix well by inverting four to five times.
- Add 16 mL isopropanol to each tube.
- Mix well by inverting.
- Incubate tube at least 1 h at −80°C. Ideally, isopropanol should freeze with formation of white globules within the tube (see Note 6).
- Centrifuge the tube 20 min at 4°C, 10,000 × *g*.
- Carefully remove supernatant and add 900 μL 1 × TE buffer.
- Incubate 15 min at 37°C with shaking to suspend pellet.

Phenol–chloroform purification and second DNA precipitation

- Transfer DNA from 3C tubes in 2 mL tubes and add same volume (900 μL) phenol–chloroform–isoamyl alcohol (10:9:1) mix to the tubes. Add 1 mL phenol–chloroform–isoamyl alcohol to control tubes.
- Vortex at top speed for 30 s.

- Centrifuge 5 min at 16,000 × *g*.
- Transfer 400 μL of the top phase of each condition to two new tubes (for control tubes as well).
- Add 40 μL sodium acetate 3 *M* to each tube, mix by inverting, and then add 1.1 mL of ice-cold 100% ethanol.
- Mix well by inverting and incubate at least 30 min at −80°C.
- Centrifuge 20 min at 4°C, 16,000 × *g*.
- Carefully remove each supernatant.
- Dry samples and controls by incubating the open tubes for 10 min at 37°C on a dry block heater.
- Suspend DNA in 60 μL Tris–HCl pH 7.5 with RNAse A (0.8 mg/mL final). Incubate 30 min at 37°C.
- Pool samples from the same conditions (ND controls, D controls, 3C-HpaII, and 3C-MluCI) together in a LoBind tube (one for each condition).

Assessment of the quality of the library

- On a 1% TAE agarose gel, load 9 μL of 3C-MluCI and 3C-HpaII and 15 μL of each of the controls and run to check the quality of the meta3C library preparation (Fig. 2).

Preparation for sequencing

The meta3C library has to be sheared and processed for paired-end sequencing (see Note 7) using standard manufacturer' protocols. Around 100 million paired-end reads 2 × 75 bp can be enough for a single gut microbiota sample. If the sample corresponds to a time point over a time course, the read number can be decreased (see Note 8) as the data from different time points can be combined for the original step of the analysis (assembly and binning of contigs).

2.3 Notes

1. Use filter tips at every step to prevent contamination.
2. More material can be processed in a single experiment by proportionally increasing the volumes used in the protocol. It is also possible to decrease the amount of material used when dealing with small samples such as mice feces. However, we do not recommend going below 50 mg material per library as it is harder to work with less material. Note that 250 mg of material should be sufficient to generate two to three libraries. Consequently, we recommend storing pellets containing 250 mg of material.
3. We advise generating two libraries per sample using two restriction enzymes covering different GC contents. The restriction enzymes

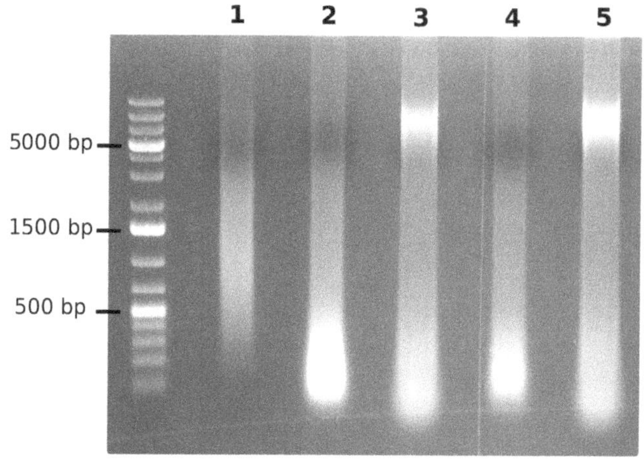

Fig. 2 Estimation on a gel of the quality of a meta3C library. (1) Undigested DNA, (2) D-MluCI, (3) 3C-MluCI, (4) D-HpaII, and (5) 3C-HpaII. Representative DNA fragment pattern expected when checking for meta3C library quality on a 1% agarose gel. After digestion with restriction enzyme, average DNA length is expected to decrease (compare lane 1 with lanes 2 and lane 4). After ligation, the DNA molecules are expected to display a higher molecular weight and the pattern therefore should reflect this change (compare lanes 2 and 4 with lanes 3 and 5, respectively).

mentioned in this protocol are routinely and successfully used in our lab, but other restriction enzymes could also be envisioned.

4. Use P1000 tips at this step as P200 tips can be clogged with beads. Some of the smallest beads will probably be carried along with the liquid. These inert beads will be discarded later during the phenol–chloroform purification step, and should not raise any concern.
5. At this step, it is possible to fill DNA restriction fragment extremities in order to integrate biotinylated bases and therefore generate conventional Hi-C libraries. This extra step will be accompanied by a pull-down of the biotinylated sequences at the later steps of the library generation, just before PCR amplification (Lieberman-Aiden et al., 2009). Other adaptations of the protocol can be envisioned as well. For each extra step, the cost/benefit will have to be carefully assessed from the final sequencing output.
6. After 1 h at −80°C, the solution should now be frozen, with a large white precipitate. If it is not the case, incubation at −80°C should be continued for another 30 min.
7. It is not necessary to generate a shotgun library of sample concomitantly to the meta3C library. Indeed, only a subset of the meta3C library reads

(10%–15%) corresponds to pairs of distant loci along the chromosomes present in the mix. The remaining reads (85%–90%) correspond to shotgun reads of nondigested or nonligated DNA segments. This proportion of shotgun reads is sufficient to generate a single read de novo assembly, as discussed in Marbouty et al. (2014). The 10%–15% pairs of reads corresponding to medium/long distant contacts will then conveniently link the contigs, resulting into a meta3C network where contigs will be nodes.

8. If several samples of the same environment/ecosystems are being studied, one may want to pool the entire reads generated from the different libraries. Some species are indeed expected to be repeatedly found in these sample, and the information recovered regarding their genomes could be pooled to increase their read coverage. This will maximize the quality of the de novo assembly step and allow the generation of a reference assembly. Once binning of contigs is achieved in this assembly, the individual data points can be reprocessed taking into account this reference assembly Therefore, the number of reads for each data point does not necessarily have to be as important as if they were single libraries.

REFERENCES

Alneberg, J., Bjarnason, B. S., de Bruijn, I., Schirmer, M., Quick, J., Ijaz, U. Z., et al. (2014). Binning metagenomic contigs by coverage and composition. *Nature Methods*, *11*(11), 1144–1146. https://doi.org/10.1038/nmeth.3103.

Ardui, S., Ameur, A., Vermeesch, J. R., & Hestand, M. S. (2018). Single molecule real-time (SMRT) sequencing comes of age: Applications and utilities for medical diagnostics. *Nucleic Acids Research*, *46*(5), 2159–2168. https://doi.org/10.1093/nar/gky066.

Beitel, C. W., Froenicke, L., Lang, J. M., Korf, I. F., Michelmore, R. W., Eisen, J. A., et al. (2014). Strain- and plasmid-level deconvolution of a synthetic metagenome by sequencing proximity ligation products. *PeerJ*, *2*, e415. https://doi.org/10.7717/peerj.415.

Burton, J. N., Liachko, I., Dunham, M. J., & Shendure, J. (2014). Species-level deconvolution of metagenome assemblies with Hi-C-based contact probability maps. *G3 (Bethesda)*, *4*(7), 1339–1346. https://doi.org/10.1534/g3.114.011825.

Dekker, J., Rippe, K., Dekker, M., & Kleckner, N. (2002). Capturing chromosome conformation. *Science*, *295*(5558), 1306–1311. https://doi.org/10.1126/science.1067799.

Dick, G. J., Andersson, A. F., Baker, B. J., Simmons, S. L., Thomas, B. C., Yelton, A. P., et al. (2009). Community-wide analysis of microbial genome sequence signatures. *Genome Biology*, *10*, R85. https://doi.org/10.1186/gb-2009-10-8-r85.

Eid, J., Fehr, A., Gray, J., Luong, K., Lyle, J., Otto, G., et al. (2009). Real-time DNA sequencing from single polymerase molecules. *Science*, *323*(5910), 133–138. https://doi.org/10.1126/science.1162986.

Flot, J.-F., Marie-Nelly, H., & Koszul, R. (2015). Contact genomics: Scaffolding and phasing (meta)genomes using chromosome 3D physical signatures. *FEBS Letters*, *589*(20 Pt. A), 2966–2974. https://doi.org/10.1016/j.febslet.2015.04.034.

Jain, M., Koren, S., Miga, K. H., Quick, J., Rand, A. C., Sasani, T. A., et al. (2018). Nanopore sequencing and assembly of a human genome with ultra-long reads. *Nature Biotechnology*, *36*(4), 338–345. https://doi.org/10.1038/nbt.4060.

Kang, D. D., Froula, J., Egan, R., & Wang, Z. (2015). MetaBAT, an efficient tool for accurately reconstructing single genomes from complex microbial communities. *PeerJ*, *3*, e1165. https://doi.org/10.7717/peerj.1165.

Lieberman-Aiden, E., van Berkum, N. L., Williams, L., Imakaev, M., Ragoczy, T., Telling, A., et al. (2009). Comprehensive mapping of long-range interactions reveals folding principles of the human genome. *Science*, *326*(5950), 289–293. https://doi.org/10.1126/science.1181369.

Lim, M. Y., Song, E.-J., Kim, S. H., Lee, J., & Nam, Y.-D. (2018). Comparison of DNA extraction methods for human gut microbial community profiling. *Systematic and Applied Microbiology*, *41*(2), 151–157. https://doi.org/10.1016/j.syapm.2017.11.008.

Mahmoud, M., Zywicki, M., Twardowski, T., & Karlowski, W. M. (2017). Efficiency of PacBio long read correction by 2nd generation Illumina sequencing. *Genomics*, *41*(2), 151–157. https://doi.org/10.1016/j.ygeno.2017.12.011.

Marbouty, M., Baudry, L., Cournac, A., & Koszul, R. (2017). Scaffolding bacterial genomes and probing host-virus interactions in gut microbiome by proximity ligation (chromosome capture) assay. *Science Advances*, *3*(2), e1602105. https://doi.org/10.1126/sciadv.1602105.

Marbouty, M., Cournac, A., Flot, J.-F., Marie-Nelly, H., Mozziconacci, J., & Koszul, R. (2014). Metagenomic chromosome conformation capture (meta3C) unveils the diversity of chromosome organization in microorganisms. *eLife*, *3*, e03318. https://doi.org/10.7554/eLife.03318.

Marie-Nelly, H., Marbouty, M., Cournac, A., Flot, J.-F., Liti, G., Parodi, D. P., et al. (2014). High-quality genome (re)assembly using chromosomal contact data. *Nature Communications*, *5*, 5695. https://doi.org/10.1038/ncomms6695.

Mikheyev, A. S., & Tin, M. M. Y. (2014). A first look at the Oxford Nanopore MinION sequencer. *Molecular Ecology Resources*, *14*(6), 1097–1102. https://doi.org/10.1111/1755-0998.12324.

CHAPTER NINE

Genome-Wide Mapping of Yeast Retrotransposon Integration Target Sites

Anastasia Barkova, Amna Asif-Laidin, Pascale Lesage[1]
INSERM U944, CNRS UMR 7212, Université Paris Diderot, Sorbonne Paris Cité, Institut Universitaire d'Hématologie, Hôpital St. Louis, Paris Cedex 10, France
[1]Corresponding author: e-mail address: pascale.lesage@inserm.fr

Contents

Methods in Enzymology, Volume 612
ISSN 0076-6879
https://doi.org/10.1016/bs.mie.2018.08.002

Abstract

Transposable elements (TEs) are present in virtually all organisms. TE integration into genomes contributes to their structure and evolution, but can also be harmful in some cases. Deciphering where and how TE integration is targeted is fundamental to understand their intricate relationship with their host and explore the outcome of TE mobility on genome evolution and cell fitness. In general, TEs display integration site preference, which differs between elements. High-throughput mapping of de novo insertions by deep sequencing has recently allowed identifying genome-wide integration preferences of several TEs. These studies have provided invaluable clues to address the molecular determinants of integration site preference. Here, we provide a step-by-step methodology to generate massive de novo insertion events and prepare a library of genomic DNA for next-generation sequencing. We also describe a primary bioinformatic procedure to map these insertions in the genome. The whole procedure comes from our recent work on the integration of Ty1 in *Saccharomyces cerevisiae*, but could be easily adapted to the study of other TEs.

1. INTRODUCTION

Transposable elements (TEs) are mobile DNA sequences that play a fundamental role in shaping genomes. TEs can be deleterious in the short-term as their integration into the host genome can inactivate or deregulate gene expression. The homologous recombination between distant copies of TEs can also induce large chromosomal rearrangements (Levin & Moran, 2011; Sultana, Zamborlini, Cristofari, & Lesage, 2017). However, TEs have also been shown to contribute to genetic innovation (Chuong, Elde, & Feschotte, 2016). Hence, the cell must protect its genome from TE activity, while enabling their mobility when genome diversity can be beneficial to its fate. One way to achieve this balance is to target TE insertions into regions where these elements are harmless. The sequencing of many genomes thus revealed that almost each TE family accumulates in specific regions of the genome (Sultana et al., 2017). However, their nonrandom distribution in the host genome at steady-state could be misleading since only neutral or beneficial insertions would be maintained in a long-term evolution process. It is therefore important to locate de novo insertion events to determine whether a given TE really displays integration site preferences and to analyze the underlying molecular mechanism.

The model yeasts *Schizosaccharomyces pombe* and *Saccharomyces cerevisiae* contain several families of LTR-retrotransposons (Tf1 and Tf2 in *S. pombe*, Ty1 to Ty5 in *S. cerevisiae*). These TEs replicate by reverse transcription of their

RNA genome into a cDNA copy that is stably integrated into the host-cell genome by their own integrase. During evolution, mechanisms have emerged to avoid disruption of yeast functions by targeting integration into noncoding regions of the genome (Levin & Moran, 2011; Sultana et al., 2017). Our knowledge of the molecular mechanisms underlying these preferences of integration comes from many studies performed in the earliest days of yeast genetics as well as recent work employing genome deep-sequencing technologies. At first, detection and mapping of de novo yeast Tf or Ty insertion events were based on classical molecular genetic approaches. In particular, functional elements containing a retrotransposition marker and expressed from strong inducible promoters were engineered on plasmids, to recover large number of insertion events. Then, upon digestion of genomic DNA, the junction between the retrotransposon and genomic DNA sequences could be visualized on Southern-blots or Sanger-sequenced after an inverse PCR (Behrens, Hayles, & Nurse, 2000; Chalker & Sandmeyer, 1990; Gai & Voytas, 1998; Ji et al., 1993; Zou, Kim, & Voytas, 1996; Zou, Ke, Kim, & Voytas, 1996). Although these approaches provided important insights into the sites targeted by yeast LTR-retrotransposons, they were laborious and the low number of recovered events was generally not sufficient to obtain a complete panorama of all the targets in the genome. More recently, improved methodologies based on ligation-mediated PCR or inverse PCR amplification allowed specific enrichment of DNA fragments corresponding to de novo insertions from the bulk of host genomic DNA. These approaches, combined with next-generation sequencing (NGS) using 454 Roche or Illumina technologies, have offered the possibility to characterize thousands to millions of integration sites of retroviruses and TEs, including Tf1 in *S. pombe* and Ty1, Ty3, and Ty5 in *S. cerevisiae* (Baller, Gao, Stamenova, Curcio, & Voytas, 2012; Baller, Gao, & Voytas, 2011; Bridier-Nahmias et al., 2015; Chatterjee et al., 2014; Guo & Levin, 2010; Jacobs et al., 2015; Qi et al., 2012). With all these studies, it has thus been possible to draw a precise profile of the integration preferences of yeast retrotransposons and the molecular mechanisms involved.

In *S. pombe*, Tf1 preferentially integrates into nucleosome-free promoters of Pol II-transcribed genes with an enrichment of stress-responsive genes (Chatterjee et al., 2014; Guo & Levin, 2010). The DNA-binding factor Sap1 and its replication fork barrier activity collaborate to determine Tf1 target site selection (Hickey et al., 2015; Jacobs et al., 2015). In *S. cerevisiae*, Ty5 integrates into nucleosome-free regions at subtelomeres and silent mating-type loci (HM) through an interaction between its integrase and the Sir4 heterochromatin protein (Baller et al., 2011; Xie et al., 2001). In contrast, Ty3 targets the transcription start site of RNA polymerase III (Pol III)-transcribed genes through an interaction between its integrase and the Pol III transcription factors TFIIB and TFIIIC (Qi et al., 2012; Qi & Sandmeyer, 2012). As for Ty1, its integration occurs in a narrow 1-kb window upstream of Pol III-transcribed genes (Bachmann, Eby, & Boeke, 2004; Devine & Boeke, 1996) and targets nucleosomal DNA, near the H2A/H2B interface (Baller et al., 2012; Mularoni et al., 2012). The primary determinant of Ty1 targeting to these sites remained elusive until we identified a physical interaction between Ty1 integrase and the AC40 subunit of Pol III (Bridier-Nahmias et al., 2015). High-throughput sequencing (HTS) of de novo Ty1 insertion events was not only essential to highlight the fundamental role of AC40 in Ty1 targeting to regions upstream of Pol III-transcribed genes but was also necessary to identify chromosome ends as new preferred targets of integration in the absence of AC40/integrase interaction (Bridier-Nahmias et al., 2015). This suggests that in addition to AC40, other factor(s) likely contribute(s) to Ty1 integration site selection.

The goal of this chapter is to provide a turnkey strategy that can be used by anyone aiming to address the role of additional factors in integration site selection in yeasts. For this purpose, we have taken into account the methodologies to sequence genome-wide de novo yeast retrotransposon insertion sites, recently published by several labs including ours, to set up our own protocol for Ty1 studies (Table 1). The proposed methodology combines ligation-mediated PCR and NGS using the Illumina technology.

Below, we describe step-by-step how to recover sufficient independent de novo Ty1 insertion events, prepare high-quality genomic DNA for library construction and sequencing, and finally align reads against the yeast genome to map de novo Ty1 insertions. An overview of the whole procedure is shown Fig. 1. We believe that this strategy could be easily adapted to address the molecular basis of other retrotransposon integration targeting mechanisms with minor modifications.

2. MATERIALS

2.1 Plasmid

Endogenous Ty1 copies have a low integration rate of approximately 10^{-5} to 10^{-7} per element per generation (Curcio, Lutz, & Lesage, 2014). In order to recover a large number of de novo integration events into the cell population, it is necessary to increase Ty1 integration frequency. To this end,

Table 1 Methodologies to Sequence Genome-Wide De Novo Yeast Retrotransposon Insertion Sites

Element	Authors	Library Preparation	Sequencer and Number of Reads
Ty1	Baller et al. (2012)	• p*GAL1*-Ty1*his3AI*-SCUF 2μ *URA3* plasmid with 6 nt unique tag in the 5′ LTR • Ura^-His^+ selected colonies • Digestion of genomic DNA with *Aci*I or *Taq*I • Ligation of adapters by PCR	• 454 Life Sciences sequencer • For BY4741: approx. 18,000 reads out of which approx. 2000 were unambiguous • For YPH499: approx. 90,000 reads out of which approx. 6000 were unambiguous
	Mularoni et al. (2012)	• p*GAL1*-Ty1-H3 2μ optimized *URA3* plasmid with an "ssb" sequence of 25 bp in 3′ LTR • Ura^- selected colonies • Digestion of genomic DNA with *Mse*I and *Bgl*II • Ligation of adapters by PCR	• Illumina GAII sequencer • Total reads: 7,990,112; nonredundant insertion sites, 38,733
	Bridier-Nahmias et al. (2015)	• p*GAL1*-Ty1*his3AI*-SCUF 2μ *LEU2* plasmid with 6 nt unique tag in the 5′ LTR • His^+ selected colonies • Digestion of genomic DNA with *Mse*I • Ligation of adapters by PCR	• Illumina Miseq sequencer • Unambiguous sequences: 90,289 and 225,856
Ty3	Qi et al. (2012)	• p*GAL1*-Ty3*his3*-ppt low-copy *URA3* plasmid with a HIS3 gene flanked by unique 60 bp tags • Ura^- His^+ selected colonies • Ty3 insertions amplified by inverse PCR (iPCR)	• Illumina GAIIx sequencer • 1,243,106 reads for two experiments, 370,000 aligned, but not uniquely

Continued

Table 1 Methodologies to Sequence Genome-Wide De Novo Yeast Retrotransposon Insertion Sites—cont'd

Element	Authors	Library Preparation	Sequencer and Number of Reads
Ty5	Baller et al. (2011)	• p*GAL1*-Ty5*his3AI* centromeric *URA3* plasmid • His^+ selected colonies • Digestion of genomic DNA with *Aci*I/*Taq*I and *Msp*I/*Hin*pII • Ligation of adapters by PCR	• 454 Life Sciences sequencer • Total of ~337,000 sequence reads, 64,000 were unambiguous (40% of 160,000 high quality)
Tf1	Guo and Levin (2010)	• p*Nmt1*-Tf1*neo* *URA3* plasmid with a unique tag in U5 • Ura^-G418^+ selected colonies • Digestion of genomic DNA with *Mse*I or HpyCH4IV • Ligation of adapters by PCR	• 454 Life Sciences sequencer • 599,760 reads out of which 73,152 mapped to unique positions
	Chatterjee et al. (2014)	• p*Nmt1*-Tf1*neo* *URA3* plasmid with a random 8 nt tag in U5 of 5′ LTR • Ura^-G418^+ selected colonies • Digestion of genomic DNA with *Mse*I • Ligation of adapters by PCR	• Illumina GAIIx sequencer • 1,061,068 independent integration events in WT and mutant
	Jacobs et al. (2015)	• p*Nmt1*-Tf1*neo* *URA3* plasmid • Ura^- $G418^+$ selected colonies • Digestion of genomic DNA with *Mse*I • Ligation of adapters by PCR	• Illumina Miseq sequencer

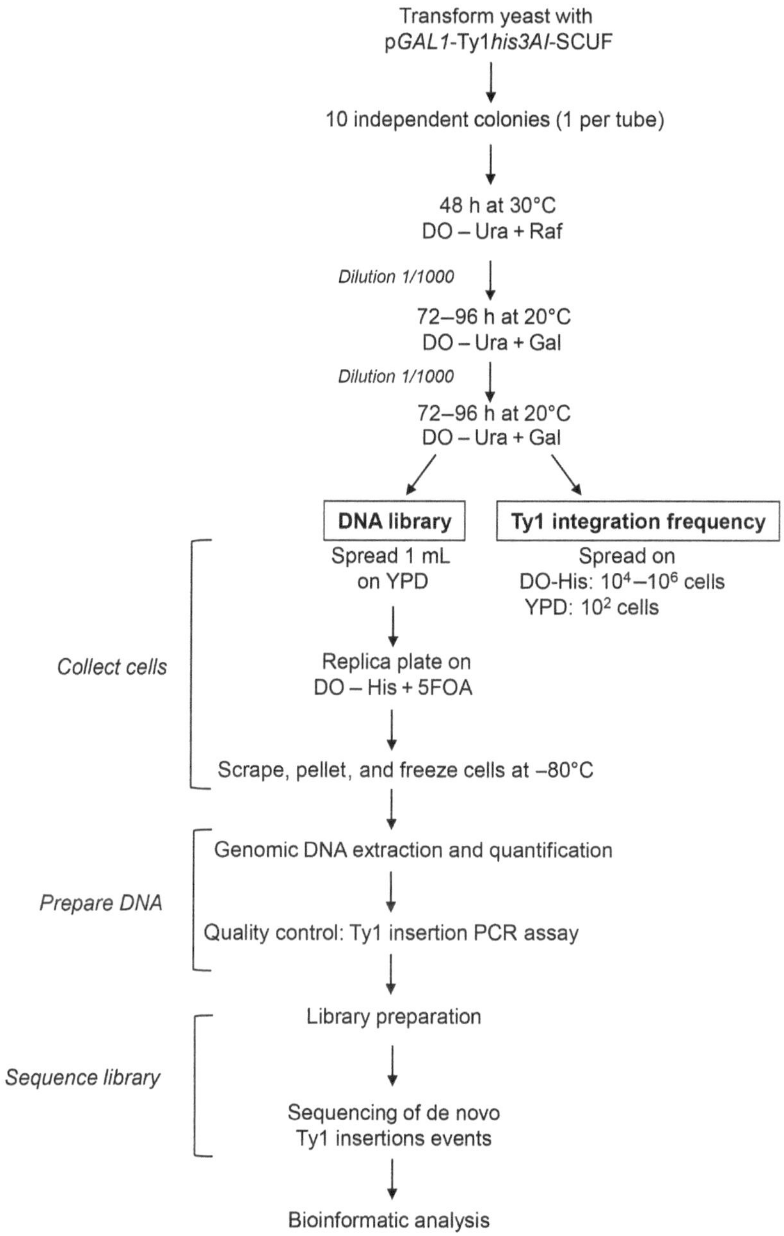

Fig. 1 Overview of the whole experimental procedure: From yeast transformation to sequencing data analysis.

we use a high-copy number (2μ) plasmid, p*GAL1*-Ty1*his3AI*-SCUF, which encodes the functional Ty1-H3 element from the inducible strong *GAL1* promoter (Boeke, Garfinkel, Styles, & Fink, 1985). The Ty1 sequence has been modified to contain two markers important for the recovery of de novo Ty1 insertions (Fig. 2). First, the 3′ untranslated region contains the retrotransposition indicator gene (RIG) *his3AI*, which consists in the *HIS3* gene in antisense orientation relative to Ty1 transcription and interrupted by an artificial intron in the sense orientation (Curcio & Garfinkel, 1991). This construct allows functional protein synthesis from

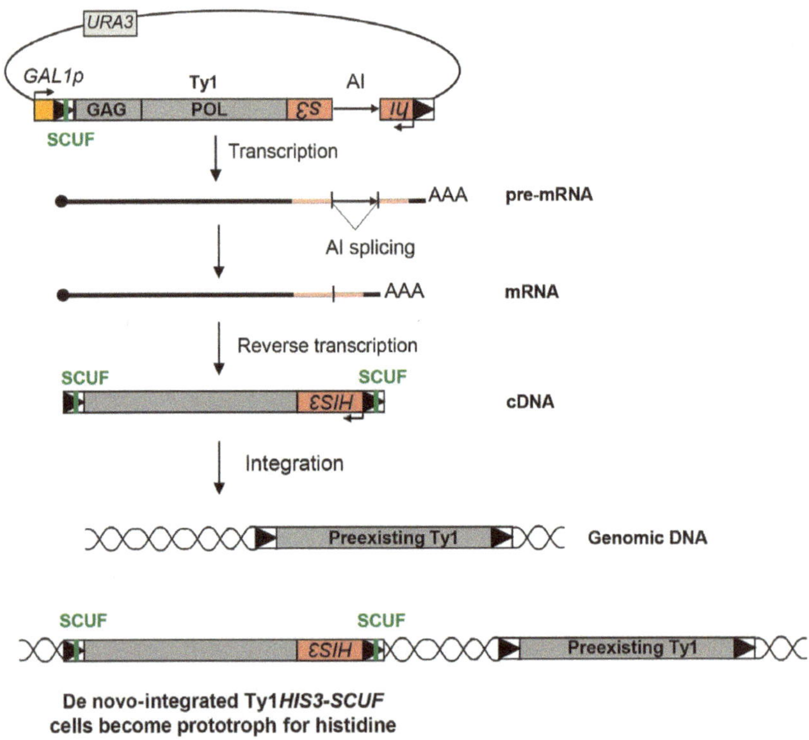

Fig. 2 p*GAL1*-Ty1*his3AI*-SCUF structure and retrotransposition assay. A 2μ plasmid carries a functional Ty1 element expressed from the galactose inducible *GAL1* promoter (*GAL1*p, *yellow box*) and the *URA3* selection marker (*light gray box*). Ty1 contains in its 3′ untranslated region a *HIS3* gene in an antisense orientation interrupted by an intron, which is in a spliceable orientation in the Ty1 transcript (*his3AI, pink box* interrupted by a *black arrow*). Transcription from the 5′ LTR, splicing of AI and reverse transcription result in a Ty1 cDNA bearing a functional *HIS3* gene. Integration of Ty1*HIS3* cDNA in the yeast genome confers histidine prototrophy to His$^-$ transformants. The 5′ LTR harbors the SCUF sequence, which is copied in the 3′ LTR by reverse transcription, allowing Ty1 de novo insertion events to be distinguished from endogenous Ty1 elements.

HIS3 only upon reverse transcription and integration of Ty1*HIS3* cDNA, thus conferring histidine prototrophy to His^- cells. Therefore, cells that have undergone at least one retrotransposition event can be selected on medium lacking histidine (see Note *N-2.1*). Second, because it is important to discriminate de novo integration events from endogenous Ty1 elements, six nucleotides in the 5′ LTR of this engineered Ty1 were substituted downstream of the initiation codon of *GAG* ORF (Baller et al., 2012). This tag (referred to as SCUF) does not affect Gag amino acids sequence or the efficiency of Ty1 retrotransposition. Importantly, SCUF is copied by reverse transcription in the 3′ LTR, allowing to distinguish during the sequencing step the 3′ LTR/genomic DNA junction of de novo Ty1 insertions from the 3′ LTR/genomic DNA junctions of Ty1 elements preexisting in the genome (Fig. 2 and see Note *N-2.2*). Finally, the plasmid encoding *GAL1*-Ty1*his3AI*-SCUF construct also carries the *URA3* gene that can be counter-selected in the presence of 5-fluoroorotic-acid (5FOA) (Boeke, LaCroute, & Fink, 1984). Plasmid elimination is not essential but avoids contamination of the library of Ty1-specific amplified DNA fragments by Ty1 sequences from the plasmid (see Note *N-2.3*). Galactose induction of *GAL1*-Ty1*his3AI*-SCUF retrotransposition results in a frequency of 10^{-3} to 10^{-2} His^+ colony/cell at 20°C, a temperature permissive for Ty1 retrotransposition.

2.2 Yeast Strains

We recommend using the BY4741, BY4742 haploid strains, and the BY4743 diploid strain, which are close derivatives of the S288C laboratory strain, whose genome serves as reference in SGD (https://www.yeastgenome.org) (Brachmann et al., 1998). These strains are also the parental strains for the international systematic *S. cerevisiae* gene disruption project (Table 2). Using these strains will therefore allow aligning sequence reads of de novo insertion events on the reference genome and comparing insertions

Table 2 Recommended Yeast Strains

Strain Name	Genotype
BY4743	*MAT*a/α *his3Δ1/his3Δ1 leu2Δ0/leu2Δ0 LYS2/lys2Δ0 met15Δ0/MET15 ura3Δ0/ura3Δ0*
BY4741	*MAT*a *his3Δ1 leu2Δ0 met15Δ0 ura3Δ0*
BY4742	*MAT*α *his3Δ1 leu2Δ0 lys2Δ0 ura3Δ0*

between WT and mutant strains deleted for your gene(s) of interest. Furthermore, the strains lack the *HIS3* and *URA3* genes, which are auxotrophy markers important for plasmid and de novo integration events selection.

Although strain ploidy does not significantly influence targeting pattern in wild-type cells (Baller et al., 2012; Mularoni et al., 2012), it might be relevant to compare Ty1 insertion patterns in diploid and haploid cells when working with yeast mutants. If some mutants increase the number of Ty1 insertion in open reading frames, insertions in essential genes will not be recovered in haploid cells, which may result in misleading conclusions; such insertions will likely be recovered in diploid cells.

An important issue to avoid when studying the impact of Ty1 mutations on integration site selectivity is transcomplementation of Ty1 function by endogenous elements (Curcio & Garfinkel, 1994). For this purpose, it is recommended to use *spt3Δ* or *ste12Δ* mutants, which prevent Ty1 transcription from its own promoter but not from *GAL1* promoter (Morillon, Springer, Lesage, 2000; Winston, Durbin, & Fink, 1984). These strains can be obtained from the Yeast Knock-Out collection (available via ATCC, Open Biosystems or Invitrogen).

2.3 Notes

N-2.1. Mularoni et al. used a p*GAL1*-Ty1-H3 plasmid that contains a higher performing 2μ origin to increase the plasmid copy number, but lacks a RIG. Therefore, de novo Ty1 insertions were recovered at high frequency without selection (Mularoni et al., 2012).

N-2.2. Alternative strategies have been developed to specifically mark de novo integration events of yeast retrotransposons (Table 1). Mularoni et al. put a tag in the U3 region of the 3′ LTR that was transferred to the 5′ LTR during retrotransposition of the inducible Ty1 element (Mularoni et al., 2012). For Tf1, a random sequence tag was introduced in the U5 region of the 5′ LTR, in order to record the numbers of independent integration events at single nucleotide positions in the genome of *S. pombe* (Chatterjee et al., 2014). As for SCUF, the random tag was copied in the 3′ LTR by reverse transcription of the element. For Ty3, two unique tags were introduced on both sides of the retrotransposon indicator *HIS3* gene for amplification of integration junctions between Ty3 sequences and the flanking genome by inverse PCR (Qi et al., 2012). In the case of Ty5, no tag was added, since full-length Ty5 elements are absent in the genome of *S. cerevisiae* (Baller et al., 2011).

N-2.3. It is possible to skip p*GAL1*-Ty1*his3AI*-SCUF plasmid shuffling step on 5FOA plates. In this case, however, it is necessary to eliminate sequence reads corresponding to Ty1 sequence in the plasmid from the sequencing raw data (see below).

3. RECOVERY OF INDEPENDENT INSERTION EVENTS

3.1 Equipment

- Incubator-shaker (30°C and 20°C)
- Spectrophotometer (600 nm)
- Microcentrifuge
- Sterile velvets and replicate plate for Petri dishes
- Sterile plastic spreaders

3.2 Buffers and Reagents

- Yeast culture liquid media: Yeast Synthetic Drop-out medium (DO) lacking uracil and supplemented with raffinose 2% (DO − Ura + Raf) or 2% galactose (DO − Ura + Gal)
- Agar plates: Yeast peptone agar supplemented with 2% glucose (YPD), DO agar lacking histidine and supplemented with 2% glucose (DO − His), and DO − His supplemented with 5FOA (DO − His + 5FOA). We use 50 mg of uracil and 1000 mg of 5FOA per liter of medium
- Sterile deionized water

3.3 Procedure

3.3.1 Step 1: Assess Ty1 Integration Frequency to Estimate the Total Number of Cells Necessary for Library Preparation

Under optimal conditions, p*GAL1*-Ty1*his3AI*-SCUF generates 10^{-3} to 10^{-2} His^{+} colonies/cell upon galactose induction at 20°C. However, when working in a particular mutant or growth context, it is important to establish whether these conditions modify Ty1 insertion frequency. This information is crucial for library preparation, since it will determine the number of total cells to spread on plates to recover around 70,000–100,000 His^{+} cells. Under our experimental conditions, each His^{+} cell contains at least one Ty1*HIS3* insertion, which provides a good coverage for HTS (Bridier-Nahmias et al., 2015).

1. Transform the strain(s) of interest (for example, WT and mutant strains) with p*GAL1*-Ty1*his3AI*-SCUF, select transformants on DO − Ura plates.

2. Pick four independent colonies of each yeast transformant and inoculate in 2 mL of DO − Ura + Raf medium (one colony per tube: four precultures). Grow to saturation in an incubator-shaker (180 rpm) at 30°C overnight.
3. Dilute each preculture 1000-fold in 3 mL of DO − Ura + Gal and grow cell cultures to induce retrotransposition in an incubator-shaker at 20°C for 3–4 days depending on the strain (9–10 generations) (see Note *N-3.1*).
4. Spread 10^2 cells on YPD and 10^4, 10^5, and 10^6 cells on DO − His (For WT cells, one DO − His plate with 10^5 cells is sufficient). Incubate 2–3 days at 30°C. Count colonies on plates and then determine the integration frequency by dividing the total number of histidine prototroph colonies by the total number of colonies on YPD.
5. Determine from the observed frequency the number of total cells necessary to obtain 70,000–100,000 His^+ cells.

*3.3.2 Step 2: Collect Cells Containing De Novo Ty1*HIS3 *Insertions*

1. Pick 10 independent colonies of each yeast transformant (for example, 10 of WT and 10 of mutant transformants) and inoculate in 2 mL of DO − Ura + Raf medium (one colony per tube: 10 precultures per yeast transformant). Grow to saturation the 10 cultures in an incubator-shaker (180 RPM) at 30°C overnight (see Note *N-3.2*).
2. Dilute each culture 1000-fold in 3 mL of DO − Ura + Gal and grow cell cultures to induce retrotransposition in an incubator-shaker at 20°C for 3–4 days depending on the strain (9–10 generations). Keep one culture of WT transformant grown in DO − Ura + Raf at 30°C, which genomic DNA will serve as control in step 2 of Section 4.2.3.
3. Dilute each culture one more time 1000-fold in 3 mL of DO − Ura + Gal and grow cell cultures for 3–4 days at 20°C to increase the recovery of independent de novo insertion events.
4. When cultures are saturated, spread 1 mL of (see Note *N-3.3*) of each culture on a YPD plate to allow spontaneous loss of plasmid. Incubate 24 h at 30°C. In parallel, do a frequency assay for every culture as described in Step 1 (no. 4) to further discard cultures with abnormal retrotransposition frequency (see Note *N-3.4*).
5. Replica plate the lawn of cells obtained on YPD plates on DO − His + 5FOA with sterile velvets. As the plasmid contains the *URA3* gene, 5FOA will be toxic for cells that have not lost the plasmid.

6. After 2–3 days, add 1 mL of sterile deionized water on the plates and scrape the cells (His^+ Ura^-) with a plastic spreader. Transfer the cells to 2-mL tubes. Wash the plates again with 0.5 mL of sterile deionized water and transfer the remaining cells to the same tubes.
7. Centrifuge 5 min at 6000 rpm. Wash the cells with 1.5 mL of sterile deionized water.
8. Centrifuge 5 min at 6000 rpm. Suspend the cell pellets in 1.5 mL of sterile deionized water and split the samples into two 1.5-mL tubes. Centrifuge again and remove supernatant.
9. Freeze cell pellets at −80°C until total genomic DNA extraction (see Note *N-3.5*).

3.4 Notes

N-3.1. Although the standard growth temperature for yeast is 30°C, Ty1 retrotransposition is much more efficient at 20–22°C.

N-3.2. Ten independent cultures limit the impact of insertion events occurring early in cell culture on the library diversity.

N-3.3. If Ty1 integration frequency is low in your experimental condition, it is necessary to adjust the volume of culture to spread on YPD in order to have at least 10^4 His^+ colonies per culture to scrape from DO − His plates (meaning that there are at least 10^4 integration events, independent or not). If working with a mutant that decreases Ty1 integration frequency, we recommend using larger Petri dishes (diameter, 144 mm) to spread more cells.

N-3.4. It could happen that a transformant displays a very low His^+ colony/cell frequency due to recombination between the two LTR of *GAL1*-Ty1*his3AI*-SCUF. This clone must be discarded for the library preparation.

N-3.5. Two pellets of approximately 0.5×10^4 cells are collected for each transformant. One is used for further steps and the other one is kept as a backup at −80°C until the de novo insertion libraries have been validated.

4. TOTAL GENOMIC DNA PREPARATION

To obtain total genomic DNA, we perform a phenol-chloroform extraction according to classical procedures (Adams, Gottschling, Kaiser, & Stearns, 1997) followed by fluorometric DNA quantification to calculate the exact double strand DNA (dsDNA) concentration in each sample.

Ty1 integrates mainly within a 700 base pair window upstream of tRNA genes (Baller et al., 2012; Bridier-Nahmias et al., 2015; Mularoni et al., 2012). This can be easily observed in a dedicated PCR that monitors Ty1 integration upstream of the *SUF16* tRNA gene, a known target of integration (Nyswaner, Checkley, Yi, Stephens, & Garfinkel, 2008). Resolved on agarose gel, the PCR pattern is characterized by a ladder-like banding profile of several bands corresponding to two insertions per nucleosome. This assay is useful to discard any genomic DNA from a clone that exhibits an abnormal integration profile compared to the other samples of the same transformant as it could distort your sequencing analysis (Fig. 3). This discrepancy is exceptional but could happen by spontaneous reversion of a mutant phenotype in one clone.

4.1 DNA Extraction

4.1.1 Equipment

- Fluorometer (Qubit™)
- Centrifuge
- Vortex (Disruptor Genie®, VWR)

4.1.2 Buffers and Reagents

- Phenol:Chloroform:Isoamyl Alcohol (Sigma 77617)
- Chloroform
- DNA Lysis Buffer (2% Triton X-100, 1% SDS, 100 m*M* NaCl, 10 m*M* Tris–HCl (pH 8), 1 m*M* Na_2 EDTA)
- Acid-washed glass beads 425–600 μm (Sigma G772)
- EtOH 100% and EtOH 70%
- TE + RNase (10 m*M* Tris–HCl pH 8, 1 m*M* Na_2 EDTA, RNase A 10 mg/mL)
- Qubit™ dsDNA BR assay kit (Thermo Fisher Q32853)

4.1.3 Procedure

1. Total genomic DNA is extracted from the cell pellets of the 10 independent cultures of each yeast transformant.
2. Add 300 μL glass beads, 200 μL of DNA lysis buffer, and 200 μL of phenol:chloroform:isoamyl alcohol to each cell pellet.
3. Vortex with disruptor for 5 min at 4°C (see Note *N-4.1*).
4. Centrifuge at 10,000 rpm for 10 min at 4°C.
5. Collect the aqueous phase (approx. 150 μL). Avoid the contact with white layer.

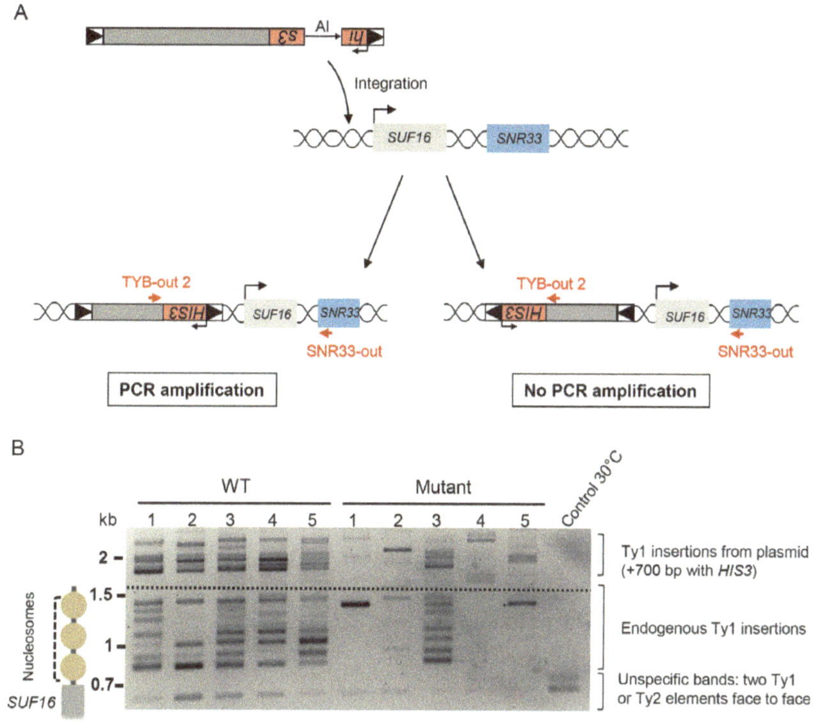

Fig. 3 Ty1 integration PCR assay at SUF16: Quality control. (A) Experimental design. PCR reactions are carried out with primers SNR33-out and TYB-out 2 (*arrowheads*). Only Ty1*HIS3* inserted in one direction are amplified by this PCR assay. (B) Example of a PCR assay resolved on agarose gel. PCR reaction was performed on extracted genomic DNA of five wild-type (WT) and five mutant yeast transformants grown in DO − Ura + Gal at 20°C and one WT yeast transformant grown in DO − Ura + Raf at 30°C (control 30°C). On the top of the gel, the bands correspond to de novo insertions of Ty1*HIS3* cDNA coming from *GAL1*-Ty1*his3AI*-SCUF retrotransposition. In the middle of the gel, the bands correspond to de novo insertions of endogenous Ty1 elements. In WT cells, the apparent periodic banding patterns observed for Ty1*HIS3* and Ty1 de novo insertion events results from two integrations per nucleosome on the first three nucleosomes upstream of *SUF16* (Baller et al., 2012; Mularoni et al., 2012). In the mutant context, the periodic pattern is altered except for mutant #3, which displays a WT profile, suggesting mutant phenotype reversion. This DNA was discarded and not used for further library construction. In all samples, one or two extra bands are detected below the minimum size expected for de novo insertion events (bottom of the gel). These bands correspond to the amplification of Ty1 and closely related Ty2 endogenous copies located face to face in the yeast genome.

6. Repeat once the extraction by adding one volume of phenol:chloroform:isoamyl alcohol. Vortex 10 s. Centrifuge at 10,000 rpm for 10 min at 4°C and collect the aqueous phase.
7. Add one volume of chloroform. Vortex 10 s. Centrifuge at 10,000 rpm for 10 min at 4°C and collect the aqueous phase.
8. Add 2.5 volume of EtOH 100% to precipitate the DNA, mix five times by inversion, and centrifuge at 10,000 rpm for 10 min at 4°C.
9. Wash the pellet with 500 µL EtOH 70%. Do not vortex to protect the pellet. Centrifuge at 10,000 rpm for 10 min at 4°C.
10. Air-dry for 5 min at 50°C.
11. Suspend the pellet in 100 µL TE + RNase. Incubate 15 min at 42°C to degrade RNAs.
12. Measure the extracted genomic DNA concentration with the fluorometer following the instructions provided by the manufacturer (see Note *N-4.2*).
13. The extracted genomic DNA can be conserved at −20°C, until the de novo insertion libraries have been validated.

4.2 Quality Control: Ty1 Integration PCR Assay at *SUF16*

4.2.1 Equipment

- Thermocycler
- Electrophoresis system

4.2.2 Buffers and Reagents

- Phusion™ High-Fidelity DNA polymerase (2 U/µL) and Green HF Buffer (Thermo Fisher F530S)
- Fresh dNTPs 10 m*M* solution (avoid thaw/freeze)
- DNase/RNase-free sterile water (Sigma W4502-1L)
- Extracted genomic DNA and PCR primers
- UltraPure™ agarose (Thermo Fisher 16500500)

4.2.3 Procedure

1. Primers used for the PCR reaction are located in Ty1 sequence upstream of *his3AI* in *GAL1*-Ty1*his3AI*-SCUF and in *SNR33*, a gene located downstream of the *SUF16* tRNA gene (Dakshinamurthy, Nyswaner, Farabaugh, & Garfinkel, 2010; Mou, Kenny, & Curcio, 2006) (Table 3). Ty1 insertion occurs with equal frequency in both orientations (Baller et al., 2012; Mularoni et al., 2012). These primers will

Table 3 Primers for PCR Assay

Primer Name	Sequence	Coordinates
TYB-out 2	5′-GTGATGACAAAACCTCTTCCG-3′	Ty1-H3: 5,500–5,200
SNR33-out	5′-TTTTAGAGTGACACCATCGTAC-3′	Chr III: 142,507–142,528

only amplify Ty1 insertions in the direct orientation upstream of *SUF16* (Fig. 3A).

2. Set up the PCR reaction for Phusion DNA Polymerase in 25 μL of total volume, for each genomic DNA of the 10 independent cultures. As control, use genomic DNA from a transformant grown in DO − Ura +Raf at 30°C. Final concentrations in the reaction mix should be: Phusion Green HF Buffer 1×, dNTPs 200 μ*M*, primers TYB-out 2 and SNR33-out 0.5 μ*M* each, Phusion 0.5 units and genomic DNA 30 ng. Complete with DNase/RNase-free water.
3. Run PCR as follows: 2 min at 98°C, 10 s at 98°C, 30 s at 61°C, 1 min at 72°C (30 cycles), 5 min at 72°C.
4. Run 1.5% agarose gel at 100 V with 12.5 μL of PCR reaction.
5. Compare the PCR pattern on gel and discard clone(s) that exhibits an abnormal profile compare to the other clones from the same yeast transformant (Fig. 3B).

4.2.4 Notes

N-4.1. Another possibility is to break cells with a MagNA Lyser™ (Roche), which provides in general higher yields of extracted genomic DNA.

N-4.2. Contrary to the commonly used spectrophotometer Nanodrop, the fluorometer will measure only dsDNA by quantifying the fluorescent molecules specifically bound to dsDNA. Importantly, we did not notice any correlation between dsDNA concentration measured by Nanodrop and fluorometer. We generally obtain a concentration of 30–100 ng of dsDNA/μL for each sample.

5. DE NOVO TY1 INSERTION DNA LIBRARY PREPARATION

Here, we describe the protocol to prepare a library of Ty1 insertion events for HTS, adapted from supplementary material of Bridier-Nahmias

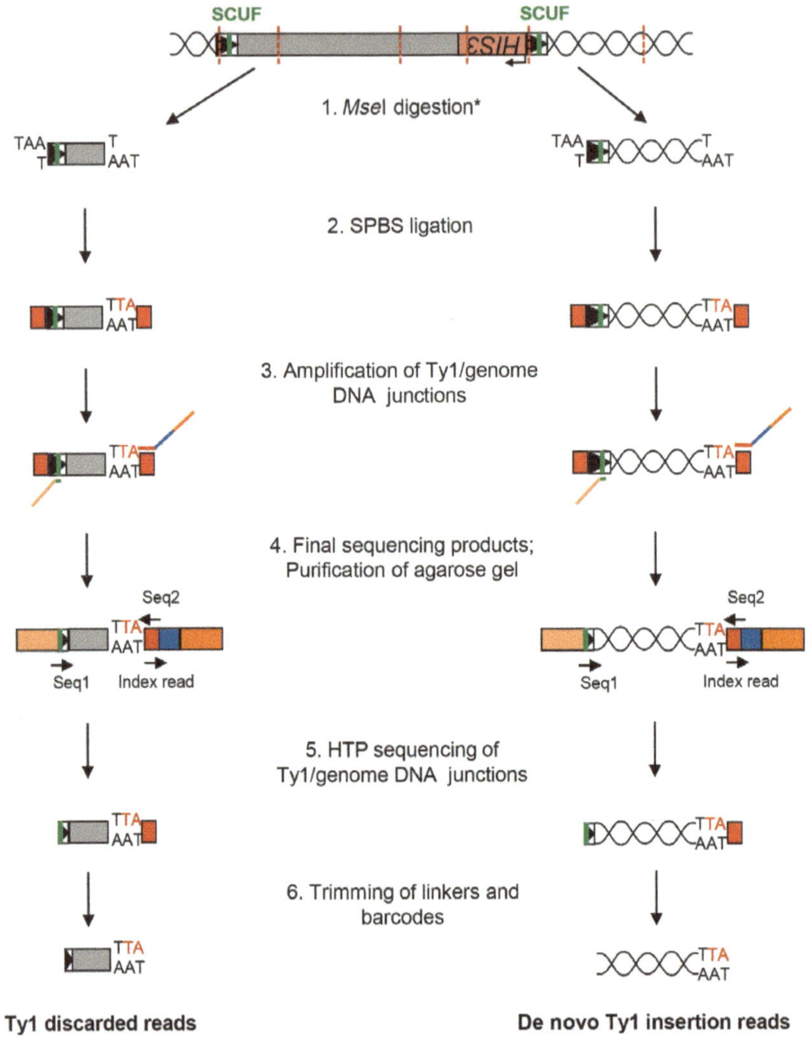

Fig. 4 Custom de novo Ty1 insertion library. After the induction of *GAL1*-Ty1*his3AI*-SCUF retrotransposition and recovery of His$^+$ cells, genomic DNA is extracted and digested with *Mse*I (T/AAT) that cuts approximately every 200 bp. The alternative is to shear DNA by sonication (indicated by a *star*). A sequence primer-binding site (SPBS, in *red*) with an *Mse*I compatible extremity (TA overhang) is ligated to *Mse*I-digested DNA. Next, Illumina adapters are added at both extremities by PCR amplification. One primer contains the 5′ Illumina adapter sequence (*light orange*) and the SCUF sequence (*green*), while the other one contains the SPBS sequence (*red*) and index tag (*blue*) to distinguish reads from different strain backgrounds (i.e., WT vs mutant) and the 3′ Illumina adapter (*orange*). After PCR amplification, DNA fragments between 300 and 700 bp are purified on agarose gel and sequenced with Seq1 and Seq2 primers with Illumina sequencer. Then, reads that passed quality control, and corresponding to 3′ LTR-SCUF, are aligned to the yeast genome.

et al. (2015) and presented in Fig. 4. Only total genomic DNA of clones that have passed the different quality controls (integration frequency and PCR assay, described in Sections 3.3 and 4.2.3) will be used for the library preparation (see Note *N-5.1*).

5.1 Equipment

- Thermocycler
- Speedvac
- UV transilluminator
- Dry bath
- Fluorometer (Qubit^{TM})

5.2 Buffers and Reagents

- Extracted genomic DNA
- *Mse*I restriction enzyme (New England Biolabs)
- Nucleospin gel and PCR clean-up kit (Macherey-Nagel)
- Sequencing Primer Binding Site (SPBS)
- NaCl 100 m*M*
- T4 DNA ligase (New England Biolabs)
- Illumina adapters
- Phusion High-Fidelity DNA polymerase (2 U/μL)
- Electrophoresis system

5.3 Procedure

1. *Digest DNA*. 500 ng of each of the 10 genomic DNA extracts from the same yeast DNA transformant are pooled together and digested in five independent enzymatic reactions with *Mse*I (10 μL NEB Buffer 2, 1 μg of genomic DNA per reaction, complete with DNase/RNase-free sterile water to 100 μL) overnight at 37°C (see Note *N-5.2*).
2. *Pool the five digestions of the same strain and purify on one PCR clean-up column* (Macherey-Nagel). Elute in 50 μL of DNase/RNase-free sterile water and determine concentration with fluorometer.
3. *Prepare SPBS*. SPBS dsDNA oligonucleotide contains a TA overhang necessary for its ligation to *Mse*I-digested genomic DNA. Anneal equal amounts (5 μL) of SPBS-primer 1 and SPBS-primer 2 (100 μM) in 100 μL NaCl 100 m*M* (Table 4). First, denature primers at 100°C for 5 min in a dry bath, then let the primer mix cool slowly by placing on the bench the dry bath block heater for 3 h. Determine dsDNA concentration with a fluorometer.

Table 4 Primers Used to Construct Ty1 De novo Insertion Library

Primer Name	Sequence	Remarks
SPBS-primer 1	5′TAACATCGAGATCGGAAGAGCACACGTCTGAACTCCAGTCA3′	Underlined: complementary SPBS sequence. At the 5′ end, the TA overhang is necessary for ligation to *Mse*I-digested genomic DNA
SPBS-primer 2	5′GTGACTGGAGTTCAGACGTGTGCTCTTCCGATCTCGATGT3′	Underlined: complementary SPBS sequence
P5-SCUF[a]	5′aatgatacggcgaccaccgagACTCGAATCGCAACAGCTTAGC3′	Sequencing adapters that binds specifically to SCUF in Ty1 and has P5 Illumina adapter. Uppercase, SCUF-binding sequence; lowercase, P5 Illumina adapter sequence; underlined, six nucleotide substitutions defining SCUF
P7-index tag-SPBS[a]	5′CAAGCAGAAGACGGCATACGAGATATCACGagatcggaagagcacacgtct3′	Sequencing adapters that binds specifically to SPBS and has P7 Illumina adapter and index tag. Uppercase, SPBS-binding sequence; lowercase, P7 Illumina sequence; underlined, Illumina index tag (index tags allow multiplex sequencing of different libraries)
Seq1	GGAATCCCAACAGCTTAGCCAACATTCACCCATTTCTCA	Primer "read 1" for sequencing, complementary to 3′ LTR of Ty1. Underlined, SCUF sequence
Seq2	AGATCGGAAGAGCACACGTCTGAACTCCAGTCAC	Primer "read 2" for sequencing. Complementary to SPBS
Index read	GTGACTGGAGTTCAGACGTGTGCTCTTCCGATCT	Primer for sequencing to read the index tag. Complementary to SPBS

[a]Illumina adapters P5 and P7 are necessary for binding to the Illumina flow cell and are specific to one flow cell and are constantly updated, therefore our primer sequences are only given as information.

4. *PCR-mediated ligation of Illumina-sequencing adapters and amplification of Ty1/ genomic DNA junctions.* Use 650 ng of digestion product and 600 ng (3,3 μL) (see Note *N-5.3*) of annealed SPBS in 200 μL reaction (10 μL NEB T4 DNA ligase, 20 μL Ligase Buffer 10×, complete with DNase/ RNase-free sterile water). Divide the reaction mixture in 10 PCR tubes (20 μL per tube) and incubate at 16°C overnight in a thermocycler.
5. *Pool and purify the ligation* mixture on a PCR clean-up column (Macherey-Nagel) and elute in 30 μL. Use one column for the 10 tubes containing genomic DNA of the same yeast transformant. Determine dsDNA concentrations with a fluorometer.
6. *PCR amplification of Ty1 insertion sites and binding of sequencing adapters.* For each yeast transformant genomic DNA, perform 10 PCR reactions in parallel with Phusion DNA polymerase to amplify 50 ng of purified ligation product in a final volume of 50 μL per tube. Final concentrations should be: Phusion HF buffer 1×, dNTPs 200 μ*M*, primers (P5-SCUF and P7-index tag-SPBS) 0.5 μ*M* each, Phusion 0.5 units and complete with DNase/RNase-free water to 50 μL (Table 4). Run PCR as follows: 30 s 98°C, 10 s 98°C, 30 s 68°C, 30 s 72°C (14–18 cycles), 5 min 72°C (see Note *N-5.4*).
7. *Concentrate PCR-amplified DNA.* For each yeast transformant, pool the 10 PCR reactions in one 1.5-mL tube and use a Speedvac at 30°C for approximately 1 h 30 min. Concentrate the DNA to 50 μL final volume.
8. *Gel purification of the DNA-sequencing library.* Load the 50 μL of DNA on a 1.5% agarose gel, migrate at 100 V. Stain gel with EtBr. Cut out the large smear of DNA fragments present between 300 and 700 bp and purify DNA fragments on gel clean-up column (Macherey-Nagel).
9. The prepared library is ready to be sequenced. Measure DNA concentration with fluorometer. The amount of DNA required for sequencing will depend on the Illumina sequencer.

5.4 Notes

N-5.1. Some companies specialized in NGS, such as Fasteris (Switzerland), have now developed strategies for de novo transposon insertion sequencing. These companies have reliable and reproducible protocols for the sequencing library preparation.

N-5.2. When the genomic DNA is cut with a single restriction enzyme, all the sequenced fragments start with the same first nucleotides. This will not generate enough diversity in the DNA extremities leading to sequencing errors. To overcome this problem, one possibility is to digest DNA with a cocktail of restriction enzymes (Baller et al., 2011) or to shear it by sonication, a strategy that we have privileged.

N-5.3. SPBS-primer 1 and SPBS-primer 2 have a molecular weight of 11,073 and 10,430 g/mol, respectively.

N-5.4. It is recommended to perform as few PCR cycles as possible for each reaction in order to avoid PCR amplification errors. For more information on PCR biases, see Van Dijk, Jaszczyszyn, and Thermes (2014).

6. DATA ANNOTATION AND ANALYSIS

The following section describes how to filter Illumina sequencing raw data to obtain reads specific for de novo Ty1 insertion events and map them to reference genome. All the tools and parameters proposed are those we use in the lab and are given as example. Everyone can choose to run the tools they prefer or set different options.

6.1 Tools

FastQC: http://bioinformatics.babraham.ac.uk/projects/fastqc
Cutadapt: http://cutadapt.readthedocs.io/en/stable/index.html
BWA: http://bio-bwa.sourceforge.net
ggplot2: http://ggplot2.tidyverse.org

6.2 Procedure

1. Reads are recovered into Fastq format from Illumina sequencer.
2. Make a quality control of the reads with FastQC.
3. If FastQC reports low-quality reads, trim sequence ends with Cutadapt, since sequencing errors come generally from sequences extremities.
4. Remove the sequencing adapters sequences with Cutadapt.
5. Run again FastQC to compare the processed sequence data file with the raw one. Make sure that you reached a good quality level (see Note *N-6.1*).
6. Remove reads starting with 5′ LTR-SCUF and ending in Ty1 *GAG* sequences (see Note *N-6.2*).
7. Map the reads with the aligner of your choice (Bowtie 2, BWA, …). We chose to align our reads using BWA-MEM (Li & Durbin, 2009) to the yeast reference genome (S288C release 64-2-1 available at SGD) for paired-end reads with default options. This will generate a SAM file.

Users can set alignment options depending on their sequenced data and the quality they desire.

8. Only properly aligned paired-end reads, SAM FLAG 83 (reverse) and 99 (forward) are retained (see Note *N-6.3*).
9. SAM file needs to be cleaned from extra information, thus only data for the FLAG with orientation (reverse or forward), chromosome, start and end position, and read sequence are conserved.
10. Duplicated reads are discarded: same orientation, same start and end positions (see Note *N-6.4*).
11. The number of reads obtained here is the number of de novo insertion events in the sequenced yeast population. Those reads represent approx. 5% of the total aligned reads. This cleaned file can be used for further downstream analysis.
12. To count the number of reads for every genomic position, the reads with same start position and orientation are added up. This number is indicated in a new column.
13. ggplot2 is used to make graphic visualization. The y-coordinate is the number of reads at each position and the x-coordinate the position of the read on the chromosome.

6.3 Notes

N-6.1. Raw reads quality has to be checked before the alignment. Sequencer can make mistakes particularly at the end of the reads. Trimming is done to cut the low-quality reads. Sometimes there can be also contamination with adapters or others sequences. FastQC is an easy and fast way to control quality.

N-6.2. Some reads correspond to the 5′ LTR-SCUF and *GAG* Ty1 sequence (Fig. 4). These reads must be filtered from your dataset with a homemade script.

N-6.3. SAM file has a precise nomenclature. SAM FLAG represents the quality score of the mapped read. More information is available at http://samtools.sourceforge.net.

N-6.4. During sequencing library preparation, PCR amplification of Ty1/genomic DNA junctions (step 4 in Section 5.3) may overproduce a subpopulation of DNA fragments. This will bias interpretation of sequencing data by increasing the number of reads corresponding to these fragments (i.e., false hotspot of Ty1 integration). To minimize PCR artifact, only one read among the duplicated reads (same orientation, same start, and end position) is therefore retained for further analysis.

7. CONCLUDING REMARKS

HTS can provide valuable data to identify TE target-site preferences. However, it is essential to get high-quality de novo insertions DNA libraries in order to perform thorough and conclusive bioinformatics analysis. Every step in the library preparation has to be checked and validated before starting downstream processes. Here, we have described our procedure to map genome-wide Ty1 integration sites. We chose to select for de novo Ty1 insertion events based on the RIG *his3*AI. Yet, the necessity to express the *HIS3* gene upon integration may favor selection of insertions in open chromatin. However, this bias should be marginal since there is no significant difference in integration profiles from unselected and selected Ty1 de novo insertion events in wild-type strains (Baller et al., 2012; Mularoni et al., 2012). In addition, in the absence of AC40/integrase interaction, we recovered de novo Ty1*HIS3* insertion events in chromosome ends, which are poorly enriched in euchromatin, indicating that new targets can be identified by selecting His^+ cells (Bridier-Nahmias et al., 2015). Nevertheless, we cannot exclude that some mutants may massively redirect Ty1 integration to heterochromatic regions, where *HIS3* might be silenced. Thus, future development will be to create libraries of de novo insertion events without any selection bias. Likewise, using diploid cells instead of haploid cells would be relevant when working with mutants to recover potential de novo insertions into essential genes. Combined together, these modifications should permit to recover all Ty1 de novo insertion events.

Previously, Chatterjee et al. added random tags in Tf1 in order to discriminate between true Tf1-independent integration events from clonal expansion or PCR amplification artifacts in the library (Chatterjee et al., 2014). Using random tags is not straightforward in the case of Ty1 because the LTR contains *GAG* sequences that need to be conserved at the amino acids levels. One possibility might be to add random tags in U3 of the 3′ LTR, which will be copied in the 5′ LTR by reverse transcription but should not impair *HIS3* expression upon Ty1*HIS3* integration.

Finally, we would like to emphasize that assaying Ty1 integration frequency before starting the whole procedure is crucial. It is not only important to determine the number of total cells to collect to get the right amount of His^+ cells, in a new yeast mutant or physiological context. It is also essential to highlight an important decrease in Ty1 integration frequency that might reveal a defect in step(s) of Ty1 replication upstream of integration.

Different assays are already available to study Ty1 transcription, translation, and reverse transcription (Curcio et al., 2014). In this case, genome-wide mapping of de novo insertions might not be necessary.

ACKNOWLEDGMENTS

We are grateful to Jill Keeney and Amandine Bonnet for their constructive comments on the manuscript.

P.L. acknowledges support from intramural funding from Centre National de la Recherche Scientifique-CNRS, the Université Paris Diderot and the Institut National de la Santé et de la Recherche Médicale-INSERM, and from grants from the Fondation ARC pour la Recherche sur le Cancer (PJA 20151203412), the Agence Nationale de la Recherche through the initiatives d'excellence (Idex ANR-11-IDEX-0005-02), the Labex Who am I (ANR11-LABX-0071), and the generic call projects ANR-13-BSV3-0012 and ANR-17-CE11-0025. A.A.-L. is supported by a postdoctoral fellowship from Fondation pour la Recherche Médicale (SPF20170938755) and A.B. by Ph.D. fellowships from the Ministère de l'Enseignement Supérieur et de la Recherche and by Fondation pour la Recherche Médicale.

REFERENCES

Adams, A., Gottschling, D. E., Kaiser, C. A., & Stearns, T. (1997). *Methods in yeast genetics: A cold spring harbor laboratory course manual*. Cold Spring Harbor, NY: Cold Spring Harbor Laboratory Press.

Bachmann, N., Eby, Y., & Boeke, J. D. (2004). Local definition of Ty1 target preference by long terminal repeats and clustered tRNA genes. *Genome Research*, *14*(7), 1232–1247. https://doi.org/10.1101/gr.2052904.

Baller, J. A., Gao, J., Stamenova, R., Curcio, M. J., & Voytas, D. F. (2012). A nucleosomal surface defines an integration hotspot for the Saccharomyces cerevisiae Ty1 retrotransposon. *Genome Research*, *22*(4), 704–713. https://doi.org/10.1101/gr.129585.111.

Baller, J. A., Gao, J., & Voytas, D. F. (2011). Access to DNA establishes a secondary target site bias for the yeast retrotransposon Ty5. *Proceedings of the National Academy of Sciences of the United States of America*, *108*(51), 20351–20356. https://doi.org/10.1073/pnas.1103665108.

Behrens, R., Hayles, J., & Nurse, P. (2000). Fission yeast retrotransposon Tf1 integration is targeted to 5′ ends of open reading frames. *Nucleic Acids Research*, *28*(23), 4709–4716. https://doi.org/10.1093/nar/28.23.4709.

Boeke, J. D., Garfinkel, D. J., Styles, C. A., & Fink, G. R. (1985). Ty elements transpose through an RNA intermediate. *Cell*, *40*(3), 491–500. https://doi.org/10.1016/0092-8674(85)90197-7.

Boeke, J. D., LaCroute, F., & Fink, G. R. (1984). A positive selection for mutants lacking 5′ phosphate decarboxylase activity in yeast: 5 Fluoro-orotic acid resistance. *Molecular and General Genetics*, *197*, 345–346.

Brachmann, C. B., Davies, A., Cost, G. J., Caputo, E., Li, J., Hieter, P., et al. (1998). Designer deletion strains derived from Saccharomyces cerevisiae S288C: A useful set of strains and plasmids for PCR-mediated gene disruption and other applications. *Yeast*, *14*(2), 115–132. https://doi.org/10.1002/(SICI)1097-0061(19980130)14:2<115::AID-YEA204>3.0.CO;2-2.

Bridier-Nahmias, A., Tchalikian-Cosson, A., Baller, J. A., Menouni, R., Fayol, H., Flores, A., et al. (2015). An RNA polymerase III subunit determines sites of retrotransposon integration. *Science*, *348*(6234), 585–588. https://doi.org/10.1126/science.1259114.

Chalker, D. L., & Sandmeyer, S. B. (1990). Transfer RNA genes are genomic targets for de novo transposition of the yeast retrotransposon Ty3. *Genetics*, *126*(4), 837–850. https://doi.org/10.1146/annurev.ge.24.120190.002423.

Chatterjee, A. G., Esnault, C., Guo, Y., Hung, S., McQueen, P. G., & Levin, H. L. (2014). Serial number tagging reveals a prominent sequence preference of retrotransposon integration. *Nucleic Acids Research*, *42*(13), 8449–8460. https://doi.org/10.1093/nar/gku534.

Chuong, E. B., Elde, N. C., & Feschotte, C. (2016). Regulatory evolution of innate immunity through co-option of endogenous retroviruses. *Science*, *351*(6277), 1083–1087.

Curcio, M. J., & Garfinkel, D. J. (1991). Single-step selection for Ty1 element retrotransposition. *Proceedings of the National Academy of Sciences of the United States of America*, *88*(3), 936–940. https://doi.org/10.1073/pnas.88.3.936.

Curcio, M. J., & Garfinkel, D. J. (1994). Heterogeneous functional Ty1 elements are abundant in the Saccharomyces cerevisiae genome. *Genetics*, *136*(4), 1245–1259. https://doi.org/10.1042/BJ20030755.

Curcio, M. J., Lutz, S., & Lesage, P. (2014). The Ty1 LTR-retrotransposon of budding yeast, Saccharomyces cerevisiae. *Mobile DNA*, 927–964. https://doi.org/10.1128/microbiolspec.MDNA3-0053-2014.

Dakshinamurthy, A., Nyswaner, K. M., Farabaugh, P. J., & Garfinkel, D. J. (2010). BUD22 affects Ty1 retrotransposition and ribosome biogenesis in Saccharomyces cerevisiae. *Genetics*, *185*(4), 1193–1205. https://doi.org/10.1534/genetics.110.119115.

Devine, S. E., & Boeke, J. D. (1996). Integration of the yeast retrotransposon Ty1 is targeted to regions upstream of genes transcribed by RNA polymerase III. *Genes and Development*, *10*(5), 620–633. https://doi.org/10.1101/gad.10.5.620.

Gai, X., & Voytas, D. F. (1998). A single amino acid change in the yeast retrotransposon Ty5 abolishes targeting to silent chromatin. *Molecular Cell*, *1*(7), 1051–1055. https://doi.org/10.1016/S1097-2765(00)80105-7.

Guo, Y., & Levin, H. L. (2010). High-throughput sequencing of retrotransposon integration provides a saturated profile of target activity in Schizosaccharomyces pombe. *Genome Research*, *20*(2), 239–248. https://doi.org/10.1101/gr.099648.109.

Hickey, A., Esnault, C., Majumdar, A., Chatterjee, A. G., Iben, J. R., McQueen, P. G., et al. (2015). Single-nucleotide-specific targeting of the Tf1 retrotransposon promoted by the DNA-binding protein sap1 of Schizosaccharomyces pombe. *Genetics*, *201*(3), 905–924. https://doi.org/10.1534/genetics.115.181602.

Jacobs, J. Z., Rosado-Lugo, J. D., Cranz-Mileva, S., Ciccaglione, K. M., Tournier, V., & Zaratiegui, M. (2015). Arrested replication forks guide retrotransposon integration. *Science (New York, N.Y.)*, *349*(6255), 1549–1553. https://doi.org/10.1126/science.aaa3810.

Ji, H., Moore, D. P., Blomberg, M. A., Braiterman, L. T., Voytas, D. F., Natsoulis, G., et al. (1993). Hotspots for unselected Ty1 transposition events on yeast chromosome III are near tRNA genes and LTR sequences. *Cell*, *73*(5), 1007–1018. https://doi.org/10.1016/0092-8674(93)90278-X.

Levin, H. L., & Moran, J. V. (2011). Dynamic interactions between transposable elements and their hosts. *Nature Reviews. Genetics*, *12*(9), 615–627. https://doi.org/10.1038/nrg3030.

Li, H., & Durbin, R. (2009). Fast and accurate short read alignment with Burrows-Wheeler transform. *Bioinformatics*, *25*(14), 1754–1760. https://doi.org/10.1093/bioinformatics/btp324.

Morillon, A., Springer, M., & Lesage, P. (2000). Activation of the Kss1 invasive-filamentous growth pathway induces Ty1 transcription and retrotransposition in Saccharomyces cerevisiae. *Molecular and Cellular Biology*, *20*(15), 5766–5776. https://doi.org/10.1128/MCB.20.15.5766-5776.2000.

Mou, Z., Kenny, A. E., & Curcio, M. J. (2006). Hos2 and Set3 promote integration of Ty1 retrotransposons at tRNA genes in Saccharomyces cerevisiae. *Genetics*, *172*(4), 2157–2167. https://doi.org/10.1534/genetics.105.054072.

Mularoni, L., Zhou, Y., Bowen, T., Gangadharan, S., Wheelan, S. J., & Boeke, J. D. (2012). Retrotransposon Ty1 integration targets specifically positioned asymmetric nucleosomal DNA segments in tRNA hotspots. *Genome Research*, *22*(4), 693–703. https://doi.org/10.1101/gr.129460.111.

Nyswaner, K. M., Checkley, M. A., Yi, M., Stephens, R. M., & Garfinkel, D. J. (2008). Chromatin-associated genes protect the yeast genome from Ty1 insertional mutagenesis. *Genetics*, *178*(1), 197–214. https://doi.org/10.1534/genetics.107.082602.

Qi, X., Daily, K., Nguyen, K., Wang, H., Mayhew, D., Rigor, P., et al. (2012). Retrotransposon profiling of RNA polymerase III initiation sites. *Genome Research*, *22*(4), 681–692. https://doi.org/10.1101/gr.131219.111.

Qi, X., & Sandmeyer, S. (2012). In vitro targeting of strand transfer by the Ty3 retroelement integrase. *Journal of Biological Chemistry*, *287*(22), 18589–18595. https://doi.org/10.1074/jbc.M111.326025.

Sultana, T., Zamborlini, A., Cristofari, G., & Lesage, P. (2017). Integration site selection by retroviruses and transposable elements in eukaryotes. *Nature Reviews. Genetics*, *18*(5), 292–308. https://doi.org/10.1038/nrg.2017.7.

Van Dijk, E. L., Jaszczyszyn, Y., & Thermes, C. (2014). Library preparation methods for next-generation sequencing: Tone down the bias. *Experimental Cell Research*, *322*(1), 12–20. https://doi.org/10.1016/j.yexcr.2014.01.008.

Winston, F., Durbin, K. J., & Fink, G. R. (1984). The SPT3 gene is required for normal transcription of ty elements in S. cerevisiae. *Cell*, *39*(3 Pt. 2), 675–682. https://doi.org/10.1016/0092-8674(84)90474-4.

Xie, W., Gai, X., Zhu, Y., Zappulla, D. C., Sternglanz, R., & Voytas, D. F. (2001). Targeting of the yeast Ty5 retrotransposon to silent chromatin is mediated by interactions between integrase and Sir4p. *Molecular and Cellular Biology*, *21*(19), 6606–6614. https://doi.org/10.1128/MCB.21.19.6606-6614.2001.

Zou, S., Ke, N., Kim, J. M., & Voytas, D. F. (1996). The Saccharomyces retrotransposon Ty5 integrates preferentially into regions of silent chromatin at the telomeres and mating loci. *Genes and Development*, *10*(5), 634–645. https://doi.org/10.1101/gad.10.5.634.

Zou, S., Kim, J. M., & Voytas, D. F. (1996). The Saccharomyces retrotransposon Ty5 influences the organization of chromosome ends. *Nucleic Acids Research*, *24*(23), 4825–4831. http://doi.org/6e0307.

CHAPTER TEN

Genome-Wide Quantitation of Protein Synthesis Rates in Bacteria

Grace E. Johnson, Gene-Wei Li[1]
Department of Biology, Massachusetts Institute of Technology, Cambridge, MA, United States
[1]Corresponding author: e-mail address: gwli@mit.edu

Contents

Abstract

Bacteria produce different amounts of their proteins in response to different conditions. The ability to accurately quantitate the rates of protein synthesis across the genome is an important step toward understanding both underlying regulation and bacterial physiology at a systems level. Ribosome profiling (deep sequencing of ribosome-protected mRNA fragments) enables accurate and high-throughput measurement of such synthesis rates. Ribosomes protect RNAs from nuclease digestion; thus, by collecting and sequencing protected footprints, one can obtain information on the position of every ribosome at the time of cell collection. Assuming ribosomes go on to translate

Methods in Enzymology, Volume 612
ISSN 0076-6879
https://doi.org/10.1016/bs.mie.2018.08.031

full-length proteins, the density of ribosomes across an ORF can be used to determine protein synthesis rates. Here we outline a step-by-step protocol and discuss the steps where variability and bias may be introduced, including ways to minimize it.

1. INTRODUCTION

Quantitating protein synthesis rates is important for understanding the physiological state of a cell. While mRNA levels provide a convenient proxy for measuring gene expression, they do not directly reflect protein output due to extensive posttranscriptional regulation. In bacteria, the components of multiprotein complexes provide a common example of genes for which mRNA levels do not accurately reflect the amount of proteins synthesized. The proteins that makeup a complex are often transcribed together as polycistronic operons, and although the mRNA level for each component is typically the same, differential translation ensures that subunits are synthesized in proportion with their stoichiometry within the complex (Li, Burkhardt, Gross, & Weissman, 2014). More broadly, translation efficiency of endogenous mRNAs varies by more than 100-fold in *Escherichia coli*, further highlighting the importance of measuring the rates of protein synthesis directly (Li et al., 2014).

Ribosome profiling takes advantage of high-throughput sequencing technology to allow for global measurement of protein synthesis rates. As ribosomes protect mRNA from nuclease digestions, collecting and sequencing protected fragments yield a snapshot of the location of every ribosome at the time of cell harvesting (Ingolia, Brar, Rouskin, McGeachy, & Weissman, 2012; Ingolia, Ghaemmaghami, Newman, & Weissman, 2009; Oh et al., 2011) (Fig. 1A). If each ribosome is in the process of translating a full-length protein, the density of ribosomes over a gene, or footprint reads per unit length, is representative of the synthesis rate of that protein. Additionally, for proteins that are stable and therefore share the same half-life as the cell doubling time, their steady-state abundance is proportional to the synthesis rate, making this method a simple tool to estimate proteome composition.

Relating ribosome density and synthesis rate requires two key assumptions. One, that there is little ribosome drop-off over the length of a transcript, and two, that the average elongation rates across genes are similar (Li, 2015). For bacteria in steady-state growth, a superb agreement between ribosome density in the first and second halves of transcripts indicates both that there is little ribosome drop-off and that potential ribosome pause

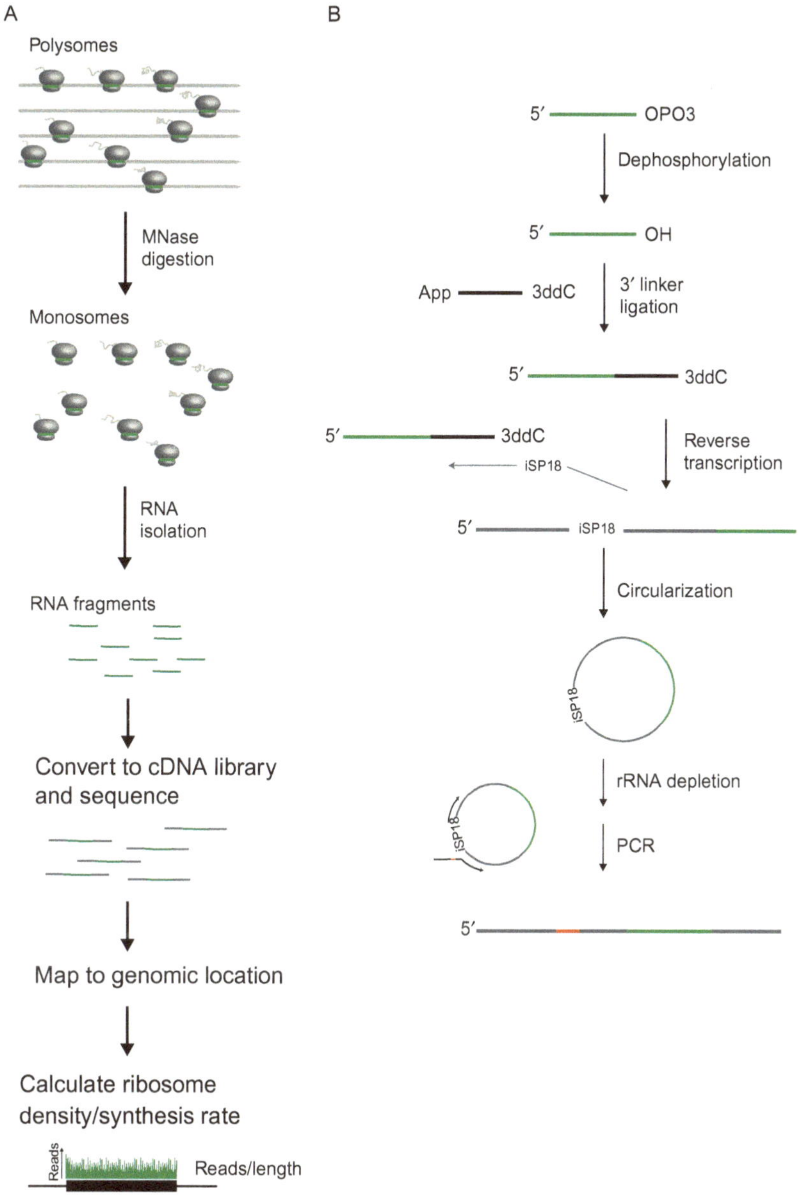

Fig. 1 (A) Overview of ribosome profiling. Polysomes are harvested and digested to monosomes with MNase. Monosomes are isolated and RNA fragments collected. These RNA fragments are then converted into cDNA libraries and sequenced, and mapped reads used to calculate ribosome density and protein synthesis rate. (B) Overview of Illumina sequencing library preparation. Briefly, following size selection, RNA is dephosphorylated, and a preadenylated DNA linker ligated to the 3′ end. RNA is then reverse transcribed and cDNA circularized. After rRNA is depleted, final libraries are generated through PCR amplification.

sequences or codon-dependent transit time does not substantially change average elongation rate over the length of the transcript (Li et al., 2014; Li, Oh, & Weissman, 2012; Subramaniam, Zid, & O'Shea, 2014). Small residual variations can be corrected for computationally.

Here, we outline a ribosome profiling protocol that can be used to measure protein synthesis rates in *E. coli* and other bacteria. We also discuss key considerations that must be made to ensure collection of high quality, quantitative information.

2. CELL COLLECTION, RIBOSOME FOOTPRINTING, AND MONOSOME ISOLATION

2.1 Overview

The first step of ribosome profiling involves collecting cells and preparing cell lysates from which polyribosomes can be prepared. Care must be taken during this step to ensure that the positions of ribosomes accurately represent a snapshot of the cells' translational state. During cell harvesting, cultures are filtered and cells rapidly collected and flash-frozen to minimize the time during which translation can be altered. Further, cell collection is performed at 37°C with equipment that has also been prewarmed to this temperature to minimize perturbations to cells that could induce changes to translation. Cells are then lysed in the presence of a translational inhibitor, chloramphenicol, and kept frozen to ensure minimal movement of ribosomes.

Several factors may contribute to artifacts and irreproducible results. If cells are not under steady-state growth, gene expression and translation are sensitive to both the exact time point of harvesting and the history of the culture, e.g., dilution factor and inoculum size. Further, cells transitioning between growth phases often experience sequential depletion of carbon sources and amino acids. The differences in the limiting nutrients can both directly and indirectly affect the kinetics of translation, leading to substantial ribosome drop-off and/or pausing that invalidate the assumptions required for estimating the rates of protein synthesis (Li et al., 2012; Mohammad, Woolstenhulme, Green, & Buskirk, 2016; Subramaniam et al., 2013). To ensure steady-state growth, we recommend careful measurement of a growth curve, and allowing for at least six cell doublings (and ideally 10) of exponential growth to minimize residual effects from the saturated culture. Another source of irreproducibility can occur during cell harvesting using vacuum filtration, in which cells must be rapidly scraped off from the filter as soon as medium is passed through. Repeated scraping from the same filter should be avoided to minimize the amount of time that cells are depleted of nutrients.

Polyribosomes collected from lysed cells are then treated with a ribonuclease to generate monosomes from which protected RNA fragments can be obtained. Ribosome profiling in yeast typically relies on RNase I for such footprinting, a nuclease that is inhibited by the *E. coli* 30S ribosomal subunit, and thus cannot be used for bacterial footprinting (Datta & Burma, 1972). Instead, Micrococcal nuclease (MNase) is used. MNase activity is biased by sequence, cleaving faster upstream of A and T residues, and contributing to more variability in footprint size than is typically seen with RNase I (Dingwall, Lomonossoff, & Laskey, 1981). Longer ribosome footprints are also due to slow translation elongation at Shine–Dalgarno-like sequences, which interact with the anti-Shine–Dalgarno sequence of the 16S rRNA for a few rounds of amino acid incorporation (Li et al., 2012; O'Connor, Li, Weissman, Atkins, & Baranov, 2013). Therefore, a wider range of fragments should be selected for inclusion in final cDNA libraries. MNase digestion must be run long enough to ensure polysomes are sufficiently collapsed to monosomes, but terminated before overdigestion results in cleavage of rRNA and loss of footprint integrity (Li et al., 2012).

2.2 Reagents

- 1 *M* Tris pH 8.0 (FisherSci, cat. no. AM9856)
- Ammonium Chloride (Sigma, cat. no. A9434)
- Magnesium Chloride (ThermoFisher, cat. no. AM9530G)
- Triton X-100 (VWR, cat. no. 97062-208)
- Nonidet P-40 Substitute (NP-40, Sigma-Aldrich, cat. no. 11332473001)
- Chloramphenicol (VWR, cat. no. 97061-244)
- RNase-free DNase I (Roche, cat. no. 04716728001)
- DEPC-Treated, DNase- and RNase-free water (Fisher, cat. no. BP561-1)
- Tris (1 *M*), pH 7.0 (FisherSci, cat. no. AM9851)
- Nuclease S7 (Micrococcal nuclease) (Roche, cat. no. 10107921001)
- SUPERase•In RNase Inhibitor (ThermoFischer, cat. no. AM2696)
- Calcium chloride dihydrate (VWR, cat. no. 97061-904)
- EGTA (Sigma, cat. no. E3889)
- Sucrose, RNase/DNase free (VWR, 97061-428)

2.3 Equipment

- 2-L filter flask
- GE Nitrocellulose-Mixed Esters of Cellulose Membrane Filters, E02WP09025 (GE)
- Ultraware microfiltration assembly (90 mm) (Fisher Scientific)
- Flat head stainless steel lab spatulas
- Stainless steel reagent digger
- 50-mL conical tubes
- Liquid nitrogen
- 10 mL mixer mill canister with 12 mm diameter ball charge (Retsch)
- Mixer mill (Qiagen Tissuelyser 2)
- 4°C microcentrifuge
- Seton Open-top Polyclear centrifuge tubes (Seton)
- SW41 rotor (Beckman)
- Ultracentrifuge
- BioComp Instruments 153 Gradient Station ip (BioComp)
- BioRad Econo UV Monitor (BioRad)
- Cryovials

2.4 Buffer Preparation

Lysis buffer

- Prepare immediately before use

Ingredient	Final Conc.	Stock Conc.	Volume
Tris pH 8.0	20 m*M*	1 *M*	30 μL
NH_4Cl[a]	100 m*M*	1 *M*	150 μL
$MgCl_2$	10 m*M*	1 *M*	15 μL
Triton X-100[a]	0.4%	25%	24 μL
NP-40[a]	0.1%	20%	7.5 μL
Chloramphenicol[b]	1 m*M*	155 m*M* (50 mg/mL)	10 μL
RNase-free DNase	100 U/mL	10 U/μL	15 μL
DEPC H_2O			1248.5 μL
Total volume			1500 μL

[a]Prepare in advance and store at room temperature.
[b]Prepare in advance and store at −20°C.

1 *M* $CaCl_2$

- Prepare in advance and store at room temperature

0.5 *M* EGTA

- Prepare in advance and store at room temperature

10%/55% sucrose solution

Ingredient	Final Conc.	Stock Conc.	Volume (ex)
Sucrose	10%/55%		1.5 g/8.25 g
Tris pH 8.0	20 m*M*	1 *M*	300 μL
NH_4Cl	100 m*M*	1 *M*	1.5 mL
$MgCl_2$	10 m*M*	1 *M*	150 μL
Chloramphenicol	1 m*M*	155 m*M*	97.5 μL
DEPC H_2O			to 15 mL
Final volume	15 mL		15 mL

- Prepare without chloramphenicol at least 1 h before use and allow sucrose to dissolve at room temperature
  - Can be prepared the day before use
- Add chloramphenicol immediately before use and mix well

2.5 Procedure

2.5.1 Cell Harvesting *(Fig. 2)*

Estimated time: Variable; depends on cell growth

1. Inoculate medium from a single colony and grow overnight at 37°C.
2. Back dilute overnight culture to $OD_{600} = 0.001$ in 200–250 mL prewarmed medium and shake at 37°C.
   a. >6 doublings before collection
3. Bring all equipment needed for cell harvesting to 37°C.
4. Collect cells when density reaches $OD_{600} = 0.3$.
   a. It is important to ensure cells are still in steady-state growth at this OD in selected growth medium
5. Rapidly filter liquid culture. When liquid has filtered through, immediately turn off vacuum and remove top of filtration apparatus. Scrape cells off filter in a single motion in ≪1 s using the long edge of a prewarmed reagent digger. Transfer immediately to a 50-mL conical tube prefilled with and stored in liquid N_2.

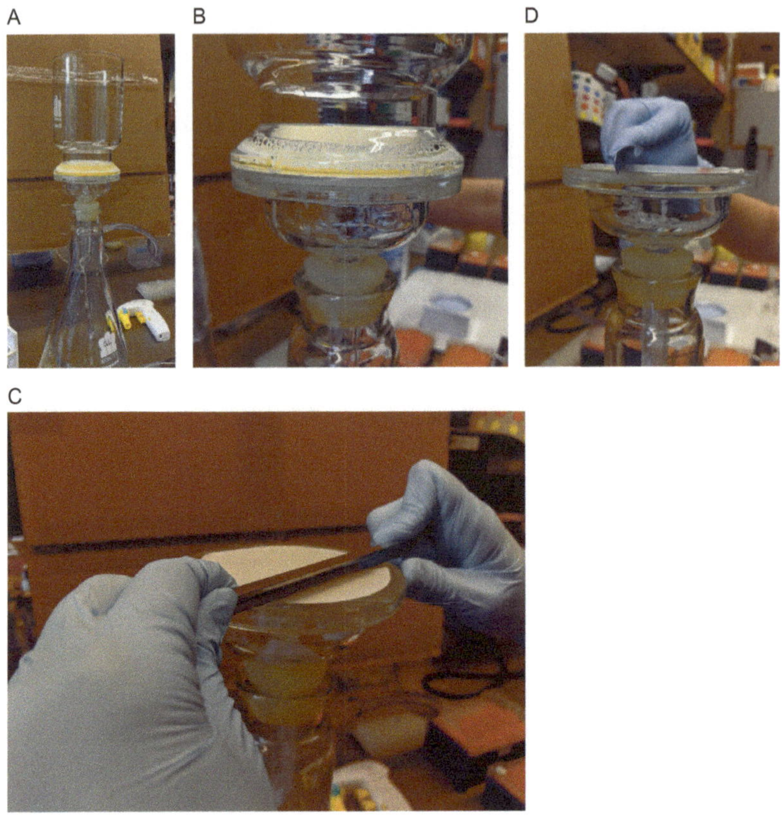

Fig. 2 (A) Cell filtration setup. (B) When liquid is almost finished filtering, immediately turn off vacuum, remove top, and (C) scrape cells in one rapid motion. (D) A side view demonstrating the angle at which the reagent digger should be held.

6. Use a prechilled flat head spatula to scrape cells into tube, keeping cells submersed in liquid N_2 at all times. The frozen pellet should be at least 10 μL in volume.
7. Poke holes through cap of tube. Cap tube and tilt to dispense liquid N_2.
8. Store tube at −80°C.

2.5.2 *Cell Lysis*

Estimated time: 1 h

1. Pellet lysis buffer by slowly pipetting 650 μL into a 50-mL conical tube filled with liquid N_2. Wait for each pellet to sink before pipetting next pellet to prevent pellets from clumping.

2. Clean 10 mL mixer mill canister with Millipore water and EtOH. Use compressed air to blow dry. Make sure no residual water droplets are left. Fully cool in liquid N_2.
3. Combine the lysis pellets with the flash-frozen cells in the canister.
4. Mixer mill five times at 15 Hz for 3 min. Cool in liquid N_2 between each round.
5. Open the canister and keep in a shallow water bath to thaw. Transfer lysate to 1.5-mL tube as soon as thawed and keep on ice.

2.5.3 Footprinting

Estimated time: 1 h 40 min

1. Centrifuge lysate at 20,000 rpm 10 min 4°C.
2. Carefully transfer the supernatant to a new tube. The pellet will be very soft. Avoid transferring viscous DNA to the new tube.
3. Dilute the supernatant 1:100 in 10 m*M* Tris pH 7.0.
4. Measure A_{260} on Nanodrop and calculate the concentration of the undiluted sample. Blank with 1:100 dilution of lysis buffer.
   a. You should have ~1 mg RNA.
5. To 0.5 mg RNA add:
   a. 750 U MNase
   b. 100 U SUPERase•In
   c. 1 μL 1 *M* $CaCl_2$
   d. Lysis buffer to 200 μL
6. Incubate at 25°C for 1 h.
7. Quench the reaction by adding 2.4 μL of 0.5 *M* EGTA.
8. Leave on ice.

2.5.4 Monosome Isolation

Estimated time: 3.5 h

1. Layer 55% sucrose (~6 mL) below 10% sucrose (~7 mL) in Seton tubes for SW41 rotor. Use marker block provided with gradient maker.
2. Make the gradient using preprogramed 10–50 short program on gradient station.
3. Prechill the gradients, tubes, buckets, rotor, and the ultracentrifuge to 4°C.
4. Load the samples onto gradients at 4°C.
5. Balance with lysis buffer.
6. Spin in ultracentrifuge at 35,000 rpm for 2.5 h at 4°C.

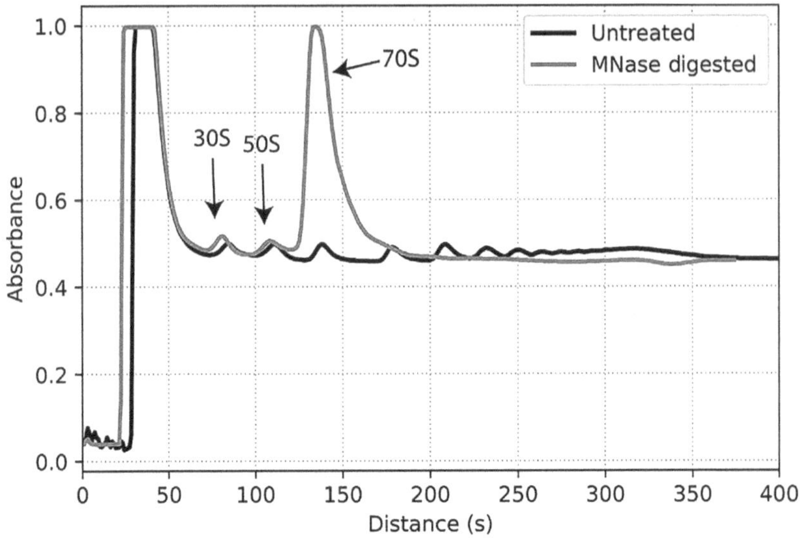

Fig. 3 Traces for an undigested polysome- (*black*) and MNase-treated (*gray*) sample. Peaks corresponding to the 30S, 50S, and 70S monosome are marked.

7. Fractionate at 0.2 mm/s. Measure A_{260} every second using BioRad Econo UV Monitor (Fig. 3).
8. Collect monosomes (~1.5 mL) in 2 mL cryovials.
9. Flash-freeze and store at −80°C.

3. LIBRARY PREPARATION

3.1 Overview

mRNA footprints must be converted into a cDNA library compatible for Illumina sequencing in a way that minimizes introduction of biases that can skew the final library composition. Here, cDNA libraries are generated by first ligating a DNA adapter to the 3′ end of the RNA fragments. Following reverse transcription from the ligated adapter, cDNA is circularized and the final library generated via PCR amplification (Fig. 1B).

Efficient ligation is critical for minimizing bias introduced during library preparation. It has been demonstrated that T4 RNA ligases have sequence and structure preferences, and that the efficiency of ligation is largely dependent on the stability of the structure formed between the RNA fragment and adapter (Zhuang, Fuchs, Sun, Zheng, & Robb, 2012). Here, several steps are taken to ensure effective and maximally efficient ligation. First, to reduce background ligation and circularization of RNA fragments that can prevent productive ligation, a truncated form of T4 RNA ligase that cannot adenylate the 5′ end of RNA/DNA is used in conjunction with a preadenylate DNA linker (linker-1) (Hafner et al., 2008; Ho, Wang, Lima, & Shuman, 2004; Yin, Ho, & Shuman, 2003). The 3′ end of the linker is blocked to prevent linker circularization. Second, a ligation reaction is performed with a control oligo (o199) in parallel to measure ligation efficiency. Ligated libraries should be used only if ligation is >80% efficient, though >90% efficiency is ideal. The control oligo additionally provides a control for all other library preparation reactions and helps to identify failed reactions that require further troubleshooting.

3.2 Reagents

- Acid phenol chloroform, pH 4.5 (ThermoFisher, cat. no. AM9720)
- 20% SDS (ThermoFisher, cat. no. AM9820)
- Chloroform (Sigma, cat. no. C2432)
- Sodium Acetate (3 *M*), pH 5.5, RNase Free (ThermoFisher, cat. no. AM9740)
- Isopropanol, molecular biology grade (FisherSci, cat. no. BP26181)
- Ethanol, 200 Proof (VWR, cat. no. AC61509)
- Tris (1 *M*), pH 7.0 (FisherSci, cat. no. AM9851)
- Novex 15% TBE-Urea Gel, 10 well (Thermo, cat. no. EC6885)
- Novex TBE-Urea Sample Buffer 2 × (Thermo, cat. no. LC6876)
- 10 × TBE Buffer (FisherSci, cat. no. AM9863)
- 10 bp DNA ladder (Invitrogen, cat. no. 10821015)
- SYBR gold nucleic acid gel stain (Invitrogen, cat. no. S11494)
- GlycoBlue (Invitrogen, cat. no. AM9516)
- 5 *M* NaCl (FisherSci, cat. no. AM9759)
- T4 Polynucleotide Kinase (NEB, cat. no. M0201L)
- SUPERase•In RNase Inhibitor (ThermoFischer, cat. no. AM2696)
- PEG 8000, 50% (Teknova, cat. No. P4180)
- T4 RNA ligase 2 truncated K227Q and buffer (NEB, cat. no. M0351S)
- DEPC-Treated, DNase, RNase-free water (Fisher, cat. no. BP561-1)
- Novex 10% TBE-Urea Gel, 10 well (Thermo, cat. no. EC6875BOX)
- Deoxynucleotide (dNTP) solution set (NEB, cat. no. N0446S)
- SuperScript III Reverse Transcriptase (Invitrogen, cat. no. 18080085)
- DTT (Sigma, cat. no. 10708984001)
- Sodium hydroxide pellets (Sigma, cat. no. 795429)
- 1 *M* Tris pH 8.0 (FisherSci, cat. no. AM9856)

- 0.5 *M* EDTA, pH 8.0 (Ambion, cat. no. AM9260G)
- CircLigase ssDNA Ligase 5000 U (Illumina, cat. no. CL4115K)
- 20 × SSC Buffer, molecular grade (Promega, cat. no. V4261)
- Tween-20, reagent grade (VWR, cat. no. 97062-332)
- Dynabeads® MyOne™ Streptavidin C1 (Thermo Scientific, cat. no. 65001)
- Phusion polymerase (NEB, cat. no. M0530L)
- 6 × DNA loading dye (Thermo, cat. no. R0611)
- Novex TBE Gels, 8%, 10 well (Invitrogen, cat. no. EC6215BOX)

3.3 Common Oligos

- o199-P: 5′_AUGUACACGGAGUCGACCCGCAACGCGA3phos_3′
- Dephosphorylated o199: 5′_AUGUACACGGAGUCGACCCGCAA CGCGA_3′
- Linker-1: 5′_App/CTGTAGGCACCATCAAT/3ddC_3′
- oCJ485: /5Phos/AGATCGGAAGAGCGTCGTGTAGGGAAAGAG TGT/iSp18/CAAGCAGAAGACGGCATACGAGATATTGATGG TGCCTACAG
- o1055: 5′_5Biosg/TCATCTCCGGGGGTAGAGCACTGTTTCG_3′
- o1056: 5′_5Biosg/GGCTAAACCATGCACCGAAGCTGCGGCAG_3′
- o1057: 5′_5Biosg/AAGGCTGAGGCGTGATGACGAGGCACT_3′
- o1058: 5′_5Biosg/CGGTGCTGAAGCAACAAATGCCCTGCTT_3′
- o231: 5′_CAAGCAGAAGACGGCATACGA_3′
- Indexing primer: 5′_AATGATACGGCGACCACCGAGATCTACA CGATCGGAAGAGCACACGTCTGAACTCCAGTCACNNNN NNACACTCTTTCCCTACAC_3′

  (NNNNNN corresponds to custom index.)

3.4 Equipment

- Thermomixer
- Table top microcentrifuge
- Nonstick RNase-free tubes, 1.5 mL
- 4°C Microcentrifuge
- Nanodrop
- Standard PAGE Equipment
- DynaMag-2 Magnet (ThermoFisher)
- PCR strip tubes
- Needle, 18G × 1–1/2

- 0.5-mL tubes
- 2-mL screw cap tubes
- Spin-X cellulose acetate column (Costar)

3.5 Buffer Preparation

80% EtOH
- Prepare in advance and store at room temperature

10 m*M* Tris pH 7.0
- Prepare in advance and store at room temperature

20 μ*M* o199-P
- Prepare in advance and store at −20°C

10 m*M* dNTPs
- Prepare in advance and store at −20°C

25 μ*M* oCJ485
- Prepare in advance and store at −20°C

1 *M* NaOH
- Prepare in advance and store at room temperature

rRNA subtraction oligo mix
- Prepare in advance and store at −20°C
- 77 μL 100 μ*M* o1055, 4 μL 100 μ*M* o1056, 17 μL 100 μ*M* o1057, 2 μL 100 μ*M* o1058

1 *M* Tris pH 7.5
- Prepare in advance and store at room temperature
- 7:3 Tris pH 7.0:Tris pH 8.0

1 × B&W Buffer
- Prepare fresh day of rRNA subtraction
- 5 μL 1 *M* Tris 7.5
- 1 μL 0.5 *M* EDTA
- 200 μL 5 *M* NaCl
- 10 μL 1% Tween
- 784 μL H_2O

2 × B&W Buffer
- Prepare fresh day of rRNA subtraction
- 10 μL 1 *M* Tris 7.5
- 2 μL 0.5 *M* EDTA
- 400 μL 5 *M* NaCl
- 10 μL 1% Tween
- 578 μL H_2O

3.6 Procedure

Note: To minimize degradation of RNA, work in an RNase-free lab space; use 70% Ethanol or RNAseZap (Invitrogen) to clean lab bench and equipment before use. Thaw and keep RNA samples on ice when possible. Work to convert RNA to cDNA as quickly as possible, and avoid storing the RNA for longer than 1 week.

3.6.1 RNA Isolation

Estimated time: 3 h

1. For each sample, prewarm 1.5 mL acid phenol chloroform to 65°C.
2. Add 80 μL 20% SDS to ~1.4 mL fractionated monosomes and split into two tubes.
3. Add 0.75 mL prewarmed acid phenol chloroform to each tube.
4. Incubate at 65°C for 5 min at 1400 rpm in a thermomixer.
5. Chill on ice for 5 min.
6. Spin at 20,000 × *g* from 2 min at room temperature.
7. Transfer top aqueous layer to fresh tube.
8. Add 0.7 mL acid phenol chloroform and incubate at room temperature for 5 min, vortexing occasionally.
9. Repeat steps 6–7.
10. Add 600 μL chloroform and vortex 30 s at room temperature.
11. Spin at 20,000 × *g* for 1 min at room temperature.
12. Transfer top aqueous layer to fresh tube.
13. Precipitate by adding 78 μL 3 *M* NaOAc and 0.75 mL isopropanol.
14. Chill samples at −80°C for ≥30 min.
15. Spin at 20,000 × *g* for 20 min at 4°C.
16. Remove supernatant and wash pellet in 750 μL 80% EtOH.
17. Spin for 5 min at 4°C.
18. Remove supernatant; air-dry ≥5 min.
19. Suspend in 11 μL 10 m*M* Tris pH 7, pool two tubes together for a total volume of 22 μL.

3.6.2 Size Selection

Estimated time: 4 h

1. Prerun a 15% TBE-Urea PAGE gel in 1 × TBE at 200 V for 1 h.
2. Measure the concentration of a 1:5 dilution of total RNA in 10 m*M* Tris pH 7 on Nanodrop.
3. Dilute samples to 20 μg of RNA in 5 μL 10 m*M* Tris pH 7.

4. Add 5 µL 2 × TBE-Urea sample loading buffer to each sample.
5. Set up o199-P control oligo by combining 1 µL 20 µ*M* o199-P, 3 µL 10 m*M* Tris pH 7, and 5 µL 2 × loading buffer.
6. Set up ladder.
7. Denature all samples at 80°C for 2 min and return to ice.
8. Run samples on prerun gel at 200 V for 65 min.
   a. *Note*: If you are running multiple samples, leave at least one well empty between samples to minimize cross-contamination
9. Stain with SYBR gold for 2 min and photograph gel (Fig. 4A).
10. Excise between ~15 and 45 nt (Fig. 4B).
    a. Also excise o199-P control band
11. Extract from gel as described in Section 3.7.1.
12. Precipitate as described in Section 3.7.2.
13. Suspend in 15 µL 10 m*M* Tris pH 7.

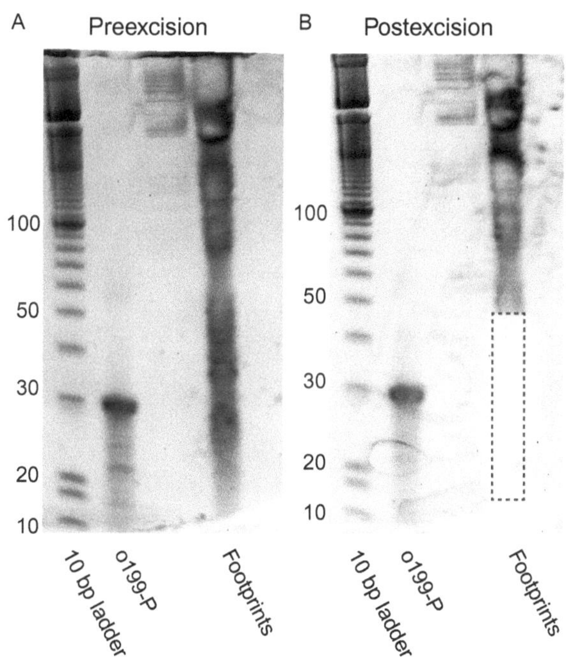

Fig. 4 Size selection gel. (A) RNA footprints are run on a 15% TBE-Urea PAGE gel at 200 V for 1 h and stained with SYBR gold. In addition to mRNA fragments, distinct rRNA fragments can be observed. Also run are a 10-bp ladder (Lane 1), and phosphorylated o199 control oligo (Lane 2). (B) Fragments between 15 and 45 nucleotides are excised and extracted.

3.6.3 Dephosphorylation

Estimated time: 2.5 h

1. Make a master mix with 2 μL T4 PNK buffer without ATP and 1 μL SUPERase•In.
2. Add 3 μL master mix to 15 μL sample.
   a. Include a reaction with o199 control gel extracted in Section 3.6.2
3. Add 2 μL T4 PNK.
4. Incubate at 37°C for 1 h.
5. Heat inactivate at 75°C for 10 min.
6. Precipitate as described in Section 3.7.2.
7. Suspend in 11 μL 10 m*M* Tris pH 7.

3.6.4 Ligation and Size Selection

Estimated time: 7 h

Note: SUPERase•In is omitted from the reaction, as we have observed that it results in decreased ligation efficiency. However, this should only be done if you are confident that the reaction is free of RNase contamination.

1. Measure the concentration of RNA.
2. Prepare 2 pmol RNA in 5 μL 10 m*M* Tris 7. Also prepare 2 pmol of the following controls:
   a. o199 dephosphorylation reaction from Section 3.6.3
   b. Dephosphorylated o199 to control for failed dephosphorylation
3. Denature at 80°C for 2 min and return to ice.
4. Make a master mix containing: (Make sure well mixed)
   a. 10 μL PEG 8000, 50%
   b. 2 μL 10 × T4 RNA ligase 2 buffer
   c. 1 μL H_2O
5. Combine 13 μL master mix with 1 μL 100 μ*M* linker-1, 1 μL T4 RNA ligase 2 truncated K227Q, and 5 μL RNA.
6. Incubate at 25°C for 2.5 h.
7. Precipitate as described in Section 3.7.2 and suspend in 6 μL 10 m*M* Tris pH 7.
8. Prerun a 10% TBE-Urea gel for 1 h at 200 V.
9. Add 6 μL TBE-urea loading buffer and denature at 80°C for 2 min followed by ice for 2 min.
10. Run on 10% TBE-Urea gel for 50 min at 200 V. Excise 35–65 bases (Fig. 5).

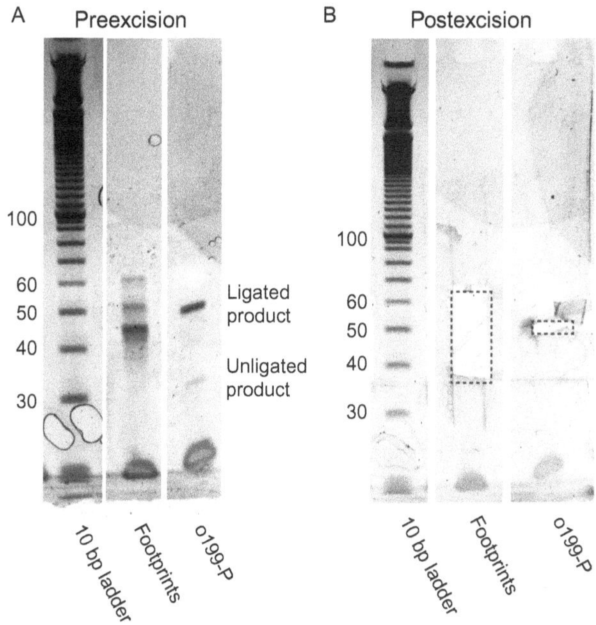

Fig. 5 Ligation size selection gel. (A) Products of the ligation reaction with the library and o199 control are run on a 10% TBE-Urea gel at 200V for 50min and stained with SYBR gold. The efficiency of ligation can be determined by comparing the intensity of unligated to ligated o199. Discrete bands in the footprint lane correspond to contaminants from rRNA fragments. (B) Ligated products (35–65nt) are excised and extracted. *Note*: This image was edited and lanes containing unrelated samples cut such that only pertinent information is displayed.

11. Gel purify and precipitate as described in Sections 3.7.1 and 3.7.2.
12. Suspend in 10μL 10m*M* Tris pH 7.0 and transfer to a new tube.

3.6.5 Reverse Transcription and Size Selection

Estimated time: 4h

1. Combine 10μL RNA, 1μL 10m*M* dNTPs, 1μL 25μ*M* oCJ485, and 1.5μL H_2O.
2. Denature at 65°C for 5min.
3. Make a master mix with 4μL 5× First Strand Buffer, 1μL SUPERase•In, and 1μL 0.1*M* DTT.
4. Add 6μL master mix to each sample and mix by pipetting.
5. Add 1μL SuperScript III.
6. Incubate at 50°C for 30min.

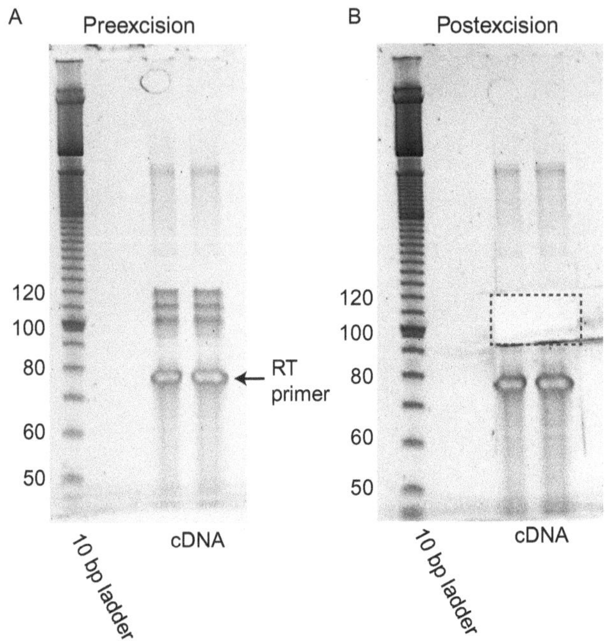

Fig. 6 RT size selection gel. (A) RT product is run over two lanes of a 10% TBE-Urea gel for 80 min at 200 V and stained with SYBR gold. (B) RT product (~90–120 nt) is excised and extracted.

7. Hydrolyze RNA by adding 2.3 μL 1 *M* NaOH and incubate at 95°C for 15 min.
8. Add 23 μL TBE-urea loading buffer and denature at 80°C for 2 min followed by ice for 2 min.
9. Prerun 10% TBE-Urea gel for 1 h at 200 V.
10. Run each sample over two wells on 10% TBE-Urea gel at 200 V for 80 min (Fig. 6A).
11. Excise 90–120 bases, avoiding excision of RT primer (Fig. 6B).
12. Gel purify and precipitate (DNA) as described in Sections 3.7.1 and 3.7.2.
13. Suspend in 15 μL 10 m*M* Tris pH 8 and transfer to a new tube.

3.6.6 Circularization

Estimated time: 2.5 h

1. Make a master mix containing the following: 1 μL 1 m*M* ATP, 2 μL 10 × CircLigase Buffer, 1 μL 50 m*M* $MnCl_2$
2. Add 4 μL master mix to 15 μL DNA

3. Add 1 μL CircLigase
4. Incubate at 60°C for 1 h
5. Spike in an additional 1 μL CircLigase into each sample
6. Incubate at 60°C for 1 h
7. Inactive at 80°C for 10 min and return to ice

3.6.7 rRNA Subtraction

Estimated time: 3 h

Note: The oligos used here are specific for *E. coli* rRNA contaminants and may not work in other bacterial species. When performing ribosome profiling on a new species for the first time, perform a low-depth sequencing run to identify the most common rRNA contaminants. Then design oligos to remove these contaminants; oligos should have the same sequence as the RNA contaminant.

1. Make a master mix containing the following: 1 μL rRNA subtraction oligo mix, 1 μL 20 × SSC, 3 μL H_2O.
2. Add 5 μL master mix to 5 μL DNA.
3. Incubate at 98°C for 75 s.
4. Ramp down to 37°C over an hour.
5. Hybridize by incubating at 37°C for 20 min.
6. Prepare 25 μL MyOne Streptavidin C1 Dynabeads per sample as follows (prepare as master mix):
   a. Wash three times with 25 μL 1 × B&W buffer
   b. Suspend in 10 μL 2 × B&W buffer
   c. Incubate at 37°C
7. Add 10 μL beads to 10 μL hybridization reaction.
8. Incubate at 37°C for 15 min.
9. Recover elute to a new tube.
10. Precipitate as described in Section 3.7.2.
11. Suspend in 10 μL 10 m*M* Tris pH 8.

3.6.8 PCR and Size Selection

1. Make a master mix containing the following: 16.7 μL 5 × HF buffer, 1.7 μL 10 m*M* dNTPs, 0.4 μL 100 μ*M* o231, 0.4 μL 100 μ*M* indexing primer, 59.2 μL H_2O, 0.8 μL HF Phusion.
2. Add 5 μL DNA to 79.2 μL of master mix.
3. Aliquot 17 μL PCR mix into four different PCR strip tubes.
4. Run the following PCR reaction, removing each strip after 6, 8, 10, and 12 cycles, respectively.

Initial	98°C	30 s
Denature	98°C	10 s
Annealing	60°C	10 s
Extension	72°C	5 s

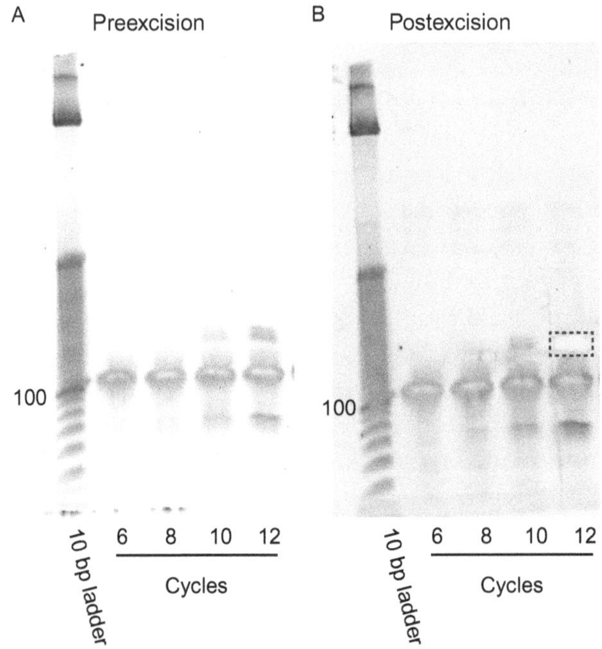

Fig. 7 PCR Gel. (A) PCR reactions run 6, 8, 10, and 12 cycles are run on an 8% TBE-PAGE gel at 180 V for 1 h and stained with SYBR gold. (B) The product amplified with the cycle number yielding the most product without excessive generation of nonspecific products is excised and extracted.

5. Add 3.5 μL 6 × DNA loading dye to each sample.
6. Prepare ladder: 1 μL 10 bp ladder, 16 μL 10 m*M* Tris 8, 3.5 μL 6 × DNA loading dye.
7. Prerun an 8% TBE-PAGE Gel at 180 V for 1 h.
8. Run samples on 8% TBE-PAGE Gel at 180 V for 50 min (Fig. 7A).
9. Excise band at ~175 bp (Fig. 7B).
10. Gel extract and precipitate as described in Sections 3.7.1 and 3.7.2.
11. Suspend in 11 μL 10 m*M* Tris pH 8.

3.6.9 Bioanalysis and Illumina Sequencing

1. Run samples on a high sensitivity DNA bioanalyzer chip to assess concentration, size, and purity.
2. Sequence using Illumina HiSeq or NextSeq.

3.7 Common Procedures

3.7.1 Gel Extraction

1. Poke a hole into the bottom of a 0.5-mL tube with a needle and place in a 2-mL screw cap tube.
2. Transfer the gel slice into 0.5-mL tube.
3. Spin at 20,000 × *g* for 3 min.
4. Collect remaining gel pieces and add 0.5 mL 10 m*M* Tris 7 (for RNA) or 10 m*M* Tris 8 (for DNA).
5. Shake at 14,000 rpm at 70°C for 10 min on a thermomixer.
6. Transfer to Spin-X cellulose acetate column. Cut the tip of the pipette tip off before transferring to ensure remaining gel pieces do not block tip.
7. Spin at 20,000 × *g* for 3 min and transfer eluate to new 1.5-mL tube.

3.7.2 Precipitation

1. Bring volume to 500 μL by adding 10 m*M* Tris 7 (for RNA) or 10 m*M* Tris 8 (for DNA).
2. Add the following:
   a. For RNA: 55 μL 3 *M* NaOAc (pH 5.5), 2 μL glycoblue, 550 μL isopropanol
   b. For DNA: 32 μL, 5 *M* NaCl, 1 μL 0.5 *M* EDTA, 2 μL glycoblue, 550 μL isopropanol
3. Vortex.
4. Incubate at −80°C for 30 min (or overnight).
5. Spin at 20,000 × *g* for 30 min at 4°C.
6. Remove supernatant and wash pellet in 750 μL 80% EtOH at 4°C.
7. Spin for 5 min at 4°C.
8. Remove supernatant; air-dry ≥5 min.
   a. *Note*: It is important that pellets are fully dry, especially when RNA/DNA will be immediately used in an enzymatic reaction. Suspended pellet can also be transferred to a new tube to minimize salt contamination.

4. ANALYSIS

4.1 Overview

Protein synthesis rates can be estimated from the density of reads mapping to a transcript. As described earlier, ribosome density is correlated with synthesis rate assuming no ribosome drop-off and constant elongation rate across transcripts. Here we provide quality control procedures to confirm that the data collected are mostly consistent with these assumptions, as well as methods to correct local variations that may be caused by small differences in elongation rate or ribosome processivity. Transcripts with strong ribosome pause sites or other nontypical features should be considered individually, as described below.

4.2 Software

- Cutadapt v.1.16
- Bowtie 1.2.2

4.3 Procedure

4.3.1 Alignment

1. Remove adapters from reads using cutadapt with the command below, where m specifies the minimum allowed read length, u trims the 5′ end of reads to remove nucleotides added by nontemplate addition during RT, a specifies the adapter sequence, and o the output file.
   a. -m 15 -u 1 -a CTGTAGGCACCATCAAT -o $OUTPUTNAME $INPUT_FASTQ
2. Discard reads that are not between 20 and 42 nt.
3. Use bowtie to align reads to a reference file containing rRNA sequences using the command below and output unmapped reads using –un. Use –v 2 to allow two mismatches, and –m to only report reads that align to a single genomic location.
   a. bowtie *rRNA_bowtie_reference* -v 2 -m 1 $INPUT_FASTQ > $OUTPUT_BOWTIE –un $OUTPUT_UNMAPPED
4. Use bowtie to align unaligned reads from step 2 to reference genome as above. Use the reads aligned here for quantification of protein synthesis rates.

4.3.2 Quality Control Procedures

1. Ribosome drop-off can be assessed via a metagene analysis. First, for each gene, calculate the average number of reads in every 50-codon window normalized to the average number of reads in the first 50 codons. Exclude genes with fewer than 128 total reads per transcript. For a sequencing depth of 100 million reads, approximately 75% of genes will be above this cutoff for *E. coli* grown in a rich medium. Fit the median value calculated for each position across all genes to an exponential decay function to assess the read drop-off over the length of the transcript. This is often referred to as metagene profile. For *E. coli*, around 20% drop-off typically occurs (Li et al., 2014), but this value can differ between species.
2. If a substantial amount of ribosome drop-off is observed, the growth medium and harvesting condition should be carefully examined to avoid nutrient depletion. For example, cell cultures transitioning out of exponential phase may start to consume amino acids, leading to prolonged ribosome pausing and subsequent rescue and release. Signatures of pausing and ribosome drop-off at specific codons can be determined by aligning ribosome density profiles by the codon of interest. Another source of substantial drop-off relates to slow cell harvesting, in which cells are depleted for all nutrients. It is therefore important to carefully characterize the growth curve and move as quickly as possible during harvesting.
3. Confirm that reads map primarily to protein coding regions of genes, and few reads map to UTRs.

4.3.3 Quantifying Protein Synthesis Rates

1. The ribosome density profile for each gene is first normalized to the metagene profile to correct for small density decreases as a function of gene length. Another correction is then applied to account for small local variations in elongation rates. For example, slow elongation at Shine–Dalgarno-like sequences can lead to elevated ribosome density without affecting protein synthesis rates. The effects of such sequences can be first quantitated by averaging across all incidences, and the magnitude of the effects used to correct for each incidence.
2. Calculate protein synthesis rates as the sum of corrected ribosome density profile to an ORF divided by the length of the ORF. Exclude reads from the first and last five codons, as the read density at these codons is affected by translation initiation and termination.

3. Compare protein synthesis rates between samples by normalizing ribosome density by the total number of mapped reads, excluding reads that map to rRNA.

4.3.4 *E. coli* Genes Requiring Further Consideration

1. Genes that include a frameshift signal (*prfB, dnaX*). One can calculate ribosome density by considering only the length of transcript following the frameshift.
2. Selenoproteins (FdhF, FdoG, FdnG). As the selenocysteine codon terminates most ribosomes, ribosome density can be calculated using the length of the transcript following this codon.
3. Proteins without a stop codon (e.g., *arfA*). Consider the last codon in the transcript the stop codon.
4. Genes with known ribosome stall sites (*secM, tnaC*). Exclude the ribosome density in the region surrounding the stall site.
5. Genes with nearly identical sequence (*tufA/tufB, gadA/gadB, ynaE/ydfK, ldrA/ldrC, ybfD/yhhI, tfaR/tfaQ, rzoD/rzoR, pinR/pinQ*). Consider these pairs as the same protein and calculate synthesis by averaging ribosome density over both coding sequences.

5. CONCLUSION

Ribosome profiling can be used to make global measurements of protein synthesis rates. To ensure such measurements are as accurate as possible and truly reflect the translational state of the cell, care must be taken to prepare samples in a way that minimizes variability and bias. When cells are collected, they must be rapidly flash-frozen to ensure ribosome location does not change, and thus that footprints represent the location of ribosomes. When cDNA is prepared, the efficiencies of reactions must be optimized to ensure that minimal bias is introduced into the final library. Lastly, data processing is critical for confirming that key assumptions required to relate ribosome density to synthesis rate hold.

Quantitative measurements of protein synthesis rates have the ability to uncover interesting new biology. Protein synthesis accounts for 50% of energy consumption in exponentially growing bacterial cells and ribosome profiling can thus reveal how cells allocate this important resource. Determining differences in protein synthesis rates in different conditions or across species can help elucidate how bacterial cells regulate their proteome in response to different constraints and can help to uncover novel forms of translational regulation.

REFERENCES

Datta, A. K., & Burma, D. P. (1972). Association of ribonuclease I with ribosomes and their subunits. *The Journal of Biological Chemistry*, *247*(21), 6795–6801.

Dingwall, C., Lomonossoff, G. P., & Laskey, R. A. (1981). High sequence specificity of micrococcal nuclease. *Nucleic Acids Research*, *9*(12), 2659–2673.

Hafner, M., Landgraf, P., Ludwig, J., Rice, A., Ojo, T., Lin, C., et al. (2008). Identification of microRNAs and other small regulatory RNAs using cDNA library sequencing. *Methods (San Diego, Calif.)*, *44*(1), 3–12. https://doi.org/10.1016/j.ymeth.2007.09.009.

Ho, C. K., Wang, L. K., Lima, C. D., & Shuman, S. (2004). Structure and mechanism of RNA ligase. *Structure (London, England: 1993)*, *12*(2), 327–339. https://doi.org/10.1016/j.str.2004.01.011.

Ingolia, N. T., Brar, G. A., Rouskin, S., McGeachy, A. M., & Weissman, J. S. (2012). The ribosome profiling strategy for monitoring translation *in vivo* by deep sequencing of ribosome-protected mRNA fragments. *Nature Protocols*, 7(8), 1534–1550. https://doi.org/10.1038/nprot.2012.086.

Ingolia, N. T., Ghaemmaghami, S., Newman, J. R. S., & Weissman, J. S. (2009). Genome-wide analysis in vivo of translation with nucleotide resolution using ribosome profiling. *Science*, *324*(5924), 218–223. https://doi.org/10.1126/science.1168978.

Li, G.-W. (2015). How do bacteria tune translation efficiency? *Current Opinion in Microbiology*, *24*, 66–71. https://doi.org/10.1016/j.mib.2015.01.001.

Li, G.-W., Burkhardt, D., Gross, C., & Weissman, J. S. (2014). Quantifying absolute protein synthesis rates reveals principles underlying allocation of cellular resources. *Cell*, *157*(3), 624–635. https://doi.org/10.1016/j.cell.2014.02.033.

Li, G.-W., Oh, E., & Weissman, J. S. (2012). The anti-Shine–Dalgarno sequence drives translational pausing and codon choice in bacteria. *Nature*, *484*(7395), 538–541. https://doi.org/10.1038/nature10965.

Mohammad, F., Woolstenhulme, C. J., Green, R., & Buskirk, A. R. (2016). Clarifying the translational pausing landscape in bacteria by ribosome profiling. *Cell Reports*, *14*(4), 686–694. https://doi.org/10.1016/j.celrep.2015.12.073.

O'Connor, P. B. F., Li, G.-W., Weissman, J. S., Atkins, J. F., & Baranov, P. V. (2013). rRNA:mRNA pairing alters the length and the symmetry of mRNA-protected fragments in ribosome profiling experiments. *Bioinformatics*, *29*(12), 1488–1491. https://doi.org/10.1093/bioinformatics/btt184.

Oh, E., Becker, A. H., Sandikci, A., Huber, D., Chaba, R., Gloge, F., et al. (2011). Selective ribosome profiling reveals the cotranslational chaperone action of trigger factor in vivo. *Cell*, *147*(6), 1295–1308. https://doi.org/10.1016/j.cell.2011.10.044.

Subramaniam, A. R., DeLoughery, A., Bradshaw, N., Chen, Y., O'Shea, E., Losick, R., et al. (2013). A serine sensor for multicellularity in a bacterium. *eLife*, *2*. https://doi.org/10.7554/eLife.01501.

Subramaniam, A. R., Zid, B. M., & O'Shea, E. K. (2014). An integrated approach reveals regulatory controls on bacterial translation elongation. *Cell*, *159*(5), 1200–1211. https://doi.org/10.1016/j.cell.2014.10.043.

Yin, S., Ho, C. K., & Shuman, S. (2003). Structure-function analysis of T4 RNA ligase 2. *The Journal of Biological Chemistry*, *278*(20), 17601–17608. https://doi.org/10.1074/jbc.M300817200.

Zhuang, F., Fuchs, R. T., Sun, Z., Zheng, Y., & Robb, G. B. (2012). Structural bias in T4 RNA ligase-mediated 3′-adapter ligation. *Nucleic Acids Research*, *40*(7), e54. https://doi.org/10.1093/nar/gkr1263.

CHAPTER ELEVEN

On the Prowl: An In Vivo Method to Identify RNA Partners of a sRNA

Marie-Claude Carrier, Claire Morin, Eric Massé[1]
Department of Biochemistry, RNA Group, Université de Sherbrooke, Sherbrooke, QC, Canada
[1]Corresponding author: e-mail address: eric.masse@usherbrooke.ca

Contents

Abstract

Bacterial cells dispose of numerous strategies to regulate gene expression. Small regulatory RNAs (sRNA) are pervasive molecules that allow gene expression regulation with exquisite precision. These molecules can bind mRNAs and negatively or positively modify their stability and interfere with translation. However, many features of sRNAs render identification of new targets or RNA interacting partners increasingly complex. In this chapter, we present a detailed procedure of MAPS, an in vivo technique based on the copurification of any type of RNA bound to an MS2-tagged sRNA. By focusing on the interaction between two RNAs rather than the outcome of this interaction, MAPS has proven useful in identifying unprecedented sRNA–RNA interactions. Below, we describe how to prepare MAPS samples and how to analyze RNA sequencing data files to determine enrichment ratios of different RNAs in an experimental condition vs a control condition. MAPS can be applied to most sRNAs of *Escherichia coli* and *Salmonella* spp., and can be easily optimized to more distant bacterial species.

Methods in Enzymology, Volume 612
ISSN 0076-6879
https://doi.org/10.1016/bs.mie.2018.09.006

ABBREVIATIONS

3′ETS^{leuZ} 3′ external transcribed spacer of tRNA^{leuZ}
DTT dithiothreitol, 1 m*M*
LB Luria–Bertani broth
MAPS MS2 affinity purification coupled to RNA sequencing
MS2-MBP MS2-maltose-binding protein fusion protein
OD optical density
ORF open reading frame
PbAc lead acetate
pBAD *araB* promoter
PCI phenol–chloroform–isoamyl alcohol
PMFS phenylmethylsulfonyl fluoride
RNAseq RNA sequencing
SAM file sequence alignment/map file format
sRNA small regulatory RNA
tRF tRNA fragment
UTR untranslated region

1. INTRODUCTION

Bacterial cells have developed many strategies to overcome the rapid changes of their environment physicochemical properties. Small regulatory RNAs (sRNAs) are of particular interest when it comes to understanding how bacteria can thrive in harsh conditions (Holmqvist & Wagner, 2017). Indeed, these regulators of gene expression are able to block translation of many mRNA transcripts while promoting translation of others. sRNAs can also modulate mRNA stability directly or indirectly by interacting with different regions or features of an mRNA (Gottesman & Storz, 2011; Papenfort & Vanderpool, 2015). Over the years, it has become evident that sRNAs were even more versatile than previously thought, as much in their characteristics as in their mechanisms of action. For example, it was well accepted that, to achieve regulation, sRNAs interact in the 5′ untranslated region (UTR) or in the first few codons of a target mRNA (Wagner & Romby, 2015). However, this concept has been shattered following the identification of sRNAs targeting mRNAs deep in the coding sequence as it is the case in the DsrA-*rbsD* and MicC-*ompD* sRNA-target pairs (Lalaouna, Morissette, Carrier, & Massé, 2015; Wroblewska & Olejniczak, 2016).

One of the main challenges in the sRNA field is to identify the targetome of a sRNA, i.e., all the mRNAs regulated by a sRNA of interest. Previously, targets were identified by expressing the sRNA and looking at RNA or protein levels. Because sRNA molecules act through a vast array of mechanisms of action (Dutta & Srivastava, 2018), resulting in diversified outcomes, these classical techniques became less attractive. One way of overcoming this challenge is to directly identify the sRNA–RNA interactions, effectively setting aside the target regulation itself. To achieve that, we developed the MAPS technique, standing for MS2 affinity purification coupled to RNA sequencing (RNAseq) (Lalaouna, Carrier, et al., 2015). This in vivo technique relies solely on interaction between molecules to identify RNA partners of a specific sRNA.

Using MAPS, new canonical targets of sRNAs were identified (Lalaouna, Carrier, et al., 2015; Lalaouna, Morissette, et al., 2015; Lalaouna, Prévost, Laliberté, Houé, & Massé, 2018), but our research also shed light on noncanonical targets (Lalaouna, Carrier, et al., 2015; Lalaouna, Morissette, et al., 2015), new mechanisms of action (unpublished data), and unveiled unforeseen types of RNA-based regulators (Lalaouna, Carrier, et al., 2015). Indeed, in a surprising twist, the ultimate regulators, sRNAs, were shown to also be subjected to posttranscriptional regulation in the form of sequestration by other RNA molecules, transforming predators into preys. Following MAPS experiments performed on RyhB and RybB sRNAs, a tRNA-derived RNA fragment (tRF) was shown to base-pair to both sRNAs in vivo. Further investigation unveiled the mechanism of action of the first functional tRF in *Escherichia coli*. This tRF, called 3′ETS^{leuZ} (external transcribed spacer of *leuZ* tRNA), sequesters RyhB and RybB to prevent sRNA transcriptional noise from regulating targets in noninducing conditions.

Over the years, MAPS has proven to be a straightforward and powerful technique to pinpoint putative RNA interacting partners of sRNAs and to narrow the biomolecular screening needed to identify novel interactions. Here, we describe the procedures needed to complete a MAPS assay on a sRNA of interest. The experiment is split in three main steps: cloning the tagged sRNA, preparing the samples, and processing the data. We will also briefly discuss validation of new interacting partners using biomolecular and biochemical approaches.

2. MS2-sRNA: EXPERIMENTAL DESIGN CONSIDERATIONS

A crucial step of the MAPS assay is to construct the tagged sRNA chimera, a step during which many aspects have to be considered. The tag we use for the MAPS consists of two 19-nucleotide-long repeats of the MS2

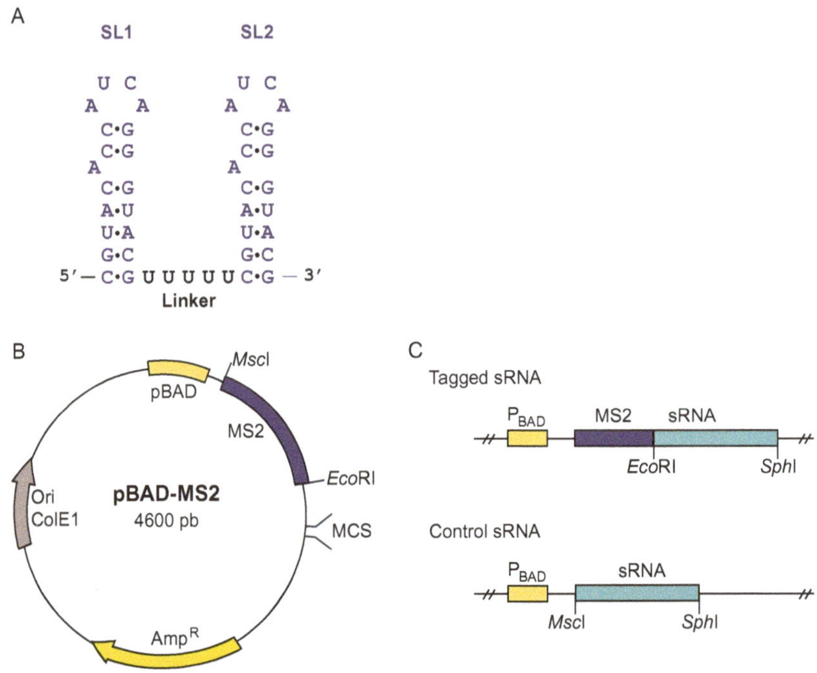

Fig. 1 Tagging a sRNA with the MS2 RNA aptamer. (A) Sequence and structure of the double MS2 RNA aptamer used for the MAPS assay. The stem-loops are connected via a linker. (B) Simplified representation of the pBAD-MS2 vector. The sequence encoding the MS2 RNA aptamer is flanked by restriction sites *Msc*I and *Eco*RI. (C) Ligation of a sRNA at the *Eco*RI/*Sph*I sites results in a chimeric MS2-sRNA construct. Ligation of the sRNA at the *Msc*I/*Sph*I sites results in the loss of the MS2 sequence and yields the sRNA control vector.

RNA aptamer linked by a 5-nucleotide-long linker (Fig. 1A). The aptamer can be inserted anywhere upstream of the transcriptional terminator of a sRNA, assuming the sRNA of interest is not submitted to processing by RNases (Davis & Waldor, 2007; Mandin & Gottesman, 2010; Papenfort, Espinosa, Casadesús, & Vogel, 2015; Papenfort et al., 2009). However, introducing such an aptamer in molecules relying on their sequence and structure could alter their functionality (Małecka, Stróżecka, Sobańska, & Olejniczak, 2015; Wu et al., 2017). When deciding where to fuse the MS2 aptamer in a sRNA, one should always take into consideration the nucleotides involved in mRNA targeting, but also consider the possible alteration to the secondary structure of the sRNA. Moreover, a subset of sRNAs interact with protein chaperones such as Hfq or ProQ, mainly relying on these chaperones for stability and functionality (Attaiech, Glover, & Charpentier, 2017;

Smirnov et al., 2016; Smirnov, Wang, Drewry, & Vogel, 2017; Updegrove, Zhang, & Storz, 2016). If introduction of the MS2 aptamer was to alter these protein-binding sites, degradation of the MS2-sRNA transcript could occur. Thus, it is strongly recommended to perform control experiments on a newly constructed MS2-sRNA and compare the results to that of the sRNA cloned without a tag. We favor performing Northern blot analysis to verify sRNA stability and expression levels of the construct and to determine the optimal expression conditions. Moreover, we suggest using techniques such as Northern blotting or reporter gene assays to validate sRNA activity on a previously characterized target.

For a sRNA that is not processed or that does not present peculiar characteristics, we find fusing the MS2 RNA aptamer to the 5′ end of the sRNA quite efficient. To facilitate the cloning process, we constructed a pBAD-MS2 plasmid (Fig. 1B), deriving from the pNM12 vector (Majdalani, Cunning, Sledjeski, Elliott, & Gottesman, 1998). In pBAD-MS2, the sequence encoding the double MS2 RNA aptamer is flanked by restriction sites *Msc*I and *Eco*RI. For cloning a tagged sRNA, one should use *Eco*RI as the 5′ restriction site. To clone the untagged control sRNA, the use of *Msc*I as the 5′ restriction site will remove the MS2 sequence. This method allows cloning of both experimental and control plasmids using the same vector (Fig. 1C). In both cases, transcription of the sRNA is under the control of the P_{BAD} promoter, induced by the presence of arabinose.

3. SAMPLES PREPARATION

Because MAPS is an in vivo technique, we must take into consideration working with living cells, in constant adaptation to their immediate environment. Moreover, bacterial cells are constantly and rapidly renewing their RNA pool, actively degrading and synthesizing RNA molecules. RNases are important players in the sRNA-based regulation of gene expression. Indeed, sRNAs can promote RNA degradation by facilitating the recruitment of specific RNases, such as RNase E, to a transcript (Mohanty & Kushner, 2016; Waters et al., 2017). The endoribonuclease RNase E is generally involved in the initial cleavage of RNAs targeted for either processing or degradation (Carpousis, 2007). Its unfolded C-terminal domain acts as a scaffolding platform on which accessory proteins assemble to form the RNA degradosome complex, facilitating RNase E activity (Bandyra & Luisi, 2018; Carpousis, 2007).

In our MAPS experiments, to prevent full degradation of negative sRNA targets and insure recovery of the whole targetome of a sRNA, we often use strains harboring the *rne131* allele. In such strains, the C-terminal scaffolding domain of RNase E is missing, preventing assembly of the RNA degradosome (Lopez, Marchand, Joyce, & Dreyfus, 1999). However, since the N-terminal catalytic domain of RNase E is left untouched, the enzyme remains capable of degrading RNA molecules, only with hindered efficiency.

Another precaution we use to avoid as much degradation as possible is to slow down the metabolism of the cell. Following the growth step of the protocol, we always keep the sample in ice-cold buffers. Slowing down the metabolism also helps in recovering RNA–RNA duplexes, as no cross-linking step is performed in this experiment.

3.1 Harvesting the Cell

The procedure described later has been optimized to recover samples from the model bacterium *E. coli* in rich medium and during the exponential phase of growth. sRNA constructs are expressed from derivatives of the pNM12 vector (pBAD-MS2-sRNA and pBAD-sRNA). With that in mind, performing MAPS assays in different conditions or bacterial species requires adaptation of samples preparation procedure.

3.1.1 Equipment

- Agitating water bath
- Spectrophotometer
- 4°C centrifuge (50 mL conic tubes)
- 4°C centrifuge (1.5 mL microtubes)

3.1.2 Buffer and Reagents

- Growth medium (LB broth)
- Arabinose, 20% w/v in ultrapure water
- Buffer A: 20 m*M* Tris–HCl (pH 8), 150 m*M* KCl, 1 m*M* $MgCl_2$, 1 m*M* dithiothreitol (DTT), 1 m*M* phenylmethylsulfonyl fluoride (PMSF). DTT and PMSF should be added to Buffer A immediately before use. Store Buffer A without DTT and PMSF at 4°C
- Liquid nitrogen

3.1.3 Procedure

1. In 100 mL of fresh LB medium supplemented with ampicillin (0.05 mg/mL), dilute an overnight culture 1/1000. Grow the cultures at 37°C with agitation.

2. At an OD_{600nm} of 0.5, induce the expression of the sRNA or MS2-sRNA construct by addition of 0.1% arabinose and let the strains grow for an extra 10 min.
3. Transfer the flasks to an icy-water bath for 10 min. Centrifuge the culture at 4°C for 15 min at 2900 × *g*. Discard the supernatant.
4. Wash cells by suspending pellets from each starting flask in a total of 1 mL of Buffer A. Centrifuge 5 min at 16,000 × *g* in 1.5 mL microtubes and discard the supernatant. Freeze the pelleted cells in liquid nitrogen. Keep at −80°C.

3.1.4 Notes

1. Various concentration of the inducer (e.g., arabinose) and inducing time (here, 10 min) should be tested prior to performing the MAPS experiment. It is preferable to overexpress the sRNA during a growth phase or a growth condition similar to that naturally inducing the expression of the endogenous sRNA.
2. We recommend taking an RNA sample after the ice-water bath as an input control if Northern blots are to be performed at later steps.

3.2 Cell Lysis Procedure

As mentioned previously, the MAPS technique does not involve cross-linking of the RNA–RNA complexes. To recover these native complexes, cells must be broken down gently and with minimal addition of chemicals reagents. The method that we describe here relies on the mechanical disruption of cell membranes by pushing the cellular suspension through a narrow valve, under high pressure, using a pressure cell press (French Press). Other methods have been successfully used to break cells for MAPS experiments such as using glass beads in a bead beater. However, we find the French Press to be the most efficient method to recover a maximal amount of RNAs and to keep RNA–RNA complexes intact.

Again, it is essential to keep samples on ice at all time and only use ice-cold solutions. Even though cells will eventually be lysed, RNases can still degrade RNA molecules in the lysate.

3.2.1 Equipment

- French press with a >3.5-mL lysis cell.
- 4°C centrifuge (1.5 mL microtubes).

3.2.2 Buffer and Reagents

- Buffer A (see Section 3.1.2)

3.2.3 Procedure

1. Thaw the pellets on ice and then suspend each pellet in 2 mL of Buffer A.
2. Using an ice-cold lysis cell, break the bacteria at 430 psi, three times per sample. Keep the samples on ice. Wash and chill the lysis cell between each sample.
3. Clear the lysates by centrifugation at 4°C, 30 min at 16,000 × *g* and transfer the lysates (soluble fraction) to clean tubes and keep on ice.

3.2.4 Notes

1. The volume in which the pellets are suspended can be adjusted according to experimental conditions. For example, cells harvested at higher OD_{600nm} should be resuspended in 3 mL to ensure proper cell lysis. Similarly, the number of times each sample is passed through the French Press can be adjusted.

3.3 MS2 Affinity Purification

Biomolecular exploitation of the specific binding of the MS2 bacteriophage coat protein to the MS2 RNA aptamer for the study of RNA started approximately two decades ago, when Bertrand et al. described a technique consisting in coupling fluorophores to the MS2 protein to track tagged mRNAs in living cells (Bertrand et al., 1998). In the present procedure, we exploit the high affinity between MS2 RNA and MS2 protein in a slightly different manner. We use an MS2-MBP (MS2-maltose-binding protein) fusion protein immobilized on amylose beads as a bait, with which we trap an MS2-tagged sRNA in complex with its cellular interacting partners. We then use the high affinity of maltose for the MBP to recover the samples (Fig. 2).

3.3.1 Equipment

- 1.5 mL Bio-Spin disposable chromatography columns and collecting tubes.
- 4°C and room temperature centrifuge (1.5 mL microtubes).

3.3.2 Buffer and Reagents

- Amylose resin
- Purified MS2-MBP fusion protein (Batey & Kieft, 2007)
- Buffer A (see Section 3.1.2)
- Elution buffer: Buffer A supplemented with 15 m*M* maltose
- Phenol–water, pH 6.0
- 25:24:1 (v/v/v) phenol–chloroform–isoamyl alcohol (PCI)

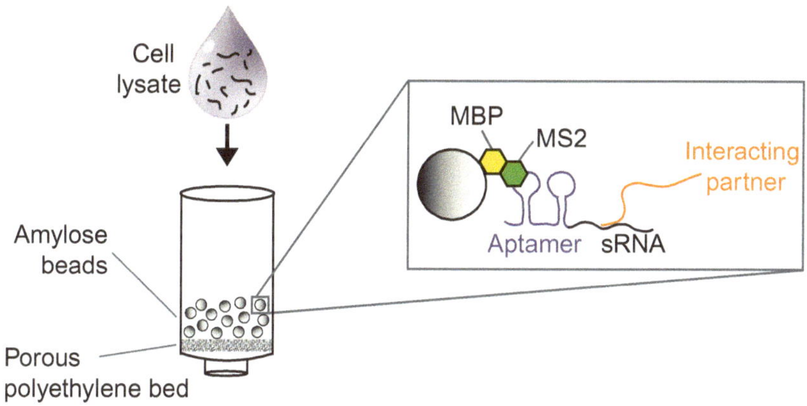

Fig. 2 MS2 affinity purification. In a chromatography column, the amylose beads sit on a porous polyethylene filter (30 μm). The MS2-MBP fusion protein is immobilized on the amylose beads via the MBP moiety. After loading the samples and washing the column, MS2-sRNAs bound to their targets are captured.

- Glycogen
- Ethanol, 95% v/v
- Ice-cold ethanol, 75% v/v

3.3.3 Procedure

1. In a Bio-Spin disposable chromatography column, add 75 μL of amylose resin.
2. Wash the column three times with 1 mL of Buffer A then seal the bottom of the column.
3. Dilute 100 pmol of MS2-MBP coat protein in 1 mL Buffer A and transfer to the column. Let stand for 5 min then remove the seal and let the column drain.
4. Wash the column two times with 1 mL of Buffer A. Load the bacterial lysate (previously cleared samples) onto the column, 1 mL at a time, and let the column drain.
5. Wash the column five times with 1 mL of Buffer A. The number of washes should be optimized.
6. Elute RNAs in a clean RNase-free collecting tube with 1 mL of Elution Buffer and aliquot the eluted samples in two tubes. Add one volume of PCI to each tube. Mix thoroughly and centrifuge 10 min at 16,000 × *g* at room temperature.

7. Transfer the aqueous phase in clean microtubes containing 20 µg of glycogen. Add two volumes of EtOH (95%). Mix well and precipitate overnight at −80°C. For each initial sample, you will have two microtubes.
8. Centrifuge 30 min at 16,000 × *g* at 4°C. Remove the supernatant carefully. Wash the samples with 500 µL of cold 75% EtOH followed by centrifugation at 16,000 × *g* and 4°C for 5 min. Remove the supernatant and let the pellets dry. Suspend the pellets of each sample in a total 86 µL of ultrapure water.

3.3.4 Notes

1. Amount of resin and of MS2-MBP protein used needs to be optimized according to the experiment. For example, if cell harvesting is performed at high OD_{600nm}, increasing the volume of resin and the amount of MS2-MBP protein used in a column will prevent quick saturation of the bait protein.
2. Column drainage should be carried out by gravity only. It might be necessary to push the first drop using a clean syringe.
3. Addition of glycogen to the RNA before adding EtOH greatly facilitates recovery of the RNA pellet after precipitation. Indeed, as glycogen is insoluble in EtOH, its precipitation traps the RNA molecules and forms a visible pellet.
4. It is strongly recommended to verify the quality of your samples after affinity purification by performing Northern blot analysis. Compare output samples to input samples (see Section 3.1.4, note 2).

3.4 RNAseq Sample Preparation

3.4.1 Equipment

- Agilent Nano Chip
- Bioanalyzer 2100
- MiSeq (Illumina)

3.4.2 Buffer and Reagents

- 10 × TURBO™ DNase Buffer
- TURBO™ DNase
- 25:24:1 (v/v/v) PCI
- Ethanol, 95% v/v
- ScriptSeq™ v2 RNA-Seq Library Preparation Kit from Illumina

3.4.3 Procedure

1. Add 10 μL of 10× TURBO™ DNase Buffer and 4 U of TURBO™ DNase to each sample (see Section 3.3.3, step 8). Incubate 30 min at 37°C.
2. Add 100 μL of PCI to each tube and mix. Centrifuge 10 min at 16,000 × *g* at room temperature.
3. Add 2.5 volumes of EtOH (95%). Mix thoroughly and precipitate overnight at −80°C. Centrifuge 30 min at 16,000 × *g* and 4°C. Remove the supernatant. Let the RNA pellets dry completely and then suspend in 6 μL of ultrapure H_2O.
4. Analyze the RNA using an Agilent Nano Chip in a Bioanalyzer 2100.
5. Prepare the cDNA libraries with the ScriptSeq™ v2 RNA-Seq Library Preparation Kit from Illumina then sequence them using Illumina MiSeq. Sequencing in both directions is recommended.

4. PROCESSING THE RNAseq DATA FILES

Processing the RNAseq data obtained through Illumina is a crucial step in identifying sRNA interacting partners. There are many different procedures that can be followed and many different softwares can be used to analyze RNAseq data. The procedure that we describe below only uses the freely available Galaxy Project platform (Afgan et al., 2016) and has the major advantage of being simple and straightforward, allowing anyone to reproduce the bioinformatical pipeline presented here. In this pipeline, many parameters should be optimized according to your dataset and to the stringency at which you want to analyze your data. A detailed visual schematization of the analysis workflow is available (Carrier, Laliberté, & Massé, 2018; Lalaouna, Prévost, Eyraud, & Massé, 2017).

4.1 Visualization of Reads Alignment on the Genome

This step allows for a visual representation of the RNAseq reads on the genome of interest and is useful to compare the experimental data set (MS2-sRNA) to the control data set (sRNA). Each dataset should be subjected independently to the data processing pipeline presented later.

4.1.1 Procedure

1. Upload the FastQ sequencing files downloaded through the Illumina platform.
2. Execute the *FastQ Groomer* tool (Sanger & Illumina 1.8+) (Blankenberg et al., 2010).

3. Run a *FastQC* analysis to visualize the quality of the sequences. If necessary, run a *FastQ Quality Trimmer* (Andrews, 2010; Blankenberg et al., 2010).
4. Use *Map with Bowtie for Illumina* and select the paired-end analysis. Use file from step 3 as input. Select your reference genome. Choose a value for the Seed parameter and set the Maximum number of mismatch allowed in the seed (Langmead, Trapnell, Pop, & Salzberg, 2009).
5. Perform a *Genome Coverage*. Select the Bowtie output data sets (SAM files) previously obtained (Quinlan & Hall, 2010). Select Report region with zero coverage. Download the output file and add the following header before the first line: track type=bedGraph name="NAME_EXPERIMENT" description="BedGraph format" visibility=full.
6. Go to the UCSC Microbial Genome Browser (http://microbes.ucsc.edu/) and select your genome of interest (Chan, Holmes, Smith, Tran, & Lowe, 2012). Add your custom tracks (step 5). You can now scan the genome and visually compare alignment of reads at different genomic positions (Fig. 3A).

4.2 Identification of Enriched mRNAs With the MS2-sRNA

After visualizing the alignment of all the RNAseq reads on the genome, it is interesting to assign each read to a gene. This permits identification of putative interacting partners according to the enrichment ratio of reads

A

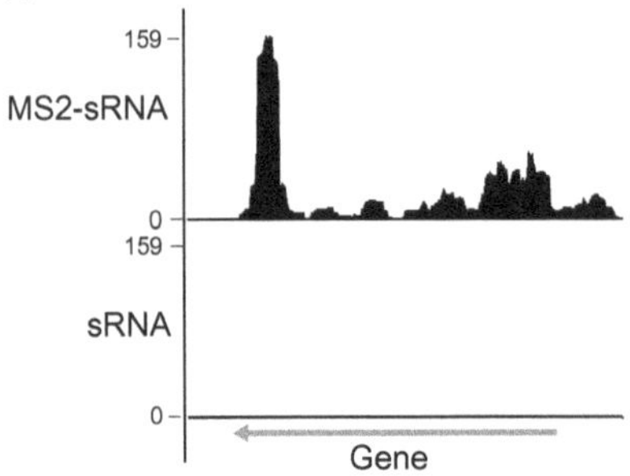

B

MS2-sRNA			sRNA		
Name	# Reads	Length	Name	# Reads	Length
Gene1	3030	1088	Gene1	11	1088
Gene2	2352	355	Gene2	692	355
Gene3	1122	1415	Gene3	306	1415
Gene4	637	419	Gene4	32	419
Gene5	2087	1103	Gene5	329	1103
Gene6	223	136	Gene6	1	1
Gene7	258	24	Gene7	224	24
Gene8	1649	736	Gene8	81	736

↓

Normalize read counts
Calculate enrichment ratio

Fig. 3 Key output of the RNAseq data processing. (A) Visual representation of RNAseq reads aligned on the reference genome (here, *E. coli* K12). Both experiment and control tracks are manually set to the same *Y* axis value. (B) Example of the table obtained after comparing the experimental data to the control data. The file obtained through the Galaxy Project platform was opened using Microsoft Excel.

for a specific gene in the experimental data set (MS2-sRNA) vs the control (sRNA or MS2) data set. The following procedure will allow this assignation, as long as the annotated genome is available. A detailed visual schematization of the analysis workflow is available (Carrier et al., 2018; Lalaouna et al., 2017).

Three different processed files will be required to complete this procedure. First, the two SAM files from Section 4.1.1, step 4. The last file required is the Gene Data Bank of the bacterial strain of interest. This file is usually found on the National Center for Biotechnology Information (NCBI) website (https://www.ncbi.nlm.nih.gov/).

4.2.1 Procedure

1. Run the *Convert SAM to interval* tool with default parameters on each SAM files.
2. Run the *Remove beginning of a file* tool. These data sets are ready to use.
3. Upload the Gene Data Bank file. Set the file type as Interval.
4. Run the *Compute an expression on every row* tool with the Add Interval parameter c3-c2. Then, modify the attributes to set the Database/Build as the reference genome. This data file is ready to use.
5. Run the *Join* tool. In the parameters, join your sample data set (step 2) to the Gene Data Bank data set (see step 4).
6. Execute the *Group* tool for each output files.
7. *Join two datasets side by side on a specified field* tool with outputs from step 6. Join the MS2-sRNA sample using column 1 with the sRNA control sample using column 1.
8. Download the output file and open it with a spreadsheet software (e.g., Microsoft Excel). The three columns represent gene name, number of reads, and gene length for the MS2-sRNA experiment (first three columns) and for the control (last three columns) (Fig. 3B).
9. Relativize the number of reads using the total number of reads and the total number of reads mapped for each sample.
10. Calculate the enrichment ratio for each gene between the MS2-sRNA experiment and the sRNA control.

4.2.2 Notes

1. Since an enrichment ratio is calculated between the experimental vs the control condition, relativizing the data according to gene size is irrelevant here.

2. Parameters for step 6:
   a. Group by column: column 23.
   b. Ignore case by grouping: no.
   c. Ignore lines beginning with these characters: select all characters except for the dot (.).
   d. Operation: insert Operation 1, Count Distinct on column 5.
   e. Operation: insert Operation 2, Mean on column 24.
3. Parameters for step 7:
   a. Keep lines of first input that do not join with second input: yes
   b. Keep lines of first input that are incomplete: no
   c. Fill Empty columns: yes
   d. Fill column by: Single fill value
   e. Fill value: 1

5. FOLLOWING THE RNAseq: BIOMOLECULAR CHARACTERIZATION OF INTERACTING PARTNERS

After identification of putative interacting partners of a sRNA of interest using the MAPS assay, it is important to remember that an RNA is not always regulated by an sRNA simply because they interact with each other. Recently, it became clear that formation of a sRNA–RNA duplex can result in much more than sRNA-based regulation. The three tRNAs encoded in the *glyW-cysT-leuZ* operon were among the most enriched RNAs following the MAPS of RyhB and RybB. One could have assumed that the polycistronic RNA was regulated by the sRNAs. However, investigation revealed that both sRNAs were rather sponged by a tRF segment derived from the operon, the $3'ETS^{leuZ}$. This is a convincing example illustrating validation requirements for new interacting partners.

Following MAPS assay, we use techniques to analyze RNA and protein levels in the absence or presence of the sRNA. Northern blotting is a straightforward qualitative technique to determine RNA levels in different conditions. However, if no change of RNA levels is detected, this does not overrule sRNA-based regulation. We suggest using a technique permitting evaluation of protein levels such as the use of reporter genes (e.g., *gfp*, *lacZ*, etc.) translationally fused to the gene of interest. Western blotting with a specific antibody or by tagging the protein of interest is also an efficient method to evaluate sRNA impact on translation.

Another important parameter to validate following MAPS is the base-pairing site of the sRNA. Reads alignment figures obtained through the UCSC genome browser provide clues about the base-pairing site of an sRNA on a given RNA. For instance, if all the RNAseq reads are located on, or near, the initiator codon of an mRNA, it suggests an interaction in the 5′ region. If a majority of reads were located in the middle of the open reading frame (ORF), for example, exploring the possibility of a pairing deep in the ORF would be interesting. Keep in mind, though, that the information provided through MAPS concerning the base-paring site is only suggestive. In vitro techniques such as lead acetate (PbAc) probing or in line probing must be performed to identify which nucleotides are involved in the interaction.

6. SUMMARY AND CONCLUSION

sRNAs are precise regulators using a vast array of mechanisms of action to control their targets. Their versatility, however, is creating one of the main challenges in the field: identify new sRNA-target pairs. The possible outcomes of sRNA-dependent regulation on an mRNA are numerous. Also, the fact that one sRNA can regulate different targets through independent mechanisms adds to the complexity of sRNA target identification. Moreover, it is increasingly clear that sRNAs are fine-tuning regulators: their action might have only a small effect on one specific target, but regulation of their whole targetome will influence cell adaptation. It is therefore difficult to try to identify the targetome of a sRNA using a technique based on the observation of the regulation.

To overcome this challenge, we decided to exploit idea behind in silico base-pairing predictors, that is to say try to identify the interaction itself, instead of the outcome of this interaction. MAPS was therefore developed and optimized with the aim of identifying RNA–RNA interactions in an in vivo context. Studying RNA–RNA interaction in vivo avoids disregarding essential characteristics of the RNAs involved. In contrast, most in silico predictors do not consider the secondary structures of both partners, nor do they considering cellular localization and expression patterns of both RNAs. Since MAPS samples originate from living cells, these characteristics of both RNAs are intrinsically taken into consideration in RNA–RNA base-pairing events.

MAPS is quite versatile as dozens of parameters can be optimized to fit specific experimental conditions and various model bacterium. In this chapter, we described in details MAPS procedures optimized for sampling RNAs from *E. coli*. The experiment described here uses a tagged sRNA as a bait, but any type of RNA can be tagged. For example, one could tag an mRNA of interest and perform MAPS assay to identify sRNAs bound to that specific transcript.

In all variation of the original MAPS protocol, similar precautions must be taken to avoid degradation of the samples. It is essential to work with RNase-free buffers and labware as the samples are subjected to quite a few treatments before the high-throughput sequencing. With that in mind, verification of the quality of the RNA through Northern blotting at various checkpoints is strongly recommended. Considering the in vivo nature of MAPS, as discussed earlier, RNA degradation pathways must be acknowledged when determining experimental conditions.

ACKNOWLEDGMENTS

This work was funded by an operating grant from the Canadian Institutes of Health Research (CIHR) to E.M. M.-C.C. holds an Alexander Graham Bell Doctoral scholarship from the Natural Sciences and Engineering Research Council of Canada (NSERC).

REFERENCES

Afgan, E., Baker, D., van den Beek, M., Blankenberg, D., Bouvier, D., Čech, M., et al. (2016). The Galaxy platform for accessible, reproducible and collaborative biomedical analyses: 2016 update. *Nucleic Acids Research*, *44*(W1), W3–W10. https://doi.org/10.1093/nar/gkw343.

Andrews, S. (2010). *Babraham bioinformatics—FastQC a quality control tool for high throughput sequence data.* Retrieved from http://www.bioinformatics.babraham.ac.uk/projects/fastqc/.

Attaiech, L., Glover, J. N. M., & Charpentier, X. (2017). RNA chaperones step out of Hfq's shadow. *Trends in Microbiology*, *25*(4), 247–249. https://doi.org/10.1016/j.tim.2017.01.006.

Bandyra, K. J., & Luisi, B. F. (2018). RNase E and the high-fidelity orchestration of RNA metabolism. *Microbiology Spectrum*, *6*(2), 1–16. https://doi.org/10.1128/microbiolspec.RWR-0008-2017.

Batey, R. T., & Kieft, J. S. (2007). Improved native affinity purification of RNA. *RNA*, *13*(8), 1384–1389. https://doi.org/10.1261/rna.528007.

Bertrand, E., Chartrand, P., Schaefer, M., Shenoy, S. M., Singer, R. H., & Long, R. M. (1998). Localization of ASH1 mRNA particles in living yeast. *Molecular Cell*, *2*(4), 437–445. Retrieved from http://www.ncbi.nlm.nih.gov/pubmed/9809065.

Blankenberg, D., Gordon, A., Von Kuster, G., Coraor, N., Taylor, J., Nekrutenko, A., et al. (2010). Manipulation of FASTQ data with Galaxy. *Bioinformatics*, *26*(14), 1783–1785. https://doi.org/10.1093/bioinformatics/btq281.

Carpousis, A. J. (2007). The RNA degradosome of *Escherichia coli*: An mRNA-degrading machine assembled on RNase E. *Annual Review of Microbiology*, *61*(1), 71–87. https://doi.org/10.1146/annurev.micro.61.080706.093440.

Carrier, M.-C., Laliberté, G., & Massé, E. (2018). Identification of new bacterial small RNA targets using MS2 affinity purification coupled to RNA sequencing. *Methods in Molecular Biology (Clifton, N.J.)*, *1737*, 77–88. https://doi.org/10.1007/978-1-4939-7634-8_5.

Chan, P. P., Holmes, A. D., Smith, A. M., Tran, D., & Lowe, T. M. (2012). The UCSC archaeal genome browser: 2012 Update. *Nucleic Acids Research*, *40*(D1), D646–D652. https://doi.org/10.1093/nar/gkr990.

Davis, B. M., & Waldor, M. K. (2007). RNase E-dependent processing stabilizes MicX, a Vibrio cholerae sRNA. *Molecular Microbiology*, *65*(2), 373–385. https://doi.org/10.1111/j.1365-2958.2007.05796.x.

Dutta, T., & Srivastava, S. (2018). Small RNA-mediated regulation in bacteria: A growing palette of diverse mechanisms. *Gene*, *656*, 60–72. https://doi.org/10.1016/J.GENE.2018.02.068.

Gottesman, S., & Storz, G. (2011). Bacterial small RNA regulators: Versatile roles and rapidly evolving variations. *Cold Spring Harbor Perspectives in Biology*, *3*(12), 1–16. https://doi.org/10.1101/cshperspect.a003798.

Holmqvist, E., & Wagner, E. G. H. (2017). Impact of bacterial sRNAs in stress responses. *Biochemical Society Transactions*, *45*(6), 1203–1212. https://doi.org/10.1042/BST20160363.

Lalaouna, D., Carrier, M. C., Semsey, S., Brouard, J. S., Wang, J., Wade, J., et al. (2015). A 3′ external transcribed spacer in a tRNA transcript acts as a sponge for small RNAs to prevent transcriptional noise. *Molecular Cell*, *58*(3), 393–405.

Lalaouna, D., Morissette, A., Carrier, M.-C., & Massé, E. (2015). DsrA regulatory RNA represses both hns and rbsD mRNAs through distinct mechanisms in Escherichia coli. *Molecular Microbiology*, *98*(2), 357–369. https://doi.org/10.1111/mmi.13129.

Lalaouna, D., Prévost, K., Eyraud, A., & Massé, E. (2017). Identification of unknown RNA partners using MAPS. *Methods*, *117*, 28–34. https://doi.org/10.1016/j.ymeth.2016.11.011.

Lalaouna, D., Prévost, K., Laliberté, G., Houé, V., & Massé, E. (2018). Contrasting silencing mechanisms of the same target mRNA by two regulatory RNAs in Escherichia coli. *Nucleic Acids Research*, *46*(5), 2600–2612. https://doi.org/10.1093/nar/gkx1287.

Langmead, B., Trapnell, C., Pop, M., & Salzberg, S. L. (2009). Ultrafast and memory-efficient alignment of short DNA sequences to the human genome. *Genome Biology*, *10*(3), R25. https://doi.org/10.1186/gb-2009-10-3-r25.

Lopez, P. J., Marchand, I., Joyce, S. A., & Dreyfus, M. (1999). The C-terminal half of RNase E, which organizes the Escherichia coli degradosome, participates in mRNA degradation but not rRNA processing in vivo. *Molecular Microbiology*, *33*(1), 188–199. Retrieved from http://www.ncbi.nlm.nih.gov/pubmed/10411735.

Majdalani, N., Cunning, C., Sledjeski, D., Elliott, T., & Gottesman, S. (1998). DsrA RNA regulates translation of RpoS message by an anti-antisense mechanism, independent of its action as an antisilencer of transcription. *Proceedings of the National Academy of Sciences of the United States of America*, *95*(21), 12462–12467. https://doi.org/10.1073/PNAS.95.21.12462.

Małecka, E. M., Stróżecka, J., Sobańska, D., & Olejniczak, M. (2015). Structure of bacterial regulatory RNAs determines their performance in competition for the chaperone protein Hfq. *Biochemistry*, *54*(5), 1157–1170. https://doi.org/10.1021/bi500741d.

Mandin, P., & Gottesman, S. (2010). Integrating anaerobic/aerobic sensing and the general stress response through the ArcZ small RNA. *The EMBO Journal*, *29*(18), 3094–3107. https://doi.org/10.1038/emboj.2010.179.

Mohanty, B. K., & Kushner, S. R. (2016). Regulation of mRNA decay in bacteria. *Annual Review of Microbiology*, *70*(1), 25–44. https://doi.org/10.1146/annurev-micro-091014-104515.

Papenfort, K., Espinosa, E., Casadesús, J., & Vogel, J. (2015). Small RNA-based feedforward loop with AND-gate logic regulates extrachromosomal DNA transfer in Salmonella. *Proceedings of the National Academy of Sciences of the United States of America*, *112*(34), E4772–E4781. https://doi.org/10.1073/pnas.1507825112.

Papenfort, K., Said, N., Welsink, T., Lucchini, S., Hinton, J. C. D., & Vogel, J. (2009). Specific and pleiotropic patterns of mRNA regulation by ArcZ, a conserved, Hfq-dependent small RNA. *Molecular Microbiology*, *74*, 139–158. https://doi.org/10.1111/j.1365-2958.2009.06857.x.

Papenfort, K., & Vanderpool, C. K. (2015). Target activation by regulatory RNAs in bacteria. *FEMS Microbiology Reviews*, *39*(3), 362–378. https://doi.org/10.1093/femsre/fuv016.

Quinlan, A. R., & Hall, I. M. (2010). BEDTools: A flexible suite of utilities for comparing genomic features. *Bioinformatics (Oxford, England)*, *26*(6), 841–842. https://doi.org/10.1093/bioinformatics/btq033.

Smirnov, A., Förstner, K. U., Holmqvist, E., Otto, A., Günster, R., Becher, D., et al. (2016). Grad-seq guides the discovery of ProQ as a major small RNA-binding protein. *Proceedings of the National Academy of Sciences of the United States of America*, *113*(41), 11591–11596. https://doi.org/10.1073/pnas.1609981113.

Smirnov, A., Wang, C., Drewry, L. L., & Vogel, J. (2017). Molecular mechanism of mRNA repression in trans by a ProQ-dependent small RNA. *The EMBO Journal*, *36*(8), 1029–1045. https://doi.org/10.15252/embj.201696127.

Updegrove, T. B., Zhang, A., & Storz, G. (2016). Hfq: The flexible RNA matchmaker. *Current Opinion in Microbiology*, *30*, 133–138. https://doi.org/10.1016/j.mib.2016.02.003.

Wagner, E. G. H., & Romby, P. (2015). Small RNAs in bacteria and archaea. *Advances in Genetics*, *90*, 133–208. https://doi.org/10.1016/bs.adgen.2015.05.001.

Waters, S. A., Mcateer, S. P., Kudla, G., Pang, I., Deshpande, N. P., Amos, T. G., et al. (2017). Small RNA interactome of pathogenic E. coli revealed through crosslinking of RNase E. *The EMBO Journal*, *36*, 374–387. https://doi.org/10.15252/embj.

Wroblewska, Z., & Olejniczak, M. (2016). Contributions of the Hfq protein to translation regulation by small noncoding RNAs binding to the mRNA coding sequence. *Acta Biochimica Polonica*, *63*(4), 701–707. https://doi.org/10.18388/abp.2016_1362.

Wu, P., Liu, X., Yang, L., Sun, Y., Gong, Q., Wu, J., et al. (2017). The important conformational plasticity of DsrA sRNA for adapting multiple target regulation. *Nucleic Acids Research*, *45*(16), 9625–9639. https://doi.org/10.1093/nar/gkx570.

CHAPTER TWELVE

Analysis of Bacterial Transcription by "Massively Systematic Transcript End Readout," MASTER

Irina O. Vvedenskaya, Seth R. Goldman, Bryce E. Nickels[1]
Waksman Institute and Department of Genetics, Rutgers University, Piscataway, NJ, United States
[1]Corresponding author: e-mail address: bnickels@waksman.rutgers.edu

Contents

Abstract

A systems-level view of cellular gene expression requires understanding the mechanistic principles governing each step of transcription. In this chapter, we describe a massively multiplexed method for the analysis of the relationship between nucleic acid

Methods in Enzymology, Volume 612
ISSN 0076-6879
https://doi.org/10.1016/bs.mie.2018.09.007

sequence and transcription termed "MASTER," for massively systematic transcript end readout. MASTER enables parallel measurements of transcription output from at least 4^{10} (~1,000,000) individual template sequences in vitro and in vivo. MASTER involves constructing a DNA template library of barcoded sequences, generating RNA transcripts from the library during transcription in vitro or in vivo, and analyzing the relative abundance and 5′-end sequences of the RNA transcripts by high-throughput sequencing. MASTER provides a powerful, rapid, and versatile method to identify sequence determinants of each step of transcription and to define the mechanistic basis by which these sequence determinants dictate transcription output.

ABBREVIATIONS

BSA	bovine serum albumin
CFU	colony-forming unit
dsDNA	double-stranded DNA
DTT	1,4-Dithiothreitol
ePCR	emulsion PCR
LB	lysogeny broth
MASTER	massively systematic transcript end readout
ssDNA	single-stranded DNA
UMI	unique molecular identifier

1. INTRODUCTION

1.1 Development of "Massively Systematic Transcript End Readout," MASTER

Proper control of gene expression is essential for organismal development, cellular response to environmental signals, and the prevention of disease. Transcription is the first step in gene expression and is a primary target of regulation. Transcription in all cells is carried out by members of a family of multisubunit RNA polymerases (RNAPs) that are highly conserved in sequence, structure, and function across all organisms (Cramer, 2002; Darst, 2001; Ebright, 2000).

During each phase of transcription, RNAP makes extensive interactions with nucleic acids and is responsive to sequence context. Thus, understanding how cells ensure proper control of gene expression requires knowledge of RNAP activity in all sequence contexts for each phase of transcription. To achieve this goal, we have developed a method termed "MASTER," for massively systematic transcript end readout (Vvedenskaya et al., 2015). MASTER enables the parallel analysis of transcription output from up to at least 4^{10} (~1 million) template sequence variants in vitro and in vivo.

1.2 Application of MASTER to Bacterial Transcription

Bacterial RNAP is the smallest and best-characterized member of the multisubunit RNAP family (Cramer, 2002; Darst, 2001; Ebright, 2000). Thus, bacterial RNAP provides an excellent model system for defining mechanistic principles that are relevant to the function of all multisubunit RNAPs. The focus of this chapter is on the application of MASTER for studies of transcription initiation by bacterial RNAP. In particular, we describe methods suitable for analysis of the effects of promoter sequence on transcription output by *Escherichia coli* RNAP. These procedures can also be adapted for studies of postinitiation steps of transcription.

The bacterial RNAP core enzyme (subunit composition $\alpha_2\beta\beta'\omega$) can carry out nonspecific transcription initiation and transcription elongation but must associate with a σ factor, forming the RNAP holoenzyme (subunit composition $\alpha_2\beta\beta'\omega\ \sigma$), to carry out promoter-specific transcription (Marchetti, Malinowska, Heller, & Wuite, 2017; Ruff, Record, & Artsimovitch, 2015). σ contains determinants for sequence-specific recognition of promoter DNA and, through those determinants, targets RNAP to promoters (Feklistov, Sharon, Darst, & Gross, 2014; Paget, 2015). During promoter-specific transcription initiation the RNAP holoenzyme binds promoter DNA, unwinds ~12–16 bp of promoter DNA to form an RNAP-promoter open complex containing an unwound "transcription bubble," and selects a transcription start site (Marchetti et al., 2017; Ruff et al., 2015). RNAP remains bound to the promoter and uses a "scrunching" mechanism during the synthesis of the first ~10 nt of the transcript (Kapanidis et al., 2006; Revyakin, Liu, Ebright, & Strick, 2006). During this phase of transcription, termed "initial transcription," the RNAP-promoter initial-transcribing complex can engage in tens or hundreds of abortive cycles of synthesis and release of short RNA products. This process, termed "abortive initiation," competes with, and limits, productive initiation. During initial transcription the RNAP-promoter initial-transcribing complex can also enter into a paused state without releasing the RNA (Duchi et al., 2016; Lerner et al., 2016). This process, termed "initiation pausing," can also limit productive initiation and has been proposed to provide a regulatory checkpoint. Once the RNAP-promoter initial-transcribing complex has synthesized a product of a threshold length of ~10 nt, RNAP breaks its interactions with promoter DNA, breaks some of its interactions with σ, and enters into the elongation phase of transcription. This last step of transcription initiation is termed "promoter escape."

Each step of transcription initiation, from promoter binding to promoter escape, can be affected by multiple sequence determinants. Because each sequence determinant for each step of transcription initiation is only one of several determinants of transcription output, their quantitative significance for different promoters differs. In addition, each step of transcription initiation can be affected by reaction conditions. Thus, predicting transcription output for given promoter sequence under different reaction conditions represents an immense challenge. To address this challenge, we have developed and applied MASTER to measure transcription output for up to at least 4^{10} (~1,000,000) individual promoter sequences in vitro and in vivo (Hochschild, 2015; Vvedenskaya et al., 2015; Winkelman et al., 2016).

Use of MASTER for studies of transcription initiation involves the construction of a promoter template library containing up to 4^{10} (~1,000,000) barcoded sequences, production of RNA transcripts from the template library in vitro and in vivo, and analysis of transcript 5′ ends and transcript yields (Fig. 1). The sections that follow provide a detailed protocol for each of these steps.

2. CONSTRUCTION OF MASTER TEMPLATE LIBRARY

2.1 Overview

To generate a MASTER template library, we use a procedure that provides a "self-assembling barcode," in which for each DNA molecule in the library a first randomized sequence in a region of interest is associated with a known corresponding second randomized sequence that serves as a barcode. To do this, we first synthesize three oligodeoxyribonucleotides (oligos) for use in PCR. One of these oligos, termed the "template" oligo, contains a first randomized sequence spanning the region of interest and a second, longer randomized sequence that serves as a barcode. The other two oligos, termed the "amplification" oligos, carry ~20 nt of 3′-end sequence complementary to the template oligo along with 5′-end sequence that introduces a BglI recognition sequence that is used to clone the PCR product into a plasmid vector.

The double-stranded DNA (dsDNA) generated by PCR is digested, and the digested product is isolated and inserted into a vector backbone by ligation (Fig. 1, top). The ligation mixture is transformed into cells, and recombinant plasmid DNA is isolated from ~10^6–10^7 transformants. This plasmid DNA is a MASTER template library that is negatively supercoiled. In the construction of the library, the barcodes have "self-assembled" such that for each DNA molecule in the library, the first randomized sequence in the region of interest is associated with a corresponding second randomized sequence in the barcode region.

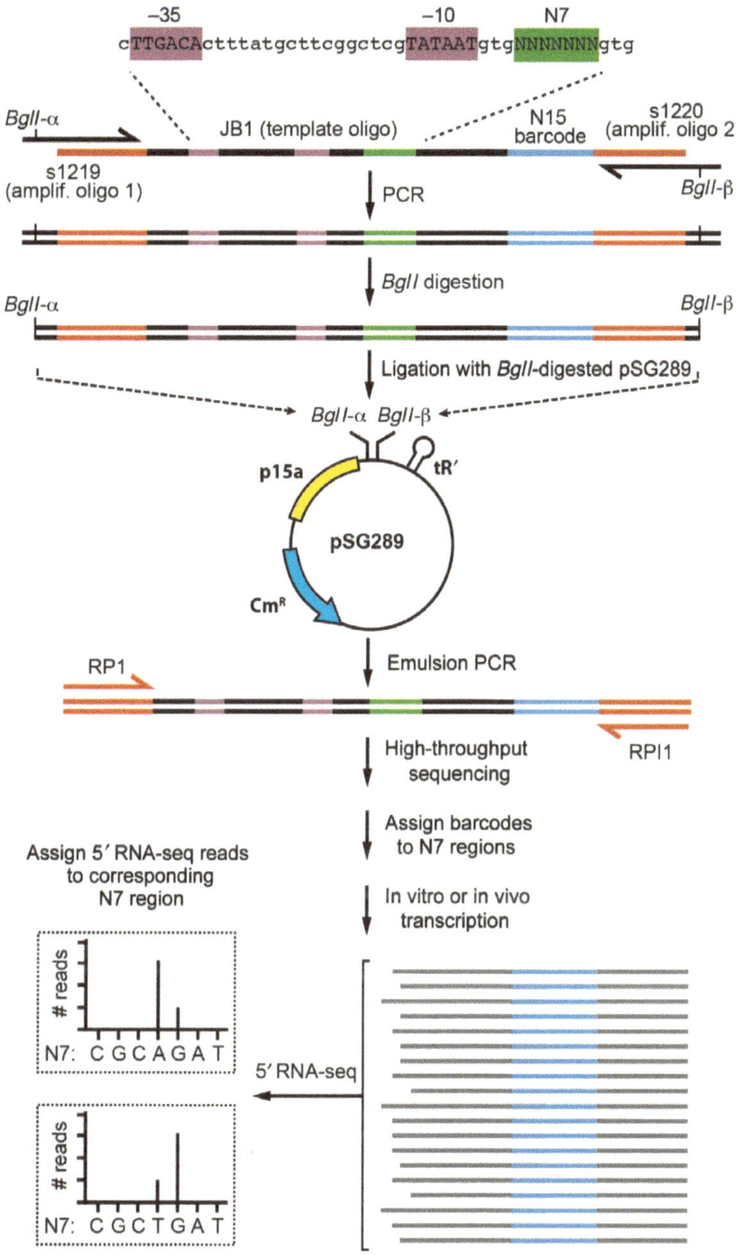

Fig. 1 Massively systematic transcript end readout, MASTER. *Top*: steps in the construction of a representative MASTER template library. A representative template oligo (JB1; Table 1) carrying the *lacCONS-N7* promoter, a 7-nt randomized region at positions 4–10 bps downstream of the *lacCONS* promoter-10 element (*green*), and a 15-nt barcode sequence (*blue*) is used as template in a PCR reaction using amplification *(Continued)*

The composition of the MASTER template library is determined through high-throughput sequencing analysis of a dsDNA product generated by emulsion PCR (ePCR) that spans the insert-containing region of the plasmid (Fig. 1, middle). The sequencing results enable us to identify each unique template sequence in the library and thereby to associate each randomized sequence in the region of interest with its corresponding randomized sequence that serves as its barcode. The sequencing results also provide a measure of the proportion of each DNA template sequence variant in the library. Use of ePCR instead of conventional PCR minimizes "template switching" during DNA synthesis that could result in the generation of amplicons in which barcodes have been switched. The dsDNA product generated by ePCR is a nonsupercoiled, linear form of the MASTER template library also suitable for use in experiments measuring transcription output in vitro.

Our goal is to obtain barcoded template libraries in which >95% of the theoretical maximum number of potential sequence variants in the region of interest are present and in which >95% of the barcodes are associated with a single sequence variant in the region of interest. Two steps of the procedure are designed to achieve this goal. First, the template oligo is designed so that the complexity of the randomized barcode region (typically 4^{15}; ~1 billion) is much greater than the complexity of the randomized region of interest (typically 4^7–4^{10}; ~16,000–1,000,000). Second, the procedure involves the transformation of a ligation mixture into bacterial cells. This step enables us to restrict the number of barcodes in the library to the total number of transformants obtained in this step. Consider the example of a library prepared from 2 million transformants where the complexity of the randomized region of interest is 4^7 (~16,000) and the complexity of the randomized barcode region is 4^{15} (~1,000,000,000). Assuming each sequence variant and each barcode has an equal chance of being represented in the final library we expect each sequence variant to be present ~120 times. Conversely, the likelihood that any of the barcodes will be represented more than once in a library of 2 million is ~0.002.

Fig. 1—Cont'd primers (s1219 and s1220, see Table 1) that introduce BglI sites. The PCR product is digested with BglI and cloned into BglI-digested pSG289 to generate a MASTER template library that contains 4^7 (~16,000) sequence variants at positions 4–10 bps downstream of the *lacCONS*-10 element. *Middle*: product generated by ePCR is sequenced to assign barcodes to template-sequence variants. The PCR primers shown in *red* (RP1 and RPI1; Table 1) carry sequences required for analysis by the Illumina sequencer. *Bottom*: 5′ RNA-seq analysis of RNA produced from the library in vitro and in vivo. The sequence of the barcode is used to assign the RNA to template-sequence variant, the sequence of the 5′ end is used to define the transcription start site, and the number of reads is used to measure transcript yield from each template-sequence variant.

2.2 Amplification of Template Oligo by PCR

2.2.1 Equipment

- Low-binding P10-P1000 barrier tips, sterile
- Low-binding 1.7 mL microtubes, nuclease-free
- Horizontal electrophoresis apparatus
- Single-channel pipettes for 2–1000 μL
- Stainless steel disposable scalpels, No. 10 blade
- Table top centrifuge
- Thermal cycler
- Thin wall 200 μL PCR tubes
- UV transilluminator, GelDoc

2.2.2 Reagents

- 6X DNA loading dye (30% glycerol (v/v), 0.25% bromophenol blue (w/v), 0.25% xylene cyanol FF (w/v))
- Ethidium bromide, 10 mg/mL
- LE Quick Dissolve agarose
- Nuclease-free water, not DEPC-treated
- Oligodeoxyribonucleotides (IDT, see Table 1; we recommend that synthesis of the "template" oligo is performed using "hand mixed" custom base solutions to ensure this solution is prepared with precisely specified dNTP ratios)
- Q5 High-Fidelity 2X Master Mix (NEB)
- QIAQuick Gel Extraction Kit (Qiagen)
- TAE buffer (40 m*M* Tris, 20 m*M* acetic acid, 1 m*M* EDTA)

2.2.3 Protocol

1. Thaw ssDNA template oligo and amplification oligos at room temperature. Thaw Q5 High-Fidelity 2X Master Mix on ice. Assemble PCR cocktail in 200 μL thin wall PCR tubes as follows.

Component	Volume (μL)
1 μ*M* template oligo (e.g., JB1)	1
100 μ*M* amplification oligo 1 (e.g., s1219)	1
100 μ*M* amplification oligo 2 (e.g., s1220)	1
Nuclease-free water	7
Q5 High-Fidelity 2X Master Mix	10
Total volume (μL)	20

Table 1 Oligonucleotides Used in MASTER

Name	Sequence (5′–3′)	Description
s1206	GUUCAGAGUUCUACAGUCCGACGAUCNNNNNNNNNNNNNNN	Standard 5′ adapter (HPLC purified)
i105	GUUCAGAGUUCUACAGUCCGACGAUC<u>CUGA</u>NNNNNNNNNNN	Barcoded 5′ adapter 1; CUGA barcode is underlined (HPLC purified)
i106	GUUCAGAGUUCUACAGUCCGACGAUC<u>GACU</u>NNNNNNNNNNN	Barcoded 5′ adapter 2; GACU barcode is underlined (HPLC purified)
i107	GUUCAGAGUUCUACAGUCCGACGAUC<u>AGUC</u>NNNNNNNNNNN	Barcoded 5′ adapter 3; AGUC barcode is underlined (HPLC purified)
i108	GUUCAGAGUUCUACAGUCCGACGAUC<u>UCAG</u>NNNNNNNNNNN	Barcoded 5′ adapter 4; UCAG barcode is underlined (HPLC purified)
JB1	GTTCAGAGTTCTACAGTCCGACGATCAGGC**TTGACA**CTTTATGCTTCGGCTCG**TATAAT**GTGNNNNNNNGTGAGCGGATAACAATNNNNNNNNNNNNNNNTGGAATTCTCGGGTGCCAAGG	Template oligo carrying *placCONS* sequence, a 7-nt randomized region, and a 15-nt randomized region; promoter −10 (TATAAT) and −35 elements (TTGACA) are in bold (PAGE purified)
RP1	AATGATACGGCGACCACCGAGATCTACACGTTCAGAGTTCTACAGTCCGA	Illumina PCR primer (HPLC purified)
RPI1	CAAGCAGAAGACGGCATACGAGATCGTGATGTGACTGGAGTTCCTTGGCACCCGAGAATTCCA	Illumina indexing PCR primer 1 (HPLC purified)
s128	CCTTGGCACCCGAGAATTCC	RT primer (HPLC purified)
s1115	CTACACGTTCAGAGTTCTACAGTCCGACGATC	Custom Illumina sequencing primer (HPLC purified)
s1219	TATAAT<u>GCCTGACCGGC</u>GTTCAGAGTTCTACAGTCCGACGATC	Amplification oligo 1; BglI-α site is underlined (standard desalting)
s1220	AATTAA<u>GCCGCTGGGCC</u>CCTTGGCACCCGAGAATTCC	Amplification oligo 2; BglI-β site is underlined (standard desalting)

2. Place the tubes into a thermal cycler and run the following program.

Step	Temp (°C)	Time (s)	Cycles
Initial denaturation	98	30	1
Denaturation	98	2	[35]
Annealing	64	5	
Extension	72	5	
Final extension	72	60	1
Hold	20	∞	1

3. Prepare a 1.5% agarose mini-gel in 1X TAE buffer. For convenience, ethidium bromide may be added to the gel solution to a final concentration 1 μg/mL prior to casting.
4. Add 7 μL of DNA loading dye to the PCR products from step 2 in Section 2.2.3, mix, and load the entire volume into a single well of the agarose gel. Electrophorese at 120 V until the bromophenol blue dye band has migrated 1/2–2/3 the length of the gel. Place the gel on a UV transilluminator. Using a new disposable scalpel quickly cut around all four sides of the PCR product band and switch off the UV light. Remove the excised band from the gel and transfer to a low-binding 1.7 mL microfuge tube.
5. Recover the PCR product from the gel slice using a gel extraction kit such as the QIAQuick Gel Extraction Kit and elute DNA with 30 μL of nuclease-free water. If using the Qiagen kit, the optional QG buffer wash is omitted and the PE buffer is allowed to stand on the column for 5 min prior to centrifugation. This latter step helps remove trace guanidine salts. Store samples at −20°C.

2.3 Digestion of PCR Product and Ligation With Plasmid Backbone

2.3.1 Equipment

- Heat block
- Qubit 2.0 fluorometer
- Shaking platform
- Sterile Petri dishes

2.3.2 Reagents

- Antarctic Phosphatase supplied with 10X Antarctic Phosphatase buffer
- BglI, restriction enzyme
- Chloramphenicol
- Ethyl alcohol, absolute
- 10X NEBuffer 3.1
- Phenol:Chloroform:IAA, pH 8
- QIAquick PCR Purification Kit (Qiagen)
- QIAprep Spin Miniprep Kit (Qiagen)
- Qubit dsDNA BR Assay Kit
- 3 *M* sodium acetate, pH 5.5
- T4 DNA ligase, supplied with 10X buffer
- XL1-Blue competent cells

2.3.3 Protocol

The cloning procedure for construction of MASTER libraries involves restriction digestion of the PCR product generated in Section 2.2 and plasmid pSG289 (Vvedenskaya et al., 2015) with BglI followed by ligation of the resultant DNA fragments. Plasmid pSG289 contains two distinct recognition sequences for BglI (BglI-α and BglI-β) upstream of the tR2 transcription terminator that enable the directional cloning of the digested PCR product (Fig. 1, middle).

(A) Restriction digest with BglI

1. Prepare a restriction digest as described later and incubate on heat block at 37°C for 90 min.

Component	Volume (μL)
pSG289 (20 ng/μL)	10
10X NEBuffer 3.1	5
Nuclease-free water	33
BglI (10 U/μL)	2
Total volume (μL)	50

2. Add 5.5 μL of 10X Antarctic Phosphatase buffer and 2 μL (10 U) of Antarctic Phosphatase, mix the reaction, and return to 37°C for an additional 30 min.
3. Purify DNA using a QIAquick PCR Purification Kit following the manufacturer's standard protocol.

4. Perform digestion of the PCR product from step 5 in Section 2.2.3 as described in step A1 in Section 2.3.3 using ~0.5 μg of linear dsDNA and without addition of phosphatase.

(B) Ligation of dsDNA library into plasmid backbone

1. Prepare a 100 μL reaction containing both the digested pSG289 (backbone) and the digested PCR product ("insert"). We recommend a molar ratio of 1:5 vector to insert. In addition, prepare a 20 μL control reaction containing only the digested pSG289. Prepare reactions on ice and transfer directly to a 16°C dry bath and incubate approximately 16 h (i.e., overnight). After incubation samples can be stored at −20°C.

Component	Volume (μL)	Volume (μL) (Control Reaction)
Vector	5	2
Insert	5	—
10X T4 DNA ligase buffer	10	2
T4 DNA ligase	5	1
Nuclease-free water	75	15
Total volume (μL)	100	20

2. Withdraw a 10 μL aliquot of each reaction and transform chemically competent *E. coli* cells (e.g., XL1-Blue). Plate each transformation on selective medium containing 25 μg/mL chloramphenicol to determine the CFU for each reaction (Fig. 2A). The CFU for the reaction containing both the backbone and insert should be at least 10 times greater than the CFU for the control reaction. If not, the ligation should be repeated from step A in Section 2.3.3.
3. Add 110 μL of nuclease-free water to the reaction containing both the backbone and insert and mix with 200 μL of phenol:chloroform:IAA (25:24:1, pH 8). Centrifuge at 21,000 × *g* for 5 min at 4°C, remove the aqueous phase, and precipitate the DNA by adding 1/10 volume of 3 *M* sodium acetate (pH 5.5) and three volumes of 100% ethanol. Allow samples to incubate for ~16 h at −20°C.
4. Pellet the DNA by centrifugation 21,000 × *g* for 30 min at 4°C. To remove all traces of salt, wash the pellet three times with 1 mL portions of 75% ethanol. Briefly air-dry the pellet and suspend in nuclease-free water. Measure the concentration of the sample using Qubit dsDNA BR Assay Kit. Store suspended DNA at −20°C.

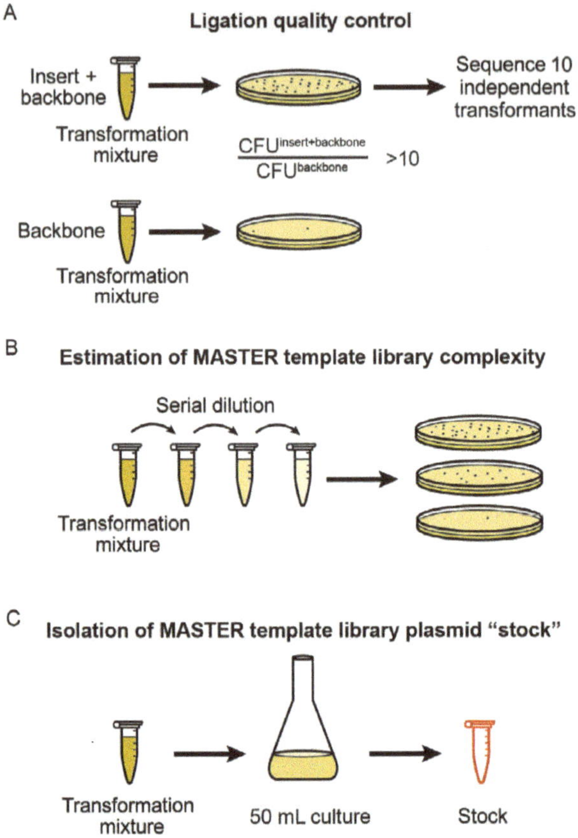

Fig. 2 Steps in the generation of plasmid-borne MASTER "stock" library.

2.4 Isolation of Recombinant Plasmid DNA From Cells

2.4.1 Equipment

- 250 mL DeLong flasks with Morton-style closure
- Electroporator
- Electroporation cuvettes with 1 mm gaps
- Rotator
- Shaking platform
- Sterile Petri dishes

2.4.2 Reagents

- LB (10 g bacto tryptone, 5 g yeast extract, and 10 g sodium chloride per 1 L)
- NEB 5-alpha competent *E. coli* (High Efficiency)

- Plasmid Midi Kit (Qiagen)
- Qubit dsDNA BR Assay Kit
- SOC outgrowth medium

2.4.3 Protocol

1. Electroporation is required to obtain the high transformation efficiency necessary to produce a high complexity library. Combine 250 ng–1 μg of ligated plasmid from step B in Section 2.3.3 with 25 μL of electrocompetent NEB 5-alpha cells in a chilled 0.1 mm gap electroporation cuvette and pulse at 1.8 kV. Add 975 μL of commercial SOC medium to the cuvette, transfer to a 1.7 mL low-binding microfuge tube, and incubate at 37°C for 1 h with gentle inversion using rotator. (Several electroporations can be performed in parallel and the resulting recovery cell suspensions pooled prior to step 2.)
2. Withdraw a 10 μL aliquot of the recovered cells suspension, dilute serially, and plate for CFU determination (Fig. 2B).
3. Inoculate the remaining recovery solution directly into 50 mL of LB containing 25 μg/mL chloramphenicol in a 250 mL DeLong flask with Morton-style closure and shake at 220 RPM for ~18 h at 37°C and 50% relative humidity.
4. Collect the entire 50 mL overnight culture by centrifugation and isolate the plasmid by Qiagen Plasmid Midi Kit. This plasmid DNA is the MASTER template library stock (Fig. 2C). The tube containing the library should be labeled with the number of CFU determined in step 2 in Section 2.4.3. Measure the DNA concentration using Qubit dsDNA BR Assay Kit and store the plasmid library at −20°C.

2.4.4 Notes

1. Use of low-retention plastic pipette tips and tubes throughout the protocol is highly recommended.
2. As a quality control measure, we recommend sequencing the region of interest of 10 individual transformants obtained at step B in Section 2.3.3.
3. The number of transformants obtained at step 2 in Section 2.4.3 determines the total complexity of the library. As the complexity of the library increases, more sequencing depth is required to assign all of the barcodes to a corresponding sequence in the region of interest.
4. Obtaining plasmid-borne libraries with complexities greater than ~10 million using the protocol above is challenging based on the limitations imposed by transformation of a ligation mixture into bacterial cells.

3. TRANSCRIPTION ASSAYS

3.1 Overview

For analysis of transcription in vitro, the experimenter has precise control of the topology of the MASTER template library as well as the reaction conditions of the assays. For in vitro experiments that require use of negatively supercoiled DNA, the stock plasmid library prepared in step 4 in Section 2.4.3 can, in principle, be used. However, as large quantities of plasmid DNA are required for these experiments, we recommend generating additional quantities of the stock plasmid library for use in transcription assays (see step A in Section 3.2.3) to avoid exhausting the stock supply. For in vitro experiments that require use of nonsupercoiled, linear DNA we use an ePCR product generated using the stock plasmid library as the template (see step B in Section 3.2.3).

For analysis of transcription in vivo, the stock plasmid library is introduced into *E. coli* cells by electroporation. To ensure isolation of transformants derived from most, if not all, of the template sequences in the library the number of transformants obtained at this step should be >10-fold the total number of distinct template sequences in the plasmid library. Transformants are pooled, cells are grown to the desired phase of growth, and total RNA is harvested. A portion of this total RNA will comprise transcripts produced during transcription from the MASTER template library in vivo.

3.2 Preparation of DNA Templates

3.2.1 Equipment

- 2.8 L Fernbach flasks
- Shaking platform
- Spectrophotometer
- Spectrophotometry cuvettes
- Thermal cycler

3.2.2 Reagents

- BSA fraction V, OmniPur 98%
- Chloramphenicol
- 5X detergent-free Phusion HF buffer
- DNA oligonucleotides (IDT, see Table 1)
- *E. coli* ElectroMax DH10B-T1^R electrocompetent cells

- Ethyl alcohol, absolute
- HiSpeed Plasmid Maxi Kit (Qiagen)
- LB (10g bacto tryptone, 5g yeast extract, and 10g sodium chloride per 1 L)
- Micellula DNA Emulsion and Purification Kit
- dNTP solution mix, 10 m*M* of each NTP
- Phusion HF DNA polymerase
- Qubit dsDNA HS Assay Kit
- Sodium acetate
- Spectinomycin

3.2.3 Protocol

(A) Generation of negatively supercoiled DNA template

1. Combine 50 ng of plasmid library stock prepared in step 4 in Section 2.4.3 with 20 μL of electrocompetent DH10B-T1^{R} cells and transform as described in Section 2.4.3. The resultant 50 mL overnight culture should contain ~20 million CFU.
2. Use 49.5 mL of overnight culture to inoculate 2.8 L baffled Fernbach flask containing 1 L of LB supplemented with chloramphenicol (25 μg/mL) (use the remaining 0.5 mL to inoculate 50 mL cultures in parallel for RNA isolation as described in step B in Section 3.3.3. Shake the flask at 175 RPM until the cells reach late exponential phase (OD_{600} ~0.8) whereupon add spectinomycin to a final concentration of 100 μg/mL to increase the copy number of the MASTER plasmid. Continue shaking for an additional 16 h.
3. Divide the culture equally among six 500 mL centrifuge bottles and collect cells by centrifugation at 13,000 × *g* at 4°C for 15 min. Use a HiSpeed Plasmid Maxi Kit to isolate supercoiled plasmids from the cell pellets. This stock of the library will serve as (i) template for ePCR to generate a linear template for in vitro transcription experiments and (ii) as a template for in vitro transcription experiments performed using supercoiled templates.

(B) Generation of nonsupercoiled, linear DNA template

1. Dilute the plasmid isolated in step A in Section 3.2.3 to ~10^9 molecules/μL in nuclease-free water and use as template in an ePCR reaction. The Micellula DNA Emulsion and Purification Kit is used with Phusion HF detergent-free buffer. The reaction mixture is prepared on ice, and emulsions are formed per the manufacturer's instructions.

Component	Final Concentration	Volume (μL)
10^9 molecules/μL template plasmid	—	1
10 μ*M* Illumina RP1 primer	0.5 μ*M*	2.5
10 μ*M* Illumina RPI 1 primer	0.5 μ*M*	2.5
0.1 mg/mL BSA	5 μg/mL	2.5
10 m*M* dNTP mix	0.4 m*M*	2
5X Phusion HF buffer (detergent-free)	1X	10
Nuclease-free water	—	27.5
Phusion HF DNA polymerase	0.04 U/μL	2
Total volume (μL)		50

2. Place the tubes containing the emulsions into a thermal cycler and run the following program.

Step	Temp (°C)	Time (s)	Cycles
Initial denaturation	95	10	1
Denaturation	95	5	
Annealing	60	5	[30]
Extension	72	15	
Final extension	72	5 min	1
Hold	4	∞	1

3. Break the emulsion and purify the DNA according to manufacturer's instructions. Concentrate the DNA by addition of 1/10 volume of 3 *M* NaOAc and three volumes of 100% ethanol and incubate ~16 h at −80°C. After precipitation, wash the DNA pellet twice with 1 mL portions of 80% ethanol chilled to −20°C before resuspension in 20 μL of nuclease-free water. Measure the DNA concentration using Qubit dsDNA HS Assay Kit. Store at −20°C.

3.3 Generation and Isolation of RNA Transcripts

3.3.1 Equipment

- Containers for gel staining (we use pipette tip box lids)
- 250 mL DeLong flasks with Morton-style closure

- Heat block
- Low-binding tubes, 0.5 mL
- Spectrophotometer
- Spectrophotometry cuvettes
- Spin-X centrifuge tube filter, 0.45 μm, RNase/DNase free
- Stainless steel disposable scalpels, No. 10 blade
- Sterile needle, 23-gauge
- Table top centrifuge
- Thermal cycler
- Vertical gel electrophoresis apparatus

3.3.2 Reagents

- Acid phenol:chloroform, pH 4.5
- BSA fraction V, OmniPur 98%
- Chloramphenicol
- Direct-zol RNA MiniPrep
- DNase I
- DTT
- 0.5 *M* EDTA, pH 8
- Elution buffer (0.3 *M* NaCl, 10 m*M* Tris–HCl pH 8, 1 m*M* EDTA)
- Ethyl alcohol, absolute
- Glycerol
- Glycogen ultrapure
- Heparin sulfate
- LB (10 g bacto tryptone, 5 g yeast extract, and 10 g sodium chloride per 1 L)
- Low Range ssRNA ladder (NEB)
- Magnesium chloride
- MICROBExpress Kit
- NTP set (ultra-pure), 100 m*M* solutions
- Potassium chloride
- QIAprep Spin Miniprep Kit (Qiagen)
- 2X RNA Gel loading dye (9.5 mL deionized formamide, 0.5 mL 0.5 *M* EDTA pH 8.0, 12.5 μL 20% SDS; bromophenol blue, xylene cyanol, and amaranth powders are added to desired color intensity)
- RNAP holoenzyme, purified as in Mukhopadhyay et al. (2003)
- RNase OUT
- Sodium acetate
- Spin-X centrifuge tube filter, 0.45 μm, RNase/DNase free
- 10% TBE-Urea gels, 1 mm × 10 wells

- TBE buffer (54 g Tris base, 27.5 g boric acid, 20 mL 0.5 *M* EDTA per 1 L)
- TRI Reagent
- Tris–HCl pH 8
- TURBO DNase
- SYBR Gold nucleic acid gel stain

3.3.3 Protocol

(A) In vitro transcription

1. *Transcription assays:* Mix template DNA from step A or B in Section 3.2.3 with RNAP holoenzyme in transcription buffer (50 m*M* Tris–HCl pH 8.0, 10 m*M* $MgCl_2$, 0.01 mg/mL BSA, 100 m*M* KCl, 5% glycerol, 10 m*M* DTT, 0.4 U/μL RNase OUT) and equilibrate at 37°C on heat block for 10 min to form open complexes. We recommend using a 1:5 ratio of template to RNAP (e.g., 10 n*M* template with 50 n*M* RNAP). Initiate a single round of transcription by addition of NTPs and heparin to final concentrations of 1 m*M* NTPs and 0.1 mg/mL heparin. After 15 min, stop the reactions by addition of 0.5 *M* EDTA to 50 m*M* final. Recover nucleic acids by ethanol precipitation, wash twice with 80% cold ethanol, and suspend in 30 μL of nuclease-free water.
2. *Separation of RNA products by gel electrophoresis*: Mix nucleic acids with an equal volume of 2X RNA loading dye, heat in thermal cycler at 95°C for 3 min. Load samples onto precast 10% TBE-urea gel next to Low Range ssRNA ladder. Visualize bands with SYBR Gold stain and excise RNA transcripts using a clean disposable scalpel.
3. *Isolation of RNA products*: Crash gel slices by centrifugal extrusion. Place the slices into 0.5 mL low-binding microfuge tubes in which three holes have been made through the bottom with a 23-gauge needle. Place this tube, in turn, into a 1.7 mL low-binding microfuge tube. Centrifuge the tube-in-tube assembly at 14,000 × *g* for 2 min or until all gel fragments have passed through the holes in the upper 0.5 mL tube and are collected in the lower 1.7 mL tube. To extract nucleic acids, add 300 μL of elution buffer to the crushed gel and incubate the slurry at 70°C for 10 min. Recover the eluate by applying the slurry to a Spin-X tube filter and centrifuging at 14,000 × *g* for 1 min. Recover the crushed gel from the Spin-X tube after centrifugation, mix with an additional 300 μL of elution buffer, and repeat the elution step. Pool the eluates, precipitate as described in step B4 in Section 3.3.3, and suspend in the specified volume of nuclease-free water.

(B) In vivo transcription

1. *Cell culture and harvest*: Use 0.5 mL of the overnight culture from step A in Section 3.2.3 to inoculate 50 mL of LB containing chloramphenicol (25 μg/mL) in a 250 mL DeLong flask and shake at 210 RPM at 37°C until cells reach mid-exponential phase (OD_{600} ~0.5). Rapidly harvest a total of 12 mL by withdrawing 6 × 2 mL aliquots of the cell suspension into 2 mL tubes and collecting the cells by centrifugation (1 min, 21,000 × *g* at room temperature). Remove the supernatants and freeze cell pellets rapidly on dry ice and store at −80°C. One of these pellets will be used for RNA isolation for 5′ RNA-seq, two pellets will be used for isolation of plasmid DNA (QIAprep Spin Miniprep Kit). This plasmid DNA will serve as template in ePCR reactions to generate a product that will be sequenced to identify and assign the barcodes present in the plasmid library. Store the remaining pellets at −80°C as a backup in case any steps need to be repeated.
2. *RNA isolation*: Suspend each cell pellet collected from 2 mL of cell suspension in 600 μL of TRI Reagent solution. Incubate samples at 70°C for 10 min to ensure complete lysis and centrifuge (10 min, 21,000 × *g* at 4°C) to remove insoluble material. Transfer the supernatant to a fresh tube, add ethanol to a final concentration of 60.5%, and divide the mixture equally between two Direct-zol spin columns (Zymo Research). Perform DNase I treatment on-column according to the manufacturer's recommendations. Elute RNA from the columns with three sequential portions of 30 μL nuclease-free water that had been heated to 70°C. To remove residual DNA, treat RNAs with 2 U of TURBO DNase at 37°C for 1 h.
3. *Sample cleanup: phenol extraction*: Following DNase treatment, add to samples an equal volume of acid phenol:chloroform (5:1, pH 4.5). Vortex samples vigorously to mix and separate phases by centrifugation at 21,000 × *g* for 5 min at 4°C. Recover the upper, aqueous phase by pipetting into a fresh tube.
4. *Sample cleanup: ethanol precipitation*: To each sample, add 1/10 volume of 3 *M* sodium acetate (pH 5.5), 1/100 volume of 10 mg/mL glycogen, and three volumes of 100% ethanol and place the samples at −80°C for a minimum of 3 h but preferably 16 h. Pellet nucleic acid by centrifugation at 21,000 × *g* for 30 min at 4°C. Remove the supernatant by pipetting and wash the pellet by adding 1 mL of cold (−20°C) 80% ethanol. Dislodge the pellet from the tube bottom by vortexing in wash solution and collect the pellet again by centrifugation as above for 5 min and

remove the supernatant by pipetting. Repeat the wash twice. After the final wash is removed, allow the pellet to air-dry for ~5 min. Suspend the pellet by addition of nuclease-free water followed by a 20 min incubation on ice to permit solubilization of the nucleic acids. A complete resuspension of nucleic acid is achieved by pipetting (no vortexing is recommended). The typical yield of total RNA isolated from 2 mL cell suspension is ~60 μg.

5. *rRNA depletion*: Due to their abundance, it may be desirable to remove ribosomal RNAs (rRNAs) from the total RNA. Deplete rRNAs from 9 μg of TURBO DNase-treated RNA using MICROBExpress Kit according to the manufacturer's protocol, precipitate overnight, wash twice with 80% cold EtOH, and suspend in 30 μL of nuclease-free water. Store RNA at −80°C for up to 6 months.

3.3.4 Notes

1. For experiments performed in vitro, assays may be performed in the presence of NTPs only (as described earlier), or in the presence of NTPs mixed with a noncanonical initiating nucleotide, NCIN, such as nicotinamide adenine dinucleotide (NAD; see Vvedenskaya et al., 2018).
2. For experiments performed in vivo, the total RNA isolated from cells can also be used to perform RNA-seq analysis of chromosomally encoded promoters to allow comparison of the behavior of natural promoters with the sequence variants in the library (see Vvedenskaya et al., 2016).

4. HIGH-THROUGHPUT SEQUENCING OF RNA 5′ ENDS (5′ RNA-Seq)

4.1 Overview

For use of MASTER to study transcription initiation we have measured transcription output by performing high-throughput sequencing of RNA 5′ ends, "5′ RNA-seq." The results provide both the 5′-end sequence and relative yields of RNA products generated from each template sequence in the library. For use of MASTER to study postinitiation steps of transcription, it may be desirable to modify the procedure to sequence RNA 3′ ends rather than RNA 5′ ends. A protocol to perform high-throughput sequencing of RNA 3′ ends can be found in Vvedenskaya et al. (2014).

Transcription initiation in vivo that involves use of a standard nucleoside triphosphate (NTP) initiating nucleotide will generate RNA products having a 5′ end that contains a triphosphate group. However, recent work has led to an emerging view that indicates bacterial RNAP can also use nonstandard initiating substrates for transcription initiation in vivo (Barvik, Rejman, Panova, Sanderova, & Krasny, 2017). In particular, RNA products generated by initiation with ~2- to 4-nt "nanoRNA" primers (Druzhinin et al., 2015; Goldman et al., 2011; Nickels, 2012; Vvedenskaya et al., 2012) or by initiation with NCINs such as NAD have also been detected in vivo (Bird et al., 2016; Cahova, Winz, Hofer, Nubel, & Jaschke, 2015; Chen, Kowtoniuk, Agarwal, Shen, & Liu, 2009; Kowtoniuk, Shen, Heemstra, Agarwal, & Liu, 2009; Vvedenskaya et al., 2018). MASTER enables the analysis of RNA products generated by NTP-mediated initiation, and RNA products generated by nanoRNA- or NCIN-mediated initiation.

RNA products generated in vivo by nanoRNA-mediated initiation contain a 5′-hydroxyl, while RNA products generated by NCIN-mediated initiation carry a 5′-NCIN "cap." The 5′ RNA-seq procedure used in MASTER enables the differential detection and quantitation of RNA transcripts on the basis of the chemical nature of the RNA 5′ ends. The procedure relies upon the ligation of single-stranded oligonucleotide adaptors to RNA 5′ ends. The ligation of single-stranded oligonucleotide adaptors requires that the RNA has a 5′-monophosphate. Thus, analysis of RNAs that do not have a 5′-monophosphate requires enzymatic processing of the RNAs to yield RNAs having a 5′-monophosphate. Accordingly, analysis of RNA products generated by NTP-mediated initiation, nanoRNA-mediated initiation, or NCIN-mediated initiation each requires a distinct enzymatic processing step to yield a 5′-monophosphate. The requirement for a distinct enzymatic processing step to yield a 5′-monophosphate enables the differential detection and quantitation of RNA transcripts on the basis of whether the RNA was generated by use of an NTP, nanoRNA, or NCIN as the initiating entity.

4.2 Enzymatic Treatments of RNA 5′ Ends

4.2.1 Equipment

- Containers for gel staining (we use pipette tip box lids)
- Heat block
- Stainless steel disposable scalpels, No. 10 blade
- Spin-X centrifuge tube filter, 0.45 μm, RNase/DNase free
- Sterile needle, 23-gauge
- Vertical gel electrophoresis apparatus

4.2.2 Reagents

- Acid phenol:chloroform, pH 4.5
- ATP, 10 m*M*
- CIP, Calf Intestinal Phosphatase
- 10X CutSmart buffer
- Elution buffer (0.3 *M* NaCl, 10 m*M* Tris–HCl pH 8, 1 m*M* EDTA)
- Ethyl alcohol, absolute
- Glycogen ultrapure
- Low Range ssRNA ladder (NEB)
- Manganese chloride
- 10X NEBuffer 2
- NudC decapping enzyme purified as in Cahova et al. (2015)
- Rai1 decapping enzyme purified as in Xiang et al. (2009)
- Rai1 buffer (10 m*M* Tris–HCl pH 7.5, 100 m*M* KOAc, 2 m*M* MgOAc, 2 m*M* DTT)
- 2X RNA gel loading dye (9.5 mL deionized formamide, 0.5 mL 0.5 *M* EDTA pH 8.0, 12.5 μL 20% SDS; bromophenol blue, xylene cyanol and amaranth powders are added to desired color intensity)
- 5′ RNA polyphosphatase supplied with 10X buffer
- RNase OUT
- RNeasy MinElute Clean Up Kit (Qiagen)
- 10% TBE-Urea gels, 1 mm × 10 wells
- T4 PNK supplied with 10X buffer
- TBE buffer (54 g Tris base, 27.5 g boric acid, 20 mL 0.5 *M* EDTA per 1 L)
- SYBR Gold nucleic acid gel stain

4.2.3 Protocol

(A) Enzymatic conversion of RNA 5′ ends to monophosphate ends suitable for ligation.

As mentioned earlier, the ligation of single-stranded oligonucleotide adaptors to RNA 5′ ends requires that the RNAs have a 5′-monophosphate. Thus, analysis of RNAs that do not have a 5′-monophosphate requires enzymatic processing of the RNAs to yield RNAs having a 5′-monophosphate. For processing of 5′-triphosphate RNA to 5′-monophosphate RNA we use the RNA processing enzyme RNA 5′-polyphosphatase (Rpp), which cleaves the phosphodiester bond between the triphosphate β and α phosphates, yielding products comprising 5′-monophosphate followed by the remainder of the RNA. For processing of 5′-hydroxyl RNA to 5′-monophosphate RNA, we use T4 PNK which catalyzes the transfer of the γ phosphate of ATP to the RNA 5′-OH. For selective processing of NCIN-capped RNA to 5′-monophosphate RNA, we use the bacterial RNA-decapping enzyme NudC or the fungal RNA-decapping enzyme Rai1. NudC processes NCIN-capped RNA to 5′-monophosphate RNA by cleaving the diphosphate group of the NCIN cap, yielding products comprising 5′-pNp (where N is the 3′ nucleoside moiety of the NCIN and "p" is a monophosphate) followed by the remainder of the RNA (Cahova et al., 2015; Hofer et al., 2016), while Rai1 processes NCIN-capped RNAs to 5′-monophosphate RNA by cleaving the phosphodiester bond connecting the NCIN cap to the remainder of the RNA, yielding products comprising 5′-monophosphate followed by the remainder of the RNA (Jiao et al., 2017).

1. *Rpp treatment*

Assemble the reaction as follows:

Component	Volume (μL)
RNA (up to 2 μg) from steps A and B in Section 3.3.3	10
10X Rpp buffer	2
RNase OUT, 40 U/μL	1
Rpp, 20 U/μL	1
Nuclease-free water	6
Total volume (μL)	20

Incubate for 30 min at 37°C.

2. *T4 PNK treatment*

Assemble the reaction as follows:

Component	Volume (μL)
RNA (up to 2 μg) from steps A and B in Section 3.3.3	10
10X T4 PNK buffer	5
10 m*M* ATP	5
RNase OUT, 40 U/μL	1
T4 PNK, 10 U/μL	1
Nuclease-free water	28
Total volume (μL)	50

Incubate for 30 min at 37°C.

3. *NudC treatment*
Assemble the reaction as follows:

Component	Volume (μL)
RNA (up to 2 μg) from steps A and B in Section 3.3.3	10
10X NEBuffer 2	2
RNase OUT, 40 U/μL	1
NudC, 55 μ*M*	1.3
Nuclease-free water	5.7
Total volume (μL)	20

Incubate for 30 min at 37°C.

4. *Rai1 treatment*
Assemble the reaction as follows:

Component	Volume (μL)
RNA (up to 2 μg) from steps A and B in Section 3.3.3	10
10X Rai1 buffer	2
100 m*M* $MnCl_2$, prepared fresh	0.2
RNase OUT, 40 U/μL	1
Rai1, 3.4 μ*M*	2
Nuclease-free water	4.8
Total volume (μL)	20

Incubate for 30 min at 37°C.

(B) Recovery of RNA products after enzymatic treatment

1. *RNA products generated in vitro*: Add 20 μL of 2X RNA loading dye. Heat samples at 95°C for 3 min and load next to ssRNA ladder on a 10% TBE-urea slab gels. Run gel in 1X TBE. Visualize RNA using SYBR Gold nucleic acid gel stain and excise RNA products of the desired size from the gel using sterile scalpel. Elute nucleic acids from the gel pieces and purify by precipitation as described in Section 3.3.3. Suspend RNAs in 10 μL of nuclease-free water.

2. *RNA products generated in vivo*: For RNA treated with T4 PNK use a RNeasy MinElute Clean Up Kit to ensure removal of excess ATP. Elute purified RNA with 10 μL of nuclease-free water. For RNA treated with Rpp, NudC, or Rai1 add 80 μL of nuclease-free water to the sample and mix with 100 μL of acid phenol/chloroform. Centrifuge the sample, recover the upper (aqueous) phase, precipitate, and recover RNA as described in step B in Section 3.3.3. Suspend RNA in 10 μL of nuclease-free water.

4.2.4 Notes

1. For analysis of RNA products containing a 5′-triphosphate or 5′-hydroxyl we recommend removal of 5′-monophosphate RNA from samples by use of 5′-phosphate-dependent exonuclease (Epicentre) prior to Rpp or PNK treatment.
2. For analysis of RNA products containing a 5′-NCIN we recommend conversion of 5′-phosphate RNA to 5′-hydroxyl RNA by use of a Phosphatase such as Calf Intestinal Phosphatase (CIP; NEB) prior to NudC or Rai1 treatment.

4.3 5′ Adapter Ligation

4.3.1 Equipment

- Containers for gel staining (we use pipette tip box lids)
- Heat block
- Stainless steel disposable scalpels, No. 10 blade
- Spin-X centrifuge tube filter, 0.45 μm, RNase/DNase free
- Sterile needle, 23-gauge
- Thermal cycler
- Vertical gel electrophoresis apparatus

4.3.2 Reagents

- ATP, 10 m*M*
- Elution buffer (0.3 *M* NaCl, 10 m*M* Tris–HCl pH 8, 1 m*M* EDTA)
- Ethyl alcohol, absolute
- Glycogen ultrapure
- Low Range ssRNA ladder (NEB)
- 50% PEG-8000
- RNA oligonucleotides (IDT, see Table 1)

- 2X RNA gel loading dye (9.5 mL deionized formamide, 0.5 mL 0.5 *M* EDTA pH 8.0, 12.5 μL 20% SDS; bromophenol blue, xylene cyanol, and amaranth powders are added to desired color intensity)
- RNase OUT
- 10% TBE-Urea gels, 1 mm × 10 wells
- TBE buffer (54 g Tris base, 27.5 g boric acid, 20 mL 0.5 *M* EDTA per 1 L)
- T4 RNA ligase 1, supplied with 10X buffer
- SYBR Gold nucleic acid gel stain

4.3.3 Protocol

1. Assemble a ligation reaction containing RNAs generated in vitro or in vivo as described and incubate in thermal cycler at 16°C for ~16 h.

Component	Final Amount	Volume (μL)
RNA from Section 4.2	—	10
10 μ*M* 5′ adapter s1206	0.33 μ*M* (in vitro) or 1 μ*M* (in vivo)	1 (in vitro) or 3 (in vivo)
50% PEG-8000	10%	6
10 m*M* ATP	1 mM	3
10X T4 RNA ligase 1 buffer	1X	3
RNase OUT	40 U	1
T4 RNA ligase 1	10 U	1
Nuclease-free water	—	5 or 3
Total volume (μL)		30

2. Stop reactions by adding 30 μL of 2X RNA loading dye, heat samples at 95°C for 3 min in thermal cycler, and separate RNAs by electrophoresis on 10% TBE-urea slab gels (equilibrated and run in 1X TBE). Incubate gels with SYBR Gold nucleic acid gel stain and visualize bands with UV transillumination. Recover RNA products from the gel as described in Section 3.3.3 and suspend in 10 μL of nuclease-free water.

4.3.4 Notes

1. The s1206 adaptor used in the ligation reactions has 15 nt of random 3′-end sequence. The use of a pool of adaptors with different 3′-end

sequences minimizes the potential for sequence-dependent effects on ligation efficiency. The use of a pool of adaptors also associates a unique molecular identifier (UMI) with each RNA molecule prior to ePCR that is used to eliminate effects of differential amplification.

2. Quantitative comparisons between two or more RNA samples subjected to different enzymatic treatments to convert RNA 5′ ends to monophosphate ends can be performed by replacing the s1206 adaptor with two or more of the barcoded 5′-adaptor oligonucleotides i105, i106, i107, and i108. The adaptors i105, i106, i107, and i108 each carry a unique 4-nt barcode sequence followed by 11 nt of random 3′-end sequence (Table 1). When using barcoded 5′-adaptor oligonucleotides for quantitative comparisons of multiple samples, the adaptor-ligated RNA products of each sample are mixed prior to the gel electrophoresis step in Section 4.3.3. For an example of the use of i105, i106, i107, and i108 to perform quantitative comparisons between samples, see Vvedenskaya et al. (2018).

4.4 Generation of cDNAs

4.4.1 Equipment

- Containers for gel staining (we use pipette tip box lids)
- Heat block
- Stainless steel disposable scalpels, No. 10 blade
- Spin-X centrifuge tube filter, 0.45 μm, RNase/DNase free
- Sterile needle, 23-gauge
- Thermal cycler
- Vertical gel electrophoresis apparatus

4.4.2 Reagents

- DTT
- Elution buffer (0.3 *M* NaCl, 10 m*M* Tris–HCl pH 8, 1 m*M* EDTA)
- Ethyl alcohol, absolute
- Glycogen ultrapure
- Low Range ssRNA ladder (NEB)
- dNTP solution mix, 10 m*M* of each NTP
- O'Gene Ruler Ultra Low Range DNA ladder
- 6X Orange DNA loading dye
- Qubit dsDNA HS Assay Kit
- Qubit ssDNA Assay Kit

- 2X RNA gel loading dye (9.5 mL deionized formamide, 0.5 mL 0.5 *M* EDTA pH 8.0, 12.5 μL 20% SDS; bromophenol blue, xylene cyanol, and amaranth powders are added to desired color intensity)
- RNase H
- RNase OUT
- SuperScript III Reverse Transcriptase supplied with 5X First-Strand buffer
- 10% TBE-Urea gels, 1 mm × 10 wells
- 10% TBE gels, 1 mm × 10 wells
- TBE buffer (54 g Tris base, 27.5 g boric acid, 20 mL 0.5 *M* EDTA per 1 L)
- SYBR Gold nucleic acid gel stain

4.4.3 Protocol

1. Combine 10 μL of adapter-ligated RNAs with 0.3 μL of oligo s128 (100 pmol/μL). Incubate the mixture for 5 min at 65°C and then cool the mixture to 4°C. Add 9.7 μL of a reverse transcription cocktail of the following composition.

RT Cocktail Component	Final Amount	Volume (μL)
5X First-Strand buffer	1X	4
10 m*M* dNTP mix	500 μ*M*	1
100 m*M* DTT	5 m*M*	1
RNase OUT	40 U	1
Nuclease-free water	—	1.7
SuperScript III Reverse Transcriptase	200 U	1
Total volume (μL)		9.7

2. Place the samples in a thermal cycler and run the following program.

Step	Temp (°C)	Time (min)	Action
RT 1	55	60	
RT 2	70	15	
Cool	25	Hold	Add 10 U of RNase H
RNase H	37	20	

3. Mix cDNAs with equal volume of 2X RNA loading dye, load on gel next to ssRNA ladder, and separate by electrophoresis on 10% TBE-urea slab gels in 1X TBE buffer. Recover products from the gel as described in Section 3.3.3 and suspend cDNA in 20 μL of nuclease-free water.
4. Quantify cDNAs using Qubit ssDNA Assay Kit, dilute, and perform ePCR as in step B in Section 3.2.3 but using only 20 cycles. Break the emulsion and purify cDNAs.
5. If residual primers are present, mix purified cDNA with 6X Orange DNA loading dye and isolate amplified cDNAs by electrophoresis using a nondenaturing 10% TBE gel. Load O'Gene Ruler Ultra Low Range DNA ladder as size standard. Isolate products by gel excision using the procedure described in Section 3.3.3. Elute cDNA from gel with 600 μL of elution buffer at 37°C for 2 h, precipitate, and suspend in 20 μL of nuclease-free water.
6. The concentration of the isolated cDNA is determined using a Qubit dsDNA HS Assay Kit.

4.5 High-Throughput Sequencing and Data Analysis

4.5.1 Overview

This section provides a general overview of the data analysis for MASTER experiments. The source code and documentation for all MASTER analyses performed in our published work is provided at: https://github.com/NickelsLabRutgers.

4.5.2 High-Throughput Sequencing

The ePCR product derived from amplification of the MASTER template library prepared in step B in Section 3.2.3 and the cDNA products derived from RNA transcripts prepared in Section 4.4.3 are sequenced using an Illumina sequencer. Sequencing primers provided by Illumina typically contain a mixture of several oligos. To avoid potential complications that may arise due to the presence of additional primers in the Illumina reagents, a custom sequencing primer can be used (e.g., primer s1115; Table 1).

4.5.3 Data Analysis: ePCR Products Derived From MASTER Template Library

The ePCR product generated in step B in Section 3.2.3 is sequenced using an Illumina sequencer. The results identify what fraction of the potential sequence variants in the region of interest are present in the library and identify the transcribed-region barcode associated with individual sequences in the region of interest. As mentioned in Section 2.1, template libraries should contain >95% of the theoretical maximum number of potential sequence variants in the region of interest. In addition, >95% of the barcode sequences present in the library should be associated with a single sequence variant in the region of interest. In addition to enabling the association of transcribed-region barcodes with a sequence variant in the region of interest, the results define the relative abundance of each sequence variant in the region of interest.

4.5.4 Data Analysis: cDNA Libraries Derived From RNA Transcripts

For cDNA libraries prepared using adaptors s1206, i105, i106, i107, or i108, the 16th base of the read is the sequence of the RNA 5′ end from which the cDNA was generated. For libraries prepared using adaptor s1206, the first 15 bases of each sequencing read provide UMI used to correct amplification bias introduced during ePCR. For cDNA libraries prepared using adaptors i105, i106, i107, or i108, the first four bases of each sequencing read provide the barcode used to identify the enzymatic treatment, and the next 11 bases provide UMI used to correct amplification bias introduced during ePCR.

For each sequencing read derived from an RNA 5′ end, the identity of the transcribed-region barcode identifies the DNA template sequence variant that produced the RNA transcript and the sequence of the RNA 5′ end defines the first nucleotide of the RNA transcript. Sequencing reads having identical RNA 5′-end sequences and identical sequence tags are counted as only a single read in the analysis.

For a given promoter sequence variant, the relative transcription output can be calculated as the total number of RNA products generated from the sequence variant divided by relative proportion of promoter sequence variants in the library (see (Vvedenskaya et al., 2015)]. In addition, for a given promoter sequence variant, the sequence of RNA 5′ ends generated from the sequence variant defines the positions where transcription start site selection occurs for this promoter sequence (see Vvedenskaya et al., 2018, 2016, 2015; Winkelman et al., 2016) and can be used to define the extent of reiterative transcription initiation (slippage synthesis) for this promoter sequence (Vvedenskaya et al., 2015).

5. CONCLUSIONS

5.1 Summary

A systems-level view of cellular gene expression requires understanding the mechanistic principles governing each step of transcription. Summarized later are examples of prior work that illustrate the broad utility of MASTER to probe the complex interplay between RNAP and nucleic acids during transcription, thereby illuminating the mechanistic principles governing this step of gene expression.

- MASTER has been used to define, comprehensively, critical sequence determinants for transcription start site selection, reiterative transcription initiation, and transcript yields for *E. coli* RNAP (Hochschild, 2015; Vvedenskaya et al., 2016, 2015; Winkelman et al., 2016).
- MASTER has been used to define promoter sequence determinants responsive to changes in DNA topology or changes in the concentrations of NTP substrates for transcription with *E. coli* RNAP (Vvedenskaya et al., 2015). The analysis of effects of DNA topology showed that the position of the transcription start site depends on DNA topology, consistent with the proposal that start site selection involves changes in the size of the transcription bubble, "DNA scrunching."
- MASTER has been used to establish that sequence-specific interactions between the *E. coli* RNAP core enzyme and the downstream part of the nontemplate strand of the transcription-bubble ("core recognition element," CRE) are a previously undocumented determinant of transcription start site selection (Vvedenskaya et al., 2016). To do this, we used MASTER to compare the transcription start site profile of wild-type RNAP to that of an RNAP derivative defective in sequence-specific interactions between RNAP and the CRE.
- MASTER has been used to define the promoter-sequence determinants for NCIN capping with NAD for *E. coli* RNAP (Vvedenskaya et al., 2018).
- MASTER has been used in parallel with a method for massively parallel protein–DNA cross-linking, termed "MASTER-XL," to determine simultaneously, for a library of 4^{10} promoter sequences, the transcription start site, the position of the RNAP leading edge, and the position of the RNAP trailing edge for *E. coli* RNAP (Winkelman et al., 2016). The results established that as the position of the transcription start site

changes, the position of the RNAP leading edge changes in lock-step register, but the position of the RNAP trailing edge does not change, consistent with the proposal that start site selection involves DNA scrunching.

5.2 Future Prospects

We anticipate MASTER (or derivatives thereof) will be used to fully define sequence determinants of each step of transcription initiation and each step of transcription that occurs postinitiation for *E. coli* RNAP. We further anticipate adapting MASTER for the analysis of other bacterial RNAPs, archaeal RNAP, and eukaryotic RNAP II.

To date, we have defined how various sequence-based features influence transcription in isolation. In future work, we anticipate the results of MASTER will enable the development of a synthetic modeling framework that allows us to integrate the relative contributions of all sequence-based features and all reaction conditions to transcription output. Such models can enable accurate predictions of how transcription of a given gene will be affected by alterations in cellular conditions, thereby providing a quantitative understanding of how cells ensure proper control of gene expression.

REFERENCES

Barvik, I., Rejman, D., Panova, N., Sanderova, H., & Krasny, L. (2017). Non-canonical transcription initiation: The expanding universe of transcription initiating substrates. *FEMS Microbiology Reviews*, *41*(2), 131–138. https://doi.org/10.1093/femsre/fuw041.

Bird, J. G., Zhang, Y., Tian, Y., Panova, N., Barvik, I., Greene, L., et al. (2016). The mechanism of RNA 5′ capping with NAD+, NADH and desphospho-CoA. *Nature*, *535*(7612), 444–447. https://doi.org/10.1038/nature18622.

Cahova, H., Winz, M. L., Hofer, K., Nubel, G., & Jaschke, A. (2015). NAD captureSeq indicates NAD as a bacterial cap for a subset of regulatory RNAs. *Nature*, *519*(7543), 374–377. https://doi.org/10.1038/nature14020.

Chen, Y. G., Kowtoniuk, W. E., Agarwal, I., Shen, Y., & Liu, D. R. (2009). LC/MS analysis of cellular RNA reveals NAD-linked RNA. *Nature Chemical Biology*, *5*(12), 879–881. https://doi.org/10.1038/nchembio.235.

Cramer, P. (2002). Multisubunit RNA polymerases. *Current Opinion in Structural Biology*, *12*(1), 89–97.

Darst, S. A. (2001). Bacterial RNA polymerase. *Current Opinion in Structural Biology*, *11*(2), 155–162.

Druzhinin, S. Y., Tran, N. T., Skalenko, K. S., Goldman, S. R., Knoblauch, J. G., Dove, S. L., et al. (2015). A conserved pattern of primer-dependent transcription initiation in *Escherichia coli* and *Vibrio cholerae* revealed by 5′ RNA-seq. *PLoS Genetics*, *11*(7), e1005348. https://doi.org/10.1371/journal.pgen.1005348.

Duchi, D., Bauer, D. L., Fernandez, L., Evans, G., Robb, N., Hwang, L. C., et al. (2016). RNA polymerase pausing during initial transcription. *Molecular Cell, 63*(6), 939–950. https://doi.org/10.1016/j.molcel.2016.08.011.

Ebright, R. H. (2000). RNA polymerase: Structural similarities between bacterial RNA polymerase and eukaryotic RNA polymerase II. *Journal of Molecular Biology, 304*(5), 687–698. https://doi.org/10.1006/jmbi.2000.4309.

Feklistov, A., Sharon, B. D., Darst, S. A., & Gross, C. A. (2014). Bacterial sigma factors: A historical, structural, and genomic perspective. *Annual Review of Microbiology, 68*, 357–376. https://doi.org/10.1146/annurev-micro-092412-155737.

Goldman, S. R., Sharp, J. S., Vvedenskaya, I. O., Livny, J., Dove, S. L., & Nickels, B. E. (2011). NanoRNAs prime transcription initiation in vivo. *Molecular Cell, 42*(6), 817–825. https://doi.org/10.1016/j.molcel.2011.06.005.

Hochschild, A. (2015). Mastering transcription: Multiplexed analysis of transcription start site sequences. *Molecular Cell, 60*(6), 829–831. https://doi.org/10.1016/j.molce.2015.12.007.

Hofer, K., Li, S., Abele, F., Frindert, J., Schlotthauer, J., Grawenhoff, J., et al. (2016). Structure and function of the bacterial decapping enzyme NudC. *Nature Chemical Biology, 12*(9), 730–734. https://doi.org/10.1038/nchembio.2132.

Jiao, X., Doamekpor, S. K., Bird, J. G., Nickels, B. E., Tong, L., Hart, R. P., et al. (2017). 5′ end nicotinamide adenine dinucleotide cap in human cells promotes RNA decay through DXO-mediated deNADding. *Cell, 168*(6), 1015–1027. e1010. https://doi.org/10.1016/j.cell.2017.02.019.

Kapanidis, A. N., Margeat, E., Ho, S. O., Kortkhonjia, E., Weiss, S., & Ebright, R. H. (2006). Initial transcription by RNA polymerase proceeds through a DNA-scrunching mechanism. *Science, 314*(5802), 1144–1147. https://doi.org/10.1126/science.1131399.

Kowtoniuk, W. E., Shen, Y., Heemstra, J. M., Agarwal, I., & Liu, D. R. (2009). A chemical screen for biological small molecule-RNA conjugates reveals CoA-linked RNA. *Proceedings of the National Academy of Sciences of the United States of America, 106*(19), 7768–7773. https://doi.org/10.1073/pnas.0900528106.

Lerner, E., Chung, S., Allen, B. L., Wang, S., Lee, J., Lu, S. W., et al. (2016). Backtracked and paused transcription initiation intermediate of *Escherichia coli* RNA polymerase. *Proceedings of the National Academy of Sciences of the United States of America, 113*(43), E6562–E6571. https://doi.org/10.1073/pnas.1605038113.

Marchetti, M., Malinowska, A., Heller, I., & Wuite, G. J. L. (2017). How to switch the motor on: RNA polymerase initiation steps at the single-molecule level. *Protein Science, 26*(7), 1303–1313. https://doi.org/10.1002/pro.3183.

Mukhopadhyay, J., Mekler, V., Kortkhonjia, E., Kapanidis, A. N., Ebright, Y. W., & Ebright, R. H. (2003). Fluorescence resonance energy transfer (FRET) in analysis of transcription-complex structure and function. *Methods in Enzymology, 371*, 144–159. https://doi.org/10.1016/S0076-6879(03)71010-6.

Nickels, B. E. (2012). A new way to start: NanoRNA-mediated priming of transcription initiation. *Transcription, 3*(6), 300–304. https://doi.org/10.4161/trns.21903.

Paget, M. S. (2015). Bacterial sigma factors and anti-sigma factors: Structure, function and distribution. *Biomolecules, 5*(3), 1245–1265. https://doi.org/10.3390/biom5031245.

Revyakin, A., Liu, C., Ebright, R. H., & Strick, T. R. (2006). Abortive initiation and productive initiation by RNA polymerase involve DNA scrunching. *Science, 314*(5802), 1139–1143. https://doi.org/10.1126/science.1131398.

Ruff, E. F., Record, M. T., Jr., & Artsimovitch, I. (2015). Initial events in bacterial transcription initiation. *Biomolecules, 5*(2), 1035–1062. https://doi.org/10.3390/biom5021035.

Vvedenskaya, I. O., Bird, J. G., Zhang, Y., Zhang, Y., Jiao, X., Barvik, I., et al. (2018). CapZyme-Seq comprehensively defines promoter-sequence determinants for RNA 5′ capping with NAD^+. *Molecular Cell*, *70*(3), 553–564. e559. https://doi.org/10.1016/j.molcel.2018.03.014.

Vvedenskaya, I. O., Sharp, J. S., Goldman, S. R., Kanabar, P. N., Livny, J., Dove, S. L., et al. (2012). Growth phase-dependent control of transcription start site selection and gene expression by nanoRNAs. *Genes & Development*, *26*(13), 1498–1507. https://doi.org/10.1101/gad.192732.112.

Vvedenskaya, I. O., Vahedian-Movahed, H., Bird, J. G., Knoblauch, J. G., Goldman, S. R., Zhang, Y., et al. (2014). Interactions between RNA polymerase and the "core recognition element" counteract pausing. *Science*, *344*(6189), 1285–1289. https://doi.org/10.1126/science.1253458.

Vvedenskaya, I. O., Vahedian-Movahed, H., Zhang, Y., Taylor, D. M., Ebright, R. H., & Nickels, B. E. (2016). Interactions between RNA polymerase and the core recognition element are a determinant of transcription start site selection. *Proceedings of the National Academy of Sciences of the United States of America*, *113*(21), E2899–E2905. https://doi.org/10.1073/pnas.1603271113.

Vvedenskaya, I. O., Zhang, Y., Goldman, S. R., Valenti, A., Visone, V., Taylor, D. M., et al. (2015). Massively systematic transcript end readout, "MASTER": Transcription start site selection, transcriptional slippage, and transcript yields. *Molecular Cell*, *60*(6), 953–965. https://doi.org/10.1016/j.molcel.2015.10.029.

Winkelman, J. T., Vvedenskaya, I. O., Zhang, Y., Zhang, Y., Bird, J. G., Taylor, D. M., et al. (2016). Multiplexed protein-DNA cross-linking: Scrunching in transcription start site selection. *Science*, *351*(6277), 1090–1093. https://doi.org/10.1126/science.aad6881.

Xiang, S., Cooper-Morgan, A., Jiao, X., Kiledjian, M., Manley, J. L., & Tong, L. (2009). Structure and function of the 5′→3′ exoribonuclease Rat1 and its activating partner Rai1. *Nature*, *458*(7239), 784–788. https://doi.org/10.1038/nature07731.

CHAPTER THIRTEEN

Transcriptomic Approaches for Studying Quorum Sensing in *Vibrio cholerae*

Roman Herzog*, Kai Papenfort*,†,1

*Faculty of Biology I, Department of Microbiology, Ludwig-Maximilians-University of Munich, Munich, Germany

†Munich Center for Integrated Protein Science (CIPSM), Munich, Germany

[1]Corresponding author: e-mail address: kai.papenfort@lmu.de

Contents

Abstract

Transcriptome analysis using RNA-sequencing (RNA-seq) has now become the standard approach to determine the transcriptional output of an organism. Various modifications to this technology have been developed over the years, usually aiming to improve the annotation of transcript borders, or to identify novel classes of RNAs, such as small regulatory RNAs (sRNAs) and antisense transcripts. RNA-seq has also led to the identification of dozens of new sRNAs in the major human pathogen, *Vibrio cholerae*. Several of these sRNAs function in the context of a cell-to-cell communication process, called quorum sensing (QS). QS is key for pathogenicity and biofilm formation of *V. cholerae* and the

Methods in Enzymology, Volume 612
ISSN 0076-6879
https://doi.org/10.1016/bs.mie.2018.09.008

sRNAs involved typically act by base pairing with multiple target mRNAs to control gene expression at the posttranscriptional level. In this chapter, we describe the use of RNA-seq technologies for the discovery and characterization of regulatory RNAs in *V. cholerae* and discuss their relevance to QS and collective functions, such as biofilm formation. We further outline possible methods for the identification and validation of sRNA target genes, which can provide crucial information as to the physiological roles of an sRNA.

1. INTRODUCTION

The major human pathogen and causative agent of cholera disease, *Vibrio cholerae*, critically depends on collective group behaviors to colonize the host's small intestine and cause disease. These collective processes include the formation of biofilms to cope with the acidic environment of the human stomach, as well as the production of virulence factors (Almagro-Moreno, Pruss, & Taylor, 2015; Silva & Benitez, 2016). Both processes, biofilm formation and virulence gene expression, are only effective when conducted simultaneously by all members of the group. In *V. cholerae*, group behavior is controlled by quorum sensing (QS), a process of interspecies communication (Ng & Bassler, 2009). QS relies on the synthesis, release, and subsequent detection of small signaling molecules, called autoinducers (AIs). Microorganisms often produce multiple AI molecules, which determine the overall QS output (Papenfort & Bassler, 2016).

In *V. cholerae*, two AIs, i.e., CAI-1 ((S)-3-hydroxytridecan-4-one) and AI-2 ((2S,4S)-2-methyl-2,3,3,4-tetrahydroxytetrahydrofuran borate), have been extensively studied. CAI-1 and AI-2 are produced by the CqsA and LuxS synthases, respectively, and both AIs are detected by membrane-bound receptor proteins (Chen et al., 2002; Higgins et al., 2007; Hurley & Bassler, 2017; Neiditch, Federle, Miller, Bassler, & Hughson, 2005; Ng et al., 2011; Surette, Miller, & Bassler, 1999). Whereas CAI-1 is recognized by CqsS, AI-2 is sensed by the LuxPQ protein complex (Fig. 1). At low cell density (LCD), when autoinducers are scarce, CqsS and LuxQ function as kinases transferring a phosphate group to the phosphorelay protein LuxU, which channels the phosphate to the response regulator LuxO (Bassler, Wright, & Silverman, 1994). Phosphorylated LuxO, together with the alternative sigma factor σ^{54}, promotes the expression of four homologous sRNAs, called Qrr1–4 (Lenz et al., 2004). Qrr1–4 are Hfq-dependent sRNAs (see below) that act in *trans* to destabilize, among others, the transcript coding for HapR, which is the main repressor of LCD functions in *V. cholerae*

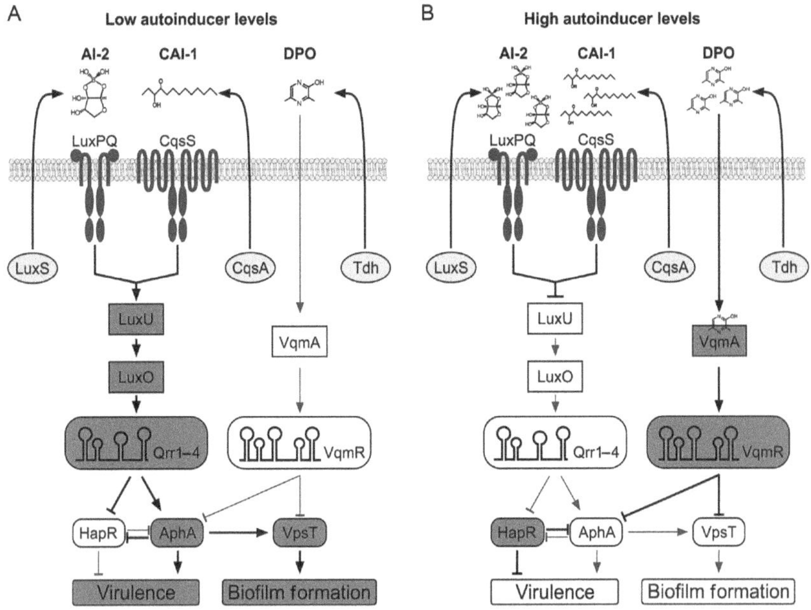

Fig. 1 Quorum Sensing pathways of *V. cholerae*. The autoinducers CAI-1 and AI-2 are synthesized by the CqsA and LuxS enzymes and sensed by the membrane-associated CqsS and LuxPQ receptors, respectively. The DPO autoinducer requires the Tdh (threonine dehydrogenase) enzyme for synthesis and binds to the VqmA receptor protein. (A) When autoinducer concentrations are low, CqsS and LuxPQ function as kinases and phosphorylate LuxU. LuxU-P channels the phosphate group to LuxO, and LuxO-P activates the transcription of the Qrr1–4 sRNAs. The Qrr sRNAs act at the posttranscriptional level to inhibit *hapR* and induce *aphA*. Under this condition virulence gene expression and biofilm formation are activated. (B) When autoinducer concentrations increase, interaction of CAI-1 and AI-2 with CqsS and LuxPQ, respectively, changes the receptors to phosphatases reducing LuxO-P levels and blocking *qrr1–4* expression. Under these conditions, AphA expression is inhibited and HapR is produced. The VqmA–DPO complex promotes expression of the VqmR sRNA. VqmR blocks VpsT production and therefore biofilm formation. In addition, HapR and AphA antagonize each other at the transcriptional level. Active factors are shown in *black*, inactivate (repressed) factors are depicted in *gray*.

(Fig. 1, left side). In addition, the Qrr2–4 sRNAs base pair with and stabilize the *aphA* mRNA encoding a key transcriptional activator of biofilm formation and virulence factor production (Rutherford, van Kessel, Shao, & Bassler, 2011; Shao & Bassler, 2012). Thus, at LCD, AphA is produced, while HapR is not.

Activation of biofilm formation by AphA relies on the induction of another transcriptional activator, called VpsT, which drives the expression of polysaccharide synthesis genes and other protein component for biofilm assembly (Casper-Lindley & Yildiz, 2004; Teschler et al., 2015; Yang, Frey, Liu, Bishar, & Zhu, 2010). Also, VpsT requires the second messenger c-di-GMP (bis-(3′–5′)-cyclic dimeric guanosine monophosphate) for full activity, thereby linking interspecies- (QS) and intracellular signaling processes (Krasteva et al., 2010). Activation of virulence genes in *V. cholerae* is a multistep process and requires several transcriptional regulators acting downstream of AphA (Almagro-Moreno et al., 2015). Specifically, AphA in concert with AphB activates the production of the membrane-bound regulators TcpP and TcpH, which together with another transmembrane regulator, ToxRS, drive the expression of ToxT. ToxT induces the expression of *ctxAB* and *tcpA*, which code for the cholera toxin and the toxin-coregulated pilus, respectively. CtxAB and TcpA are both critical for pathogenicity of *V. cholerae*. Of note, the Qrr sRNAs also inhibit the expression of factors required for Type VI secretion (Shao & Bassler, 2014), which can also affect interspecies competition during host colonization (Logan et al., 2018; Zhao, Caro, Robins, & Mekalanos, 2018).

Continued growth and accumulation of AIs in the environment trigger *V. cholerae*'s transition into a high cell density (HCD) mode of gene expression (Fig. 1, right side). Binding of CAI-1 and AI-2 to CqsS and LuxPQ, respectively, alters the receptors' function from kinases to phosphatases, leading to dephosphorylation of LuxU and LuxO (Hurley & Bassler, 2017). Unphosphorylated LuxO is inactive and consequently expression of the Qrr sRNAs ceases. Reduced production of the Qrrs results in increased HapR production and inhibition of AphA synthesis. This pattern of gene expression is reinforced by the reciprocal repression of HapR and AphA, in which both transcription factors function to repress the promoter of the other regulator (Rutherford et al., 2011). With HapR being produced, virulence gene expression and biofilm formation of *V. cholerae* are inhibited.

Recently, we identified another QS pathway in *V. cholerae*, which is independent of the canonical QS signal transduction cascade. This new system relies on the 3,5-dimethylpyrazin-2-ol (DPO) AI molecule (Papenfort et al., 2017), which is synthesized during the catabolic degradation of L-threonine by threonine dehydrogenase (Tdh) (Fig. 1). DPO is recognized by the cytoplasmic LuxR-type receptor protein, VqmA, and the active DPO–VqmA complex induces the expression of the VqmR sRNA. VqmR base pairs with and inhibits translation of the *vpsT* mRNA, encoding a major activator of biofilm formation in *V. cholerae* (Papenfort, Forstner, Cong, Sharma, & Bassler, 2015). Consequently, VqmR is a repressor of biofilm formation. VqmR also regulates other target mRNAs in *V. cholerae*, some of which also relate to QS and collective behaviors (Papenfort, Forstner, et al., 2015).

It is interesting to note that both QS pathways of *V. cholerae*, depending on either AI-2/CAI-1 or DPO, use sRNAs at crucial positions of their pathways (Fig. 1). It has been speculated that RNA regulators employ different regulatory dynamics when compared to transcription factors (Beisel & Storz, 2010; Nitzan, Rehani, & Margalit, 2017), which could also be relevant for the dynamics underlying QS regulation in *V. cholerae*. Indeed, in the related *Vibrio harveyi* species, regulation by the Qrr sRNAs can have different consequences depending on the nature of the RNA duplex formed between the sRNAs and the targets (Feng et al., 2015). In general, sRNAs can act catalytically (the target mRNA is degraded, while the sRNA remains stable in the cell), or by coupled degradation in which both target mRNA and sRNA are degraded after RNA duplex formation (Masse, Escorcia, & Gottesman, 2003). In few cases, sRNAs have also been reported to function by sequestration (Feng et al., 2015; Moller, Franch, Udesen, Gerdes, & Valentin-Hansen, 2002; Tu, Long, Svenningsen, Wingreen, & Bassler, 2010). Here, the sRNA and the target mRNA are believed to form a stable RNA duplex that is not subject to degradation by cellular ribonucleases.

Both, the Qrr sRNAs and VqmR, regulate multiple target genes, and it is evident that sRNAs, like transcription factors, can function as global regulators of gene expression (Hor, Gorski, & Vogel, 2018; Papenfort & Vogel, 2009). Similarly, QS is a multifaceted process affecting hundreds of genes with diverse physiological implications (Ball, Chaparian, & van Kessel, 2017). Therefore, genome-wide profiling technologies are required to study QS and the role of regulatory RNAs in this process. The advent of next-generation sequencing (NGS) technologies has turned transcriptome analysis and global gene expression profiling on its head and now has also been applied to study QS of *V. cholerae* and other microorganisms. In this chapter, we will revisit and update NGS-based technologies for studying QS in *V. cholerae* with an emphasis on sRNA-mediated gene regulation. We describe how to predict and validate of sRNA-target pairs, and outline possible future directions in this field of research.

2. DISCOVERY OF SMALL REGULATORY RNAS USING DIFFERENTIAL RNA-SEQ

Global expression analysis using RNA-sequencing (RNA-seq) has revolutionized the way scientists interpret and understand the transcriptional output of a cell. In general, RNA-seq has replaced hybridization-based technologies such as microarrays and tilling arrays and comes with the major advantage of single-nucleotide resolution data allowing researchers to study individual transcripts. Various different technologies allowing RNA-seq are currently available (e.g., *Solexa*, *SOLiD*, and *454*; for a comprehensive overview, see Lowe, Shirley, Bleackley, Dolan, & Shafee, 2017), some of which also allow single-molecule sequencing (e.g., nanopore sequencing and SMRT sequencing).

RNA-seq has also had a major impact on the discovery and characterization of prokaryotic regulatory RNAs (Barquist & Vogel, 2015; Hor et al., 2018). For example, RNA-seq has been used to identify RNA ligands associated with the Hfq RNA chaperone in *Salmonella enterica* (Chao, Papenfort, Reinhardt, Sharma, & Vogel, 2012; Sittka et al., 2008), *Neisseria meningitidis* (Heidrich et al., 2017), *Sinorhizobium meliloti* (Torres-Quesada et al., 2014), and *Escherichia coli* (Bilusic, Popitsch, Rescheneder, Schroeder, & Lybecker, 2014). Homologs of *hfq* are present in approximately 50% of all sequenced bacterial genomes (Sun, Zhulin, & Wartell, 2002) and typically Hfq plays a major role in posttranscriptional gene regulation in these species. The current model suggests that Hfq acts as a matchmaker promoting base pairing of sRNAs with their cognate targets; however, Hfq is also relevant for sRNA stability and can function independent of sRNAs to control mRNA turnover and translation (Santiago-Frangos & Woodson, 2018; Vogel & Luisi, 2011; Woodson, Panja, & Santiago-Frangos, 2018). Not surprisingly, mutation of *hfq* is frequently associated with complex phenotypic alterations and oftentimes impairs the virulence of pathogenic microbes (Chao & Vogel, 2010), including *V. cholerae* (Ding, Davis, & Waldor, 2004).

The identification of sRNAs using Hfq as a bait has been an important strategy to categorize sRNAs by their function, i.e., Hfq-dependent sRNAs are likely to engage base pairing with *trans*-encoded target mRNAs. However, additional classes of sRNAs exist in bacteria (Hor et al., 2018; Waters & Storz, 2009) and therefore other, unbiased approaches are required to determine the output of regulatory RNAs expressed from a bacterial genome. One such powerful approach is differential RNA-seq (dRNA-seq; Fig. 2). The dRNA-seq method takes advantage of the different chemical nature of 5′ ends present in bacterial RNAs. Specifically, the cellular RNA pool consists of primary transcripts carrying a triphosphate at the 5′ end and processed transcripts, such as transfer RNAs and ribosomal RNAs, which carry a 5′ monophosphate, or in fewer cases, a 5′ hydroxyl group. To perform a dRNA-seq experiment (see below for details), RNA is isolated using conventional methods, and the RNA is split in two. One sample remains untreated

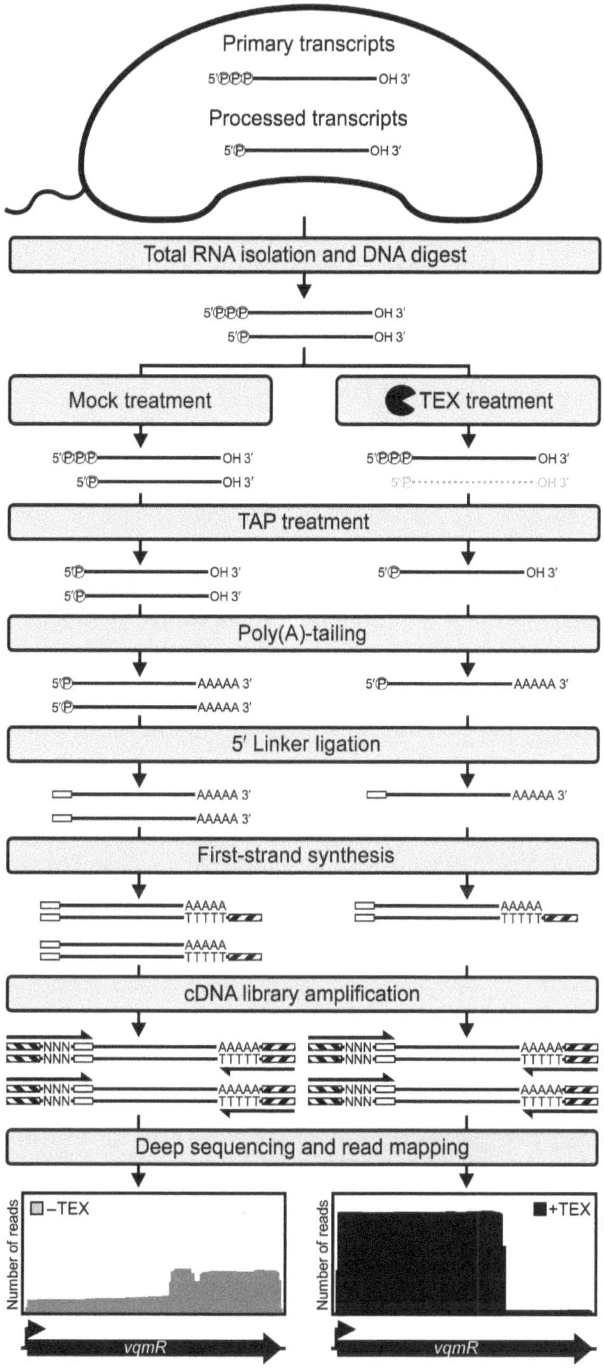

Fig. 2 See legend on next page.

(allowing the detection of all transcripts), while the other sample is treated with terminator exonuclease (TEX). TEX specifically degrades transcripts carrying a 5′ monophosphate and thus leads to the relative enrichment of primary transcripts. This strategy led to the identification of dozens of new regulatory RNAs in various species (Sharma & Vogel, 2014), which included many sRNAs acting independent of Hfq. Of note, dRNA-seq also promotes the identification of processed sRNAs (Chao et al., 2017, 2012; Kroger et al., 2012; Papenfort, Espinosa, Casadesus, & Vogel, 2015; Papenfort et al., 2009), for which there is no enrichment in the TEX-treated cDNA libraries.

2.1 RNA Preparation and DNA Digest

High quality RNA is key for reliable RNA-seq results. Therefore, the isolation of total RNA from bacterial cells is a critical step in every RNA-seq experiment. Most RNA isolation protocols use either organic extraction methods or column-based filter systems. Both have advantages and disadvantages. However, since the column-implemented silica membranes from RNA isolation kits often fail to recover small RNA fragments, we recommend the use of organic extraction protocols for transcriptome analyses. These protocols are based on five steps: Cell lysis, phase separation, RNA precipitation, washing, and elution. In the first step, cell pellets are homogenized in a phenol- and guanidine thiocyanate-containing solution to keep the RNA in solution and to inactivate RNases, respectively. The addition of chloroform separates the mixture into a lower organic phase, a solid middle phase and an aqueous, RNA-containing upper phase. RNA precipitation is performed by addition of isopropanol to the upper

Fig. 2 Workflow of the dRNA-seq experiment. Total RNA is isolated from *V. cholerae* and copurified cellular DNA is digested. RNA samples are split in two. One half is treated with water (mock), whereas the other half is treated with 5′-P-dependent terminator exonuclease (TEX), which specifically degrades processed transcripts carrying a 5′ monophosphate group (*gray, dashed line*). Next, samples are treated with tobacco acid pyrophosphatase, which converts the 5′ triphosphate groups (5′ PPP) of primary transcripts to 5′ phosphate ends (5′ P). An RNA-linker (*white box*) is ligated to the transcript's 5′ ends, and 3′ ends (3′ OH) are polyadenylated using *E. coli* poly(A) polymerase. First-strand cDNA synthesis is performed using an oligo(dT)-adapter (*striped box*). After second-strand synthesis, libraries are amplified by PCR using barcoded oligonucleotides (barcode sequences are shown as "NNN"). The cDNA libraries are sequenced and reads are mapped onto a reference genome. Primary transcripts are enriched in the +TEX-treated samples, which is exemplified by the two coverage plots for reads mapping to the *vqmR* gene (*bottom*).

phase. Subsequent washing of the pelleted RNA is required to remove residual salts, before nuclease-free water is used to dissolve the RNA. Despite careful implementation of the protocol, it is inevitable that low amounts of genomic DNA are copurified. Consequently, residual genomic DNA should be removed prior to ±TEX treatment, using DNase I. Enzymes and salts are removed in subsequent separation-, precipitation-, and washing steps. Finally, the integrity of the clean, DNA-free RNA samples should be confirmed to avoid sequencing of degraded RNA fragments. The RNA integrity number (RIN) is a reliable algorithm for RNA quality assessment and is easily calculated from Agilent 2100 Bioanalyzer measurements (Schroeder et al., 2006). To proceed with subsequent treatments, RNA samples should score a RIN of $\geq$7 on a scale from 1 (highly degraded) to 10 (highest integrity).

2.1.1 Equipment

- 2100 Bioanalyzer (Agilent)
- 5PRIME Phase Lock Gel tubes (Quantabio)
- Cooling centrifuge
- Heating block with shaking function (e.g., ThermoMixer, Eppendorf)
- RNA Nano Kit (Agilent)
- Spectrophotometer (e.g., NanoDrop 2000)

2.1.2 Buffers and Reagents

- 75% (vol./vol.) Ethanol
- Chloroform
- Ethanol–sodium acetate (3 *M*, pH 5.2) solution with 30:1 ratio
- ExtraZol (e.g., 7Bioscience)
- Propan-2-ol (Isopropanol)
- Nuclease-free water
- Phenol/Chloroform/Isoamyl alcohol solution (25:24:1 vol./vol.; e.g., Carl Roth)
- Stop-Mix (95% [vol./vol.] ethanol, 5.0% [vol./vol.] phenol)
- TURBO DNase (2 U/μL) & 10× Reaction Buffer (Thermo Fisher Scientific)

2.1.3 Procedure

1. Inoculate two or three independent clones of each strain and cultivate overnight in appropriate medium. Next day, dilute overnight cultures 1:1000 in fresh media and incubate at desired conditions.

2. Harvest cells (we recommend ≈4 OD equivalents, ≈4 × 10^9 cells) at selected growth phase(s) and immediately apply Stop-Mix (20% final conc. of the collected culture volume) to stop transcription and translation. Snap-freeze and store the mixture at −80°C, or directly continue to step 3.
3. Thaw RNA samples on ice and spin cells at 4°C (4000 rpm, 15 min for 15- or 50-mL tubes and 13,000 rpm, 2 min for 1.5- or 2-mL tubes, respectively). Discard the supernatant.
4. Suspend the cell pellet in 1 mL ExtraZol and incubate the lysates at RT for 5 min. Transfer the cell lysates to Phase Lock Gel tubes. Add 250 μL chloroform and shake vigorously for 10 s. Incubate at RT for 5 min prior to centrifugation (12°C, 13,000 rpm, 15 min). Transfer the RNA-containing aqueous phase (≈500 μL) to a fresh 1.5-mL reaction tube.
5. Precipitate the RNA using 500 μL of isopropanol, invert tubes 10 times, and let stand at RT for 30 min prior to centrifugation (4°C, 13,000 rpm, 30 min).
6. Remove the supernatant and wash (do not suspend) the RNA pellet with 350 μL of 75% ice-cold ethanol and centrifuge (RT, 13,000 rpm, 10 min). Repeat this step for a total of two washing steps.
7. Carefully remove any residual ethanol and air-dry pellets at RT for 10 min. Add 30 μL of nuclease-free water and dissolve RNA in a heating block at 60°C with vigorous shaking for 5 min. Determine the RNA concentration using a spectrophotometer.
8. Store RNA at −80°C or continue directly to step 9.
9. Use nuclease-free water to dilute 20 μg of RNA in a total volume of 89 μL. Add 10 μL of 10 × Reaction Buffer and 1 μL of TURBO DNase. Mix well and incubate at 37°C for 30–45 min to digest cellular DNA.
10. To remove the DNase enzyme, transfer the mixture to Phase Lock Gel tubes, add 100 μL PCI solution, and shake vigorously for 10 s. Centrifuge (12°C, 13,000 rpm, 15 min) and transfer the aqueous phase (≈100 μL) to a fresh 1.5-mL reaction tube.
11. For RNA precipitation, add 300 μL of the EtOH–NaOAc 30:1 solution, invert 10 times, and incubate at −20°C for 1 h to overnight. Collect RNA by centrifugation (4°C, 13,000 rpm, 30 min).
12. Wash, dry, and dissolve RNA by repeating steps 6 and 7.
13. Check RNA integrity using a RNA Nano chip on a 2100 Bioanalyzer. The electrophoresis profile should look similar to the one shown in Fig. 3A and the RIN needs to be ≥7.

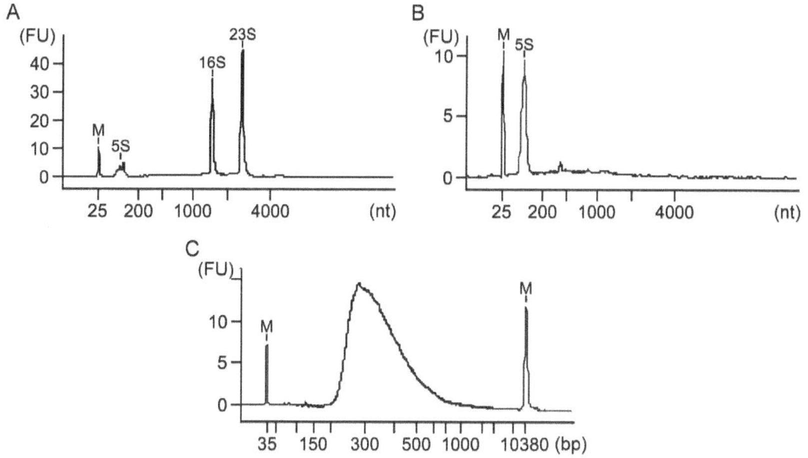

Fig. 3 Quality assessment after different steps of library preparation using a 2100 Bioanalyzer. Electrophoresis profiles obtained from (A) DNA-free total RNA, (B) 16S and 23S rRNA-depleted RNA, (C) a sequencing-ready cDNA library. Marker peaks are indicated (M).

2.1.4 Notes

1. Briefly spin the Phase Lock Gel tubes before use to collect gel at the bottom of the tubes.
2. Use nuclease-free water for the preparation of ethanol and NaOAc solutions.

2.2 Enrichment of Primary Transcripts and Generation of 5′ P Ends for Linker Ligation

The bacterial RNA pool consists of primary and processed transcripts. Primary transcripts are marked by a triphosphate group at the 5′ end (5′ PPP), whereas processed RNAs possess a 5′ monophosphate (5′ P). Less frequently, multiple processing of transcripts generates RNA species with a 5′ hydroxyl group (5′ OH). The goal of the dRNA-seq approach is to distinguish between primary and processed transcripts. For this purpose, the original RNA sample is split into two parts: One remains untreated, while the other part is treated with 5′-P-dependent TEX, an enzyme that specifically depletes transcripts carrying a 5′ P group. Thus, in the TEX-treated samples, processed transcripts such as the abundant rRNAs and tRNAs are degraded, and primary transcripts including mRNAs and sRNAs are relatively enriched. The TEX procedure is followed by treatment of both samples with tobacco acid pyrophosphatase (TAP), which converts the 5′ triphosphate ends to monophosphate and thus allows RNA linker ligation. Finally, the RNA samples are subjected to poly(A) tailing and cDNA libraries are produced. Preparation of cDNA libraries and RNA-seq of both TEX-treated and untreated samples results in a typical enrichment pattern (Fig. 2), allowing for the global annotation of transcriptional start sites (TSS) and revealing putative sRNA genes. For example, the coverage plot of a +TEX-treated library (Fig. 2, bottom) shows a distinct enrichment of reads toward the 5′ end of the *vqmR* gene, compared to the −TEX sample. Of note, processed transcripts that harbor a 5′ OH end are not substrates of TAP and will therefore not appear in the final libraries. To include these noncanonical processed RNAs in dRNA-seq experiments, samples require treatment with a T4 polynucleotide kinase prior to TAP treatment.

2.2.1 Equipment

- 2100 Bioanalyzer (Agilent)
- 5PRIME Phase Lock Gel tubes (Quantabio)
- Cooling centrifuge
- Heating block with shaking function (e.g., ThermoMixer, Eppendorf)
- RNA Nano Kit (Agilent)
- Spectrophotometer (e.g., NanoDrop 2000)

2.2.2 Buffers and Reagents

- 75% (vol./vol.) Ethanol
- EDTA solution (0.5 *M*, pH 8.0)
- Ethanol–sodium acetate (3 *M*, pH 5.2) solution with 30:1 ratio
- GlycoBlue (15 mg/mL; Thermo Fisher Scientific)
- Nuclease-free water
- Phenol/Chloroform/Isoamyl alcohol solution (25:24:1 vol./vol.; e.g., Carl Roth)
- SUPERase In RNase Inhibitor (20 U/μL; Thermo Fisher Scientific)
- Terminator 5′-Phosphate-Dependent Exonuclease (1 U/μL) and 10× reaction buffers A and B (Epicentre)
- Tobacco Acid Pyrophosphatase (10 U/μL) and 10× reaction buffer (Epicentre)

2.2.3 Procedure

1. Transfer equal amounts of DNA-free RNA into two reactions tubes ("-TEX" and "+TEX"). We use 7 μg of RNA and adjust the reaction volume to 37.5 μL using nuclease-free water.

2. Denature RNA at 90°C for 2 min. Immediately put samples back on ice and incubate for 5 min.
3. Add 0.5 μL of SUPERase In RNase Inhibitor and 5.0 μL of the 10 × TEX reaction buffer. Add 7 μL of TEX to the "+TEX" sample and treat the "−TEX" sample with the same volume of nuclease-free water. Incubate the samples at 30°C for 30 min.
4. Place tubes on ice and add 0.5 μL of the EDTA solution to stop the enzymatic reaction.
5. Add 49.5 μL of nuclease-free water to adjust to a total volume to 100 μL and transfer the mixture to Phase Lock Gel tubes. Add 100 μL of PCI solution per tube, shake vigorously for 10 s, and separate phases by centrifugation (12°C, 13,000 rpm, 15 min).
6. Transfer the aqueous phase to a fresh reaction tube and add 2 μL of GlycoBlue to increase precipitation efficiency and to visualize the RNA pellet during the following steps.
7. For RNA precipitation, add 300 μL of the EtOH–NaOAc solution, invert 10 ×, and incubate at −20°C overnight. Collect RNA pellet by centrifugation (4°C, 13,000 rpm, 30 min).
8. Remove the supernatant and wash (do not suspend) the RNA pellet with 100 μL of 75% ice-cold ethanol and centrifuge (4°C, 13,000 rpm, 10 min). Repeat this step for a total of two washing steps.
9. Carefully remove any residual ethanol and air-dry pellets at RT. Add 11 μL of nuclease-free water and dissolve RNA in a heating block at 60° C with vigorous shaking for 5 min.
10. Determine RNA concentration using a spectrophotometer. The RNA concentrations of the "+TEX" samples should be lower than in the "−TEX"-treated samples, due to the removal of processed transcripts.
11. Denature the remaining 10 μL of the ±TEX-treated RNA at 90°C for 1 min and immediately place samples back on ice for 5 min.
12. Prepare a TAP master-mix. Per sample calculate the following volumes: 2.0 μL of 10 × TAP buffer, 0.5 μL TAP enzyme, 0.5 μL SUPERase In RNase Inhibitor, and 7 μL nuclease-free water. Add 10 μL of the mix to each sample and mix well by pipetting. Incubate at 37°C for 1 h.
13. Add 80 μL nuclease-free water to adjust the total volume to 100 μL. Transfer the mixture to Phase Lock Gel tubes and add 100 μL of PCI solution. Shake vigorously for 10 s and separate phases by centrifugation (12°C, 13,000 rpm, 15 min). Transfer the aqueous phase to a fresh reaction tube and add 0.8 μL of GlycoBlue.

14. Precipitate, wash, dissolve, and measure RNA concentration, as described in steps 7–10 except dissolve in 20 μL of nuclease-free water.
15. Check TEX treatment using the RNA Nano Kit and a 2100 Bioanalyzer (Agilent).

2.2.4 Notes

1. Use nuclease-free water for the preparation of ethanol and NaOAc solutions.
2. Briefly spin Phase Lock Gel tubes before use to collect gel matrix at the bottom of the tube
3. The ±TEX samples can also be analyzed by visual inspection using polyacrylamide gels (Borries, Vogel, & Sharma, 2012).

2.3 cDNA Library Preparation

The reads obtained from dRNA-seq experiments require strand-specific information to allow mapping onto a reference genome in the correct orientation (sense or antisense). To this end, the TEX- and TAP-treated transcripts are polyadenylated at the 3′ end and RNA linkers are ligated to the 5′ ends. Oligo(dT)-adapter primers bind to the complementary 3′ poly(A) stretches and a reverse transcriptase is used to perform first-strand cDNA synthesis. The cDNA libraries are amplified in a PCR reaction using barcoded primers. These barcodes are specific nucleotide sequences allowing massive parallel sequencing on an appropriate sequencing platform and subsequent sample-specific in silico assignment of reads (demultiplexing). For a more detailed protocol of the cDNA library preparation, the reader is referred to Berezikov et al. (2006) and Borries et al. (2012).

2.3.1 Equipment

- 2100 Bioanalyzer (Agilent)
- Agencourt AMPure XP beads (Beckman Coulter)
- DNA 1000 Kit (Agilent)
- Magnetic rack
- PCR thermocycler

2.3.2 Buffers and Reagents

- *E. coli* Poly(A) Polymerase (5 U/μL) and 10× reaction buffer (New England Biolabs)
- High-fidelity DNA polymerase (e.g., Q5 Polymerase, New England Biolabs) and reaction buffer

- Illumina primers for PCR amplification
- Illumina sequencing adapter for 5′ and oligo(dT)-adapter primer for 3′ end, e.g., TrueSeq Sense primer (5′→3′ orientation):
  AATGATACGGCGACCACCGAGATCTACACTCTTTCCCT ACACGACGCTCTTCCGATCT
  Antisense primer: NNNNNN=barcode sequence for multiplexing (5′→3′ orientation)
  CAAGCAGAAGACGGCATACGAGAT-NNNNNN-GTGACT GGAGTTCAGACGTGTGCTCTTCCGATC(dT25)
- M-MLV reverse transcriptase (200 U/μL) and 5 × reaction buffer (Thermo Fisher Scientific)
- T4 RNA Ligase (10 U/μL) and 10× reaction buffer (New England Biolabs)
- dNTPs

2.3.3 Procedure

- cDNA libraries are generated as previously described (Berezikov et al., 2006; Borries et al., 2012), but omitting RNA size-fractionation prior to cDNA synthesis. Briefly, equal amounts of ±TEX- and TAP-treated RNA are incubated with poly(A) polymerase and a 5′ Illumina linker is ligated to the 5′ ends. Next, first-strand synthesis is performed using an oligo(dT)-adapter primer and the M-MLV reverse transcriptase. The incubation steps are conducted at 42°C for 20 min and 5 min at 55°C. Finally, cDNA libraries are amplified by PCR using a high-fidelity polymerase and Illumina adapter primers. The DNA is subsequently purified using AMPure XP beads (1.8 × sample volume). The quality of cDNA libraries is assessed on a DNA chip in a Bioanalyzer, and deep sequencing is performed on an Illumina HiSeq 2000 platform following standard protocols.

2.3.4 Notes

1. If adapter dimers are prominent in the electrophoresis profile of the DNA chip, a size selection step is recommended to avoid extensive sequencing of these self-ligated fragments.

3. IDENTIFICATION OF sRNA TARGET GENES

Determining the target spectrum of an sRNA is an important step in understanding its physiological roles. Given that the vast majority of Hfq-dependent sRNAs regulate multiple *trans*-encoded mRNAs

(Papenfort & Vogel, 2009; Waters & Storz, 2009), the search for putative targets requires a global view on gene expression and often involves transcriptome profiling. Such analyses have been particularly useful in the case of Hfq-dependent sRNAs because RNA-duplex formation is typically associated with changes in transcript abundance (Mohanty & Kushner, 2018), which can be measured by RNA-seq. Thus, sRNA overexpression followed by global transcriptome analysis is a promising strategy to discover target mRNAs. However, especially constitutive overexpression of an sRNA is prone to cause pleiotropic effects. For example, the regulation of a transcription factor as a primary sRNA target could indirectly affect the expression of dozens of downstream genes. Further, in some cases, constitutive sRNA overexpression from a multicopy plasmid is toxic (Sharma, Darfeuille, Plantinga, & Vogel, 2007; Sharma et al., 2011) or could increase the frequency of suppressor mutations. To circumvent these limitations, an inducible promoter (e.g., the arabinose-regulated P_{BAD} promoter) is used to tightly regulate sRNA expression from a multicopy plasmid. The rationale behind this approach is to strongly induce sRNA expression for 10–15 min (pulse-expression), which is sufficient to score changes in transcript levels of primary targets, but will not affect the regulation of downstream genes. Of note, sRNA pulse-expression followed by RNA-seq will not automatically reveal all primary targets of an sRNA because certain sRNA–mRNA interactions do not result in rapid transcript decay. Moreover, some targets may not be transcribed under the conditions tested. Fig. 4 illustrates the sRNA pulse-expression method, which we will explain in more detail in the following sections.

3.1 Generation of Inducible sRNA Overexpression Plasmids

For an sRNA pulse-expression experiment as described above a suitable sRNA expression plasmid should be generated. We recommend the use of mid- to high-copy plasmids, e.g., carrying pBR322 or p15a origins of replication (Sambrook, 2001). Further, the plasmid requires an inducible promoter, tightly controlling sRNA expression. The arabinose-dependent P_{BAD} promoter has been established as the promoter of choice for sRNA pulse-expression experiments (Papenfort et al., 2006) and is therefore used in our experiments. It is crucial, that the full-length sRNA sequence is inserted at the TSS position of the P_{BAD} promoter, since transcription of an extended or truncated sRNA variant possibly impairs or alters its regulatory potential. The Gibson assembly method (Gibson et al., 2009) allows the rapid and precise insertion of an sRNA gene into a linearized plasmid backbone carrying the P_{BAD} promoter and is described below.

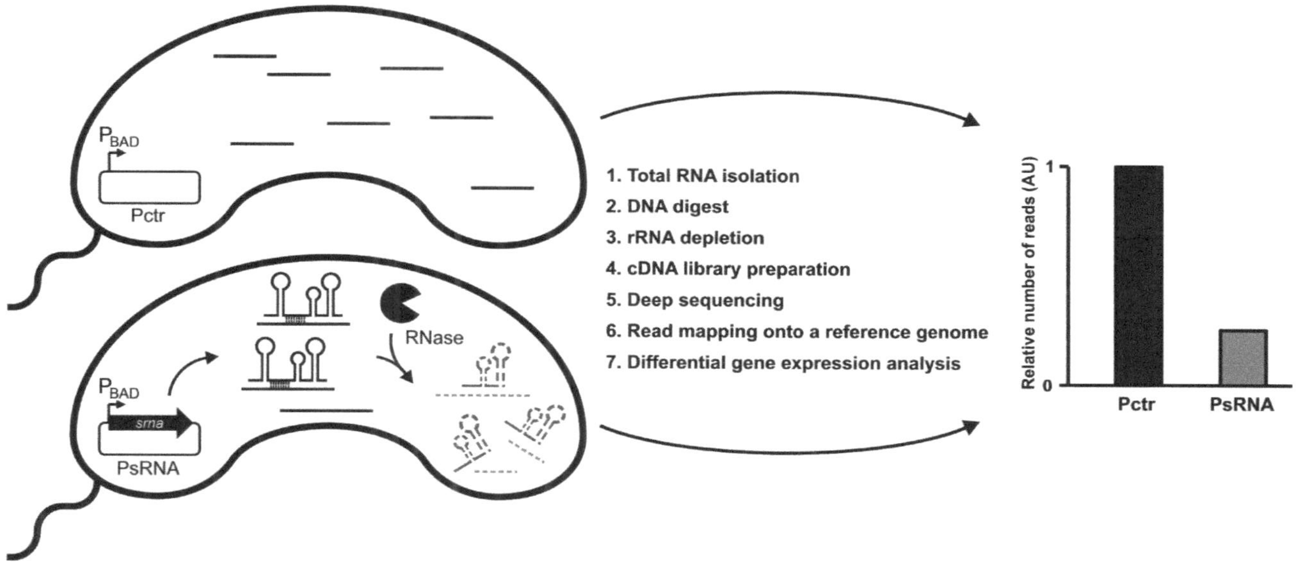

Fig. 4 Identification of sRNA targets using sRNA pulse-expression and RNA-seq. An empty control plasmid (Pctr) and an L-arabinose inducible P_{BAD} sRNA overexpression plasmid (PsRNA) are introduced into *V. cholerae* strains. Both strains are treated with L-arabinose triggering sRNA expression of the PsRNA-carrying strain. The sRNAs engage RNA duplex formation with target mRNAs, which typically results in altered transcript levels. After 10–15 min of induction, total RNA is isolated and prepared for RNA-seq. Transcripts with significantly changed read counts in the samples overexpressing the sRNA, compared to the Pctr samples, are potential sRNA targets.

3.1.1 Equipment

- Standard horizontal agarose gel electrophoresis system
- Electrophoresis power supply
- Gel documentation system
- PCR purification kit (e.g., DNA Clean & Concentrator Kit, Zymo Research)
- PCR thermocycler
- Plasmid DNA purification kit (e.g., HiYield Plasmid Mini DNA Kit, e.g., Südlabor)
- Spectrophotometer (e.g., NanoDrop 2000, Thermo)

3.1.2 Buffers and Reagents

- Appropriate antibiotics for plasmid selection
- dNTPs
- *Dpn*I restriction enzyme (20 U/μL; New England Biolabs)
- HiFi DNA Assembly Master Mix (New England Biolabs)
- High-fidelity DNA polymerase (e.g., Q5 Polymerase, New England Biolabs) and reaction buffer
- DNA oligonucleotides

3.1.3 Procedure

1. Linearize an expression plasmid carrying the inducible P_{BAD} promoter, by PCR. The primer sequences should be designed so that the −1 position of the P_{BAD} TSS is located at one end of the linearized product.
2. Amplify the sRNA of interest by PCR, using DNA oligonucleotides with 15–20 nts of sequence overlap to the ends of the linearized backbone.
3. Run PCR products on an agarose gel to ensure correct and specific amplification.
4. *Dpn*I digest all PCR products that are derived from a plasmid DNA template.
5. Purify DNA using a commercially available kit and determine the PCR product concentrations with a spectrophotometer.
6. Assemble the two fragments using the HiFi DNA Assembly Master Mix according to manufacturer's instructions. Transform the mixture into competent *E. coli* cells and plate on selective LB-agar plates.
7. Make sure that the fragments are assembled correctly via isolation and sequencing of the plasmid DNA.

8. Conjugate the sRNA expression plasmid and an empty control vector into *V. cholerae* cells. To avoid endogenous sRNA regulation, the plasmids should be introduced into an sRNA mutant strain.

3.1.4 Notes

1. For all enzymatic reactions, DNA purification, and plasmid DNA isolation, follow the manufacturers' protocols.
2. Make sure that the sRNA gene is inserted into the plasmid in the correct orientation. The DNA oligonucleotide that binds at the 5′ end of the sRNA should carry an extended sequence complementary to the end of the linearized plasmid, which displays the −1 position of the TSS of the inducible promoter.

3.2 sRNA Pulse-Expression, RNA Preparation, and DNA Digest

The timing of sRNA induction is of considerable importance, since putative target genes need to be expressed at this particular time point. If the target transcripts are not expressed, they will not be represented in the transcriptomic data. Profiling the expression of the sRNA of interest under various conditions is often helpful to identify a "sweet spot" for sRNA induction and can easily be monitored on Northern blots (see Section 4.1). Assuming that the sRNA is highest expressed when its function is needed the most, the time point that shows the highest sRNA production might also be most suitable for sRNA pulse-expression. Further, the experiment should be performed in an sRNA deletion background to avoid target regulation by the chromosomally expressed sRNA copies. For example, the VqmR sRNA is highly expressed in stationary phase and the *vpsT* mRNA is a known target of VqmR (Figs. 1 and 5). Thus, during stationary growth phase *vpsT* levels are low in *V. cholerae* wild-type cells and an additional pulse-expression of VqmR may only have mild effects. In contrast, *vpsT* levels in a *V. cholerae vqmR* mutant are increased and will be strongly affected upon VqmR overexpression. After 10–15 min of P_{BAD}-controlled sRNA induction, cells are collected and transcription is stopped. Isolation of total RNA and the depletion of genomic DNA is accomplished as described in Section 2.1. If one or more targets of the tested sRNA are already established, the experimental approach can be validated using the isolated RNA and Northern blot or qRT-PCR analyses.

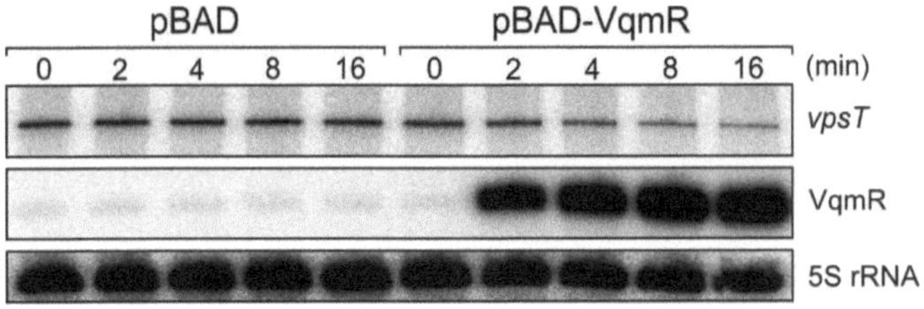

Fig. 5 Northern blot analysis of *vpsT* mRNA after VqmR pulse-expression. *V. cholerae* cells carrying either an empty control plasmid (P_{BAD}) or an L-arabinose inducible VqmR overexpression plasmid (P_{BAD}-VqmR) were cultivated to $OD_{600} = 0.2$. Cultures were treated with L-arabinose and expression levels of *vpsT* and VqmR prior induction (0 min) and 2, 4, 8, and 16 min after induction were analyzed on a Northern blot. 5S rRNA served as loading control.

3.2.1 Equipment

- 5PRIME Phase Lock Gel tubes (Quantabio)
- Cooling centrifuge
- Spectrophotometer (e.g., NanoDrop 2000)
- Heating block with shaking function (e.g., ThermoMixer, Eppendorf)

3.2.2 Buffers and Reagents

- 75% (vol./vol.) Ethanol
- Appropriate antibiotics (for plasmid maintenance)
- Inducer (e.g., L-arabinose when using the P_{BAD} promoter)
- Chloroform
- ExtraZol (7Bioscience)
- Nuclease-free water
- Phenol/Chloroform/Isoamyl alcohol solution (25:24:1 vol./vol.; e.g., Carl Roth)
- Propan-2-ol (Isopropanol)
- Stop-Mix (95% [vol./vol.] ethanol, 5.0% [vol./vol.] phenol)
- TURBO DNase (2 U/μL) & 10× Reaction Buffer (Thermo Fisher Scientific)

3.2.3 Procedure

1. Dilute overnight cultures of three independent clones each of the strains harboring the sRNA overexpression plasmid and the empty control plasmid, respectively, 1:1000 in fresh media (add appropriate antibiotics to maintain plasmid).

2. At an appropriate OD, treat all cultures with inducer to gain maximal sRNA expression (e.g., 0.2% final conc. of L-arabinose to induce a P_{BAD} promoter). After 10–15 min of induction, immediately proceed with step 3.
3. Harvest cells, isolate the RNA, and deplete chromosomal DNA by following steps 2–13 described in Section 2.1.3.

3.2.4 Notes

1. To test for potential "leakiness" of the inducible promoter, additional RNA samples should be collected immediately before addition of the inducer. Northern Blot analysis (see Section 4.1) can be used to probe cellular sRNA levels.
2. This method aims to identify only primary targets of the sRNA of interest. Thus, induced sRNA expression should not exceed 15 min.

3.3 Ribosomal RNA Depletion and cDNA Library Preparation

Ribosomal RNA (rRNA) constitutes approximately 70%–80% of the bacterial RNA pool (Westermann, Gorski, & Vogel, 2012). However, in most transcriptomic studies, rRNAs are of minor interest and are therefore depleted to enrich other RNA species (e.g., mRNAs and sRNAs) prior to cDNA library construction. Several methods have been developed to remove bacterial rRNAs, for example, size-fractionation using gel electrophoresis (Liu et al., 2009), coimmunoprecipitation of protein-bound RNAs (Sittka et al., 2008), or TEX treatment (Section 2.2). Further, there are a variety of commercially available kits, which are based on species-specific probes, which hybridize to rRNA sequences. These probes attach to magnetic beads and are easily separated in the presence of a magnet. In our experience, the Ribo-Zero Illumina Kit is suitable to remove the predominant 16S and 23S rRNA transcripts of *V. cholerae*. Of note, due to the rapid improvement of sequencing depth and the drop of sequencing costs, rRNA depletion has become an optional rather than an essential step. Nevertheless, the decision for or against rRNA depletion should be made carefully and involves multiple factors. For example, desired sequencing coverage and overall costs should be considered.

RNA sequences need to be converted into cDNA libraries in order to be applicable to most NGS platforms. To distinguish between sense and antisense reads after sequencing, the construction of cDNA libraries requires a strand-specific protocol. There are several established protocols that maintain strand-specific information, for example, poly(A)-tailing and 5′-end linker ligation (as described in Section 2.3), first-strand sequencing (Croucher et al., 2009), bisulfite-induced C to U conversions prior cDNA synthesis (He, Vogelstein, Velculescu, Papadopoulos, & Kinzler, 2008), or the incorporation of dUTPs during second-strand cDNA synthesis (Parkhomchuk et al., 2009). The latter, so-called dUTP-method, is also applied in the commercially available NEBNext Ultra II Directional RNA Library Prep Kit, which we describe here. The kit's protocol includes a fragmentation step prior to first-strand synthesis increasing the distribution of reads along transcripts. The fragments are primed with random hexamers, which enable first-strand synthesis. In contrast to standard library construction protocols, dUTP nucleotides are used instead of dTTPs for second-strand synthesis. After dA-tailing and adapter ligation, the second strand is degraded using uracil-DNA glycosylase. The first-strand cDNA sequences are recovered, enriched by PCR, and finally sequenced on an Illumina sequencing platform.

3.3.1 Equipment

- 2100 Bioanalyzer (Agilent)
- Agencourt AMPure XP beads (Beckman Coulter)
- Centrifuge
- Magnetic rack (e.g., Life technologies)
- NEBNext Multiplex Oligos for Illumina (New England Biolabs)
- NEBNext Ultra II Directional RNA Library Prep Kit (New England Biolabs)
- PCR thermocycler or heating block
- Ribo-Zero rRNA Removal Kit for Bacteria (Illumina)
- RNA Nano Kit (Agilent)
- Spectrophotometer (e.g., NanoDrop 2000)
- Vortex mixer

3.3.2 Buffers and Reagents

- Ethanol (100% and 70% [vol./vol.])
- Nuclease-free water

3.3.3 Procedure

1. Deplete rRNA using the Ribo-Zero Kit, following the manufacturer's instructions. Briefly, 1–5 μg of DNA-free total RNA is mixed with probes that specifically hybridize to microbial rRNAs. The probes bind to magnetic beads and are easily separated from the rRNA-depleted RNA fraction. To clean up the recovered RNA, an ethanol precipitation step

is required. Analyze rRNA-depleted RNA samples on a 2100 Bioanalyzer platform using the RNA Nano Kit. The electrophoresis profiles should look similar to Fig. 3B.

2. cDNA libraries are prepared using the NEBNext Ultra II Directional RNA library Prep Kit for Illumina sequencing according to the manufacturer's protocol. In brief, 1–100 ng rRNA depleted RNA is fragmented to sizes of $\approx$200 nt. These fragments are randomly primed which allows first-strand cDNA synthesis. During second-strand synthesis, dUTP is incorporated and the cDNA products are purified using AMPure XP beads. After end repair and dA-tailing, NEBNext adaptors are ligated and the USER enzyme (Uracil-Specific Excision Reagent) cleaves at the uracil-incorporation sites. AMPure XP beads are used for size-selection and purification of the cDNA libraries. The libraries are amplified using NEBNext oligos for multiplexing and purified in a final cleanup step with AMPure XP beads. Quality of cDNA libraries is assessed on a DNA chip in a 2100 Bioanalyzer. Fig. 3C illustrates an example of a typical cDNA library size distribution.

3.3.4 Notes

1. The probes of the Ribo-Zero Kit do not hybridize to the 5S rRNA of *V. cholerae* (Fig. 3B).
2. If peaks at $\approx$80 bp (primers) or 128 bp (adapter dimer) appear in the Bioanalyzer traces of a cDNA library, the final cleanup step using AMPure XP beads should be repeated.

4. VALIDATION OF sRNA-TARGET INTERACTIONS

RNA-seq data obtained from sRNA overexpression experiments as described in Section 3.2 typically reveal multiple putative target transcripts of the tested sRNA (Wagner & Romby, 2015). To validate and characterize a direct sRNA-mediated regulation of a transcript, a combination of biochemical, genetic, and computational methods is required.

4.1 Northern Blots

Northern blotting is a well-established method for sensitive and highly specific detection of transcripts using radioactively labeled probes. In contrast to other methods (such as qRT-PCR), processing of a transcript can be easily visualized. Further, the transferred RNA is covalently bound to a membrane, which allows frequent reprobing of the same Northern blot.

Prior to RNA transfer onto a membrane, RNA needs to be size-separated using denaturating agarose or polyacrylamide gels and electrophoresis. Agarose gels are preferred for separation of large transcripts (>1200 nts); however, they provide only limited resolution for smaller fragments. In contrast, polyacrylamide gels provide a superior resolution in the range of <20 to approximately 1200 nts, depending on the acrylamide–bisacrylamide concentration of the gel (Rio, Ares Jr, Hannon, & Nilsen, 2010). Consequently, polyacrylamide gels are favored for Northern blot analyses of sRNAs, which are typically between 50 and 350 nts in size, and putative target transcripts that do not exceed 1200 nts. For example, the Northern blot shown in Fig. 5 shows that the VqmR sRNA (151 nt in length) promotes the decay of the ≈750 nt *vpsT* mRNA. Expression levels of larger transcripts (e.g., transcribed operons) can either be monitored by Northern blotting using agarose gel electrophoresis, or via qRT-PCR. Due to its minimal input requirements of RNA template, qRT-PCR can be the method of choice to quantify poorly expressed transcripts. In this section, we explain the Northern blot procedure using polyacrylamide gels. For more detailed information about Northern blotting using agarose gels and qRT-PCR analyses, we refer the reader to published protocols (Lan, Tang, Un San Leong, & Love, 2009; Rio, 2015).

4.1.1 Equipment

- Vertical electrophoresis system (e.g., VWR, PerfectBlue Dual Gel System Twin L)
- Tank blotter (e.g., VWR, PerfectBlue Electroblotter)
- Microcentrifuge
- Ambion MAXIscript T7 In Vitro Kit (Thermo Fisher Scientific)
- Centrifuge
- Electrophoresis power supply
- Transparent sheet protectors
- Gel documentation system
- Heating block
- Hybridization oven
- Hybridization tubes
- Microspin G-25/G-50 Columns (GE Healthcare)
- Nylon membrane (e.g., Amersham Hybond-XL Membrane, GE Healthcare)
- PCR purification kit (e.g., DNA Clean & Concentrator, Zymo Research)

- PCR thermocycler
- Phosphorimager (e.g., Typhoon FLA 7000, GE Healthcare Life Sciences)
- Spectrophotometer (e.g., NanoDrop 2000)
- Storage phosphor screen and exposure cassette
- UV light chamber
- Whatman paper

4.1.2 Buffers and Reagents

- 2 × RNA gel loading buffer (98%[v/v] formamide, 0.02% [w/v] xylene cyanol, 0.02% [w/v] bromophenol blue, 2 m*M* EDTA)
- 10 × TBE buffer
- 19:1 Acrylamide:Bisacrylamide solution (40% [w/v])
- Alpha-^{32}P-UTP (10 mCi/mL, for riboprobes)
- APS (10% [w/v])
- DNA marker (e.g., 50 bp DNA Ladder, New England Biolabs)
- DNA polymerase (e.g., Q5 Polymerase, New England Biolabs) and 5 × reaction buffer
- dNTPs
- Gamma-^{32}P-ATP (10 mCi/mL)
- Hybridization buffer (e.g., Roti-Hybri.Quick, Carl Roth)
- Nuclease-free water
- SSC buffers (5 ×, 1 ×, 0.5 ×)
- DNA oligonucleotides
- SDS solution (10% [w/v])
- T4 Polynucleotide Kinase (10 U/μL) and 10 × reaction buffer (New England Biolabs)
- TEMED
- Urea

4.1.3 Procedure

1. Prepare a denaturing 7 *M* urea gel for RNA gel electrophoresis (Table 1). For probing of transcripts up to 1200 nts (e.g., mRNAs) use 4% PAA gels; for sRNA detection 6%–8% PAA gels are recommended.
2. Immediately pour the gel and insert the comb. After approximately 10 min the gel is polymerized and fixed in a vertical electrophoresis unit. Fill up the apparatus with 1 × TBE to avoid drying of the gel.
3. For RNA size estimation on Northern Blots, size markers need to be labeled radioactively. Start with diluting 5 μL of a DNA marker (0.5 μg/μL) with 7 μL water, denature the mixture at 95°C for 5 min,

Table 1 Composition of Denaturing Urea Gels With Different PAA Concentrations

	4% PAA 7 *M* Urea Gel	**6% PAA 7 *M* Urea Gel**
19:1 Acrylamide solution	10 mL	15 mL
Urea	42 g	42 g
10 × TBE	10 mL	10 mL
10% APS	800 μL	800 μL
TEMED	106 μL	106 μL
H_2O	ad 100 mL	ad 100 mL

and place it on ice. Add 2 μL 10 × T4 PNK reaction buffer, 1 μL T4 PNK (10 U/μL), and 3–5 μL gamma-^{32}P-ATP. Incubate for 1 h at 37°C. Bring mixture to a total volume of 50 μL and purify the labeled marker using a Microspin G-50 Column according to the manufacturer's instructions. Dilute the cleanedup marker with 150 μL 1 × RNA loading buffer.

4. Prepare 10 μg of total RNA for each sample using a spectrophotometer and adjust volumes with nuclease-free water. Add the same volume of 2 × RNA loading dye. Use ≈1 μL of a freshly labeled marker and adjust the volume to 10 μL with 1 × RNA loading buffer. Boil all samples (including the marker) for 5 min at 95°C and immediately put back on ice.
5. Remove the comb from the gel and use a pipette to flush the pockets with 1 × TBE to allow proper loading. Carefully load the samples into the wells without disturbing the gel matrix and run the gel at 250–300 V for approximately 2–3 h.
6. Build the blotting "sandwich" in the following order (bottom to top): three layers of Whatman paper, released gel (note 2), nylon membrane, three layers of Whatman paper. Make sure that the Whatman paper and the membrane are wetted with 1 × TBE before use and that trapped air bubbles between the layers are removed (for example, by rolling a glass pipette across the surface). Place the transfer sandwich in the correct orientation into the tank electroblotter and fill up with 1 × TBE. Move the chamber to 4°C and start transfer 50 V for 1 h at 50 V.
7. Disassemble the blotting sandwich, quickly dry the membrane between two Whatman paper, and cross-link the RNA to the membrane using UV light (320 nm, 120 mJ). Make sure that the RNA site is facing the UV lamp.

8. Wet the membrane in 15 mL hybridization buffer in a hybridization tube at 42°C (for oligonucleotide probes) or 63°C (for riboprobes) for at least 30 min, before adding the labeled probe.
9. Riboprobes are generated using the Ambion MAXIscript T7 In Vitro Kit following manufacturer's instructions. Oligoprobes are prepared by mixing 1 μL DNA oligonucleotide (10 μ*M* stock solution), 2 μL 10 × T4 PNK reaction buffer, 1 μL T4 PNK (10 U/μL), and 13 μL nuclease-free water, followed by incubation at 37°C for 1–1.5 h. To remove unincorporated labeled nucleotides, the ribo- and oligonucleotide probes are cleaned up using Microspin G-50 and G-25 Columns, respectively. Both probes can be stored at −20°C.
10. Boil the cleaned-up probe for 5 min and quick chill on ice. Start hybridization by adding 2 or 4 μL of an oligo- or riboprobe, respectively, into the hybridization buffer. Perform hybridization for 2–12 h.
11. Discard hybridization buffer and perform three washing steps (15 min each) using SSC buffers of different concentrations in the following order: 5 ×, 1 ×, and 0.5 × SSC. Always use a washing volume of approximately 50 mL and add SDS to a final concentration of 0.1% (w/v). All washing steps are performed at 42°C.
12. Dry Northern blot membrane between two Whatman papers, move it to a sheet protector and seal the foil. Expose the blot to a storage phosphor screen in an exposure cassette.
13. Detect the signals using a phosphorimager (e.g., Typhoon FLA 7000, GE Healthcare Life Sciences).

4.1.4 Notes

1. The total volume of each sample that is loaded on the gel should not exceed 30 μL. If the RNA concentration is insufficient to prepare 10 μg of total RNA, we recommend to either load less RNA (e.g., 5 μg total RNA) or to increase the RNA concentration via precipitation methods or by using a vacuum concentrator system.
2. After electrophoresis, release the gel from one glass plate by slowly turning one spacer. Use a flexible material with a sticky surface to remove the gel from the second glass plate and transfer it to a Whatman paper.
3. Mark the orientation of the membrane with a pencil.
4. Oligoprobes are easily prepared and sufficient to detect abundant transcripts, e.g., most sRNAs. The generation of riboprobes requires a PCR template to perform the T7 in vitro transcription. In vitro transcription incorporates multiple alpha-^{32}P-UTPs, which allows the

detection of weakly expressed transcripts (e.g., mRNAs of transcription factors) on a phosphor screen.

5. Northern Blot membranes can be reprobed several times. Therefore, previous probes often need to be removed (stripped), which can be achieved by applying boiling distilled water onto the membrane and adding SDS solution to a final concentration of $\approx 1\%$ (w/v). Seal the container with a lid and incubate on a rocking platform for 20 min. Discard the solution and repeat the "stripping" step. Seal the membrane into a foil and apply an erased phosphor screen to check for removal of the previous probe(s) using a phosphorimager.

4.2 Generation of Plasmid-Based Posttranscriptional Reporters

Transcripts that are regulated in response to sRNA overexpression cannot automatically be classified as direct sRNA targets. For example, an sRNA could inhibit the synthesis of a transcription factor, which would also result in decreased transcript levels of the genes that are controlled by this regulator. Methods like Northern blots and qRT-PCR do not allow to discriminate between direct and indirect sRNA targets. Thus, reporter assays have been established that specifically score for sRNA-mediated posttranscriptional regulation of a selected transcript (Corcoran et al., 2012). Our posttranscriptional reporter assay is based on the constitutive expression of the sRNA and the putative target transcript, which is translationally fused to a fluorescent reporter gene (e.g., *sfgfp*). Specifically, the P_{tac} promoter drives the sRNA expression from a mid-copy plasmid, whereas a P_{tetO} promoter on a low-copy pXG10 plasmid backbone (carrying a pSC101* origin of replication; Corcoran et al., 2012) controls the expression of the translational reporter fusion (Fig. 6). In the case of a direct sRNA–mRNA interaction, translation of the reporter fusion is altered, leading to a decreased (repressed targets, Fig. 6, right) or increased (activated targets) fluorescence signal. Considering that sRNA/mRNA duplex formation frequently occurs at or close to the ribosome-binding site (Waters & Storz, 2009), we recommend to narrow down the transcript sequence to the 5′ UTR and the first 60 bases of the coding sequence. Shortening of the putative target sequence can further reduce the number of "false-positive" results. To reduce *Vibrio*-specific background noise (e.g., through autoregulatory feedback loops and other sRNAs), which could affect the frequency or stability of sRNA–mRNA interactions, these assays can be performed in a heterologous host, such as *E. coli*.

4.2.1 Equipment

- Standard horizontal agarose gel electrophoresis system
- Clear bottom 96-well Microplate (e.g., Greiner 96 Flat Black)
- Electrophoresis power supply
- Gel documentation system
- Microplate reader (e.g., Tecan Spark 10 *M*)
- PCR purification kit (e.g., DNA Clean & Concentrator, Zymo Research)
- PCR thermocycler
- Spectrophotometer (e.g., NanoDrop 2000)

4.2.2 Buffers and Reagents

- 1 × PBS
- Appropriate antibiotics for plasmid cloning
- dNTPs
- *Dpn*I restriction enzyme (20 U/μL; New England Biolabs)
- HiFi DNA Assembly Master Mix (New England Biolabs)
- High-fidelity DNA polymerase (e.g., Q5 Polymerase, New England Biolabs) and reaction buffer
- DNA oligonucleotides

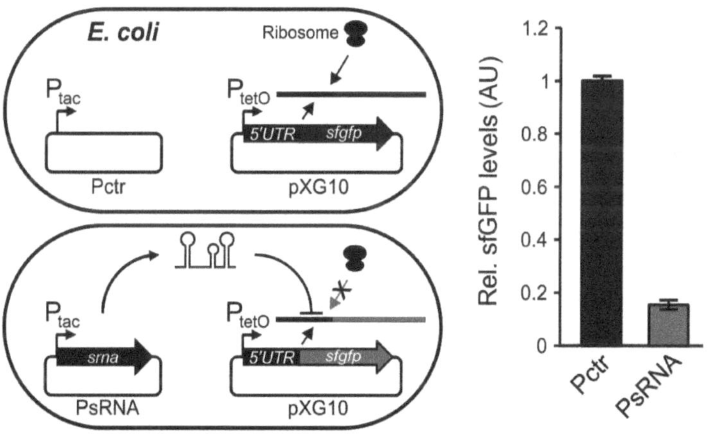

Fig. 6 A posttranscriptional reporter system to investigate direct sRNA-target interactions. *E. coli* cells carrying a pXG10-based target gene fusion were cotransformed with either an empty control plasmid (Pctr) or an sRNA overexpression plasmid (PsRNA). For inhibited target genes translation of the target gene fusion is reduced in the presence of the sRNAs. Thus, sfGFP levels are decreased compared to the cells that carry the control plasmid, and this effect can be measured, e.g., in a microplate reader or by Western blot analysis.

4.2.3 Procedure

1. The sRNA overexpression plasmid can be generated following steps 1–7 of Section 3.1.3; however, the template plasmid should harbor a constitutive promoter. We use a P_{tac} promoter implemented in a plasmid backbone. Additionally, transform an empty vector control.
2. The second plasmid is based on a pXG10 backbone that constitutively expresses a start codon-lacking *sfgfp* gene (starting at the second codon) from a P_{tetO} promoter. To construct a translational reporter fusion of a gene of interest, open the backbone by PCR between the −1 site of the TSS and the first nucleotide of the *sfgfp* gene. Next, the 5′ UTR and the coding sequence of the first 20 amino acids of the putative target gene is amplified from genomic DNA by a set of two primers that carry 15–20 nt overhangs to the ends of the linearized plasmid backbone. Follow steps 3–7 of Section 3.1.3 and transform the plasmid into *E. coli* cells that already carry the sRNA overexpressing plasmid or the empty vector control.
3. Inoculate a single colony of the *E. coli* strains harboring the translational reporter plasmid and either the sRNA overexpression plasmid (PsRNA), or the empty control plasmid (Pctr) and incubate for ~12 h. Make sure to additionally inoculate an *E. coli* strain that lacks any fluorescent proteins (autofluorescence control).
4. Dilute the cultures 1:1000 in fresh media and let the cells grow to an OD_{600} of 0.5–1. Harvest 200 μL of each culture by centrifugation (4°C, 13,000 rpm, 1 min) and wash cell pellets with 1 mL PBS.
5. Repeat the previous step and suspend each pellet in 500 μL PBS. Transfer 3 × 150 μL to 96-well plates (technical triplicates) and measure absorbance at 600 nm and sfGFP fluorescence in a microplate reader.
6. Correct the fluorescence over OD_{600} and subtract the values of the autofluorescence control.

4.2.4 Notes

1. Make sure that both plasmids carry compatible origins of replication (e.g., p15A and pSC101*).
2. Aerobic conditions can promote fluorescence measurements, since oxygen is required for chromophore formation.

4.3 In Silico Prediction of sRNA–Target mRNA Interactions

Hfq-dependent sRNAs usually act through imperfect base pairing with *trans*-encoded target mRNAs. This challenges the development of computational tools, which can predict sRNA–mRNA duplexes at a genome-wide scale. Algorithms like CopraRNA (Wright et al., 2013), IntaRNA (Mann, Wright, & Backofen, 2017), and TargetRNA2 (Kery, Feldman, Livny, & Tjaden, 2014) allow for the identification of putative sRNA-binding sites in candidate mRNA sequences. Nonetheless, experimental validation of these computationally generated predictions is usually required. After target identification, e.g., by RNA-seq (Section 3), and posttranscriptional reporter experiments (Section 4.2), the exact base pairing site should be located. Here, we recommend the easy-to-use web interface of the RNAhybrid algorithm, which determines the energetically most favorable hybridization site between an sRNA and its target sequence. For example, Fig. 7 shows the predicted duplex formation between VqmR and the *vpsT* mRNA (Papenfort, Forstner, et al., 2015).

4.3.1 Procedure

1. Open the RNAhybrid Web tool at the following URL: https://bibiserv2.cebitec.uni-bielefeld.de/rnahybrid and go to "Submission."
2. Copy and paste the 5′ UTR sequence and the sequence of the first 20 codons of the putative target gene as "target sequence(s)." Next, copy and paste (parts of) the sRNA of interest in the 5′–3′ direction as "miRNA sequence(s)." Convert both sequences into the FASTA format and click next.
3. Select desired parameters. We usually choose a value of 3 for "hits per target" and check the "Generate graphics" box. Click "nothing" as a source of approximate *P*-value estimation. Click next and start calculation.

4.4 Compensatory Base Pair Exchange Experiments

Once an sRNA–mRNA duplex formation has been predicted, it needs to be validated experimentally. To this end, we use the two plasmid posttranscriptional reporter assay described in Section 4.2 and introduce mutations into

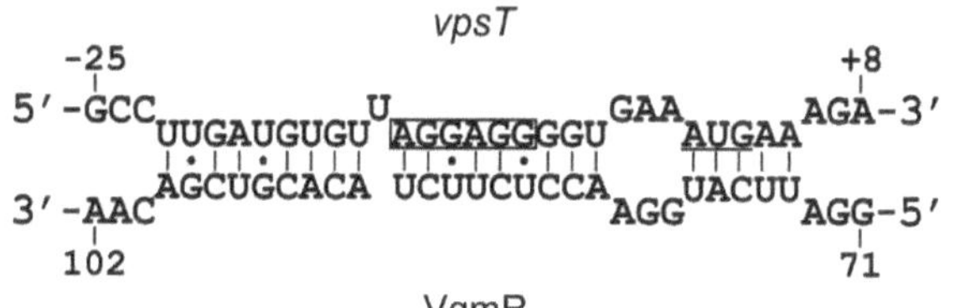

Fig. 7 Predicted base pairing between VqmR and the *vpsT* mRNA. *Numbers* indicate the distances from the translational start site for the *vpsT* mRNA, and the distance from the 5′ end for the VqmR sRNA. The ribosome-binding site is *boxed* and the AUG start codon is *underlined.*

the sRNA and mRNA sequences. Specifically, one base pair of the putative base pairing region is altered to the complementary base. An sRNA–mRNA interaction is confirmed, when the mutagenesis of one nucleotide impairs posttranscriptional regulation, while the introduction of a complementary nucleotide restores regulation.

4.4.1 Equipment

- Standard horizontal agarose gel electrophoresis system
- Black 96-well microplate
- Gel documentation system
- Microplate reader (e.g., Tecan Spark 10 *M*)
- PCR purification kit (e.g., DNA Clean & Concentrator, Zymo Research)
- PCR thermocycler
- Plasmid DNA purification kit (e.g., HiYield Plasmid Mini DNA Kit, SLG)

4.4.2 Buffers and Reagents

- 1 × PBS
- Appropriate antibiotics for plasmid selection
- dNTPs
- *Dpn*I restriction enzyme (20 U/μL; New England Biolabs)
- High-fidelity DNA polymerase (e.g., Q5 Polymerase, New England Biolabs) and 5 × reaction buffer
- DNA oligonucleotides

4.4.3 Procedure

1. Select one base pair of a predicted sRNA–mRNA duplex for site-directed mutagenesis that potentially disrupts the sRNA–mRNA interaction. For example, the G at position −13 of the mRNA shown in Fig. 8A that is predicted to base pair with the C at position 67 of the sRNA is altered to a C. In the reciprocal approach, the C67 of the sRNA is mutated into a G. The implementation of these point mutations into the mRNA and sRNA sequences of the plasmids generated in Section 4.2 is described in the following steps.
2. For each plasmid, design two complementary DNA oligonucleotides that are between 25 and 40 nt in length ($T_m > 70°C$) and carry the desired mutation in the middle. Additionally, a C or G should define the 3′ end of each primer.

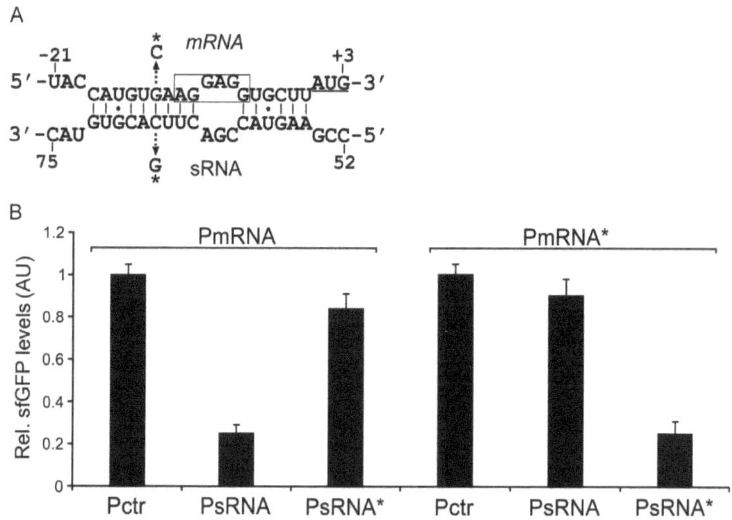

Fig. 8 Compensatory base pair exchange experiments. (A) Predicted RNA duplex formation of a hypothetical sRNA-target pair. The sRNA sequesters the ribosome-binding site (*boxed*) directly upstream of the AUG start codon (*underlined*). *Arrows* and *asterisks* indicate the mutations, which were tested in (B). (B) The posttranscriptional reporter system described in Section 4.2 shows a direct interaction between the target mRNA (PmRNA) and the sRNA (PsRNA). Altering cytosine at position 67 of the sRNA (PsRNA*) inhibits posttranscriptional regulation by the sRNA (*left part*). A compensatory mutation changing guanine at position −13 in the mRNA sequence (PmRNA*) restores repression by PsRNA*, whereas the native sRNA does not inhibit translation (*right part* of the figure).

3. Amplify the plasmids by PCR using the corresponding primer set and run products on an agarose gel.
4. Perform *Dpn*I digest, purify DNA using a commercially available kit, and transform the plasmid DNA into competent *E. coli* cells. The nicks in the mutated plasmid are repaired in vivo.
5. Isolate the plasmid DNA and confirm mutagenesis using Sanger sequencing.
6. Transform the translational reporter plasmid (PmRNA) and its mutated variant (PmRNA*) into competent *E. coli* cells.
7. Introduce a control plasmid (Pctr), the sRNA overexpressing plasmid (PsRNA), and its mutated variant (PsRNA*) into both strains generated in step 6 (resulting in six strains total).
8. Analyze sfGFP levels of these strains as described in Section 4.2.3 (step 3–6). The bar graphs shown in Fig. 8B illustrate the characteristic regulation found in compensatory base pair exchange experiments.

4.4.4 Notes

1. G-C base pairs are a more stable when compared to A-U and G-U pairs and thus make higher contributions to RNA duplex formation. Therefore, G-C base pairs are preferred targets for compensatory base pair exchanges.
2. It may be necessary to alter more than a single nucleotide to disrupt interaction of certain sRNA–mRNA pairs.

5. SUMMARY AND OUTLOOK

In this chapter, we provide a workflow from initial sRNA discovery, to target mRNA identification and validation highlighting the advantages of RNA-seq. dRNA-seq has been employed to explore the primary transcriptomes of bacteria and archaea with single-nucleotide resolution and has boosted the number of TSS and sRNAs in these organisms (Sharma & Vogel, 2014). In *V. cholerae*, dRNA-seq analysis led to the discovery of 107 new regulatory RNAs, including the VqmR sRNA (Papenfort, Forstner, et al., 2015). It has now been established that VqmR is the centerpiece of a novel QS pathway relying on the DPO autoinducer molecule and the VqmA transcription factor (Papenfort et al., 2017). DPO signaling is independent of the AI-2 and CAI-1 autoinducers; however, when detected by *V. cholerae* all three autoinducers repress biofilm formation (Fig. 1). In addition, LuxPQ and CqsS receptors are required to sense AI-2 and CAI-1, respectively. Phosphorylation of LuxO is also affected by two additional sensors, called CqsR and VpsS (Jung, Chapman, & Ng, 2015). Future experiments will show which signaling molecules affect the activity of these sensors and if there are additional sRNAs involved in QS of *V. cholerae*.

In past few years, RNA-seq based methods have largely replaced microarrays to investigate the regulatory networks of bacterial sRNA. Here, we describe the use of sRNA pulse-expression followed by RNA-seq analysis as a tool to identify sRNA target genes. However, this approach only reports on target transcripts with changing stability upon sRNA overexpression. In contrast, certain targets are only translationally inhibited by the sRNA. For example, pulse-expression of the Qrr4 sRNA in *V. harveyi* did not reveal a change in *luxO* transcript stability (Shao, Feng, Rutherford, Papenfort, & Bassler, 2013), although the Qrrs bind to the *luxO* mRNA and block translation (Feng et al., 2015; Tu et al., 2010). To detect these cases of translational inhibition, the pulse-expression RNA-seq approach can be coupled with ribosome profiling analysis, a method that yields genome-wide profiles of protein synthesis (Guo et al., 2014; Wang et al., 2015).

Besides sRNA pulse-expression, several alternative methods have been described to discover target genes of bacterial sRNAs. For instance, in the MAPS (MS2 affinity purification coupled with RNA-sequencing) approach an sRNA of interest is fused to a MS2 aptamer tag, purified from cell lysates and bound interaction partners are identified using RNA-seq (Carrier, Lalaouna, & Masse, 2016) (see chapter "On the prowl: An in vivo method to identify RNA partners of a sRNA" by Carrier, Morin, & Massé). A related method is GRIL-seq (Global small noncoding RNA target identification by ligation and sequencing), which is based on the in vivo coexpression of an RNA ligase that fuses the sRNA to its targets. The generated chimeric RNAs can be analyzed by RNA-seq and the method comes with the advantage that the sRNA of choice does not require genetic manipulation, e.g., by adding an aptamer tag (Han, Tjaden, & Lory, 2016).

Most sRNAs require an RNA-binding "matchmaker" protein, e.g., Hfq, to facilitate base pairing of the sRNA with its targets, a feature that is harnessed in a related technology, called RIL-seq (RNA interaction by ligation and sequencing). Here, sRNA-target pairs are cross-linked on Hfq, trimmed by ribonucleases, and ligated using T4 RNA Ligase (Melamed et al., 2018, 2016). Generation of cDNA and sequencing of the chimeric RNA allow for the global identification of sRNA-target interactions under selected conditions. Of note, RIL-seq is not limited to Hfq and can also be applied to other RNA-binding proteins such as RNase E (Waters et al., 2017) (see chapter "Transcriptome-wide analysis of protein–RNA and RNA–RNA interactions in pathogenic bacteria" by Tree, Gerdes, & Tollervey). We are yet to apply similar technologies to *V. cholerae* to determine the global RNA interactome of this important pathogen, which could also involve other RNA chaperones mediating sRNA-target interactions, such as ProQ (Holmqvist, Li, Bischler, Barquist, & Vogel, 2018; Smirnov et al., 2016).

ACKNOWLEDGMENTS

We thank Kathrin Fröhlich for comments on the manuscript. Work in the Papenfort lab is supported by DFG (Exc114-2, GRK2062, SPP2002, PA2820/1 and TRR174), the Human Frontiers Science Program (CDA00024/2016-C), GIF (G-2411-416.13/2016), and the European Research Council (StG-758212).

REFERENCES

Almagro-Moreno, S., Pruss, K., & Taylor, R. K. (2015). Intestinal colonization dynamics of Vibrio cholerae. *PLoS Pathogens*, *11*(5), e1004787.

Ball, A. S., Chaparian, R. R., & van Kessel, J. C. (2017). Quorum sensing gene regulation by LuxR/HapR master regulators in Vibrios. *Journal of Bacteriology*, *199*(19), e00105-17.

Barquist, L., & Vogel, J. (2015). Accelerating discovery and functional analysis of small RNAs with new technologies. *Annual Review of Genetics*, *49*, 367–394.

Bassler, B. L., Wright, M., & Silverman, M. R. (1994). Multiple signalling systems controlling expression of luminescence in Vibrio harveyi: Sequence and function of genes encoding a second sensory pathway. *Molecular Microbiology*, *13*(2), 273–286.

Beisel, C. L., & Storz, G. (2010). Base pairing small RNAs and their roles in global regulatory networks. *FEMS Microbiology Reviews*, *34*(5), 866–882.

Berezikov, E., Thuemmler, F., van Laake, L. W., Kondova, I., Bontrop, R., Cuppen, E., et al. (2006). Diversity of microRNAs in human and chimpanzee brain. *Nature Genetics*, *38*(12), 1375–1377.

Bilusic, I., Popitsch, N., Rescheneder, P., Schroeder, R., & Lybecker, M. (2014). Revisiting the coding potential of the E. coli genome through Hfq co-immunoprecipitation. *RNA Biology*, *11*(5), 641–654.

Borries, A., Vogel, J., & Sharma, C. M. (2012). Differential RNA sequencing (dRNA-Seq): Deep-sequencing-based analysis of primary transcriptomes. In *Tag-based next generation sequencing*. Wiley-VCH Verlag GmbH & Co.

Carrier, M. C., Lalaouna, D., & Masse, E. (2016). A game of tag: MAPS catches up on RNA interactomes. *RNA Biology*, *13*(5), 473–476.

Casper-Lindley, C., & Yildiz, F. H. (2004). VpsT is a transcriptional regulator required for expression of vps biosynthesis genes and the development of rugose colonial morphology in Vibrio cholerae O1 El Tor. *Journal of Bacteriology*, *186*(5), 1574–1578.

Chao, Y., Li, L., Girodat, D., Forstner, K. U., Said, N., Corcoran, C., et al. (2017). In vivo cleavage map illuminates the central role of RNase E in coding and non-coding RNA pathways. *Molecular Cell*, *65*(1), 39–51.

Chao, Y., Papenfort, K., Reinhardt, R., Sharma, C. M., & Vogel, J. (2012). An atlas of Hfq-bound transcripts reveals 3' UTRs as a genomic reservoir of regulatory small RNAs. *The EMBO Journal*, *31*(20), 4005–4019.

Chao, Y., & Vogel, J. (2010). The role of Hfq in bacterial pathogens. *Current Opinion in Microbiology*, *13*(1), 24–33.

Chen, X., Schauder, S., Potier, N., Van Dorsselaer, A., Pelczer, I., Bassler, B. L., et al. (2002). Structural identification of a bacterial quorum-sensing signal containing boron. *Nature*, *415*(6871), 545–549.

Corcoran, C. P., Podkaminski, D., Papenfort, K., Urban, J. H., Hinton, J. C., & Vogel, J. (2012). Superfolder GFP reporters validate diverse new mRNA targets of the classic porin regulator, MicF RNA. *Molecular Microbiology*, *84*(3), 428–445.

Croucher, N. J., Fookes, M. C., Perkins, T. T., Turner, D. J., Marguerat, S. B., Keane, T., et al. (2009). A simple method for directional transcriptome sequencing using Illumina technology. *Nucleic Acids Research*, *37*(22), e148.

Ding, Y., Davis, B. M., & Waldor, M. K. (2004). Hfq is essential for Vibrio cholerae virulence and downregulates sigma expression. *Molecular Microbiology*, *53*(1), 345–354.

Feng, L., Rutherford, S. T., Papenfort, K., Bagert, J. D., van Kessel, J. C., Tirrell, D. A., et al. (2015). A qrr noncoding RNA deploys four different regulatory mechanisms to optimize quorum-sensing dynamics. *Cell*, *160*(1–2), 228–240.

Gibson, D. G., Young, L., Chuang, R. Y., Venter, J. C., Hutchison, C. A., 3rd, & Smith, H. O. (2009). Enzymatic assembly of DNA molecules up to several hundred kilobases. *Nature Methods*, *6*(5), 343–345.

Guo, M. S., Updegrove, T. B., Gogol, E. B., Shabalina, S. A., Gross, C. A., & Storz, G. (2014). MicL, a new sigmaE-dependent sRNA, combats envelope stress by repressing synthesis of Lpp, the major outer membrane lipoprotein. *Genes & Development, 28*(14), 1620–1634.

Han, K., Tjaden, B., & Lory, S. (2016). GRIL-seq provides a method for identifying direct targets of bacterial small regulatory RNA by in vivo proximity ligation. *Nature Microbiology, 2*, 16239.

He, Y., Vogelstein, B., Velculescu, V. E., Papadopoulos, N., & Kinzler, K. W. (2008). The antisense transcriptomes of human cells. *Science, 322*(5909), 1855–1857.

Heidrich, N., Bauriedl, S., Barquist, L., Li, L., Schoen, C., & Vogel, J. (2017). The primary transcriptome of Neisseria meningitidis and its interaction with the RNA chaperone Hfq. *Nucleic Acids Research, 45*(10), 6147–6167.

Higgins, D. A., Pomianek, M. E., Kraml, C. M., Taylor, R. K., Semmelhack, M. F., & Bassler, B. L. (2007). The major Vibrio cholerae autoinducer and its role in virulence factor production. *Nature, 450*(7171), 883–886.

Holmqvist, E., Li, L., Bischler, T., Barquist, L., & Vogel, J. (2018). Global maps of ProQ binding in vivo reveal target recognition via RNA structure and stability control at mRNA 3′ ends. *Molecular Cell, 70*(5), 971-982 e976.

Hor, J., Gorski, S. A., & Vogel, J. (2018). Bacterial RNA biology on a genome scale. *Molecular Cell, 70*(5), 785–799.

Hurley, A., & Bassler, B. L. (2017). Asymmetric regulation of quorum-sensing receptors drives autoinducer-specific gene expression programs in Vibrio cholerae. *PLoS Genetics, 13*(5), e1006826.

Jung, S. A., Chapman, C. A., & Ng, W. L. (2015). Quadruple Quorum-Sensing Inputs Control *Vibrio cholerae* Virulence and Maintain System Robustness. *PLOS Pathogens, 11*(4), e1004837.

Kery, M. B., Feldman, M., Livny, J., & Tjaden, B. (2014). TargetRNA2: Identifying targets of small regulatory RNAs in bacteria. *Nucleic Acids Research, 42*(Web Server issue), W124–W129.

Krasteva, P. V., Fong, J. C., Shikuma, N. J., Beyhan, S., Navarro, M. V., Yildiz, F. H., et al. (2010). Vibrio cholerae VpsT regulates matrix production and motility by directly sensing cyclic di-GMP. *Science, 327*(5967), 866–868.

Kroger, C., Dillon, S. C., Cameron, A. D., Papenfort, K., Sivasankaran, S. K., Hokamp, K., et al. (2012). The transcriptional landscape and small RNAs of Salmonella enterica serovar Typhimurium. *Proceedings of the National Academy of Sciences of the United States of America, 109*(20), E1277–E1286.

Lan, C. C., Tang, R., Un San Leong, I., & Love, D. R. (2009). Quantitative real-time RT-PCR (qRT-PCR) of zebrafish transcripts: Optimization of RNA extraction, quality control considerations, and data analysis. *Cold Spring Harbor Protocols, 2009*(10), 1–12. pdb prot5314.

Lenz, D. H., Mok, K. C., Lilley, B. N., Kulkarni, R. V., Wingreen, N. S., & Bassler, B. L. (2004). The small RNA chaperone Hfq and multiple small RNAs control quorum sensing in *Vibrio harveyi* and *Vibrio cholerae*. *Cell, 118*(1), 69–82.

Liu, J. M., Livny, J., Lawrence, M. S., Kimball, M. D., Waldor, M. K., & Camilli, A. (2009). Experimental discovery of sRNAs in Vibrio cholerae by direct cloning, 5S/tRNA depletion and parallel sequencing. *Nucleic Acids Research, 37*(6), e46.

Logan, S. L., Thomas, J., Yan, J., Baker, R. P., Shields, D. S., Xavier, J. B., et al. (2018). The Vibrio cholerae type VI secretion system can modulate host intestinal mechanics to displace gut bacterial symbionts. *Proceedings of the National Academy of Sciences of the United States of America, 115*, E3779–E3787.

Lowe, R., Shirley, N., Bleackley, M., Dolan, S., & Shafee, T. (2017). Transcriptomics technologies. *PLoS Computational Biology, 13*(5), e1005457.

Mann, M., Wright, P. R., & Backofen, R. (2017). IntaRNA 2.0: Enhanced and customizable prediction of RNA-RNA interactions. *Nucleic Acids Research*, *45*(W1), W435–W439.

Masse, E., Escorcia, F. E., & Gottesman, S. (2003). Coupled degradation of a small regulatory RNA and its mRNA targets in Escherichia coli. *Genes & Development*, *17*(19), 2374–2383.

Melamed, S., Faigenbaum-Romm, R., Peer, A., Reiss, N., Shechter, O., Bar, A., et al. (2018). Mapping the small RNA interactome in bacteria using RIL-seq. *Nature Protocols*, *13*(1), 1–33.

Melamed, S., Peer, A., Faigenbaum-Romm, R., Gatt, Y. E., Reiss, N., Bar, A., et al. (2016). Global mapping of small RNA-target interactions in bacteria. *Molecular Cell*, *63*(5), 884–897.

Mohanty, B. K., & Kushner, S. R. (2018). Enzymes involved in posttranscriptional RNA metabolism in Gram-negative bacteria. *Microbiology Spectrum*, *6*(2), 1–16.

Moller, T., Franch, T., Udesen, C., Gerdes, K., & Valentin-Hansen, P. (2002). Spot 42 RNA mediates discoordinate expression of the E. coli galactose operon. *Genes & Development*, *16*(13), 1696–1706.

Neiditch, M. B., Federle, M. J., Miller, S. T., Bassler, B. L., & Hughson, F. M. (2005). Regulation of LuxPQ receptor activity by the quorum-sensing signal autoinducer-2. *Molecular Cell*, *18*(5), 507–518.

Ng, W. L., & Bassler, B. L. (2009). Bacterial quorum-sensing network architectures. *Annual Review of Genetics*, *43*, 197–222.

Ng, W. L., Perez, L. J., Wei, Y., Kraml, C., Semmelhack, M. F., & Bassler, B. L. (2011). Signal production and detection specificity in vibrio CqsA/CqsS quorum-sensing systems. *Molecular Microbiology*, *79*, 1407–1417.

Nitzan, M., Rehani, R., & Margalit, H. (2017). Integration of bacterial small RNAs in regulatory networks. *Annual Review of Biophysics*, *46*, 131–148.

Papenfort, K., & Bassler, B. L. (2016). Quorum sensing signal-response systems in Gram-negative bacteria. *Nature Reviews. Microbiology*, *14*(9), 576–588.

Papenfort, K., Espinosa, E., Casadesus, J., & Vogel, J. (2015). Small RNA-based feedforward loop with AND-gate logic regulates extrachromosomal DNA transfer in Salmonella. *Proceedings of the National Academy of Sciences of the United States of America*, *112*(34), E4772–E4781.

Papenfort, K., Forstner, K. U., Cong, J. P., Sharma, C. M., & Bassler, B. L. (2015). Differential RNA-seq of Vibrio cholerae identifies the VqmR small RNA as a regulator of biofilm formation. *Proceedings of the National Academy of Sciences of the United States of America*, *112*(7), E766–E775.

Papenfort, K., Pfeiffer, V., Mika, F., Lucchini, S., Hinton, J. C., & Vogel, J. (2006). SigmaE-dependent small RNAs of Salmonella respond to membrane stress by accelerating global omp mRNA decay. *Molecular Microbiology*, *62*(6), 1674–1688.

Papenfort, K., Said, N., Welsink, T., Lucchini, S., Hinton, J. C., & Vogel, J. (2009). Specific and pleiotropic patterns of mRNA regulation by ArcZ, a conserved, Hfq-dependent small RNA. *Molecular Microbiology*, *74*(1), 139–158.

Papenfort, K., Silpe, J. E., Schramma, K. R., Cong, J. P., Seyedsayamdost, M. R., & Bassler, B. L. (2017). A Vibrio cholerae autoinducer-receptor pair that controls biofilm formation. *Nature Chemical Biology*, *13*(5), 551–557.

Papenfort, K., & Vogel, J. (2009). Multiple target regulation by small noncoding RNAs rewires gene expression at the post-transcriptional level. *Research in Microbiology*, *160*(4), 278–287.

Parkhomchuk, D., Borodina, T., Amstislavskiy, V., Banaru, M., Hallen, L., Krobitsch, S., et al. (2009). Transcriptome analysis by strand-specific sequencing of complementary DNA. *Nucleic Acids Research*, *37*(18), e123.

Rio, D. C. (2015). Northern blots: Capillary transfer of RNA from agarose gels and filter hybridization using standard stringency conditions. *Cold Spring Harbor Protocols, 2015*(3), 306–313.

Rio, D. C., Ares, M., Jr., Hannon, G. J., & Nilsen, T. W. (2010). Polyacrylamide gel electrophoresis of RNA. *Cold Spring Harbor Protocols, 2010*(6), 1–6. pdb prot5444.

Rutherford, S. T., van Kessel, J. C., Shao, Y., & Bassler, B. L. (2011). AphA and LuxR/HapR reciprocally control quorum sensing in vibrios. *Genes & Development, 25*(4), 397–408.

Sambrook, J. (2001). In J. Sambrook & D. W. Russell (Eds.), *Molecular cloning: A laboratory manual* (pp. 251–254). Cold Spring Harbor, NY: Cold Spring Harbor Laboratory. (Chapter 3).

Santiago-Frangos, A., & Woodson, S. A. (2018). Hfq chaperone brings speed dating to bacterial sRNA. *Wiley Interdisciplinary Reviews. RNA, 9*(4), e1475.

Schroeder, A., Mueller, O., Stocker, S., Salowsky, R., Leiber, M., Gassmann, M., et al. (2006). The RIN: An RNA integrity number for assigning integrity values to RNA measurements. *BMC Molecular Biology, 7*, 3.

Shao, Y., & Bassler, B. L. (2012). Quorum-sensing non-coding small RNAs use unique pairing regions to differentially control mRNA targets. *Molecular Microbiology, 83*(3), 599–611.

Shao, Y., & Bassler, B. L. (2014). Quorum regulatory small RNAs repress type VI secretion in Vibrio cholerae. *Molecular Microbiology, 92*(5), 921–930.

Shao, Y., Feng, L., Rutherford, S. T., Papenfort, K., & Bassler, B. L. (2013). Functional determinants of the quorum-sensing non-coding RNAs and their roles in target regulation. *The EMBO Journal, 32*(15), 2158–2171.

Sharma, C. M., Darfeuille, F., Plantinga, T. H., & Vogel, J. (2007). A small RNA regulates multiple ABC transporter mRNAs by targeting C/A-rich elements inside and upstream of ribosome-binding sites. *Genes & Development, 21*(21), 2804–2817.

Sharma, C. M., Papenfort, K., Pernitzsch, S. R., Mollenkopf, H. J., Hinton, J. C., & Vogel, J. (2011). Pervasive post-transcriptional control of genes involved in amino acid metabolism by the Hfq-dependent GcvB small RNA. *Molecular Microbiology, 81*(5), 1144–1165.

Sharma, C. M., & Vogel, J. (2014). Differential RNA-seq: The approach behind and the biological insight gained. *Current Opinion in Microbiology, 19C*, 97–105.

Silva, A. J., & Benitez, J. A. (2016). Vibrio cholerae biofilms and cholera pathogenesis. *PLoS Neglected Tropical Diseases, 10*(2), e0004330.

Sittka, A., Lucchini, S., Papenfort, K., Sharma, C. M., Rolle, K., Binnewies, T. T., et al. (2008). Deep sequencing analysis of small noncoding RNA and mRNA targets of the global post-transcriptional regulator, Hfq. *PLoS Genetics, 4*(8), e1000163.

Smirnov, A., Forstner, K. U., Holmqvist, E., Otto, A., Gunster, R., Becher, D., et al. (2016). Grad-seq guides the discovery of ProQ as a major small RNA-binding protein. *Proceedings of the National Academy of Sciences of the United States of America, 113*(41), 11591–11596.

Sun, X., Zhulin, I., & Wartell, R. M. (2002). Predicted structure and phyletic distribution of the RNA-binding protein Hfq. *Nucleic Acids Research, 30*(17), 3662–3671.

Surette, M. G., Miller, M. B., & Bassler, B. L. (1999). Quorum sensing in Escherichia coli, Salmonella typhimurium, and Vibrio harveyi: A new family of genes responsible for autoinducer production. *Proceedings of the National Academy of Sciences of the United States of America, 96*(4), 1639–1644.

Teschler, J. K., Zamorano-Sanchez, D., Utada, A. S., Warner, C. J., Wong, G. C., Linington, R. G., et al. (2015). Living in the matrix: Assembly and control of Vibrio cholerae biofilms. *Nature Reviews. Microbiology, 13*(5), 255–268.

Torres-Quesada, O., Reinkensmeier, J., Schluter, J. P., Robledo, M., Peregrina, A., Giegerich, R., et al. (2014). Genome-wide profiling of Hfq-binding RNAs uncovers extensive post-transcriptional rewiring of major stress response and symbiotic regulons in Sinorhizobium meliloti. *RNA Biology, 11*(5), 563–579.

Tu, K. C., Long, T., Svenningsen, S. L., Wingreen, N. S., & Bassler, B. L. (2010). Negative feedback loops involving small regulatory RNAs precisely control the Vibrio harveyi quorum-sensing response. *Molecular Cell*, *37*(4), 567–579.
Vogel, J., & Luisi, B. F. (2011). Hfq and its constellation of RNA. *Nature Reviews. Microbiology*, *9*(8), 578–589.
Wagner, E. G., & Romby, P. (2015). Small RNAs in bacteria and archaea: Who they are, what they do, and how they do it. *Advances in Genetics*, *90*, 133–208.
Wang, J., Rennie, W., Liu, C., Carmack, C. S., Prevost, K., Caron, M. P., et al. (2015). Identification of bacterial sRNA regulatory targets using ribosome profiling. *Nucleic Acids Research*, *43*(21), 10308–10320.
Waters, S. A., McAteer, S. P., Kudla, G., Pang, I., Deshpande, N. P., Amos, T. G., et al. (2017). Small RNA interactome of pathogenic E. coli revealed through crosslinking of RNase E. *The EMBO Journal*, *36*(3), 374–387.
Waters, L. S., & Storz, G. (2009). Regulatory RNAs in bacteria. *Cell*, *136*(4), 615–628.
Westermann, A. J., Gorski, S. A., & Vogel, J. (2012). Dual RNA-seq of pathogen and host. *Nature Reviews. Microbiology*, *10*(9), 618–630.
Woodson, S. A., Panja, S., & Santiago-Frangos, A. (2018). Proteins that chaperone RNA regulation. *Microbiology Spectrum*, *6*(4), 1–13.
Wright, P. R., Richter, A. S., Papenfort, K., Mann, M., Vogel, J., Hess, W. R., et al. (2013). Comparative genomics boosts target prediction for bacterial small RNAs. *Proceedings of the National Academy of Sciences of the United States of America*, *110*(37), E3487–E3496.
Yang, M., Frey, E. M., Liu, Z., Bishar, R., & Zhu, J. (2010). The virulence transcriptional activator AphA enhances biofilm formation by Vibrio cholerae by activating expression of the biofilm regulator VpsT. *Infection and Immunity*, *78*(2), 697–703.
Zhao, W., Caro, F., Robins, W., & Mekalanos, J. J. (2018). Antagonism toward the intestinal microbiota and its effect on Vibrio cholerae virulence. *Science*, *359*(6372), 210–213.

CHAPTER FOURTEEN

In Vitro Study of the Major *Bacillus subtilis* Ribonucleases Y and J

Liliana Mora, Saravuth Ngo, Soumaya Laalami, Harald Putzer[1]
CNRS UMR8261—Université Paris Diderot, Sorbonne Paris Cité, Institut de Biologie Physico-Chimique, Paris, France
[1]Corresponding author: e-mail address: putzer@ibpc.fr

Contents

Abstract

The metabolic instability of mRNA is fundamental to the adaptation of gene expression. In bacteria, mRNA decay follows first-order kinetics and is primarily controlled at the steps initiating degradation. In the model Gram-positive organism *Bacillus subtilis*, the major mRNA decay pathway initiates with an endonucleolytic cleavage by the membrane-associated RNase Y. High-throughput sequencing has identified a large number of potential mRNA substrates but our understanding of what parameters affect cleavage in vivo is still quite limited. In vitro reconstitution of the cleavage event is thus instrumental in defining the mechanistic details, substrate recognition, the role of auxiliary factors, and of membrane localization in cleavage. In this chapter, we describe not only the purification and assay of RNase Y but also RNase J1/J2 which shares a similar low-specificity endoribonucleolytic activity with RNase Y. We highlight

Methods in Enzymology, Volume 612
ISSN 0076-6879
https://doi.org/10.1016/bs.mie.2018.08.004

potential problems in the set-up of these assays and include methods that allow purification of full-length RNase Y and its incorporation in multilamellar vesicles created from native *B. subtilis* lipids that might best mimic in vivo conditions.

1. INTRODUCTION

Low-specificity endoribonucleases are keys in initiating bacterial mRNA decay. The conservation of a similar although vague cleavage specificity for the three major decay-initiating ribonucleases Y, J, and E constitutes an impressive case of convergent evolution. All prokaryotic *phyla* whose genomes have been sequenced contain at least one enzyme related to RNases E, J, or Y and sometimes all three of them (Laalami & Putzer, 2011; Laalami, Zig, & Putzer, 2014). High-density sequence applications and other transcriptomic approaches have been crucial to appreciate the global impact of these enzymes. While RNase E has been characterized in detail (Chao et al., 2017; Kime, Clarke, Romero, Grasby, & McDowall, 2014), we will focus here on the in vitro analysis of *Bacillus subtilis* RNase Y and RNase J1/J2. RNase Y is an endoribonuclease with a global effect on mRNA stability whose activity can be stimulated by a monophosphorylated 5′ end of the RNA substrate (Shahbabian, Jamalli, Zig, & Putzer, 2009). Transcriptomic studies have shown that the decay/processing of the majority of expressed cellular mRNAs in *B. subtilis* is prone to be initiated by RNase Y (Durand, Gilet, Bessieres, Nicolas, & Condon, 2012; Laalami et al., 2013; Lehnik-Habrink, Schaffer, et al., 2011). A very recent transcriptomic study (H. Putzer, unpublished) finds that the cellular localization of RNase Y also plays a significant role in the overall pattern of its activity and that the level of many mRNAs appears to depend on the binding of RNase Y to the inner side of the cell membrane.

RNases J1/J2 are the other major ribonucleases to influence gene expression on a large scale in *B. subtilis*. The transcriptome/proteome of mutant strains has highlighted the global impact of RNases J1/J2 on protein expression and the steady-state levels of many mRNAs and degradation intermediates (Durand et al., 2012; Mäder, Zig, Kretschmer, Homuth, & Putzer, 2008). In addition, RNase J1 is a dual activity enzyme with an *endo*- and a 5′ exoribonucleolytic activity (Even et al., 2005; Mathy et al., 2007), which complicates its analysis. The current model for involvement of RNase Y and RNase J1/J2 in the major mRNA decay pathway in *B. subtilis* is shown in Fig. 1.

Over the last years, the existence of a potential degradosome-like complex has been discussed for *B. subtilis* (Redder, 2018). Many proteins have

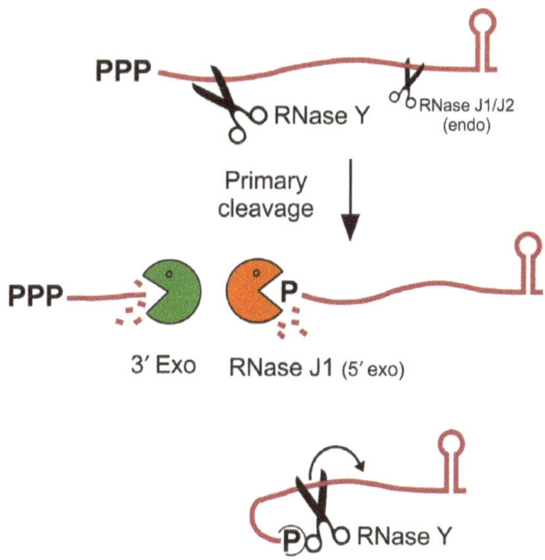

Fig. 1 The major direct entry pathway for mRNA degradation in *B. subtilis*. RNase Y and to a much lesser and unknown extent RNases J1/J2 initiate mRNA degradation through an initial endonucleolytic cleavage event. The cleavage products can be further degraded both from the 3′ end by 3′ exonucleases and from the 5′ end by the 5′ exonuclease activity of RNase J1.

been shown to potentially interact with RNase Y but no convincing RNase Y-based complex has been isolated without chemical crosslinking, in contrast to the *E. coli* RNase E-based degradosome. At present, the most promising accessory factors influencing RNase Y activity are a three-protein complex (YlbF, YmcA, and YaaT), that has been shown in a recent RNAseq analysis (DeLoughery, Lalanne, Losick, & Li, 2018) to be required for a number of RNase Y-mediated mRNA cleavages. In order to understand how these enzymes work and recognize the many different substrates revealed by high-throughput techniques, we will have to analyze their influence on RNase Y activity in vitro. In this chapter, we describe the purification and assay of different forms of RNase Y and RNases J1/J2 and discuss parameters that might explain differences observed between in vivo and in vitro cleavage behavior.

2. RNase Y

The characterization of RNase Y as a key endoribonuclease provided the first evidence that endonucleolytic cleavage plays a major role in the mRNA metabolism in *B. subtilis*. RNase Y shares important functional features with *E. coli* RNase E, notably a similar cleavage specificity and a preference for 5′ monophosphorylated substrates (Shahbabian et al., 2009). Other similarities with RNase E include a localization at the membrane (Hunt, Rawlins, Thomaides, & Errington, 2006) and the potential to interact with other proteins to form a degradosome-like complex (Commichau et al., 2009; DeLoughery et al., 2018; Lehnik-Habrink, Newman, et al., 2011). However, no such complexes have so far been isolated under native conditions. For practical reasons, in vitro assays of RNase Y have been carried out with a truncated enzyme lacking the N-terminal 25 amino acids that anchor the nuclease at the inside of the cell membrane (Shahbabian et al., 2009). Here, we describe the purification of tagged and native untagged versions of the enzyme. The subcellular localization of RNase Y is important for its activity on a large number of mRNA substrates (Laalami et al., manuscript to be submitted). In addition, the oligomerization state of the enzyme and its interaction with potential partner proteins in the cell might also be conditioned by its tethering to the cell membrane.

We therefore describe protocols for purification not only of a cytoplasmic version of the ribonuclease but also the full-length protein in a soluble form. In order to test the latter under more realistic conditions, insertion into membrane vesicles should be envisaged. At present, commercially available lipid extracts cannot mimic the branched-chain fatty acids present in the *B. subtilis* membrane. We therefore provide a protocol to constitute membrane vesicles based on a native lipid preparation from *B. subtilis*.

RNase Y cleaves single-stranded RNA with very low-sequence specificity (generally quite AU rich) but clearly prefers proximity to RNA secondary structure (Shahbabian et al., 2009).

2.1 Purification of RNase Y

2.1.1 Purification of RNase Y (Δ 1–25) Without Affinity Tag

Native RNase Y lacking the N-terminal transmembrane domain (56.2 kDa) is overexpressed as a C-terminal Intein fusion protein (vector pKYB1, New England Biolabs) from BL21 Star Codon + cells (Thermo Fisher Scientific) containing plasmid pHMD3 (Shahbabian et al., 2009). A 500 mL culture (LB + 10 μg/mL chloramphenicol + 25 μg/mL neomycin) is grown to an OD_{600} of 0.5 at 37°C. RNase Y expression is induced by the addition of 20 μ*M* IPTG and incubation of the culture O/N at 15°C. All subsequent steps are performed at 4°C. Cells are pelleted (10,000 × *g*) and can be stored frozen at −80°C until further use.

The fusion protein is bound to a chitin column and the target protein is cleaved and eluted from the column as described in Standard Protocol 2.

Fractions containing eluted RNase Y are measured by the Bradford protein assay (Bio-Rad), analyzed by SDS-PAGE, pooled and stored in aliquots at −20°C or −80°C (long-term storage). Yields of RNase Y purified by this method are around 1 mg of protein/500 mL culture.

2.1.2 Purification of RNase Y (Δ 1–25) With 6x His-Tag

C-terminal 6x His-tagged RNase Y (Δ 1–25, 57 kDa) is overexpressed in BL21 (DE3) Star Codon + cells (Thermo Fisher Scientific) containing plasmid pHMD2. This plasmid contains the *rny* open reading frame starting at amino acid 26 and harboring 6 histidine residues after the RNaseY C-terminal lysine residue. The sequence coding for the 6 histidine residues is inserted between the sites NdeI-BamHI of the pKYB1T vector (Liponska et al., 2018). A 100 mL culture (LB + 10 μg/mL chloramphenicol + 25 μg/mL neomycin) is grown to an OD_{600} of 0.5 at 37°C. RNase Y expression is induced by the addition of 0.3 m*M* IPTG and incubation of the culture O/N at 15°C. All subsequent steps are performed at 4°C. Cells are pelleted (10,000 × *g*) and can be stored frozen at −80°C until further use. The His-tagged protein is purified on a 1 mL Ni-NTA agarose resin according to Standard Protocol 3. Average yields are about 0.5 mg of RNase Y per 100 mL culture.

RNase Y can be further purified to separate the dimeric form from oligomeric forms by gel filtration on a Superdex 200 26/60 column (GE Healthcare). To do this, the protein is concentrated to ~2.5 mg/mL in buffer A (20 m*M* Hepes pH 8, 500 m*M* NaCl, 10% glycerol) before loading on the column and the protein peak eluting at ~150 mL at a flow rate of 0.5 mL/min is collected (optical density read at 230 n*M*) and concentrated with Amicon Ultra filters (10 kDa MWCO) and stored at −80°C.

2.1.3 Purification of Full-Length RNase Y With 6x His-Tag

Full-length RNase Y with a C-terminal 6x His-tag (59.8 kDa) is overexpressed in BL21 (DE3) Rosetta cells containing plasmid pHMD27. This plasmid has the full *rny* open reading frame and the sequence coding for the six C-terminal histidine residues inserted between the sites NdeI-BamHI of the vector pKYB1T (Liponska et al., 2018). Two 800 mL cultures (LB + 10 μg/mL chloramphenicol + 25 μg/mL neomycin) are grown at 37°C to OD_{600} = 0.5–0.6. Expression of RNase Y is induced by the addition of 0.5 m*M* IPTG followed by incubation of the culture at 18°C overnight.

All subsequent steps are carried out at 4°C. Cells are pelleted at 5200 × *g* for 30 min, washed in cold LB and pelleted again.

The cell pellet is suspended in 30 mL sonication buffer (25 m*M* Tris–HCl pH 7.6, 0.5 *M* NaCl, 30 μg DNase I, one EDTA-free protein inhibitor tablet (Pierce). Cells are broken by sonication (4 × 1 min (pulse mode—70% duty, 30 s pause) and the lysate is cleared by centrifugation (30 min, 10,000 × *g*).

The pellet is suspended in 10 mL of homogenization buffer (25 m*M* Tris–HCl pH 7.6, 0.5 *M* NaCl, 1% Triton X-100, 10% glycerol, 10 m*M* imidazole) supplemented with one EDTA free protein inhibitor tablet (Pierce) and incubated on ice for 1 h.

The protein is solubilized with a Potter homogenizer with a PTFE pestle, which is manually moved up and down 15–20 times. The sample is then centrifuged (10,000 × *g* for 30 min) and the supernatant loaded on a 1 mL Ni-NTA agarose resin (Protino, Macherey-Nagel) previously equilibrated in homogenization buffer. The column connected to a peristaltic pump and UV detector is first washed with 10 mL of the same buffer and then with 30 mL of buffer where the 1% Triton X-100 has been replaced with 0.2% reduced Triton X-100 (Sigma). The bound RNase Y is eluted with the reduced Triton buffer containing 250 m*M* imidazole. The use of reduced Triton X-100 at 0.2% permits detecting RNase Y at 280 n*M* which is otherwise impossible due to the low-absorption coefficient of RNase Y (no tryptophan residues). The peak fractions are pooled and dialyzed against storage buffer (25 m*M* Tris–HCl pH 7.6, 0.5 *M* NaCl, 0.1% reduced Triton X-100, 10% glycerol), aliquoted and stored at −80°C.

2.2 Multilamellar Vesicle Preparation

The lipid extraction protocol from *B. subtilis* is based on and adapted from a recently published study (Nickels et al., 2017), and multilamellar vesicles (MLVs) are prepared by adapting previously published procedures (Khemici, Poljak, Luisi, & Carpousis, 2008; Zhou & Lutkenhaus, 2003) and using the *B. subtilis* native lipids.

2.2.1 Lipid Extraction From B. subtilis

A *B. subtilis* 168 wild-type strain 1A2 (BGSC) is grown in SMS minimal medium (Spizizen, 1958) with 0.75% fructose, supplemented with 0.1% L-glutamine, 0.2 m*M* L-tryptophan, 22 mg/mL ferric ammonium citrate (CAF). Three liters of culture are inoculated to an $OD_{600} = 0.05$ and grown overnight at 37°C (final $OD_{600} = \sim 1.2$). Cells were pelleted by centrifugation (5200 × *g*, 15 min, 4°C), washed three times in 1% NaCl, and stored frozen at −20°C until further use. Cell pellets were lyophilized overnight and stored at −80°C. The total output is ~1 g of dried cells. 200 mg of the lyophilized cells were crushed to a coarse powder with a spatula and transferred to a separation funnel. The following solvents were added in the given order with vigorous shaking in a separating funnel before addition of the next solvent: 60 mL chloroform, 120 mL methanol, and 48 mL H_2O. The funnel was kept under a hood at room temperature for 2 days with occasional shaking. 60 mL chloroform and 30 mL H_2O were added, the mixture was vigorously shaken and left for 1 day under the hood to allow for phase separation. The lower phase was recovered and the upper phase extracted with 30 mL chloroform. The lipids from the combined chloroform phases were recovered by rotary evaporation under reduced pressure (Buchi) for 10 min at 30°C. The lipid precipitate was dissolved in 5 mL chloroform and recovered by another round of evaporation. Total lipid yield is ~40 mg.

2.2.2 MLV Preparation

40 mg of total lipids from *B. subtilis* (see Section 2.2.1) are resuspended in 700 μL of 25 m*M* Tris–HCl pH 7.6, 40% glycerol under vigorous vortexing. The suspension is diluted 50-fold by adding 35 mL 25 m*M* Tris–HCl pH 7.6, 50 m*M* NaCl, and incubated 30 min on ice.

The lipids are centrifuged (163,000 × *g*, 20 min, 4°C) and resuspended in 15 mL 25 m*M* Tris–HCl pH 7.6, 50 m*M* NaCl supplemented with 1 tablet of EDTA-free protease inhibitor (Pierce). The suspension is incubated for 3 h at 65°C with regular vortexing and not-well-suspended particles are removed by centrifugation (5000 × *g*, 10 min, 4°C). Aliquots of 1 mL, saturated with argon to avoid lipid oxidation are stored at −20°C.

2.2.3 Binding of RNase Y to MLVs

To verify the integration of RNase Y into MLV, approximately 1°μg of RNase Y (16°pmol) is mixed with 70 μL MLV (prepared as described in Section 2.2.2) and incubated for 20 min at 30°C. The MLVs are pelleted by centrifugation at 15000 × *g* for 2 min at room temperature. The protein in the pellet was separated by SDS-PAGE (10%) and visualized by Coomassie blue staining. For in vitro activity assays RNase Y was bound to the MLVs in the same proportions and added to the reaction in a scaled-up volume (100 μL).

2.3 Assay of RNase Y Activity In Vitro

The endonucleolytic activity of RNase Y can be tested on any RNA substrate that is thought to be cleaved by the enzyme. However, due to the complex RNA recognition pattern that is not fully understood it is judicious to start out with a substrate where an RNase Y cleavage has been shown to occur in vivo. A suitable in vitro transcript (preferentially $< \sim 350$ nt) should be designed not only to cover the region where cleavage has been shown or is suspected to occur, but it should also take into account potential alternative folding of the RNA. This might require, for example, changing the length of a transcript in order to favor formation of a certain secondary structure. This is clearly illustrated by reproducing RNase Y cleavage of the *yitJ* S-box riboswitch in vitro where the formation of the secondary structure necessary for the cleavage can be favored by shortening the in vitro transcript (Fig. 2).

The in vitro cleavage sites of an RNA substrate are best identified by end labeling the RNA, either on the 5′ or the 3′ end. Continuous labeling might be useful to visualize all cleavage products. Since RNase Y cleavage has been shown to occur faster on a 5′ monophosphorylated RNA (at least in the case of *yitJ*) it is probably best to start to set-up pilot assays with a 5′ P [^{32}P]-labeled substrate. The procedures to radioactively label an RNA in vitro at different positions are given in Standard Protocol 1.

RNase Y cleavage assays are carried out in a reaction volume of 10 μL containing 20 m*M* HEPES pH 8, 8 m*M* $MgCl_2$, 100 m*M* NaCl, 0.24 U/μL RNasin (Promega). About 0.1 pmol of the labeled RNA substrate (~10 n*M*,

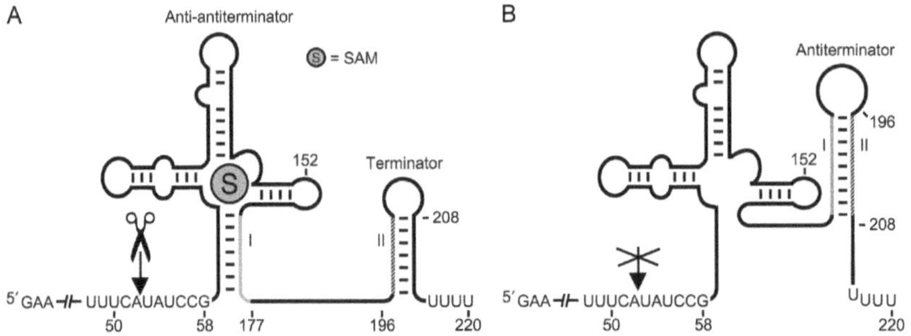

Fig. 2 The *yitJ* SAM riboswitch is cleaved in vivo by RNase Y upstream of the SAM binding secondary structure at position 52 (A). In the absence of SAM, the alternative antiterminator structure folds and no cleavage occurs (B). In vitro, cleavage of the 220 nt RNA substrate by RNase Y can be observed only when SAM is added to the reaction. However, a shorter transcript ending at nucleotide 196 alleviates the need to add SAM because the alternative structure cannot form anymore (Shahbabian et al., 2009).

preincubated for 3 min at 80°C) are incubated with 10 pmol of RNase Y (1 μ*M*) for 10 min at 30°C, and the reaction is stopped by the addition of 5 μL of 3x gel loading buffer (87.5% formamide, 0.05% xylene cyanol, 0.05% bromophenol blue, and 5 m*M* EDTA). Half of the reaction is heated for 3 min at 70°C and directly loaded on 5% denaturing polyacrylamide gels (7 *M* urea). Cleavage products <50 nt are best resolved on a 20% gel.

This protocol can be adapted to optimize the reaction for different substrates and enzyme preparations. For example, full-length RNase Y preparations included in MLVs will require larger reaction volumes (up to 100 μL). The same is true when other cofactors that potentially alter RNase Y activity are included in the assay. Given the importance of RNA folding on the endonucleolytic activity, it might also be useful to vary the $MgCl_2$ concentration in the reaction between 4 and 20 m*M*. For yet unknown reasons the enzyme is always in large excess to the substrate.

3. RNase J

B. subtilis has two RNase J orthologs, RNase J1 and RNase J2 that have a very similar endonucleolytic cleavage activity (Even et al., 2005). They occur in vivo as a heterooligomeric complex and can behave synergistically to alter cleavage site preference and to increase cleavage efficiency at specific sites (Mathy et al., 2010). However, the transcriptome and proteome of a strain that does not express RNase J2 show no profound differences compared to a wild-type strain (Mäder et al., 2008). Since RNase J1 has a strong 5′ exonucleolytic activity (Mathy et al., 2007) the in vitro analysis of its endonucleolytic activity is complicated and depends on the RNA substrate chosen. Notably an initial 5′ triphosphorylated transcript cannot be degraded exonucleolytically from the 5′ end (Li de la Sierra-Gallay, Zig, Jamalli, & Putzer, 2008). To what extent the enzyme can switch to exonucleolytic mode after endonucleolytic cleavage is still not clear but might also be substrate dependent. In this respect, it is also important to note that an affinity tag can significantly influence the 5′ exonuclease activity of RNase J1. We found that a C-terminal 6x His-tagged protein has a 4- to 10-fold higher exonuclease activity compared to both the native untagged protein and an N-terminal 6x His-tagged version of RNase J1. We therefore prefer to work with RNase J1 proteins that carry their affinity tag at the N-terminal end.

Later, we describe several procedures that allow purification of untagged, 6x His-tagged versions of RNase J1 as well as the heterooligomeric RNase J1/J2 complex.

3.1 Purification of RNase J1

3.1.1 Purification of RNase J1 Without Affinity Tag

Native RNase J1 (61.5 kDa) is overexpressed as a C-terminal Intein fusion protein (vector pKYB1, New England Biolabs) from BL21 (DE3) Star Codon+ cells (Thermo Fisher Scientific) containing plasmid pHMJ1 (Even et al., 2005). A 500 mL culture (LB + 10 μg/mL chloramphenicol + 25 μg/mL neomycin) is grown to an OD_{600} of 0.5 at 37°C. RNase J1 expression is induced by the addition of 20 μ*M* IPTG and incubation of the culture O/N at 15°C. All subsequent steps are performed at 4°C. Cells are pelleted (10,000 × *g*) and can be stored frozen at −80°C until further use.

The fusion protein is bound to a chitin column and the target protein is cleaved and eluted from the column as described in Standard Protocol 2.

Fractions containing eluted RNase J1 are measured by the Bradford protein assay (Bio-Rad), analyzed by SDS-PAGE, pooled and stored in aliquots at −20°C or −80°C (long-term storage). Yields of RNase J1 purified by this method are <1 mg of protein.

3.1.2 Purification of RNase J1 With N-Terminal 6x His-Tag

RNase J1 with a N-terminal 6x His-tag (62.5 kDa) is overexpressed in BL21 (DE3) Star Codon+ cells containing plasmid pHMJ6 (Liponska et al., 2018).

A 100 mL culture (LB + 10 μg/mL chloramphenicol + 25 μg/mL neomycin) is grown to an OD_{600} of 0.5 at 37°C. RNase J1 expression is induced by the addition of 0.3 m*M* IPTG incubation of the culture O/N at 15°C. All subsequent steps are performed at 4°C. Cells are pelleted (10,000 × *g*) and can be stored frozen at −80°C until further use. The His-tagged protein is purified on a 1 mL Ni-NTA agarose resin according to Standard Protocol 3. Average yields are about 2 mg of RNase J1 per 100 mL culture.

RNase J1 can be further purified by gel filtration on a Superdex 200 26/60 column (GE Healthcare). The protein is concentrated to ~2.5 mg/mL in buffer A (20 m*M* Hepes pH 8, 500 m*M* NaCl, 10% glycerol) and the protein peak eluting with an apparent MW of 150 kDa is collected and concentrated with Amicon Ultra filters (10 kDa MWCO) and stored at −80°C.

3.1.3 Purification of the RNase J1/J2 Complex

400 mL BL21 (DE3) Codon+ cells containing the overexpression plasmid pET28-BsubYmfA-YkqCHis (Mathy et al., 2010) are grown at 37°C to $OD_{600} = 0.3$ and induced by addition of IPTG (0.5 m*M*) for 3 h. This leads to the tandem overexpression of native RNase J2 and RNase J1-C-6x His. All further steps are at <4°C. The cell pellet is resuspended in 10 mL buffer containing 50 m*M* NaH_2PO_4, 300 m*M* NaCl, 10% glycerol, 0.1% Triton-X100, to which is added 10 μg/mL DNase I and an EDTA-free antiprotease tablet (Pierce). The suspension is sonicated 4x 1 min (pulse mode—70% duty, 30 s pause) and the lysate centrifuged for 30 min at 15000 × *g*. Imidazole–HCl, pH 7.8, is added to the supernatant to give a final concentration of 1 m*M*. The resulting mix is applied to 1 mL Cobalt-Talon resin (Clontech) in a small column, attached to a peristaltic pump, and previously equilibrated with approximately 15 mL cell resuspension buffer at 10 mL/h. The resin is then washed with 10 mL 50 m*M* NaH_2PO_4, 300 m*M* NaCl, 20 m*M* imidazole, and eluted in 10 mL buffer containing 50 m*M* NaH_2PO_4, 300 m*M* NaCl, 150 m*M* imidazole. 1 mL fractions are collected and analyzed by SDS-PAGE for protein purity. Peak fractions are pooled and dialyzed against 50 m*M* NaH_2PO_4 pH 7.0, 450 m*M* NaCl, 15% glycerol and applied to a Superdex 200 (26/60) gel-filtration column (GE Healthcare) equilibrated in the same buffer. The protein peak was concentrated to 1 mg/mL using Amicon Ultra 30 kDa filters and stored frozen.

3.2 Assay of RNase J Activity In Vitro

The endonucleolytic activity of RNase J1 or the RNase J1/J2 complex can be assayed under exactly the same experimental conditions described for RNase Y (point 2.3). We have previously also shown that the cleavage specificity can be very similar between RNase J1 and RNase Y using the *thrS* leader (Even et al., 2005) or the *yitJ* riboswitch (Shahbabian et al., 2009), unpublished data). However, it is important to use a 5′ triphosphorylated substrate in the case of RNase J1 because the quite robust 5′ exonuclease activity will otherwise dominate. On the other hand, downstream endonucleolytic cleavage products generated on original 5′ PPP transcripts, which carry a 5′ P moiety, are relatively stable, most likely due to the fact that they generally occur immediately upstream of a secondary structure. This apparently impairs binding of a new molecule of RNase J1 but likely also limits the potential switching from *endo-* to exonucleolytic mode on the same substrate.

The 5′ exonuclease activity of RNase J1 can also be tested under the same experimental conditions on a 5′-P or 5′-OH transcript. If other factors are to be tested to see whether they have an influence on the activity of the RNase J1, it is probably better to work with the native protein or the N-terminal tagged version rather than the C-terminal His-tagged RNase J1, which appears to have a significantly increased 5′ exonuclease activity.

4. CONCLUSION

High-throughput sequencing data and a number of single case studies have clearly established RNase Y and RNase J1/J2 as the major players in mRNA metabolism in *B. subtilis*. However, our understanding of how these enzymes act at the molecular level, including substrate recognition and their potential interaction with other partners remains rather obscure. In order to make further progress on the mechanistic aspects of these enzymes we will have to reconstitute the reaction in vitro. The protocols presented here can be taken as a good starting point to assay the activities of *B. subtilis* RNase Y and RNase J1/J2 on any potential or known substrate. It will allow studying the effect of primary sequence and secondary structure on the enzyme activity and also test the potential effect of other proteins that might interact with RNase Y. Such studies might well require working with enzyme preparations that best mimic in vivo conditions. For this purpose we have included protocols that allow purification of full-length RNase Y and association with MLVs created from native *B. subtilis* lipids. Such an approach obviously complicates the setup of in vitro assays but could turn out to be necessary to faithfully reconstitute the activity pattern of RNase Y.

ACKNOWLEDGMENTS

We thank Jacqueline Plumbridge for critical reading of the manuscript. CNRS and University Paris Diderot, Sorbonne Paris Cité and the "Initiative d'Excellence" program from the French State (Grant DYNAMO, ANR-11-LABX-0011) provided funding to the Research Unit UMR8261.

APPENDICES

Standard Protocol 1: T7 In Vitro Transcription and Radioactive RNA Labeling

Protocol 1A: Uniformly Labeled RNA

Uniformly [^{32}P]-labeled RNAs are synthesized with either PCR generated (0.3 μg) or linearized plasmid templates (1 μg) bearing T7 RNA polymerase promoters. A 30 μL reaction will contain 6 μL of 5x transcription buffer (Promega), 2.25 μL of 100 m*M* DTT, 1 μL of 40 U/mL RNasin (Promega), 3 μL of NTP mix (2.5 m*M* ATP, CTP, GTP), 1.5 μL 500 μ*M* (UTP), 4.5 μL α-[^{32}P]-UTP (10 μCi/μL) template DNA, 1.5 μL of 20 U/μL T7 RNA polymerase (Promega), and nuclease-free water to 30 μL. The 5x reaction buffer should be diluted with the other components of the reaction before addition of the template DNA to avoid precipitation by spermidine in the reaction buffer. Reactions are incubated at 37°C for 1.5 h. 2 μL RQ DNase (1 U/μL, Promega) are added and incubation at 37°C continued for 15 min. 20 μL H_2O is added to bring the volume to 50 μL. RNAs are separated from unincorporated nucleotides by purification on G50 spin columns (GE Healthcare). T7 RNA polymerase-transcribed RNAs are generally initiated with between one and three G-residues, with the first nucleotide bearing a 5′-triphosphate group. RNAs bearing 5′-monophosphate or 5′-hydroxyl groups can be synthesized by adding an eight-fold excess of GMP or guanosine, respectively, to the in vitro transcription reaction from 50 or 100 m*M* stock solutions (i.e., 4 m*M* final concentration).

Protocol 1B: Synthesis [^{32}P] 5′ End-Labeled RNA With a 5′ P Moiety

A 30 μL reaction is set-up containing 6 μL of 5x transcription buffer (Promega) 2.25 μL of 100 m*M* DTT, 1 μL of 40 U/mL RNasin (Promega), 3 μL of NTP mix (2.5 m*M* ATP, CTP, GTP, UTP), template DNA (as in Protocol 1A), 1.5 μL of 20 U/μL T7 RNA polymerase (Promega), and nuclease-free water to 30 μL. The 5x reaction buffer should be diluted with the other components of the reaction before addition of the template DNA to avoid precipitation by spermidine in the reaction buffer. Reactions are incubated at 37°C for 1.5 h. 2°μL RQ DNase (1 U/μL, Promega) is added and incubation at 37°C continued for 15 min. The RNA is precipitated by addition of 1/10 vol. 3 *M* NaOAc pH 4.7, 1 μL glycogen (25 μg/μL) and 3 vol. EtOH (96%), centrifuged for 10 min at >10,000 × *g*, and the pellet is washed once with 70% EtOH. The RNA air-dried pellet is taken up in 8 μL H_2O.

Dephosphorylation of the Transcript Is Performed in 10 μL Reaction Volume

1 μL 10x phosphatase reaction buffer and 1 μL Antarctic Phosphatase (5 U, Biolabs) are added to the RNA and incubated 15 min at 37°C followed by 5 min at 70°C to denature the phosphatase.

Labeling Reaction

To the dephosphorylation reaction (10 μL) is added 2 μL 10x T4 kinase reaction buffer, 0.5 μL RNasin (40 U/μl), 4.5 μL γ-[^{32}P] ATP (10 μCi/μl) and 1 μL T4 polynucleotide kinase (10 U), H_2O to 20 μL. Incubate for 2 h at 37°C. Add 20 μL H_2O and separate unincorporated radioactivity by purification on G25 spin columns (GE Healthcare). Precipitate the RNA as described above for its synthesis and dissolve the pellet in 7 μL gel loading buffer and load on an 8% denaturing polyacrylamide gel (7 M urea). After migration, the gel is covered by Saranwrap and exposed for 30 min at room temperature. The film is developed and superposed on the gel to excise the labeled RNA. The gel slice is incubated overnight with shaking at 37°C in 300 μL elution buffer (0.1% SDS, 1 m*M* EDTA, 50 m*M* NaAc pH 4.7, 0.1 vol. phenol/H_2O). The gel slice is pelleted by centrifugation for 10 min at 13,000 × *g*. The supernatant is precipitated as described earlier and the radioactive RNA is dissolved in 10 μL H_2O. Generally, 1 μL is used per assay. However, depending on the specific activity, the RNA can be further diluted (up to 20-fold).

Protocol 1C: Synthesis of [^{32}P] 5′ End-Labeled RNA With a 5′ PPP Moiety

A 30 μL reaction is set up to contain 6 μL of 5x transcription buffer (Promega) 2.25 μL of 100 m*M* DTT, 1 μL of 40 U/mL RNasin (Promega), 3 μL of NTP mix (2.5 m*M* ATP, CTP, UTP), 3.75 μL 100 μ*M* GTP), 7.5 μL γ-[^{32}P]-GTP (10 μCi/μL), template DNA, 1.5 μL of 20 U/μL T7 RNA polymerase (Promega), and nuclease-free water to 30 μL. Reactions are incubated at 37°C for 1.5 h. 2°μL RQ DNase (1 U/μL, Promega) are added and incubation at 37°C continued for 15 min. Add 20 μL H_2O to bring the volume to 50 μL. RNAs are separated from unincorporated nucleotides by purification on G50 spin columns (GE Healthcare).

Protocol 1D: Synthesis of [^{32}P] 3′ End-Labeled RNA

The transcript is synthesized as described in Protocol 1B. The RNA (5–10 pmol carrying a 5′ P or 5′ PPP moiety) is labeled on the 3′ end in a 25 μL reaction containing 2.5 μL 10x T4 RNA ligase buffer, 2.5 μL 5 m*M* ATP, 5 μL [α-^{32}P]-pCp (3.33 pmol/μL; PerkinElmer), 2.5 μL DMSO and 2.5 μL T4 RNA ligase (10 U/μL, Biolabs). The reaction is incubated overnight at 16°C, precipitated and the labeled RNA purified on a denaturing gel as described in Protocol 1B.

Standard Protocol 2: Native Purification of Intein Fusion Proteins

500 mL culture (LB) are grown at 37°C to OD_{600} = 0.5. Recombinant protein expression is induced by the addition of 0.3 m*M* IPTG (can be

reduced to 20 μ*M* for less expression) and overnight incubation at 15°C. Further steps are carried out at 4°C. Cells are pelleted (10,000 × *g*) and frozen at −80°C until further use. The frozen cell pellet is resuspended in 10 mL of buffer A (20 m*M* HEPES pH 8, 500 m*M* NaCl, 10% glycerol) supplemented with 10 μg/mL DNase I, one EDTA free protein inhibitor tablet (Pierce), kept in ice and sonicated 4x 1 min (pulse mode—70% duty, 30 s pause).

The lysate is centrifuged for 30 min at 10,000 × *g*. The supernatant is loaded on a 2 mL chitin column (~3.2 mL of slurry, Biolabs) attached to a peristaltic pump, and previously equilibrated with 40 mL of buffer A (20 vol.) at 20 mL/h. The column is then washed with 40 mL buffer B (buffer A + 1 *M* NaCl), 12 mL of buffer A and 6 mL of cleavage buffer (buffer A + 50 m*M* DTT freshly made). When the elution buffer has completely entered the chitin beads, the flow is stopped and the column incubated at 4°C overnight.

The cleaved protein is eluted with 5 mL elution buffer (buffer A + 0.1 mg/mL acetylated BSA) and 1 mL fractions are collected. Fractions are analyzed by SDS-PAGE for protein purity, pooled and aliquoted for storage.

Standard Protocol 3: Native Purification of His-Tagged Proteins on Ni-NTA Agarose

A 100 mL culture (LB) is grown at 37°C to an OD_{600} = 0.5. Expression of the recombinant protein is induced by the addition of 0.3 m*M* IPTG and overnight incubation at 15°C (can be reduced to 0.5 to 3 h at 37°C in many cases). All further steps are at <4°C. The cell pellet is suspended in 3 mL buffer A (20 m*M* Hepes pH 8, 500 m*M* NaCl, 10% glycerol) supplemented with 10 μg/mL DNase I, 1 tablet of EDTA-free protease inhibitor (Pierce). The suspension is kept in ice and sonicated 4x 1 min (pulse mode—70% duty, 30 s pause). The lysate is cleared by centrifugation for 10 min at 12,000 × *g* and the supernatant gently mixed with 1 mL of Ni-NTA agarose resin (Protino, Macherey-Nagel) for 1 h on a wheel in a closed column. The column is connected to a peristaltic pump and the resin is washed (0.5 mL/min) 3x with 30 mL wash buffer 1 (20 m*M* Hepes pH 8, 1 *M* NaCl, 10% glycerol, 20 m*M* Imidazole) and 1x with 30 mL wash buffer 2 (20 m*M* Hepes pH 8, 500 m*M* NaCl, 10% glycerol, 20 m*M* Imidazole). The bound protein is eluted with 5x 1 mL elution buffer (20 m*M* Hepes pH 8, 500 m*M* NaCl, 10% glycerol, 300 m*M* Imidazole). The peak fractions are analyzed for purity by SDS-PAGE, pooled and dialyzed against the relevant storage buffer.

REFERENCES

Chao, Y., Li, L., Girodat, D., Forstner, K. U., Said, N., Corcoran, C., et al. (2017). In vivo cleavage map illuminates the central role of RNase E in coding and non-coding RNA pathways. *Molecular Cell*, *65*(1), 39–51. https://doi.org/10.1016/j.molcel.2016.11.002.

Commichau, F. M., Rothe, F. M., Herzberg, C., Wagner, E., Hellwig, D., Lehnik-Habrink, M., et al. (2009). Novel activities of glycolytic enzymes in *Bacillus subtilis*: Interactions with essential proteins involved in mRNA processing. *Molecular & Cellular Proteomics*, *8*(6), 1350–1360. https://doi.org/10.1074/mcp.M800546-MCP200. doi: M800546-MCP200 [pii].

DeLoughery, A., Lalanne, J. B., Losick, R., & Li, G. W. (2018). Maturation of polycistronic mRNAs by the endoribonuclease RNase Y and its associated Y-complex in *Bacillus subtilis*. *Proceedings of the National Academy of Sciences of the United States of America*, *115*(24), E5585–E5594. https://doi.org/10.1073/pnas.1803283115.

Durand, S., Gilet, L., Bessieres, P., Nicolas, P., & Condon, C. (2012). Three essential ribonucleases-RNase Y, J1, and III-control the abundance of a majority of Bacillus subtilis mRNAs. *PLoS Genetics*, *8*(3), e1002520. https://doi.org/10.1371/journal.pgen.1002520.

Even, S., Pellegrini, O., Zig, L., Labas, V., Vinh, J., Brechemmier-Baey, D., et al. (2005). Ribonucleases J1 and J2: Two novel endoribonucleases in *B. subtilis* with functional homology to *E. coli* RNase E. *Nucleic Acids Research*, *33*(7), 2141–2152.

Hunt, A., Rawlins, J. P., Thomaides, H. B., & Errington, J. (2006). Functional analysis of 11 putative essential genes in *Bacillus subtilis*. *Microbiology*, *152*(Pt. 10), 2895–2907.

Khemici, V., Poljak, L., Luisi, B. F., & Carpousis, A. J. (2008). The RNase E of *Escherichia coli* is a membrane-binding protein. *Molecular Microbiology*, *70*(4), 799–813. doi:MMI6454 [pii]. https://doi.org/10.1111/j.1365-2958.2008.06454.x.

Kime, L., Clarke, J. E., Romero, A. D., Grasby, J. A., & McDowall, K. J. (2014). Adjacent single-stranded regions mediate processing of tRNA precursors by RNase E direct entry. *Nucleic Acids Research*, *42*(7), 4577–4589. https://doi.org/10.1093/nar/gkt1403.

Laalami, S., Bessieres, P., Rocca, A., Zig, L., Nicolas, P., & Putzer, H. (2013). Bacillus subtilis RNase Y activity in vivo analysed by tiling microarrays. *PLoS One*, *8*(1), e54062. https://doi.org/10.1371/journal.pone.0054062.

Laalami, S., & Putzer, H. (2011). mRNA degradation and maturation in prokaryotes: The global players. *Biomolecular Concepts*, *2*, 491–506. https://doi.org/10.1515/bmc.2011.042.

Laalami, S., Zig, L., & Putzer, H. (2014). Initiation of mRNA decay in bacteria. *Cellular and Molecular Life Sciences*, *71*(10), 1799–1828. https://doi.org/10.1007/s00018-013-1472-4.

Lehnik-Habrink, M., Newman, J., Rothe, F. M., Solovyova, A. S., Rodrigues, C., Herzberg, C., et al. (2011). RNase Y in Bacillus subtilis: A natively disordered protein that is the functional equivalent of RNase E from *Escherichia coli*. *Journal of Bacteriology*, *193*(19), 5431–5441. https://doi.org/10.1128/JB.05500-11.

Lehnik-Habrink, M., Schaffer, M., Mader, U., Diethmaier, C., Herzberg, C., & Stulke, J. (2011). RNA processing in Bacillus subtilis: Identification of targets of the essential RNase Y. *Molecular Microbiology*, *81*(6), 1459–1473. https://doi.org/10.1111/j.1365-2958.2011.07777.x.

Li de la Sierra-Gallay, I., Zig, L., Jamalli, A., & Putzer, H. (2008). Structural insights into the dual activity of RNase J. *Nature Structural & Molecular Biology*, *15*(2), 206–212.

Liponska, A., Jamalli, A., Kuras, R., Suay, L., Garbe, E., Wollman, F. A., et al. (2018). Tracking the elusive 5′ exonuclease activity of Chlamydomonas reinhardtii RNase J. *Plant Molecular Biology*, *96*(6), 641–653. https://doi.org/10.1007/s11103-018-0720-2.

Mäder, U., Zig, L., Kretschmer, J., Homuth, G., & Putzer, H. (2008). mRNA processing by RNases J1 and J2 affects *Bacillus subtilis* gene expression on a global scale. *Molecular Microbiology, 70*(1), 183–196. doi:MMI6400 [pii]. https://doi.org/10.1111/j.1365-2958.2008.06400.x.

Mathy, N., Benard, L., Pellegrini, O., Daou, R., Wen, T., & Condon, C. (2007). 5′-to-3′ exoribonuclease activity in Bacteria: Role of RNase J1 in rRNA maturation and 5' stability of mRNA. *Cell, 129*(4), 681–692.

Mathy, N., Hebert, A., Mervelet, P., Benard, L., Dorleans, A., Li de la Sierra-Gallay, I., et al. (2010). Bacillus subtilis ribonucleases J1 and J2 form a complex with altered enzyme behaviour. *Molecular Microbiology, 75*(2), 489–498. doi:MMI7004 [pii]. https://doi.org/10.1111/j.1365-2958.2009.07004.x.

Nickels, J. D., Chatterjee, S., Mostofian, B., Stanley, C. B., Ohl, M., Zolnierczuk, P., et al. (2017). Bacillus subtilis lipid extract, a branched-chain fatty acid model membrane. *Journal of Physical Chemistry Letters, 8*(17), 4214–4217. https://doi.org/10.1021/acs.jpclett.7b01877.

Redder, P. (2018). Molecular and genetic interactions of the RNA degradation machineries in Firmicute bacteria. *Wiley Interdisciplinary Reviews RNA, 9*(2), e1460. https://doi.org/10.1002/wrna.1460.

Shahbabian, K., Jamalli, A., Zig, L., & Putzer, H. (2009). RNase Y, a novel endoribonuclease, initiates riboswitch turnover in *Bacillus subtilis*. *The EMBO Journal, 28*(22), 3523–3533. doi:emboj2009283 [pii]. https://doi.org/10.1038/emboj.2009.283.

Spizizen, J. (1958). Transformation of biochemically deficient strains of *Bacillus subtilis* by deoxyribonucleate. *Proceedings of the National Academy of Sciences of the United States of America, 44*, 407–408.

Zhou, H., & Lutkenhaus, J. (2003). Membrane binding by MinD involves insertion of hydrophobic residues within the C-terminal amphipathic helix into the bilayer. *Journal of Bacteriology, 185*(15), 4326–4335.

CHAPTER FIFTEEN

Mapping 5′-Ends and Their Phosphorylation State With EMOTE, TSS-EMOTE, and nEMOTE

Peter Redder[1]
Laboratoire de Microbiologie et Génétique Moléculaires, Centre de Biologie Integrative, University of Toulouse, Toulouse, France
[1]Corresponding author: e-mail address: peter.redder@univ-tlse3.fr

Contents

Methods in Enzymology, Volume 612
ISSN 0076-6879
https://doi.org/10.1016/bs.mie.2018.08.011

Abstract

Knowing the exact 5′-end of a specific RNA molecule has long been of great interest, since this information can reveal transcription start sites, which subsequently pinpoints promoter regions and potential binding sites for transcriptional regulators. Mapping of the 5′-end can also identify endoribonucleolytic cleavage sites, for the maturation of ribosomal RNA or events which initiate degradation of all or parts of the RNA molecule. If the 5′ nucleotide of an RNA is triphosphorylated, then the nucleotide represents the original, unaltered 5′-end of the RNA. Similarly, if an endoribonuclease cleaves the RNA with a mechanism that leaves a monophosphate on the 5′-end downstream cleavage product, then it is useful to search exclusively for monophosphorylated 5′-ends to identify the cleavage sites of the endoribonuclease. This chapter details three variations of the EMOTE (exact mapping of transcriptome ends) protocol, which can identify RNA 5′-ends on a transcriptome-wide scale, and simultaneously provide information about the phosphorylation state of the detected ends. EMOTE, TSS-EMOTE (transcription start specific EMOTE), and nEMOTE (nonphosphorylated EMOTE) reveals mono-, tri-, and nonphosphorylated 5′-ends, respectively, and the protocols can be performed individually or in parallel on the same RNA sample.

1. INTRODUCTION

Investigations of the 5′-end of RNA have classically been done with primer extension experiments or nuclease protection assays, and more recently with variations of 5′ RACE (rapid amplification of cDNA ends) all of which are still in frequent use. Most modern protocols for 5′-RACE are derived from RLM-RACE (RNA ligase-mediated amplification of cDNA ends) (Fromont-Racine, Bertrand, Pictet, & Grange, 1993; Liu & Gorovsky, 1993) and call for ligation of a synthetic oligo to the 5′-ends of all RNA in the sample, whereupon reverse transcription (RT) is carried with a primer that is specific for the RNA which is under investigation. The resulting cDNA will have a known sequence at each end (RT primer at the 5′-end and synthetic oligo at the 3′-end), and it can therefore easily be amplified by PCR. The length of this (these) PCR product(s) already provides an idea of the position of the 5′-end(s), but to determine the exact 5′ nucleotide, the PCR product is cloned into a vector and transformed into *Escherichia coli*. Plasmid DNA is prepared from several colonies, and the sequence of their inserts is determined with Sanger sequencing, whereupon the extreme 5′-nucleotide of the original RNA molecule can be identified as the base immediately after the sequence of the synthetic oligo. While the RLM-RACE methods are accurate down to the individual nucleotide, they suffer from three drawbacks:

(1) The cloning step in the classic RLM-RACE protocols is somewhat cumbersome and limits the number of individual examined RNA molecules, because each colony has to be isolated, plasmid DNA extracted and sequenced separately. This typically means that only 5–30 clones are sequenced, which may translate into even fewer original RNA molecules, since the PCR step might have amplified a single ligation product to become the insert in multiple vectors. Most promoters give rise to heterogeneity in the transcription start site, in the sense that the RNA polymerase sometimes starts transcription one or more nucleotides upstream or downstream of the main site (see, for example, Prados, Linder, & Redder, 2016). This has classically been observed as minor bands in primer extension autoradiographs, but only a few individual RNA molecules are examined with RLM-RACE, so this method is sometimes inadequate for judging the extent of heterogeneity.

(2) Ligation to the 5′-end of RNA can only take place if the 5′-end is monophosphorylated. This can be ensured by treating triphosphorylated (or capped) RNAs with a pyrophosphate, and by treating nonphosphorylated 5′-ends with polynucleotide kinase to add a phosphate to the 5′-end. However, if the RNA sample contains a mixture of RNA species with different phosphorylation states, then the enzymatic treatments will remove all information regarding these phosphorylation states. If several RNA molecules have identical 5′ nucleotides but different 5′ phosphorylation states, then this information will be lost. PABLO (phosphorylation assay by ligation of oligonucleotides), an alternative technique, can be used to identify and quantify the phosphorylation states but requires a priori knowledge about the exact 5′-end of the RNA (Celesnik, Deana, & Belasco, 2008).

(3) Finally, the most severe limitation of RLM-RACE is that only a single locus can be examined at a time, as defined by the primer used in the RT step. For many projects this will not be a limiting factor, but when 5′-end data are required on a transcriptome-wide scale, then transcriptome-wide methods must be employed. Unfortunately, most standard RNAseq protocols will not provide an accurate mapping of the 5′-ends and will additionally lose information about the phosphorylation states of the sequences RNA. Specialized protocols are therefore needed to obtain these data.

The following protocols will describe how to detect 5′-ends on a transcriptomic scale, while simultaneously gauging the phosphorylation state of each detected 5′-end. The three protocols (EMOTE, TSS-EMOTE, and nEMOTE) will be presented as part of the same workflow, but each part can also be performed individually (if, for example, only information about nonphosphorylated RNAs is of interest). Certain steps can be skipped if only one of the protocols are used, and these optional steps will be clearly highlighted.

1.1 Preparing the RNA Sample

High quality RNA (i.e., RNA that has neither been degraded during cell lysis, isolation, or storage) is key to obtaining usable data, and extreme care must therefore be taken right from the moment of harvesting the bacterial cultures. This chapter will not describe the details of how to ensure good quality RNA, since the methods may vary greatly between different projects, different laboratories, and different organisms. However, a short list of useful advice which will improve the chances of success are listed below:

Use only certified RNase-free microtubes and pipette tips (always filtered). These are commercially available and usually only slightly more expensive than their normal counterparts.

Always work with gloves and a lab coat dedicated to RNA work. A lab coat that has been worn while purifying plasmid with a miniprep kit is highly suspect due to the high concentration of RNase A found in such kits.

Ideally have a dedicated work-space for RNA work with a dedicated set of micropipettes. If this is not possible, then clean the bench and the pipettes thoroughly with an RNase removal agent such as RNase Away (Molecular BioProducts) before each use. This also applies to microcentrifuges, which are often shared by the whole laboratory and are therefore often contaminated by RNase A from miniprep kits.

RNA degradation is a rapid process in bacteria, and it is therefore crucial to immediately inhibit the cellular ribonucleases upon harvesting. Several commercial buffers exist for harvesting and protecting the RNA (for example, RNAlater from Invitrogen) but we have found that in our hands an inexpensive solution is to pipet 8 mL bacterial culture into 40 mL ice-cold Ethanol–Acetone mix (1:1 volume). The cells are then pelleted by centrifuging 10 min at 4000 × g (or max speed) at 4°C, the supernatant is decanted, and 1 mL fresh Ethanol–Acetone is added (the pellet will not dissolve in Ethanol–Acetone) whereupon the tube can be stored for up to a month at −80°C with no visible degradation of the RNA. The RNA can be recovered by removing the supernatant (after a centrifugation to ensure the cell pellet sticks to the bottom of the tube), suspending the cell pellet in 1 mL TE buffer which is transferred to a microtube and centrifuged at max speed for 2 min. The supernatant is removed, and the pellet is suspended in the buffer of your choice. The Ethanol–Acetone method has the additional advantage that it immediately kills bacterial pathogens, thus permitting subsequent steps to be performed outside a level 2 biosafety laboratory.

Once RNA has been isolated, it is highly recommended to verify the quality (nondegradation) by Bioanalyzer (Agilent), which furthermore serves to obtain an accurate measurement of the RNA concentration in the sample.

2. EMOTE

The detection of monophosphorylated 5′-ends of RNA is the simplest of the three EMOTE protocols described here, and it forms the basis for the two other protocols. However, the vast majority of monophosphorylated 5′-ends in the cell can be found on the three ribosomal RNAs, 16S, 23S, and 5S, and it is thus desirable to remove the ribosomes before initiating the EMOTE protocol. Several commercial kits are available for depleting the ribosomes from bacterial RNA samples, and when working with *Staphylococcus aureus*, we have had good results with the Ribo-Zero rRNA Removal Kit (Bacteria) from Illumina. *Note that the ribosome depletion step is not strictly necessary for the nEMOTE and TSS-EMOTE protocols, where the treatment with XRN-1 will remove the vast majority of 16S and 23S rRNA (but not the 5S rRNA).*

2.1 Ribosome Depletion

2.1.1 Equipment

The equipment needed depends on which kit is chosen for the removal of rRNA; however, the following items are usually required:

Adjustable heating block (or water bath)

Magnetic stand for microtubes

2.1.2 Reagents

Ribosome depletion kit, for example, Ribo-Zero (Illumina)

2.1.3 Procedure

1. Deplete the ribosomes from 8 μg total RNA, by following the instructions of the manufacturer of the ribosome depletion kit (for example, RiboZero from Illumina) and suspend the ribosome-depleted RNA in 20 μL 0.1 × TE. *Note*: In our laboratory, we usually use 8 μg total RNA as starting material for EMOTE; however, we have successfully performed the protocol with as little as 1 μg RNA. The following protocol(s) remain unchanged.

2.2 Ligation Step

2.2.1 Equipment

Heating block at 80°C
Heating block (or incubator) at 37°C
Ice-water

2.2.2 Reagents

RNase-free water
100 μ*M* BioRp8 **RNA oligo**: Biotin-dG-CGGCACCAACCGAGG VVVVVVVACAGA
10 × T4 RNA ligase buffer (New England Biolabs)
10 m*M* ATP
T4 RNA ligase 1 (New England Biolabs)
RNasin Plus (Promega)
3 *M* Sodium Acetate RNase-Free (Invitrogen)
Glycogen RNase-Free (Roche)
96% Ethanol (99% is equally good)
75% Ethanol
0.1 × TE (1 m*M* Tris–HCl pH 8, 0.1 m*M* EDTA)

2.2.3 Procedure

1. Mix 9 μL of each ribosome-depleted sample with 1 μL 100 μ*M* BioRp8 oligo. Heat to 80°C for 2 min, then flash-cool in ice-water.
2. Make a ligation master mix with 2 μL 10 × T4 RNA ligase buffer, 2 μL 10 m*M* ATP, 2 μL (30 U) T4 RNA ligase 1, 1 μL (40 U) RNasin Plus and 3 μL RNase-free water for each RNA sample to be processed.

3. Add 10 μL ligation master mix to the RNA + oligo. Mix gently by pipetting and incubate for 3 h at 37°C.
4. Add 70 μL RNase-free water, 10 μL 3 *M* Sodium Acetate and 1 μL (20 mg/mL) Glycogen. Vortex and add 300 μL 96% ethanol. Vortex again and store at −80°C overnight.
5. Centrifuge at maximum speed for 45 min at 4°C, and then remove supernatant.
6. Wash the pellet twice in 750 μL cold 75% ethanol and suspend in 20 μL 0.1 × TE. Store samples at − 80°C or pass directly to the reverse transcription step (Section 2.3).

2.3 Reverse Transcription Step

2.3.1 Equipment

Heating block (or water bath) at 80°C
Heating block (or water bath) at 42°C
Heating block (or water bath) at 65°C

2.3.2 Reagents

RNase-free water
ProtoScript II Reverse Transcriptase (New England Biolabs)
5 × ProtoScript II Reverse Transcriptase Reaction Buffer (New England Biolabs)
10 m*M* dNTP (2.5 m*M* each)
10 m*M* DTT
20 μ*M* DROAA primer: GGCATTCCTGCTGAACCGCTCTTCC GATCTNNNNNNNNAA
RNasin Plus (Promega)
GeneJET PCR Purification Kit (Thermo Scientific) [QIAquick PCR Purification Kit (Qiagen) or similar will also work]

2.3.3 Procedure

1. Mix 1 μL 20 μ*M* DROAA primer with 9 μL ligated RNA (from step 6 of Section 2.2.3) and heat for 2 min at 80°C. Then let the mix cool slowly to room temperature.
2. Prepare Reverse Transcription mix with (per reaction) 1 μL RNase-free water, 4 μL 5 × ProtoScript II Reverse Transcriptase Reaction Buffer, 2 μL 10 m*M* DTT, 1 μL 10 m*M* dNTP, 1 μL RNasin Plus, 1 μL (200 U) ProtoScript II Reverse Transcriptase.

3. Add 10 μL Reverse Transcription mix to each sample of RNA with DROAA primer. Leave at room temperature for 10 min and then for 50 min at 42°C.
4. Heat inactivate the reverse transcriptase by heating to 65°C for 30 min.
5. Add 180 μL water and use a GeneJET PCR Purification Kit according to the manufacturer's instructions (without the optional added isopropanol). Elute in 50 μL water.

2.4 PCR Amplification and Addition of EMOTE Barcodes and Illumina Adaptors

2.4.1 Equipment

PCR Thermo cycler

Thin-walled 200 μL PCR tubes

Standard agarose gel apparatus and gel visualization equipment.

2.4.2 Reagents

MilliQ water (or other high purity water)

1.5 μL 10 m*M* dNTP (2.5 m*M* each)

100 n*M* Primers D6A, D6B, etc. (note the low concentration!)

10 μ*M* Primer B-PE-PCR20

10 μ*M* Primer A-PE-PCR10

5 × Q5 Reaction Buffer (New England Biolabs)

Q5 Hot Start High-Fidelity DNA Polymerase (New England Biolabs)

D6A, D6B, etc., primers (examples below)

D6A: CTCTTTCCCTACACGACGCTCTTCCGATCTNTACAC GGCACCAACCGAGG

D6B: CTCTTTCCCTACACGACGCTCTTCCGATCTNGTATC GGCACCAACCGAGG

D6C: CTCTTTCCCTACACGACGCTCTTCCGATCTNCGTCC GGCACCAACCGAGG

D6D: CTCTTTCCCTACACGACGCTCTTCCGATCTNAAGTC GGCACCAACCGAGG

D6E: CTCTTTCCCTACACGACGCTCTTCCGATCTNACACC GGCACCAACCGAGG

D6F: CTCTTTCCCTACACGACGCTCTTCCGATCTNGGTAC GGCACCAACCGAGG

A-PE-PCR10: AATGATACGGCGACCACCGAGATCTACACTC TTTCCCTACACGACG

B-PE-PCR20: CAAGCAGAAGACGGCATACGAGATCGGTCTC GGCATTCCTGCTGAACCGC

2.4.3 *Procedure*

1. In a separate PCR tube for each reverse transcriptase reaction, prepare a mix of 10 μL purified RT-reaction (from step 5 in Section 2.3.3) and 1.5 μL 100 n*M* primer D6A, D6B, etc., a different primer for each reverse transcriptase reaction). Please note the low concentration of these primers.
2. Make a PCR master mix with (per reverse transcriptase reaction) 25 μL water, 10 μL 5 × Q5 Reaction Buffer, 1.5 μL dNTP, 1.5 μL 10 μ*M* Primer B-PE-PCR20, 0.5 μL (1 U) Q5 Hot Start High-Fidelity DNA Polymerase.
3. Add 38.5 μL PCR master mix to each of the PCR tubes from step 1. Mix well by pipetting up and down.
4. Place the PCR tubes in the thermo cycler and launch the following program:

   2 min @98°C; [10s @98°C; 20s @50°C; 2 min @ 72°C] × 31 cycles; 5 min @72°C; then 4°C

   Important. *At the end of cycle 5, pause the thermocycler and quickly add 1.5 μL of 10 μM A-PE-PCR10 to each PCR tube. Then replace the PCR tubes in the thermocycler and continue the program to run the remaining 26 cycles.*

2.5 Adjusting Relative DNA Concentrations, Concentration, and Size-Selection of PCR Products

2.5.1 *Equipment*

Standard agarose gel apparatus and gel visualization equipment
2-mL microtube
Microcentrifuge
Adhesive tape
Scalpel
NanoDrop spectrophotometer (or equivalent) for measuring DNA concentration

2.5.2 *Reagents*

Xylene cyanol FF agarose gel loading buffer, for example, 6 × Orange DNA Loading Dye (Thermo Scientific) [if possible avoid loading buffers with Bromophenol blue, since it will stain the area of the agarose gel which is of interest for EMOTE].

Molecular marker for DNA sizes (for example, GeneRuler 1 kb Plus DNA Ladder from Thermo Scientific).

Fluorescent DNA staining agent, for example, SYBR Safe DNA Gel Stain (Invitrogen).

GeneJET PCR Purification Kit (Thermo Scientific) [QIAquick PCR Purification Kit (Qiagen) or similar will also work].

GeneJET Gel Extraction Kit (Thermo Scientific) [QIAquick Gel Extraction Kit (Qiagen) or similar will also work]. Note furthermore that the gel extraction and PCR purification kits are identical in composition for most brands.

Isopropanol

2.5.3 Procedure

1. Migrate 10 μL of each PCR reaction (from step 4 in Section 2.4.3) on a 0.8% agarose gel.
2. If the yields of all the PCR reactions are similar, i.e., the smears on the agarose gel appear to visually be of similar intensity, then the remaining 40 μL from each PCR should be added to an equal volume (in this case 40 μL) of Binding Buffer (GeneJET PCR Purification Kit). If one or more of the smears appear to significantly more intense than the other smears, then less than 40 μL of that (these) PCR reaction(s) should be used. Note that an exact equilibration between the samples is not required, but it is advantageous for subsequent sequencing and analyses steps to avoid significant overrepresentation of a single sample (or two).
3. Mix all PCR reactions + binding buffer together in a single 2-mL microtube and vortex.
4. Add 1/2 volume of isopropanol for every volume of PCR reaction and vortex thoroughly.
5. Load 750 μL of the mix from step 4 onto a single DNA binding column from the GeneJET PCR Purification Kit.
6. Centrifuge according to manufacturer's protocol (usually 14,000 × *g* for 1 min).
7. Discard flow-through.
8. Repeat steps 5–7 until all the mix has been through the DNA binding column.
9. Wash column according to manufacturer's instructions and then elute in 50 μL water.
10. Use adhesive tape to combine several of the teeth on the comb used for casting agarose gels, in a way that results in a single well with room for a 50 μL sample. Be sure to have untaped teeth on both sides of the tape. Then cast a 1% agarose gel using the taped comb.

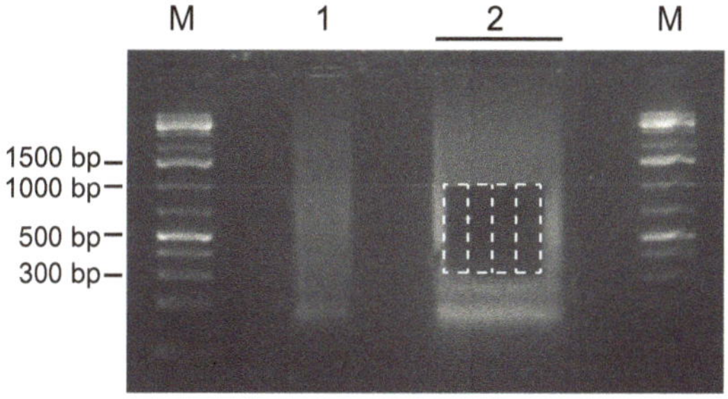

Fig. 1 Example of an agarose gel where the relevant gel-piece has been cut out. Lanes 1 and 2 were loaded with the same mix of PCR products (see Section 2.5.3, step 11), but with 10 μL in lane 1 and 50 μL in lane 2 (note that the teeth of the comb were taped to make lane 2 much wider). *M*: size-markers. The *white lines* indicate the separation into four smaller agarose pieces (see Section 2.5.3, step 13). Gel photo kindly provided by Céline Pélissier.

11. Add 10 μL 6 × Orange DNA Loading Dye to the mixed and concentrated PCR products and load 50 μL in the large well of the agarose gel, and then load the remaining 10 μL in an adjacent untaped well. Load a DNA molecular marker on either side of the sample wells, leaving a gap of one empty well between marker and sample (see Fig. 1).
12. Migrate the DNA for approximately 40 min at 7 V/cm.
13. Cut the agarose gel with a scalpel to excise the DNA of the 50 μL well in the size-range from 300 to 1000 bp. Cut the gel piece into four smaller pieces and transfer each to a 2-mL microtube (it is preferable to cut the gel piece in the direction of the electrophoresis migration, so that each of the four resulting gel pieces contains an equal sampling of DNA sizes from 300 to 1000 bp) (see Fig. 1).
14. Extract the DNA from the agarose gel using a standard kit such as GeneJET Gel Extraction Kit. However, be sure to add the optional isopropanol (1 mL/g of agarose gel) to ensure optimal recovery of DNA molecules smaller than 500 base pairs. Elute in 50 μL water.
15. Determine the DNA concentration using a NanoDrop spectrophotometer and send a 20 μL aliquot of the size-selected and purified PCR product to be sequenced with Illumina technology at a sequencing platform or company. Note that the DNA concentrations are usually not very high (typically between 20 and 60 ng/μL); however, this is more than enough for Illumina sequencing.

3. TSS-EMOTE

TSS-EMOTE (transcription start specific exact mapping of transcriptome ends) maps the 5′-end of RNAs which are triphosphorylated, a phosphorylation state which can only be achieved if the 5′ nucleotide was the first nucleotide in a de novo transcription. Triphosphorylated ends cannot be ligated directly to a synthetic oligo, and the EMOTE protocol (see Section 2) therefore has to be modified in order to permit such a ligation. The first step is to remove as much as possible of the monophosphorylated RNA in the sample, since these RNAs will contribute to the background noise of the TSS-EMOTE data. A pyrophosphatase will then be used to convert all triphosphorylated RNA 5′-ends to their equivalent monophosphorylated species, whereupon ligation with a synthetic oligo can take place (equivalent to step 2.2).

3.1 Removal of 5′ Monophosphorylated RNA Species by XRN-1

3.1.1 Equipment

Ice
Heating block (or water bath) at 37°C
Microcentrifuge

3.1.2 Reagents

RNase-free water
RNasin Plus (Promega)
XRN-1 (New England Biolabs)
10X NEBuffer 3 (New England Biolabs). This buffer is provided with the XRN-1 enzyme and should be used exclusively for RNA to avoid contamination with RNases.
Phenol:Chloroform:Isoamyl Alcohol 25:24:1 Saturated with 10 m*M* Tris, pH 8.0, 1 m*M* EDTA (Sigma)
2 mL Phase Lock Gel tubes (VWR)
3 *M* Sodium Acetate RNase-Free (Invitrogen)
Glycogen RNase-Free (Roche)
96% Ethanol (99% is equally good)
75% Ethanol
1 × TE (10 m*M* Tris–HCl pH 8, 1 m*M* EDTA)
0.1 × TE (1 m*M* Tris–HCl pH 8, 0.1 m*M* EDTA)

3.1.3 Procedure

1. Dilute 8 μg RNA in RNase-free water to reach 30 μL and place on ice. It is also possible to use an aliquot of ribosome depleted RNA from Section 2.1.
2. Prepare XRN-1-mix. For one sample of RNA, mix 35 μL RNase-free water, 8 μL 10 × NEBuffer 3, 2 μL (80 U) RNasin Plus, and 5 μL (5 U) XRN-1.
3. Add 50 μL XRN-mix into each RNA-tube. Pipet gently up and down to mix, and then incubate for 4 h at 37°C.
4. Prepare the Phase Lock Gel tubes by centrifugation for 30 s at 14,000 × *g*.
5. Add 220 μL 1 × TE and 300 μL Phenol:Chloroform:Isoamyl Alcohol, then vortex.
6. Add vortexed samples to Phase Lock Gel tubes and shake gently to mix (no vortex).
7. Centrifuge at 14,000 × *g* for 5 min and add another 300 μL Phenol: Chloroform:Isoamyl Alcohol to each tube, and shake to mix (no vortex).
8. Centrifuge at 14,000 × *g* for 5 min and then transfer the aqueous phase to a clean RNase-free microtube.
9. Add 10% volume Sodium Acetate (final concentration should be 0.3 *M*), 1 μL glycogen (20 μg/μL). Vortex and add three times the volume of 96% Ethanol (for example, 900 μL ethanol for 300 μL sample + Sodium Acetate). Vortex again and put at −80°C overnight.
10. Centrifuge at maximum speed for 45 min at 4°C and then remove supernatant.
11. Wash the pellet twice in 750 μL cold 75% ethanol and suspend it in 20 μL 0.1 × TE. Store samples at − 80°C or pass directly to the ligation step (Section 3.2).

3.2 Ligation Step With and Without RppH

The pyrophosphatase RppH converts 5′ triphosphorylated RNAs to their monophosphorylated equivalents. Thus, when RppH is added to a ligation reaction, it allows the triphosphorylated RNAs to first be converted to monophosphorylated species and then be ligated to a synthetic RNA oligo (BioRp8 in this protocol) by the T4 RNA ligase 1. Most (possibly all) bacteria naturally express pyrophosphatases, and the 5′ triphosphorylated transcripts (i.e., full-length transcripts) are therefore slowly but continuously converted to monophosphorylated RNAs, generating a population of such RNAs in all total RNA samples. The pretreatment with XRN-1

(Section 3.1) greatly diminishes the number of RNA molecules with monophosphorylated 5′-ends in the sample. However, XRN-1 (and equivalent enzymes such as Terminator from Illumina) is not 100 % efficient unless high excess of this relatively expensive enzyme is used, and there are even RNA molecules with structures that protects against XRN-1 degradation, even though they are monophosphorylated (notably 5S rRNA and tRNAs).

In TSS-EMOTE, the RppH pyrophosphatase is used to convert all triphosphorylations at the 5′-ends to monophosphorylations, thus permitting ligation to the synthetic BioRp8 oligo. However, the mere existence of an BioRp8 + RNA ligation product is not enough to conclude that the original RNA was triphosphorylated, since the XRN-1-treated sample will contain a low but detectable amount of 5′ monophosphorylated RNA that has not been digested by XRN-1. The TSS-EMOTE protocol addresses this problem by splitting the XRN-1-treated sample into two aliquots. One aliquot is used for a ligation with RppH added, to ligate BioRp8 to both triphosphorylated and monophosphorylated RNAs in the sample. The second aliquot is used for a ligation without RppH, and any ligation products in this aliquot therefore represent the baseline monophosphorylated RNA molecules present in the sample. Comparison between the data for the −RppH and the +RppH aliquots can then reveal which RNAs were triphosphorylated in the original RNA sample (see Section 5).

3.2.1 Equipment

Heating block at 80°C
Heating block (or incubator) at 37°C
Ice-water

3.2.2 Reagents

100 μ*M* BioRp8 oligo
10 × T4 RNA ligase buffer (New England Biolabs)
10 m*M* ATP
RppH (New England Biolabs)
T4 RNA ligase 1 (New England Biolabs)
RNasin Plus (Promega)
3 *M* Sodium Acetate RNase-Free (Invitrogen)
Glycogen RNase-Free (Roche)
96% Ethanol (99% is equally good)
75% Ethanol
0.1 × TE (1 m*M* Tris–HCl pH 8, 0.1 m*M* EDTA)

3.2.3 Procedure

1. For each RNA sample, prepare two separate RNase-free microtubes, each with 9 μL XRN-treated RNA and 1 μL 100 μ*M* BioRp8. Mark the tubes with the name of the RNA sample and label one tube −RppH and the other tube +RppH.
2. Heat the tubes to 80°C for 2 min, then flash-cool in ice-water.
3. Prepare 10 μL of-RppH mix or + RppH mix for each RNA sample with the following compositions.

   "−RppH mix": 3 μL RNase-free water, 2 μL RNA ligase buffer, 2 μL 10 m*M* ATP, 1 μL RNasin Plus, 2 μL (20 U) RNA ligase 1.

   "+RppH mix": 1 μL RNase-free water, 2 μL RNA ligase buffer, 2 μL 10 m*M* ATP, 1 μL RNasin Plus, 2 μL (20 U) RNA ligase 1, 2 μL (10 U) RppH.
4. Add 10 μL of either ×RppH mix or +RppH mix to each tube (the two master mixes are quite viscous to pipet, so it is advisable to make some extra to ensure that there is enough).
5. Incubate 60 min at 37°C, then add 1 μL 10 m*M* ATP to each tube and incubate overnight at 16°C.
6. Next day, add 70 μL RNase-free water, 10 μL 3 *M* Sodium Acetate, and 0.5 μL Glycogen. Mix and add 300 μL ethanol. Vortex and store overnight at −80°C.
7. Centrifuge at maximum speed for 45 min at 4°C and then remove supernatant.
8. Wash the pellet once in 900 μL cold 75% ethanol and suspend it in 20 μL 0.1 × TE. Store samples at − 80°C or pass directly to the reverse transcription step (Section 3.3).

3.3 Reverse Transcription (Same as Section 2.3)

The reverse transcription step in the TSS-EMOTE protocol is identical to the reverse transcription step of the EMOTE protocol (see Section 2.3).

3.4 PCR Amplification and Addition of EMOTE Barcodes and Illumina Adaptors (Same as Section 2.4)

The PCR amplification and addition of EMOTE barcodes and Illumina adaptors step in the TSS-EMOTE protocol are identical to the equivalent step of the EMOTE protocol (see Section 2.4).

3.5 Adjusting Relative DNA Concentrations, Concentration, and Size-Selection of PCR Products (Same as Section 2.5)

Adjusting relative DNA concentrations, concentration, and size-selection of PCR products step in the TSS-EMOTE protocol are identical to the equivalent step of the EMOTE protocol (see Section 2.5).

4. nEMOTE

nEMOTE (nonphosphorylated exact mapping of transcriptome ends) maps the 5′-end of RNAs which have no phosphate at their 5′-ends. This phosphorylation state can arise from a complete dephosphorylation of a mono-, di-, or triphosphorylated RNA, but is more likely to be the result of an endoribonucleolytic cleavage event between the phosphate group and the 5′ OH of the following nucleotide. One origin of such cleavages is inherent to RNA, where the 2′ OH performs a nucleophilic attack on the phosphodiester bond to form a 2′ 3′ cyclic phosphate and cause the RNA to cleave itself. Additionally, there are a range of bacterial toxin ribonucleases (part of toxin–antitoxin systems) that enzymatically cleave RNA to generate 5′ nonphosphorylated products (see, for example, Kirkpatrick et al., 2016). Finally, the ubiquitous eukaryal endoribonuclease A (RNase A) also generates 5′ OH RNA ends, which necessitates extreme prudence regarding RNase A contamination when performing the nEMOTE assay.

Nonphosphorylated 5′-ends cannot be ligated directly to a synthetic oligo, and the EMOTE protocol (see Section 2) therefore has to be modified in order to permit such a ligation. The first step is to remove as much as possible of the monophosphorylated RNA in the sample (similar to TSS-EMOTE in Section 3.1), since these RNAs will contribute to the background noise of the nEMOTE data. T4 Polynucleotide Kinase will then be used to convert all nonphosphorylated RNA 5′-ends to their equivalent monophosphorylated species, whereupon ligation with a synthetic oligo can take place (equivalent to step 2.2).

4.1 XRN-1 Treatment (Same as Section 3.1)

The XRN-1 treatment step in the nEMOTE protocol is identical to the equivalent step of the TSS-EMOTE protocol (see Section 3.1).

4.2 Ligation With and Without Polynucleotide Kinase (PNK)

T4 Polynucleotide Kinase (PNK) converts 5′ nonphosphorylated RNAs to their monophosphorylated equivalents. Thus, when PNK is added to a ligation reaction, it allows the nonphosphorylated RNAs to first be converted to monophosphorylated species and then be ligated to a synthetic RNA oligo (BioRp8 in this protocol) by the T4 RNA ligase 1. The pre-treatment with XRN-1 (Section 3.1) greatly diminishes the number of RNA molecules with monophosphorylated 5′-ends in the sample. However, XRN-1 (and equivalent enzymes such as Terminator from Illumina) is not 100 % efficient unless high excess of this relatively expensive enzyme is used, and there are even RNA molecules with structures that protects against XRN-1 degradation, even though they are monophosphorylated (notably 5S rRNA and tRNAs).

In nEMOTE, the PNK enzyme is used to convert all non-phosphorylations at the 5′-ends to monophosphorylations, thus permitting ligation to the synthetic BioRp8 oligo. However, the mere existence of a BioRp8 + RNA ligation product is not enough to conclude that the original RNA was nonphosphorylated, since the XRN-1-treated sample will contain a low but detectable amount of 5′ monophosphorylated RNA that has not been digested by XRN-1. The nEMOTE protocol addresses this problem by splitting the XRN-1-treated sample into two aliquots. One aliquot is used for a ligation with PNK added, to ligate BioRp8 to both nonphosphorylated and monophosphorylated RNAs in the sample. The second aliquot is used for a ligation without PNK, and any ligation products in this aliquot therefore represent the baseline monophosphorylated RNA molecules present in the sample. Comparison between the data for the −PNK and the +PNK aliquots can then reveal which RNAs were nonphosphorylated in the original RNA sample (see Section 5). Throughout the ligation step it is important to be extra careful to avoid RNase A-contamination, and to avoid heating the RNA in a buffer containing divalent cations, since this will catalyze RNA autocleavage and generate new nonphosphorylated 5′-ends.

4.2.1 Equipment

Ice

Heating block (or water bath) at 37°C

Microcentrifuge

4.2.2 Reagents

100 μ*M* BioRp8 oligo
10 × T4 Polynucleotide Kinase Reaction Buffer (New England Biolabs)
10 m*M* ATP (New England Biolabs)
T4 Polynucleotide Kinase (New England Biolabs)
T4 RNA ligase 1 (New England Biolabs)
RNasin Plus (Promega)
3 *M* Sodium Acetate RNase-Free (Invitrogen)
Glycogen RNase-Free 20 mg/mL (Roche)
96% Ethanol (99% is equally good)
75% Ethanol
0.1 × TE (1 m*M* Tris–HCl pH 8, 0.1 m*M* EDTA)

4.2.3 Procedure

1. Make an initial master mix with the following composition for each RNA sample: 8 μL RNase-free water, 4 μL 10 × T4 Polynucleotide Kinase Reaction Buffer, 2 μL (80 U) RNasin Plus, 4 μL 10 m*M* ATP, 4 μL T4 RNA ligase 1, 2 μL 100 μ*M* BioRp8 oligo (it is recommended to make a little extra of the master mix).
2. Mark two microtubes with "master-mix −PNK" and "master-mix +PNK," respectively.
3. Transfer half (12 μL per RNA sample) of the initial master mix to each of the two tubes labeled "master-mix −PNK" and "master-mix +PNK".
4. Add 2 μL RNase-free water per RNA sample to the tube labeled "master-mix −PNK" and 2 μL (20 U) T4 Polynucleotide Kinase to the tube labeled "master-mix +PNK." Mix by gentle pipetting.
5. Label two microtubes per RNA sample with sample name as well as −PNK and +PNK, respectively.
6. Aliquot 14 μL "master-mix −PNK" to each of the microtubes labeled "−PNK" and 14 μL "master-mix +PNK" to each microtube labeled "+PNK."
7. Add 6 μL XRN-1-treated RNA sample (from Section 4.1) to each of the microtubes with the corresponding sample name (i.e., two tubes per RNA sample). Mix by gentle pipetting.
8. Centrifuge briefly and incubate 3 h at 37°C.
9. Add 70 μL RNase-free water, 10 μL 3 *M* Sodium Acetate, and 0.5 μL Glycogen (20 mg/mL) to each tube. Vortex and quickly add 300 μL 96% ethanol. Vortex again and store at −80°C overnight.

10. Centrifuge at maximum speed for 45 min at 4°C, then remove supernatant.
11. Wash the pellet once in 900 μL cold 75% ethanol and suspend in 20 μL 0.1 × TE. Store samples at −80°C or pass directly to the reverse transcription step (Section 4.3).

4.3 Reverse Transcription (Same as Section 2.3)

The reverse transcription step in the nEMOTE protocol is identical to the reverse transcription step of the EMOTE protocol (see Section 2.3).

4.4 PCR Amplification and Addition of EMOTE Barcodes and Illumina Adaptors (Same as Section 2.4)

The PCR amplification and addition of EMOTE barcodes and Illumina adaptors step in the nEMOTE protocol are identical to the equivalent step of the EMOTE protocol (see Section 2.4).

4.5 Adjusting Relative DNA Concentrations, Concentration, and Size-Selection of PCR Products (Same as Section 2.5)

The Adjusting relative DNA concentrations, concentration, and size-selection of PCR products step in the nEMOTE protocol are identical to the equivalent step of the EMOTE protocol (see Section 2.5).

5. TIPS AND ADVICE ON THE ANALYSES OF EMOTE DATA

The main focus of this chapter is to describe the exact protocols used to obtain data with the EMOTE, TSS-EMOTE, and nEMOTE protocols (Sections 2, 3, and 4). Since these protocols can be used to answer a wide variety of biological questions in an equally large variety of settings, it is difficult to provide an exact method for analyses of the EMOTE data. However, this section will nevertheless attempt to provide helpful hints and advice about the approaches it is possible to take when analyzing the data.

5.1 What Does a Read (i.e., the Raw Data) Look Like?

The data provided by the Illumina sequencing are in the form of a ".fastq" file, with the sequence of each read and corresponding information about the quality of the sequencing read.

When using the oligos and primers described in the protocols above, an EMOTE read that has passed all quality control steps will have the organization shown in Fig. 2. Since all quality control criteria have to be fulfilled before a sequencing read is considered "valid," it is in principle immaterial in which order the criteria are checked. However, the order that was used to design the EMOTE-conv software pipeline will be presented here (Yasrebi & Redder, 2015): (1) The first quality control step is to verify the existence of the recognition sequence (15 nt), starting at the sixth nucleotide of the read. (2) Then it is verified that nucleotide 2–5 of the read corresponds to an EMOTE barcode used in the project. (3) The presence of a correct control sequence at nucleotide 28–32 of the read is then confirmed. (4) The absence of "T" in nucleotides 21–27 is then verified (see Section 6.1.3). (5) Finally, the reads where the mapping sequences (nucleotides 33–50) have ambiguous base calling are also removed.

5.2 How to Map the Reads Onto the Genome

As can be seen in Fig. 2, the first section of the Illumina read (from nucleotide 1–32) does not originate from the examined organism. It is therefore not possible to map the entire read directly onto the genome with a standard short read aligner (for example, Langmead, Trapnell, Pop, & Salzberg, 2009). The mapping sequence first needs to be separated from the rest of the Illumina read, but in such a way that information about the EMOTE barcode and UMI (Unique Molecular Identifier) is not lost. Once each read has been mapped onto the genome, it should be assigned to the RNA sample where it originated (via the EMOTE barcode) and corrected for PCR amplification biases (via the UMI). This somewhat complex process has been simplified to a series of perl scripts in the software pipeline EMOTE-conv (Yasrebi & Redder, 2015). The EMOTE-conv pipeline will generate four data tables from the .fastq file (in addition to two tables with

nBBBB<u>CGGCACCAACCGAGG</u>vvvvvvv<u>ACAGA</u>555555555555555555

Fig. 2 The layout of a 50-nucleotide Illumina read from an EMOTE experiment. n signifies a random nucleotide (see Section 6.3.2). BBBB indicates the four-nucleotide EMOTE barcode (see Section 6.3.3). The recognition sequence has been *single-underlined*. The seven "v" nucleotides indicate the UMI sequence (see Section 6.1.3). The control sequence (see Section 6.1.4) has been *double-underlined*. The *row* of the number 5 signifies the "mapping sequence," i.e., the nucleotide sequence of the 5′-end of the original RNA molecule, before ligation to BioRp8.

information about the statistics of the sorting and mapping). Two of the four tables are for the positive strand of the genome (UnambPosTable and AmbPosTable) and two tables for the negative strand (UnambNegTable and AmbNegTable), where positive and negative strands are defined by the genome reference sequence. The separation of the two strands into individual tables is based on the fact that reads that align with positive strand cannot possibly originate from the same RNA molecule as a read that aligns with the negative strand.

Most of the mapping sequences will map unambiguously (i.e., only a single time) onto the reference genome, but some mapping sequences will map to two or more locations on the reference genome (for example, transcripts from multicopy transposons). All reads with unambiguously aligned mapping sequences will contribute to the UnambPosTable or UnambNegTable and all reads with mapping sequences that align to two or more places on the genome contribute to the AmbPosTable and AmbNegTable, depending on whether the mapping sequence aligns to the positive or negative strand.

So, what data are available in the EMOTE-conv tables?

Each row in the EMOTE-conv tables represents a genomic position where a 5′-nucleotide of an RNA was detected in the assay (Fig. 3). In other words, if, for example, the mapping sequence ACTTTATCGTTAAAACGT aligns perfectly with nucleotides 2432–2450 in the genomic sequence, then an entry will be generated with the row-name "2432," since the nucleotide at the very 5′-end of the original RNA corresponds to nucleotide 2432 of the reference genome. Genomic positions where no corresponding 5′-ends were detected are not included in the table (since these would simply be empty rows).

Each EMOTE barcode in the data is assigned two columns in the EMOTE-conv tables (Fig. 3). One column showing how many individual reads with the barcode in question align with their 5′ nucleotide of the mapping sequence to the exact position on the genome indicated by the row name (see previous paragraph). A second column (identified by "_Q" after the name) indicates the number of different UMI sequences found among the mapping reads, once again applied to the specific genomic position found in the row name (the number of different UMIs will always be equal or smaller than the number of reads). For more detailed information about EMOTE-conv, please refer to the help file in the EMOTE-conv package (http://sourceforge.net/projects/emoteconv) (Redder, 2015; Yasrebi & Redder, 2015).

Genomic_pos	Strand	5′_base	D6A	D6B	D6C	D6A_Q	D6B_Q	D6C_Q	GeneSymbol_startPos_endPos_strand	GeneAlias_desc
312	Plus	G	0	1	0	0	1	0		
313	Plus	G	68	185	407	35	76	128		
314	Plus	A	7	1	7	3	1	4		
315	Plus	U	0	0	2	0	0	2		
316	Plus	A	5	9	10	3	2	4		
320	Plus	C	0	1	0	0	1	0		
333	Plus	A	0	0	6	0	0	1		
344	Plus	A	0	1	0	0	1	0		
467	Plus	C	1	0	0	1	0	0		
483	Plus	U	0	1	0	0	1	0		
487	Plus	U	0	1	0	0	1	0		
490	Plus	A	0	0	2	0	0	1		
494	Plus	U	0	1	0	0	1	0		
496	Plus	U	0	4	0	0	1	0		
499	Plus	G	0	0	1	0	0	1		
507	Plus	G	0	0	4	0	0	1		
567	Plus	A	0	2	0	0	1	0	dnaA_517_1878_plus	SA0001_chromosome replication initiator DnaA
649	Plus	A	0	0	3	0	0	2	dnaA_517_1878_plus	SA0001_chromosome replication initiator DnaA
684	Plus	A	11	5	13	5	3	4	dnaA_517_1878_plus	SA0001_chromosome replication initiator DnaA
685	Plus	U	0	3	0	0	1	0	dnaA_517_1878_plus	SA0001_chromosome replication initiator DnaA

Fig. 3 An example of an EMOTE-conv table. Genomic_pos: Genomic position of the first nucleotide in the detected RNA molecule (i.e., the first nucleotide of the mapping sequence). Strand: indicates which strand the RNA molecule corresponds to. 5′_base: The nucleotide at the 5′-end of the RNA (A, C, G, or U). Three samples were included in this example, using oligos D6A, D6B, and D6C (see Sections 2.4 and 6.3). Columns D6A, D6B, and D6C show the number of reads that correspond to that particular genomic position, for each of the three samples. Columns D6A_Q, D6B_Q, and D6C_Q show the number of unique UMI sequences found within the reads that correspond to that particular genomic position, for each of the three samples. GeneSymbol_startPos_endPos_strand and GeneAlias_desc are columns dedicated to show the annotated features at each of the genomic positions included in the table (annotations are taken from the Genbank .gff file of the reference genome, and are not dependent on the EMOTE data) (Yasrebi & Redder, 2015). The data shown are from an EMOTE experiment, but only the *rows* corresponding to genomic positions from 300 to 700 from the positive strand are included in this figure (i.e., from the

5.3 Interpreting Data From EMOTE, TSS-EMOTE, and nEMOTE Experiments

Interpretation of data generated by the EMOTE protocol (Section 2) is the most straight forward of the three protocols described in this chapter. For each position on the reference genome, each Illumina read with a unique UMI (see Section 6.1.3) corresponds to a detected RNA molecule with that particular 5′-end. Additionally, in order to be detected with the EMOTE protocol, the original RNA 5′-end must have been monophosphorylated. Likewise, if no 5′-ends are detected, then either RNA molecules with such a monophosphorylated 5′-end are below detection level in the RNA sample, or some kind of bias in the protocol (most probably the ligation step) have prevented its detection. However, such bias should be identical between different RNA samples originating from isogenic (or almost isogenic) strains, since a given 5′-end, which maps to the same nucleotide, is identical in its nucleotide sequence, even though it originates from different samples.

For example, if it is possible to detect a 5′-end in RNA from a wild-type strain, but the same 5′-end is undetectable in RNA from an endoribonuclease mutant, then it is likely that the mutated endoribonuclease is responsible for generating the 5′-end in question. Naturally, the likelihood of this interpretation being true goes up the more individual RNA molecules are detected in the wild-type RNA (see also Khemici, Prados, Linder, & Redder, 2015; Redder, 2015).

In TSS-EMOTE, each detected 5′-end still corresponds to an actual RNA molecule with such a 5′-end (when using the UMI to correct for PCR bias). However, the data from TSS-EMOTE add the additional complexity that the detection of a 5′-end not necessarily means that it

---

UnambPosTable). This includes a transcription start site at genomic position 313 (with minor RNA polymerase initiation sites at positions 314 and 316), which is detectable by EMOTE due to the action of cellular pyrophosphatases. The *dnaA* coding sequence was annotated to start at genomic position 517 (see the *two columns* to the *right*). Genomic position 684 appears to either be a site for endoribonucleolytic cleavage or possibly a second transcription start site. The data from the remaining genomic positions are most likely background noise, but further statistical analyses may show that some of these signals are significant. *Data were adapted from Khemici, V., Prados, J., Linder, P., & Redder, P. (2015). Decay-initiating endoribonucleolytic cleavage by RNase Y is kept under tight control via sequence preference and sub-cellular localisation.* PLoS Genetics, 11*(10), e1005577. https://doi.org/10.1371/journal.pgen.1005577.*

was triphosphorylated in the original RNA sample, due to the inherently incomplete removal of monophosphorylated RNAs in the XRN-1 digestion (Section 3.1). Consequently, TSS-EMOTE samples are split into two pools after the XRN-1 treatment, and ligated with and without the presence of RppH, respectively. The underlying idea is that the data from ligation without RppH serve as a "baseline" to indicate how much monophosphorylated RNA remains undigested by XRN-1. In our analyses we have employed two different strategies to use this "baseline" information. The simplest strategy is to require that the baseline is at zero for a given position, otherwise it is disregarded for further analyses. Such a strategy is easy to implement, but will naturally eliminate all RNAs with 5′-ends that can be detected even as a single copy in the "-RppH" data, regardless of whether the "+RppH" data contain thousands of detected molecules with an identical 5′-end. A second, and more sophisticated, strategy is therefore to use beta-binomial statistics to determine the natural baseline noise in the data and calculate the probability that the signal from the "+RppH" data is significantly stronger than the signal from the "−RppH" data, for the same genomic position. Such a beta-binomial analysis was, for example, employed to determine the transcription start sites in a transcriptome-wide TSS-EMOTE assay of four different bacteria (Prados et al., 2016).

nEMOTE has until now mainly been used to map cleavage sites of endoribonucleases that cleave between the phosphate group and the 5′ OH of the RNA (Kirkpatrick et al., 2016). For such an analysis, it is necessary to combine a strategy from TSS-EMOTE to ensure that the detected 5′-end is nonphosphorylated (and not a monophosphorylated RNA that survived the XRN-1 treatment) with a comparison between a wild-type and an endoribonuclease mutant, similar to the analysis described above for EMOTE data. Analyses of nEMOTE data are therefore the most complicated of the three.

6. ADAPTING THE EMOTE OLIGOS TO SPECIFIC NEEDS

The EMOTE, TSS-EMOTE, and nEMOTE protocols were each developed to address specific biological questions (Khemici et al., 2015; Kirkpatrick et al., 2016; Linder, Lemeille, & Redder, 2014; Prados et al., 2016). However, these biological questions will almost certainly differ, to a smaller or larger extent, from the questions that will be asked in the future, and it is therefore important to be able to adjust and modify the EMOTE, TSS-EMOTE, and nEMOTE to fit the needs of a particular experiment. The following sections explain the design features of each of the oligos, in order to help the reader to adapt the oligo sequences to their specific needs.

6.1 BioRp8

BioRp8 oligo: Biotin-dG-CGGCACCAACCGAGGVVVVVVVACAGA

T4 RNA ligase 1 is able to ligate an RNA 3′ OH to a single stranded 5′ monophosphorylated polynucleotide (RNA or DNA). BioRp8 therefore needs to be an RNA oligo which unfortunately increases its price and decreases its "shelf-life." It is possible that a few ribonucleotides at the 3′-end of BioRp8 is enough to permit ligation and that the rest of the oligo can be made of deoxyribonucleotides, but we have not tested this.

6.1.1 The 5′ Blocking Biotin

Biotin-dG-CGGCACCAACCGAGGVVVVVVVACAGA

The 5′-end of the oligo blocked by a biotin-deoxyguanine. This is necessary for the nEMOTE protocol, where a free 5′ OH would be phosphorylated by the Polynucleotide Kinase, resulting in ligations between oligos to form oligo-concatemers. This 5′ block does not need to be biotin-deoxyguanine, but can be anything the protects the 5′ OH group. The 5′ block is not needed for the EMOTE and TSS-EMOTE protocols.

6.1.2 The Recognition Sequence

Biotin-dG-CGGCACCAACCGAGGVVVVVVVACAGA

The 15 ribonucleotides CGGCACCAACCGAGG constitute the recognition sequence, which serves a dual purpose. (i) it is used for hybridization with the D6A, D6B, etc., primers (Section 6.1.3) during the PCR amplification (see Section 2.4). (ii) it is also used as the first sorting criteria for analyses of the EMOTE reads (i.e., if a read does not contain the sequence, then it is rejected). The recognition sequence can be changed to whatever sequence is convenient, if the following guidelines are followed: (a) Avoid the use of uridines (see Section 6.2.3). (b) Maintain a high melting temperature of the duplex with the D6A, D6B, etc., primers to ensure specificity of the PCR amplification. (c) Avoid palindromic sequences to prevent hairpin formation, both within the recognition sequence and in combination with the control sequence (see below). (d) If possible avoid that the recognition sequence can be found in the genome of the examined organism, or at the very least choose a sequence which occurs rarely overall and not at all in the ribosomal RNAs.

6.1.3 The Unique Molecular Identifier

Biotin-dG-CGGCACCAACCGAGGVVVVVVVACAGA

Each of the seven ribonucleotides VVVVVVV corresponds to a randomly incorporated G, C, or A ribonucleotide, which creates a UMI (a sort of "social security number") for each BioRp8 molecule. The resulting 2187 potential combinations are obviously not enough to truly uniquely identify each of the multitude of BioRp8 molecules used in an EMOTE experiment, but the diversity generated is enough for the vast majority of EMOTE applications (see Section 5). If additional diversity is needed, then BioRp8 can simply be redesigned with a longer stretch of randomly incorporated ribonucleotides. A key point in the design of the UMI is that no uridines are included (same rule as for the recognition sequence; Section 6.1.2).

6.1.4 The Control Sequence

Biotin-dG-CGGCACCAACCGAGGVVVVVVVACAGA

During the PCR amplification in step 2.4, the D6A, D6B, etc., primers ideally only hybridize to the complementary sequence of the recognition sequence (see Section 6.1.2). However, since the template for the PCR is randomly primed cDNA, its diversity is very high, and the D6A, D6B, etc., primers may occasionally hybridize elsewhere, on the interior sequence of a cDNA. The result of such a "false" priming is a double-stranded DNA molecule with D6A, D6B, etc., sequence at one end and an Illumina PE 2.0 adaptor sequence at the other (see also Section 6.5), providing a perfect template for further amplification in the following cycles of the PCR. These "false" PCR products will have the correct Illumina adaptors at either end and are perfectly valid substrates for Illumina sequencing. The resulting "false" reads will contain the recognition sequence (since this sequence is included in the D6A, D6B, etc., primers), and are therefore not be rejected by the first step of in silico sorting (see Section 5). Two factors permit the elimination of such "false" reads: (i) if the seven nucleotides immediately downstream of the recognition sequence contain any thymidines, then it cannot be a genuine ligation event, since the UMI does not contain any uridine residues (see Section 6.1.3). (ii) nucleotides 8–13 downstream of the recognition sequence should be ACAGA; if this is not the case, then the DNA molecule is not an amplification product of BioRp8 ligated to an RNA. Together, these two quality criteria greatly reduce the risks of including "false" positives in the subsequent analyses.

The control sequence should not contain any uridines, and ideally have a purine at its 3′ terminus, since this will augment the ligation frequency of

T4 RNA ligase 1. The longer the control sequence the less likely it is that "false" PCR products by coincidence contain an identical sequence. However, a longer control sequence also means a more expensive BioRp8 oligo and a shorter mapping sequence (see Section 5), so five ribonucleotides was chosen as a reasonable compromise for use in our laboratory (theoretically it should reduce "false" positives 1024-fold).

6.2 DROAA

DROAA primer: GGCATTCCTGCTGAACCGCTCTTCCGATCT NNNNNNNNAA

DROAA is a DNA oligo used to prime the reverse transcription reaction in Sections 2.3, 3.3, and 4.3.

6.2.1 Part of the Illumina Adaptor

GGCATTCCTGCTGAACCGCTCTTCCGATCT NNNNNNNNAA

The 30 nucleotides at the 5′ terminus of DROAA is part of the Illumina adaptor sequence. The full Illumina adaptor sequence is only added when the B-PE-PCR20 primer hybridizes to these 30 nucleotides and adds the final components of the PE 2.0 Illumina adaptor during the PCR reaction (see Sections 2.4, 3.4, and 4.4).

6.2.2 Random Sequence for Random Priming of the Reverse Transcription

GGCATTCCTGCTGAACCGCTCTTCCGATCTNNNNNNNNAA

The eight randomly incorporated nucleotides (i.e., a random octamer) permit the DROAA oligo to hybridize randomly all over the transcriptome. The exact length of the random stretch of nucleotides is not highly important, and we have tested 7, 8, and 9 random nucleotides with success (Peter Redder, unpublished).

6.2.3 Two 3′ Adenosines to Avoid Priming Inside the BioRp8 Oligo

GGCATTCCTGCTGAACCGCTCTTCCGATCTNNNNNNNNAA

The two adenosines at the 3′-end of DROAA prevent hybridization between DROAA and BioRp8, since the latter contains no uridines for base-pairing. BioRp8 is used in large excess in the ligation step (Sections 2.2, 3.2, and 4.2), which is carried over into the reverse transcription reaction (Sections 2.3, 3.3, and 4.3) where it is important to minimize the use of BioRp8 as template and maximize the use of sample RNA as template. A single adenosine is in principle enough to prevent priming on BioRp8, but we prefer to use two adenosines to increase specificity as much as possible. It is clear that the inclusion of nonrandom nucleotides at the 3′-end of the DROAA primer reduces the number of potential priming sites for the, but on average, DROAA should still be able to initiate reverse transcription every 16th position (on a transcriptome with 25% uridines).

6.3 D6A, D6B, etc.

Example D6A: CTCTTTCCCTACACGACGCTCTTCCGATCTNTA CACGGCACCAACCGAGG

The role of the D6A, D6B, etc., primers is to add the EMOTE barcodes to the individual samples, as well as a part of the Illumina PE 1.0 adaptor. A separate D6A, D6B, etc., oligo is required for each EMOTE barcode, and we usually work with 16 different barcodes in our laboratory, although more can easily be added (see Section 6.3.3). In this section we will use D6A as example to explain the details of the design, but all of the D6A, D6B, etc., primers are identical except for the EMOTE barcode.

Examples of 16 compatible four-nucleotide barcodes: TACA, GTAT, CGTC, AAGT, ACAC, GGTA, TCGG, CAAG, TTGA, GCTG, CCGA, CTCG, AGGA, ATTG, GACG, TGTT.

6.3.1 *Partial Illumina PE 1.0 Adaptor*

<u>CTCTTTCCCTACACGACGCTCTTCCGATCT</u>NTACACGGCAC CAACCGAGG

This sequence serves to add a part of the Illumina PE 1.0 adaptor, which will be extended to the full-length adaptor by the addition of A-PE-PCR10 to the last 25 cycles of the PCR reaction (see Section 2.4.3, step 4).

6.3.2 *Randomly Incorporated Nucleotide to Increase Diversity of the First Four Nucleotides of the Reads*

CTCTTTCCCTACACGACGCTCTTCCGATCT<u>N</u>TACACGGCAC CAACCGAGG

The algorithm used by the Illumina software to identify individual DNA clusters in the photographs of the flow-cell (corresponding to colored spots on the photograph) is based on a computer analysis of the first four nucleotides sequenced (i.e., the first four photographs in the series). If two DNA clusters are close to each other, or perhaps slightly overlap, then they can only be distinguished if their first four nucleotides are different. In a standard RNA (or DNA) sequencing project, the first four nucleotides will be close to random, and the risk that two DNA clusters with the same quadruplet end up close to each other is extremely low. However, the diversity of the first four nucleotides is highly reduced when sequencing EMOTE samples, since the EMOTE barcodes are found right at the beginning of the sequencing reads. Since EMOTE barcodes have defined sequences with only 16 combinations or fewer (see Section 6.3.3 for discussion of this), it is desirable to augment this diversity by adding a random nucleotide immediately before the EMOTE barcode. The four first nucleotides in each sequencing read will then be the random nucleotide followed by the three first nucleotides of the EMOTE barcode, thus increasing the diversity by four. For a project where less than four different EMOTE barcodes are used, it is advisable to increase the number of random nucleotides to two or three.

6.3.3 EMOTE Barcode Sequence

CTCTTTCCCTACACGACGCTCTTCCGATCTNTACACGGCAC CAACCGAGG

The EMOTE barcode sequence serves to identify which RNA sample each individual sequencing read belongs to, even though the PCR products have been mixed together (see Sections 2.5, 3.5, and 4.5). Each EMOTE barcode should obviously be unique; however, the length can be chosen according to the project. We have found that a length of four nucleotides permits the design of a useful range of different EMOTE barcodes, while ensuring that each EMOTE barcode has at least two nucleotides different than all other EMOTE barcodes in the project. A difference of two nucleotides prevents mistakenly assigning a sequencing read in the case of a single poorly called base (i.e., a sequencing mistake). Note that an EMOTE barcode is not the same as the index (sometimes referred to as a "barcode") used within the Illumina sequencing systems. The EMOTE barcodes are part of the sequencing read, whereas the Illumina indices have been removed before the sequencing read is delivered to the customer.

6.3.4 Recognition Sequence

CTCTTTCCCTACACGACGCTCTTCCGATCTNTACACGGCAC CAACCGAGG

The 3′ recognition sequence is used to hybridize to the cDNA molecules which contain the reverse complement of this sequence, i.e., the cDNA which has been reverse transcribed from an RNA molecule ligated to the BioRp8 oligo.

6.4 A-PE-PCR10

A-PE-PCR10: AATGATACGGCGACCACCGAGATCTACACTCTT TCCCTACACGACG

This primer is added for the final 25 cycles of the PCR amplification reaction (Sections 2.4, 3.4, and 4.4), where it serves to add the remainder of the adaptor sequence needed for Illumina sequencing. This is the adaptor sequence which is used for the "forward" sequencing in an Illumina sequencing.

A-PE-PCR10 does not contain any Illumina index sequences, but these can be included if the researcher needs it for their project. As the primer is shown here, the resulting Illumina reads will be classified by the Illumina software as "Undetermined," but this does not prevent further analyses, since only truly valid EMOTE reads will contain the recognition sequence, EMOTE barcode, and control sequence (see Section 5.1).

6.5 B-PE-PCR20

B-PE-PCR20: CAAGCAGAAGACGGCATACGAGATCGGTCTCG GCATTCCTGCTGAACCGC

This primer adds the remainder of the "reverse" adaptor sequence needed for Illumina sequencing.

B-PE-PCR20 does not contain any Illumina index sequences, but these can be included if the researcher needs it for their project.

REFERENCES

Celesnik, H., Deana, A., & Belasco, J. G. (2008). PABLO analysis of RNA: 5′-phosphorylation state and 5′-end mapping. *Methods in Enzymology*, *447*, 83–98. https://doi.org/10.1016/S0076-6879(08)02205-2.

Fromont-Racine, M., Bertrand, E., Pictet, R., & Grange, T. (1993). A highly sensitive method for mapping the 5′ termini of mRNAs. *Nucleic Acids Research*, *21*(7), 1683–1684.

Khemici, V., Prados, J., Linder, P., & Redder, P. (2015). Decay-initiating endoribonucleolytic cleavage by RNase Y is kept under tight control via sequence preference and sub-cellular localisation. *PLoS Genetics*, *11*(10), e1005577. https://doi.org/10.1371/journal.pgen.1005577.

Kirkpatrick, C. L., Martins, D., Redder, P., Frandi, A., Mignolet, J., Chapalay, J. B., et al. (2016). Growth control switch by a DNA-damage-inducible toxin–antitoxin system in *Caulobacter crescentus*. *Nature Microbiology*, *1*, 16008.

Langmead, B., Trapnell, C., Pop, M., & Salzberg, S. L. (2009). Ultrafast and memory-efficient alignment of short DNA sequences to the human genome. *Genome Biology*, *10*(3), R25. https://doi.org/10.1186/gb-2009-10-3-r25.

Linder, P., Lemeille, S., & Redder, P. (2014). Transcriptome-wide analyses of 5′-ends in RNase J mutants of a gram-positive pathogen reveal a role in RNA maturation, regulation and degradation. *PLoS Genetics*, *10*(2), e1004207. https://doi.org/10.1371/journal.pgen.1004207.

Liu, X., & Gorovsky, M. A. (1993). Mapping the 5′ and 3′ ends of *Tetrahymena thermophila* mRNAs using RNA ligase mediated amplification of cDNA ends (RLM-RACE). *Nucleic Acids Research*, *21*(21), 4954–4960.

Prados, J., Linder, P., & Redder, P. (2016). TSS-EMOTE, a refined protocol for a more complete and less biased global mapping of transcription start sites in bacterial pathogens. *BMC Genomics*, *17*, 849. https://doi.org/10.1186/s12864-016-3211-3.

Redder, P. (2015). Using EMOTE to map the exact 5′-ends of processed RNA on a transcriptome-wide scale. *Methods in Molecular Biology (Clifton, N.J.)*, *1259*, 69–85. https://doi.org/10.1007/978-1-4939-2214-7_5.

Yasrebi, H., & Redder, P. (2015). EMOTE-conv: A computational pipeline to convert exact mapping of transcriptome ends (EMOTE) data to the lists of quantified genomic positions correlated to related genomic information. *Journal of Applied Bioinformatics & Computational Biology*, *04*(02). https://doi.org/10.4172/2329-9533.1000118.

CHAPTER SIXTEEN

MS2-Affinity Purification Coupled With RNA Sequencing Approach in the Human Pathogen *Staphylococcus aureus*

David Lalaouna, Emma Desgranges, Isabelle Caldelari[1], Stefano Marzi[1]
Architecture et Réactivité de l'ARN, Université de Strasbourg, IBMC-CNRS, Strasbourg, France
[1]Corresponding authors: e-mail address: i.caldelari@ibmc-cnrs.unistra.fr; s.marzi@ibmc-cnrs.unistra.fr

Contents

Abstract

Staphylococcus aureus is a Gram-positive major human pathogen involved in a wide range of human infectious diseases (from minor skin infections to septicemia, endocarditis or toxic shock syndrome). The treatment of *S. aureus* infections is very challenging due to the emergence of multiple antibiotic-resistant isolates. The high diversity of clinical symptoms caused by *S. aureus* depends on the precise expression of numerous virulence factors and stress response pathways, which are tightly regulated at every level (transcriptional, posttranscriptional, translational, and posttranslational).

During the last two decades, it has become evident that small regulatory RNAs (sRNAs) play a major role in fast adaptive responses, mainly by targeting mRNA

Methods in Enzymology, Volume 612
ISSN 0076-6879
https://doi.org/10.1016/bs.mie.2018.08.022

translation. sRNAs act as antisense RNAs by forming noncontiguous pairings with their target mRNAs and their mechanisms of action vary according to the interaction site. To obtain a global and detailed view of the regulatory networks involved in the adaptive processes of *S. aureus*, we have adapted the MAPS approach to get individual sRNA targetomes. We also set up different strategies to validate MAPS results and establish sRNA regulatory activities. As this method has been first developed in Gram-negative bacteria, we provide here a protocol for its application in *S. aureus* and highlight underlying differences. Finally, we discuss several points that have been and could be further improved and provide a workflow file for the automatic analysis of the sequencing in Galaxy.

Highlights

- Detailed protocol for MAPS application in *S. aureus* to define sRNA targetome.
- Detailed procedures for both qualitative and quantitative data analysis.
- Analysis of advantages and drawbacks of MAPS.
- Comparison with MAPS applied to Gram-negative bacteria.
- Discussions of critical steps.
- Downloadable Galaxy Workflow for complete MAPS analysis.

1. INTRODUCTION

1.1 *Staphylococcus aureus*: A Major Human Opportunistic Pathogen

Staphylococcus aureus is an opportunistic pathogen, which is the major cause of nosocomial and community infections. This versatile bacterium can invade many tissues and then causes a wide spectrum of infections (cutaneous abscesses, endocarditis, septic shock, etc.). Moreover, the emergence of multidrug resistant strains makes urgent the exploration of new antimicrobial therapies. Its pathogenesis depends on the production of numerous wall-associated or secreted virulence factors (Dale, Doherty-Kirby, Lajoie, & Heinrichs, 2004; Goldmann & Medina, 2017; Novick & Geisinger, 2008; Oliveira, Borges, & Simoes, 2018; Penades, Chen, Quiles-Puchalt, Carpena, & Novick, 2015). They contribute to defense mechanisms against the host immune system, participate in adhesion, invasion, and dissemination in host tissues and also in the acquisition of nutrients, for example, by producing siderophores. The expression of these factors is finely regulated at several levels and requires many actors such as two-component systems, transcription factors, and regulatory RNAs.

All these macromolecules form complex networks that allow *S. aureus* to sense and respond to a wide variety of signals by coordinating the expression of virulence genes dynamically (Tomasini et al., 2014).

1.2 sRNA-Mediated Regulation in Bacteria

Small regulatory RNAs (sRNAs) have been identified and characterized in numerous bacteria, from Proteobacteria to Firmicutes (Wagner & Romby, 2015). Bacterial genomes, including the human pathogen *S. aureus*, encode a few hundred sRNAs, which are mainly encoded in the core genome, while few are localized within mobile elements, pathogenic islands or plasmids (Abu-Qatouseh et al., 2010; Beaume et al., 2010; Bohn et al., 2010; Geissmann et al., 2009; Pichon & Felden, 2005). These pleiotropic and powerful regulators are involved in every aspect of cell physiology (e.g., primary or secondary metabolism, motility, iron starvation). In *S. aureus*, sRNAs also have a crucial role in virulence (Le Pabic, Germain-Amiot, Bordeau, & Felden, 2015; Manna et al., 2018; Nitzan et al., 2015; Quereda & Cossart, 2017; Romilly et al., 2014), and potentially in antibiotic resistance (Eyraud, Tattevin, Chabelskaya, & Felden, 2014; Howden et al., 2013; Qin, McCausland, Cheung, & Otto, 2016).

Bacterial sRNAs are commonly short (ranging from 50 to 300 nt), non-coding RNA molecules. Some exceptions exist, notably in *S. aureus*, where large coding sRNAs are found (e.g., RNAIII, a 516 nt long sRNA coding for the δ-hemolysin) (Novick et al., 1993; Vanderpool, Balasubramanian, & Lloyd, 2011). Most sRNAs act in *trans* via short, and imperfect base pairings in the vicinity of the ribosome binding site (RBS) and thus interfere with the translational initiation of multiple target mRNAs. Many *S. aureus* sRNAs carry a consensus UCCC motif hybridizing with the Shine-Dalgarno sequence of target mRNAs (Bohn et al., 2010; Geissmann et al., 2009; Romilly et al., 2014). Remarkably, *S. aureus* sRNAs often bind to a second site located either in the 5′ untranslated region or in the coding region of the target mRNA. As a consequence, these binding characteristics complicate the identification of their target mRNAs using in silico predictions.

Recent developments in high-throughput RNA sequencing made it possible to draw a comprehensive interaction map of sRNAs (Saliba, C Santos, & Vogel, 2017). Commonly, an RNA binding protein such as the RNA chaperone Hfq is used as bait to copurify sRNA:mRNA duplexes. More recently, Melamed et al. (2016) improved this approach by ligation of sRNA and mRNAs pairing sites (RIL-seq). Although Hfq is involved in several aspects of sRNA-dependent regulation in most studied organisms

(Kavita, de Mets, & Gottesman, 2018; Santiago-Frangos & Woodson, 2018), its role is rather limited in *S. aureus* (Bohn, Rigoulay, & Bouloc, 2007; Liu et al., 2010). As an alternative, MS2-affinity purification coupled with RNA-Seq (MAPS) approach uses a MS2-tagged sRNA to copurify its target mRNAs (Lalaouna, Carrier, et al., 2015; Lalaouna, Prevost, Eyraud, & Masse, 2017). As MAPS has already revealed the targetome of several sRNAs in *Escherichia coli*, we have decided to adapt it to the study of *S. aureus* sRNAs (Tomasini et al., 2017).

2. MAPS PROTOCOL APPLIED TO *S. aureus*

The principle of MAPS is to isolate complexes formed in vivo between sRNA and its target mRNAs using an MS2 tag fused to the 5′ end of sRNA. This tag is specifically recognized by the phage MS2 capsid protein (MS2-CP) fused to MBP (maltose binding protein) which is itself bound to an amylose resin to specifically retain complexes containing the modified MS2-sRNA. By coupling this affinity chromatography to high-throughput sequencing, RNAs-bound to a specific sRNA can be characterized in depth.

2.1 Plasmids, Strains, and Growth Conditions

Experiments are performed in the model strain HG001, a derivative of the RN1 strain (NTC8325) with a restored *rsb*U gene, an activator of SigB, and in sRNA deletion mutants obtained by homologous recombination. The genome of HG001 strain has been recently sequenced (Caldelari et al., 2017) (GenBank accession number CP018205.1). Staphylococcal strains are grown routinely in BHI (brain heart infusion) broth at 37°C under agitation. MS2-sRNA plasmids are prepared using pCN51 as template shuttle vector (Charpentier et al., 2004). The MS2 tag alone or MS2-tag fused to the 5′ end of sRNAs is cloned under the control of the agr-dependent P3 promoter (replacing the cadmium inducible promoter). This allows an accumulation of sRNAs at the late exponential phase of growth, which correspond to most sRNAs natural expression (Geissmann et al., 2009). All plasmids are constructed in *E. coli* strain DC10B (Monk, Shah, Xu, Tan, & Foster, 2012). Plasmids extracted from *E. coli* are directly transformed into *S. aureus* by electroporation.

2.2 Preparation of Crude Bacterial Extracts

For preparation of *S. aureus* crude extracts expressing MS2-RNAs or MS2 alone, cells are grown in 50 mL BHI medium (Sigma) containing 10 μg/mL of erythromycin to an OD_{600} of 5 (~6 h of culture at 37°C), immediately chilled on ice, and then pelleted by centrifugation (3,273 × *g*, 15 min, 4° C). Supernatants are carefully removed, and the pellets are resuspended in 4 mL ice-cold Buffer A (20 m*M* Tris–HCl pH 8, 150 m*M* KCl, 1 m*M* $MgCl_2$, 1 m*M* DTT). Cells are then transferred into Lysis Matrix B Tubes (MP Biomedicals) and lysed with FastPrep apparatus (MP Biomedicals). Samples are centrifuged 15 min (15,700 × *g*, 4°C). The supernatants (about 4 mL) are carefully removed and placed in new ice-cold tubes that are maintained at 4°C until subsequent use for the affinity chromatography or for the total RNA extraction by phenol–chloroform (Fig. 1).

2.3 Purification of MS2-MBP Protein

The MS2 coat protein fused to maltose binding protein (MS2-MBP) is expressed in *E. coli* BL21DE (+ pLysS) containing the pMBP-MS2 plasmid (Addgene #65104) (Macias, Bragulat, Tardiff, & Vilardell, 2008) by isopropyl-1-thio-β-D-galactopyranoside (IPTG) induction and then purified as followed. After 3 h induction with 1 m*M* IPTG, bacterial cells are lysed by sonication in Buffer AB1 (20 m*M* Tris–HCl pH 8, 200 m*M* KCl, 1 m*M* EDTA) supplemented with protease inhibitors and DNase I. Cleared lysate is loaded on an AB1-equilibrated MBPTrap HP column (GE Healthcare). Washes are performed first with Buffer AB1 and then Buffer AB2 (20 m*M* Tris–HCl pH 8, 20 m*M* KCl, 1 m*M* EDTA). For elution, 10 m*M* of maltose are added to Buffer AB2. Fractions containing MS2-MBP are pooled and loaded on a monoQ anion exchange column (GE Healthcare, monoQ 5/50 GL) equilibrated in Buffer AB2 to remove unspecifically bound nucleic acids. MS2-MBP is then eluted with a gradient of Buffer B (20 m*M* Tris–HCl pH 8, 1 *M* KCl, 1 m*M* EDTA) and dialyzed in Buffer AB2.

2.4 Affinity Purification of Regulatory Complexes Containing MS2-Tagged RNA

Affinity chromatography purification steps are performed at 4°C. For each cell extract (control or MS2-sRNA), 300 μL of amylose resin (New England BioLabs) are packed into a Poly-Prep empty chromatography column

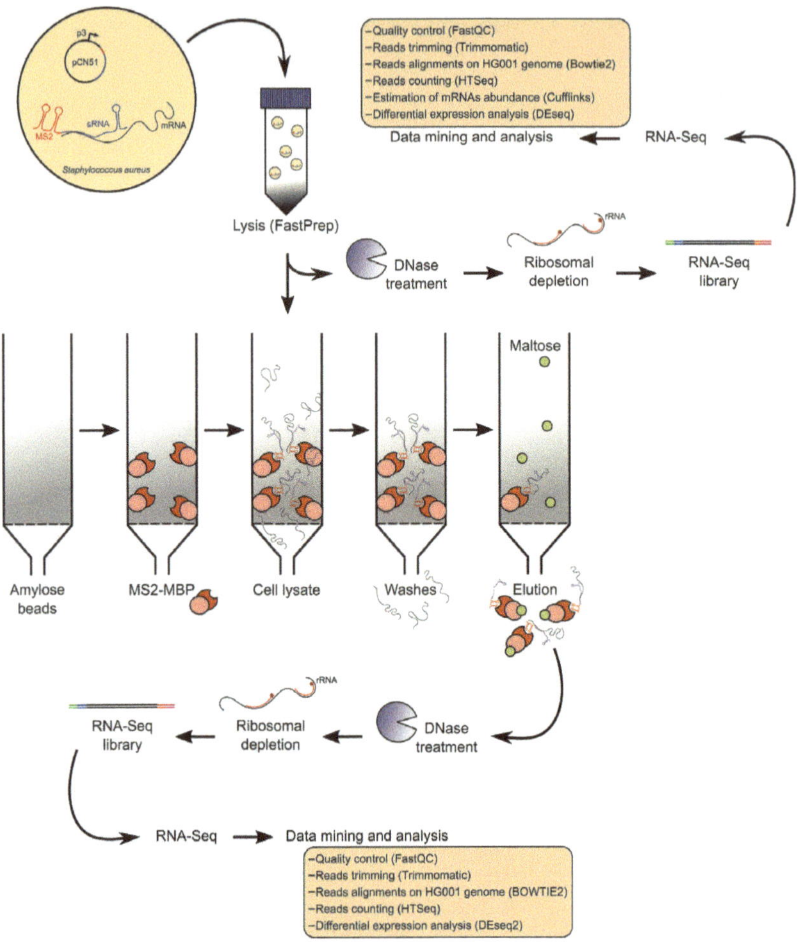

Fig. 1 Schematic representation of MAPS approach in *Staphylococcus aureus*. The MS2-sRNA construct (or MS2 alone) is expressed in vivo (under the control of a quorum sensing-dependent P3 promoter). Bacterial cells are then lysed with FastPrep apparatus. Lysates are loaded onto a column where the MS2-MBP (maltose binding protein) fusion protein is immobilized (amylose resin). The MS2 protein binds with high specificity and affinity to the MS2 aptamer and, therefore, enables the copurification of RNAs interacting with the MS2-sRNA. Several washes allow removing unspecifically bound RNAs (column, beads, MS2-MBP). RNA:RNA duplexes are eluted by the addition of maltose. After DNase I treatment and ribosomal depletion, cDNA libraries are build (ScriptSeq Complete Kit (Bacteria), Illumina) and sequenced (Mi-Seq, Illumina). *Data are analyzed using Galaxy Project (Afgan, E., Baker, D., van den Beek, M., Blankenberg, D., Bouvier, D., Cech, M., et al. (2016). The Galaxy platform for accessible, reproducible and collaborative biomedical analyses: 2016 Update.* Nucleic Acids Research, 44*(W1), W3–W10. doi:10.1093/nar/gkw343).*

(Biorad). The columns containing the amylose resin are washed three times with 10 mL Buffer A, and 3600 pmol of MS2-MBP (diluted in 6 mL of Buffer A) are then immobilized. The crude bacterial extracts (4 mL) are directly loaded, followed by three washes with 10 mL Buffer A. Finally, RNAs are eluted with 1 mL of Buffer E (20 m*M* Tris–HCl pH 8, 150 m*M* KCl, 1 m*M* $MgCl_2$, 1 m*M* DTT, 0.1% Triton X-100, 12 m*M* maltose). Eluted RNAs are extracted with phenol:chloroform:isoamylalcohol (25:24:1 (v/v), Roth) and precipitated with three volumes of cold absolute ethanol. MS2-affinity purifications are performed in duplicates.

2.5 Transcriptomics and RNA Sequencing

In order to take into account the expression levels of mRNA candidates, total RNA is purified from bacterial cell extract (Fig. 1) by phenol:chloroform:isoamylalcohol (25:24:1 (v/v), Roth) and precipitated with three volumes of cold absolute. Then both total RNA and MS2-eluted RNAs are treated with DNase I (0.1U/μL) for 1 h at 37°C to remove genomic DNA. RNAs are then purified by phenol:chloroform:isoamylalcohol and precipitated with ethanol in the presence of sodium acetate 0.3 *M* at pH 5. RNA pellets are resuspended in milliQ water. RNAs are quantified by Qubit (Life Technologies) and their integrity is assessed with Bioanalyzer (Agilent Technologies). RNAs were treated to deplete abundant rRNAs, and the cDNA libraries are prepared using the ScriptSeq complete kit (bacteria) from Illumina.

2.6 RNA Sequencing Analysis

The libraries are sequenced with Illumina Mi-seq using a V4 chemistry sequencing kit (Illumina), which preserves the information about the orientation of transcripts and produces reads of 150 nt. The reads are then analyzed using Galaxy (Afgan et al., 2016). The workflow we have developed can be downloaded from the Supplementary Material in the online version at https://doi.org/10.1016/bs.mie.2018.08.022 session and used directly in Galaxy. The utilities which are needed, available in most of the Galaxy instances, are the following: FastQC (Galaxy Version 0.62), Trimmomatic (Galaxy Version 0.32.3), Bowtie2 (Galaxy Version 2.2.6), Flagstat (Galaxy Version 1.0.1), HTSeq-count (Galaxy Version 0.6.1), and SARTools DEseq2 (Galaxy Version 0.99.2). Briefly, after having assessed the quality of the reads with FastQC, they are processed to remove adapter sequences and bad quality reads using Trimmomatic (Bolger, Lohse, & Usadel, 2014). For this step, the parameters are given by (i) ILLUMINACLIP, which allows a maximum of two mismatches between the read and the TruSeq3 adapter sequences; (ii) SLIDINGWINDOW, which scans from the 5′ end of the read and trims once the average Illumina quality score of 4 consecutive bases falls below the 20 threshold; (iii) MINLEN, which drops the reads shorter than 20 nucleotides. The reads are then aligned on the HG001 genome (GenBank accession number CP018205.1) using Bowtie2 (Langmead & Salzberg, 2012). The alignment is done using the option very-sensitive-local. In this mode, Bowtie2 uses a seed length of 20 with 0 mismatch allowed, and might "trim" or "clip" some read characters from one or both ends of the alignment if doing so maximizes the alignment score. In this way, reads belonging to MS2-sRNA can also be recovered. To obtain statistics on the alignment including the percentage of mapped reads we use Flagstat (Li et al., 2009). Finally, the number of reads mapping to each annotated feature is counted with HTSeq (Anders, Pyl, & Huber, 2015) using the interception nonempty protocol. To estimate the enrichment values for the MAPS experiment or the differential expression analysis for the transcriptomic experiment, we use DEseq2 (Varet, Brillet-Gueguen, Coppee, & Dillies, 2016). DEseq2 uses one of the most common approaches in comparative analysis of transcriptomics data. It assumes that the count data follow a negative binomial distribution testing the null hypothesis that the logarithmic fold change (LFC) between MS2-sRNA and MS2 for the recovering of a given transcript is exactly zero, i.e., that the target mRNA is not at all enriched by the purification via the sRNA. The statistical analysis process includes data normalization, graphical exploration of raw and normalized data, estimation of the dispersion of the data and test for differential enrichment for each feature between the MS2 and MS2-sRNA samples, raw *P*-value adjustment, and export of lists of features having a significant differential enrichment (threshold *P*-value = 0.05; fold change threshold = 2) between the conditions.

Since there is a great variation in the expression of the different transcripts purified with the MS2-sRNA bait, the rank obtained with the LFCs does not necessarily reflect the affinities between the sRNA and its targets. Thus, we also estimate the cellular abundance of each of the transcripts of the targetome. In combination with the enrichment values of the MAPS analysis, the in vivo affinities (designated as affinity rank) between sRNA and its targets can be then evaluated. Affinity rank is obtained by the ratios between the counted fragments (normalized for the gene length and the deepness of the sequencing—"FPKM") of the sRNA and those of each of mRNAs issued from the MAPS (Tomasini et al., 2017).

2.7 Validation of sRNA Target and Mechanism of Regulation

Putative target mRNAs identified by MAPS are subsequently validated using methods described in Fig. 2 (also reviewed by Jagodnik, Brosse, Le Lam, Chiaruttini, & Guillier, 2017).

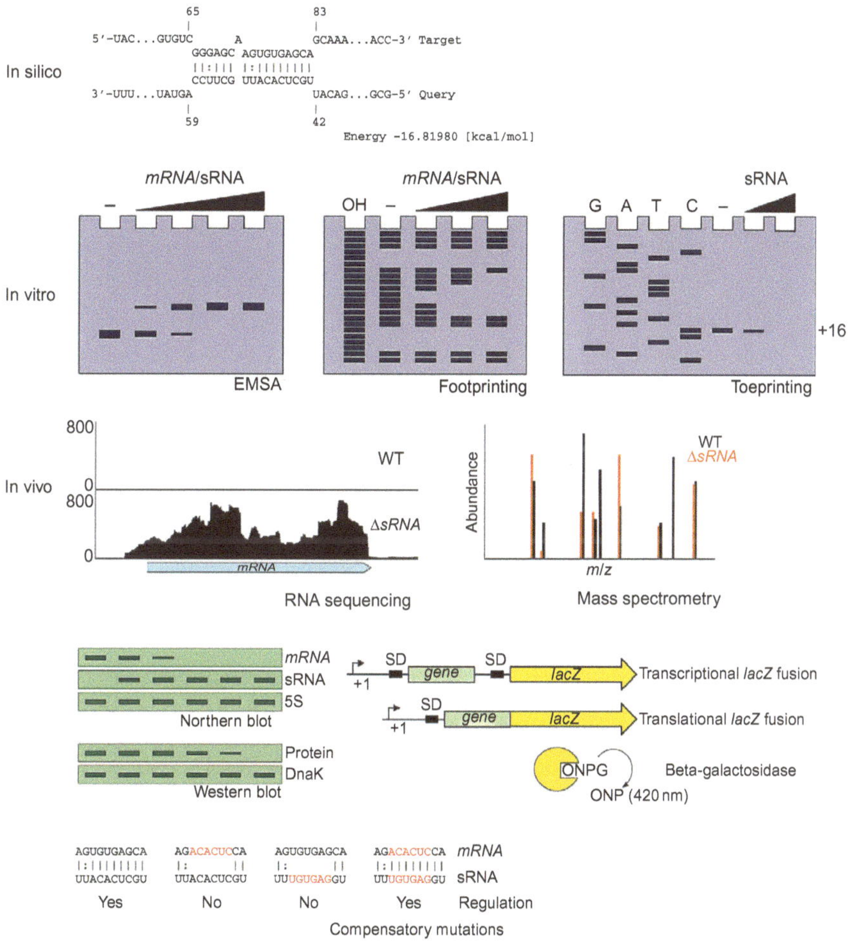

Fig. 2 Overview of methods commonly used to validate mRNA candidates identified by MAPS.

Gel retardation assays are first used to visualize direct interactions between RNAs. Briefly 5′ end radiolabelled sRNA and cold mRNAs are renatured separately at room temperature. Then, increasing concentrations of cold mRNA are mixed with labeled sRNA in a buffer containing 20 m*M* Tris–HCl pH 7.5, 10 m*M* $MgCl_2$, and 60 m*M* KCl. Complex formation is carried out at 37°C for 15 min. After incubation, 10 μL of glycerol blue is added and the samples are loaded on a 6% polyacrylamide gel under nondenaturing conditions (5 h, 300 V, 4°C). To characterize sRNAs regulatory mechanism and notably their impact on target mRNA translation, we usually perform toeprinting assays to monitor the formation of a simplified initiation ribosomal complex (Fechter et al., 2009), but also in vitro translation and in vivo β-galactosidase assays (using *lacZ* translational fusion in cells expressing or not the sRNA of interest) (see Tomasini et al., 2017). Moreover, more precise interactions can be mapped by footprinting using enzymatic hydrolysis and/or chemical modification. To prove direct interaction beyond doubt, the seed region of both sRNA and target mRNA is punctually mutated. Both mutations must individually impair sRNA-dependent regulation, but they must restore sRNA activity together (compensatory mutations). This can be performed either in vitro or in vivo.

2.8 Frequently Asked Questions

2.8.1 How to Fuse the MS2 Aptamer to Your Favorite sRNA?

This is an essential question since to modify an sRNA is tantamount to walk in a minefield. Said et al. (2009) tried to add the MS2 tag at various positions, notably just before the native transcriptional terminator of the sRNA, which is often Rho-independent (a stem–loop followed by a poly-U tail). They even replaced this latter by an exogenous terminator, originating from *vrrA* sRNA gene in *Vibrio cholerae* (Song et al., 2008). Unfortunately, they noticed the transcription of RNA fragments of variable sizes due to aberrant processing and transcriptional read-through. We, and Massé's group, chose to add the MS2 sequence at the 5′-end of sRNA genes (Bronesky et al., 2018; Lalaouna, Carrier, et al., 2015; Lalaouna, Morissette, Carrier, & Masse, 2015; Lalaouna, Prevost, Laliberte, Houe, & Masse, 2018; Tomasini et al., 2017). In this case, a major risk mentioned by Said et al. (2009) is that the addition of MS2 tag could disrupt sRNA function when the seed region is located in its 5′-end (e.g., *E. coli* RybB sRNA). However, such a limit can be circumvented by the addition of a 6-nt long spacer between the MS2 aptamer and the +1 of sRNA (Lalaouna et al., 2017). This has been validated notably using RybB sRNA (Lalaouna, Carrier, et al., 2015).

It is also possible to add the MS2 tag in other regions of the sRNA, but it is important to mention that the regulatory RNAs are composed of distinct functional regions that must be preserved which are sometimes difficult to localize:

(1) A seed region, which generally initiates the recognition with the target mRNAs by sequence complementarity. Note that several *S. aureus* sRNAs carry multiple seed sequences such as RsaA (Bronesky et al., 2018) or RNAIII (Bronesky et al., 2016). A behavior already observed for some *E. coli* sRNAs like DsrA (Lalaouna & Masse, 2016) and GcvB (Coornaert, Chiaruttini, Springer, & Guillier, 2013).

(2) A possible binding site of an RNA chaperone. Even if no RNA chaperone proteins have been so far identified for sRNA regulation in *S. aureus*, we cannot exclude that specific regions could constitute a binding site for an unidentified protein. For Gram-negative (*E. coli* and *Salmonella*) the importance of RNA chaperone proteins has been amply demonstrated. This is the case for *E. coli* Hfq, which uses two anchorage points on the sRNAs (Schu, Zhang, Gottesman, & Storz, 2015). The involvement of Hfq in sRNA-dependent regulation is still controversial in *S. aureus* (Bohn et al., 2007; Liu et al., 2010) and no homologs of ProQ, another described RNA chaperone protein (Olejniczak & Storz, 2017; Smirnov, Schneider, Hor, & Vogel, 2017), have been identified.

(3) Remaining sequences which may be involved in the secondary structure of the sRNA, which is a key determinant in its stability and functions.

(4) The above-mentioned terminator, which could also be involved in RNA:RNA interactions (Susan Gottesman, 5th Meeting of Regulating with RNA in Bacteria & Archaea).

2.8.2 How to Validate MS2 Constructs?

Addition of a MS2 aptamer (43-nt long) to regulatory RNAs (ranging from 50 to 500 nucleotides) can generate certain risks notably concerning the impact of the nucleotides of the tag on sRNA folding. Thus, MS2-sRNA constructs must be carefully validated before subsequent applications. A crucial step is to determine whether the MS2 aptamer affects cellular level and the functions of chimeric sRNA (MS2-sRNA) compared to untagged sRNA (control). Classically, MS2-sRNA level is compared to sRNA control in vivo using Northern blot analysis. For its activity, even if informative, a simple in silico prediction of the sRNA folding in the presence and the absence of the MS2 aptamer is not enough. If possible, at least one previously described target mRNA should be used to verify that MS2-sRNA is able to fulfill its regulatory role in a similar manner than untagged sRNA using Northern blot analysis. MS2-sRNA construct could be also validated using in vitro translation assays on specific target mRNAs (Tomasini et al., 2017).

2.8.3 Endogenous Expression or Overexpression of MS2 Constructs?

MS2-sRNA and cognate controls are generally overexpressed from plasmids to ensure their sufficient expression and increase the yield of MS2-affinity purification (Bronesky et al., 2018; Lalaouna et al., 2018; Lalaouna, Carrier, et al., 2015; Lalaouna, Morissette, et al., 2015; Said et al., 2009; Tomasini et al., 2017). For *S. aureus* the plasmid is a modified pCN51 (quorum sensing-dependent P3 promoter) with a copy number of 20–25 copies/cell (Bronesky et al., 2018; Charpentier et al., 2004; Tomasini et al., 2017). This is similar to the plasmid used in *S.* Typhimurium, pZ plasmid (constitutive PLlacO-1 promoter) with a copy number of 20–30 copies/cell (Lutz & Bujard, 1997; Said et al., 2009), or *E. coli,* pBAD24 plasmid (arabinose pBAD promoter) with a copy number of 30 copies/cell (Guzman, Belin, Carson, & Beckwith, 1995; Lalaouna, Carrier, et al., 2015). Although the induction from plasmids only slightly overexpressed RsaA sRNA (1.7-fold) or RyhB (2.7-fold) compared to their endogenous expression (Lalaouna, Carrier, et al., 2015; Tomasini et al., 2017), other *S. aureus* sRNAs are drastically overproduced (unpublished data). Therefore, identified mRNA candidates could be pure artifacts, for example, due to indirect copurification through associated proteins such as RNA binding proteins (e.g., Hfq or other RNA chaperones). Intriguingly, we and other noticed that some highly enriched mRNA candidates only respond to overexpressed sRNAs (unpublished data). Indeed, no significant effect was observed when a Δ*sRNA* strain is compared to WT. A possible explanation is that the overexpression of sRNAs may sequestered some RNA chaperone proteins and impair the stability and function of other RNA chaperone protein-dependent sRNAs (Moon & Gottesman, 2011).

To settle this problem, we could envisage to use chromosomally encoded MS2-sRNA constructs (under the control of native promoter). Said et al. (2009) successfully copurified Hfq protein using endogenously transcribed MS2-RybB. It remains to be determined whether this approach enables the copurification of target mRNAs in detectable amounts.

2.8.4 Constitutive or Pulsed Expression?

The advantage of pulse expression compared to constitutive expression is to restrict sRNA activity in a narrow window of time.

(1) In Gram-negative bacteria, many sRNAs negatively regulate their target mRNAs at the posttranscriptional level by recruiting ribonucleases such as RNase E and thus actively inducing mRNA decay (Lalaouna, Simoneau-Roy, Lafontaine, & Masse, 2013; Mohanty & Kushner, 2016). If sRNA genes are constitutively expressed, the level of many, if not most, target mRNAs will drastically drop. In *S. aureus*, there is no homolog of RNase E. Other RNases such as RNase III, but also RNase Y, J1, and J2, might be involved in sRNA-mediated mRNA decay (Durand, Tomasini, Braun, Condon, & Romby, 2015). To slow down sRNA-dependent mRNA decay in *E. coli*, an RNA degradosome assembly mutant (*rne131*) of RNase E is used (Lalaouna et al., 2017). The use of an *rne131* strain significantly increases the enrichment ratio of negatively regulated targets, as confirmed with MS2-DsrA (WT *rne* vs *rne131*; D. Lalaouna, unpublished data). In *S. aureus*, we work in a wild-type background, which may principally shed light on stable interactions and translation regulation instead of regulation by targeting the mRNA stability. The mutation of the nonessential RNase III could be envisioned (Lioliou et al., 2012).

(2) sRNA synthesis all along the cell growth is detrimental, enabling sRNAs to regulate target mRNAs in nonphysiological conditions, but also to meet new transcripts (e.g., no more separation in time, exponential vs stationary phase of growth). Most sRNAs are solely produced under specific conditions (e.g., iron starvation, host invasion, osmotic stress). Even transcriptional noise could be damaging for bacterial cells which use sRNAs sponge to avoid it (Lalaouna, Carrier, et al., 2015).

2.8.5 Which Genetic Background Should Be Used?

To optimize MS2-affinity purification, the endogenous sRNA gene should be deleted to avoid competition for targets.

2.8.6 What Are Suitable Controls for MS2-Affinity Purification?

Two main controls can be used to validate specific enrichment of mRNA candidates: a sRNA without MS2 tag (untagged sRNA) or a MS2 tag alone.

(1) Untagged sRNA control enables to deduct false positives resulting from unspecific interactions with the column or agarose beads. Theoretically no RNA should be purified in the presence of untagged sRNA.

Unfortunately, numerous RNAs are eluted such as ribosomal RNAs (which could represent more than 90% of copurified RNAs). Accordingly, an additional step of ribosomal depletion (Ribo-Zero rRNA Removal Kit (Bacteria), Illumina) has been added to the original protocol established by Lalaouna, Carrier, et al. (2015) (see Bronesky et al., 2018; Tomasini et al., 2017).

(2) MS2 tag control expressed in the parental strain enables to remove unspecific RNAs interacting with the column, agarose beads, and most importantly with the MS2 aptamer. To note, the choice of the transcriptional terminator is primordial. Said et al. (2009) used a transcriptional terminator originating from *V. cholerae* to lower the probability to be recognized in *S.* Typhimurium. We, and Massé's group, used sRNA gene terminators. For example, both MS2-*Rsa*I and MS2 tag constructs share the same transcriptional terminator (Tomasini et al., 2017).

Massé's group mainly used an enrichment ratio corresponding to MS2-sRNA/untagged sRNA. In addition to eliminate unspecifically bound RNAs, this enables a comparable sRNA-induced regulation in both samples. In Lalaouna, Morissette, et al. (2015), they also published a list of contaminants that bound to the MS2 aptamer in tested conditions (ratio pBAD-MS2/empty vector).

For *S. aureus*, we use an enrichment ratio corresponding to MS2-sRNA/MS2 tag. Since the MS2 tag alone is expressed in the wild-type background, the untagged sRNA is also present. Nevertheless, as we discussed above, the level of endogenous sRNA and plasmid expressed MS2-sRNA might differ. We also extracted and sequenced total RNAs before the MS2-affinity purification. The comparison of both datasets enables to determine relative level of sRNA and mRNA target candidates before and after purification.

2.8.7 How Reliable are MAPS Results?

Massé's group commonly performs MAPS only once. For our part, we perform MAPS in duplicate, adding statistical analysis, and thus robustness. However, we all agree that MAPS results are more like snapshot views. Any variation in medium, harvesting time, or even experimentalist will influence results.

2.8.8 Are Highly Enriched mRNAs Always Direct Targets?

As previously described, some of mRNA candidates could be artifacts due to unspecific interactions with the column or the MS2 aptamer. Moreover, mRNAs could be copurified by indirect interactions, notably through

RNA binding proteins or RNAs having the same targets. Hence, every mRNA candidate must be validated using in vivo or in vitro approaches (Fig. 2). Prediction of basepairing interactions might also be useful to distinguish primary from unspecific target RNAs. We often used CopraRNA and/or IntaRNA (Wright et al., 2014).

2.8.9 Is It Possible to Infer the Prioritization of Target RNAs From MAPS Data?

Performing MAPS can potentially generate a long list of RNA candidates sorted according to fold change (or enrichment ratio). Hence, we could imagine that sRNAs have a greater affinity for highly enriched target RNAs. However, it is crucial to obtain information on their relative or absolute levels before MS2-affinity purification. For example, a poorly transcribed RNA which is highly enriched will be recognized with a higher affinity than a highly transcribed one. For this purpose, we performed total RNA sequencing simultaneously with MAPS (Bronesky et al., 2018; Tomasini et al., 2017).

ACKNOWLEDGMENTS

We thank Pascale Romby for helpful advice and discussions. We are grateful to Joseph Vilardell for the gift of the plasmid expressing the MS2-MBP. RNA-Seq analyses have been done using the Roscoff (France) instance of Galaxy (http://galaxy.sb-roscoff.fr/).

FUNDING

This work was supported by the Agence Nationale de la Recherche (ANR, Grant ANR-16-CE11-0007-01, RIBOSTAPH, to Pascale Romby [P.R.]). It has also been published under the framework of the LABEX: ANR-10-LABX-0036 NETRNA to P.R., a funding from the state managed by the French National Research Agency as part of the investments for the future program. D.L. was supported by the European Union's Horizon 2020 research and innovation program under the Marie Skłodowska-Curie Grant Agreement No. 753137—SaRNAReg.

REFERENCES

Abu-Qatouseh, L. F., Chinni, S. V., Seggewiss, J., Proctor, R. A., Brosius, J., Rozhdestvensky, T. S., et al. (2010). Identification of differentially expressed small non-protein-coding RNAs in Staphylococcus aureus displaying both the normal and the small-colony variant phenotype. *Journal of Molecular Medicine*, *88*(6), 565–575. https://doi.org/10.1007/s00109-010-0597-2.

Afgan, E., Baker, D., van den Beek, M., Blankenberg, D., Bouvier, D., Cech, M., et al. (2016). The Galaxy platform for accessible, reproducible and collaborative biomedical analyses: 2016 Update. *Nucleic Acids Research*, *44*(W1), W3–W10. https://doi.org/10.1093/nar/gkw343.

Anders, S., Pyl, P. T., & Huber, W. (2015). HTSeq—A Python framework to work with high-throughput sequencing data. *Bioinformatics*, *31*(2), 166–169. https://doi.org/10.1093/bioinformatics/btu638.

Beaume, M., Hernandez, D., Farinelli, L., Deluen, C., Linder, P., Gaspin, C., et al. (2010). Cartography of methicillin-resistant S. aureus transcripts: Detection, orientation and temporal expression during growth phase and stress conditions. *PLoS One*, *5*(5), e10725. https://doi.org/10.1371/journal.pone.0010725.

Bohn, C., Rigoulay, C., & Bouloc, P. (2007). No detectable effect of RNA-binding protein Hfq absence in Staphylococcus aureus. *BMC Microbiology*, *7*, 10. https://doi.org/10.1186/1471-2180-7-10.

Bohn, C., Rigoulay, C., Chabelskaya, S., Sharma, C. M., Marchais, A., Skorski, P., et al. (2010). Experimental discovery of small RNAs in Staphylococcus aureus reveals a riboregulator of central metabolism. *Nucleic Acids Research*, *38*(19), 6620–6636. https://doi.org/10.1093/nar/gkq462.

Bolger, A. M., Lohse, M., & Usadel, B. (2014). Trimmomatic: A flexible trimmer for Illumina sequence data. *Bioinformatics*, *30*(15), 2114–2120. https://doi.org/10.1093/bioinformatics/btu170.

Bronesky, D., Desgranges, E., Corvaglia, A., François, P., Caballero, C. J., Prado, L., et al. (2018). A dual sRNA in *Staphylococcus aureus* induces a metabolic switch responding to glucose consumption. *bioRxiv*. https://doi.org/10.1101/278127.

Bronesky, D., Wu, Z., Marzi, S., Walter, P., Geissmann, T., Moreau, K., et al. (2016). Staphylococcus aureus RNAIII and its regulon link quorum sensing, stress responses, metabolic adaptation, and regulation of virulence gene expression. *Annual Review of Microbiology*, *70*, 299–316. https://doi.org/10.1146/annurev-micro-102215-095708.

Caldelari, I., Chane-Woon-Ming, B., Noirot, C., Moreau, K., Romby, P., Gaspin, C., et al. (2017). Complete genome sequence and annotation of the Staphylococcus aureus strain HG001. *Genome Announcements*, *5*(32). https://doi.org/10.1128/genomeA.00783-17.

Charpentier, E., Anton, A. I., Barry, P., Alfonso, B., Fang, Y., & Novick, R. P. (2004). Novel cassette-based shuttle vector system for Gram-positive bacteria. *Applied and Environmental Microbiology*, *70*(10), 6076–6085. https://doi.org/10.1128/aem.70.10.6076-6085.2004.

Coornaert, A., Chiaruttini, C., Springer, M., & Guillier, M. (2013). Post-transcriptional control of the Escherichia coli PhoQ-PhoP two-component system by multiple sRNAs involves a novel pairing region of GcvB. *PLoS Genetics*, *9*(1), e1003156. https://doi.org/10.1371/journal.pgen.1003156.

Dale, S. E., Doherty-Kirby, A., Lajoie, G., & Heinrichs, D. E. (2004). Role of siderophore biosynthesis in virulence of Staphylococcus aureus: Identification and characterization of genes involved in production of a siderophore. *Infection and Immunity*, *72*(1), 29–37.

Durand, S., Tomasini, A., Braun, F., Condon, C., & Romby, P. (2015). sRNA and mRNA turnover in Gram-positive bacteria. *FEMS Microbiology Reviews*, *39*(3), 316–330. https://doi.org/10.1093/femsre/fuv007.

Eyraud, A., Tattevin, P., Chabelskaya, S., & Felden, B. (2014). A small RNA controls a protein regulator involved in antibiotic resistance in Staphylococcus aureus. *Nucleic Acids Research*, *42*(8), 4892–4905. https://doi.org/10.1093/nar/gku149.

Fechter, P., Chevalier, C., Yusupova, G., Yusupov, M., Romby, P., & Marzi, S. (2009). Ribosomal initiation complexes probed by toeprinting and effect of trans-acting translational regulators in bacteria. *Methods in Molecular Biology*, *540*, 247–263. https://doi.org/10.1007/978-1-59745-558-9_18.

Geissmann, T., Chevalier, C., Cros, M. J., Boisset, S., Fechter, P., Noirot, C., et al. (2009). A search for small noncoding RNAs in Staphylococcus aureus reveals a conserved

sequence motif for regulation. *Nucleic Acids Research*, *37*(21), 7239–7257. https://doi.org/10.1093/nar/gkp668.
Goldmann, O., & Medina, E. (2017). Staphylococcus aureus strategies to evade the host acquired immune response. *International Journal of Medical Microbiology*. https://doi.org/10.1016/j.ijmm.2017.09.013.
Guzman, L. M., Belin, D., Carson, M. J., & Beckwith, J. (1995). Tight regulation, modulation, and high-level expression by vectors containing the arabinose PBAD promoter. *Journal of Bacteriology*, *177*(14), 4121–4130.
Howden, B. P., Beaume, M., Harrison, P. F., Hernandez, D., Schrenzel, J., Seemann, T., et al. (2013). Analysis of the small RNA transcriptional response in multidrug-resistant Staphylococcus aureus after antimicrobial exposure. *Antimicrobial Agents and Chemotherapy*, *57*(8), 3864–3874. https://doi.org/10.1128/aac.00263-13.
Jagodnik, J., Brosse, A., Le Lam, T. N., Chiaruttini, C., & Guillier, M. (2017). Mechanistic study of base-pairing small regulatory RNAs in bacteria. *Methods*, *117*, 67–76. https://doi.org/10.1016/j.ymeth.2016.09.012.
Kavita, K., de Mets, F., & Gottesman, S. (2018). New aspects of RNA-based regulation by Hfq and its partner sRNAs. *Current Opinion in Microbiology*, *42*, 53–61. https://doi.org/10.1016/j.mib.2017.10.014.
Lalaouna, D., Carrier, M. C., Semsey, S., Brouard, J. S., Wang, J., Wade, J. T., et al. (2015). A 3′ external transcribed spacer in a tRNA transcript acts as a sponge for small RNAs to prevent transcriptional noise. *Molecular Cell*, *58*(3), 393–405. https://doi.org/10.1016/j.molcel.2015.03.013.
Lalaouna, D., & Masse, E. (2016). The spectrum of activity of the small RNA DsrA: Not so narrow after all. *Current Genetics*, *62*(2), 261–264. https://doi.org/10.1007/s00294-015-0533-7.
Lalaouna, D., Morissette, A., Carrier, M. C., & Masse, E. (2015). DsrA regulatory RNA represses both hns and rbsD mRNAs through distinct mechanisms in Escherichia coli. *Molecular Microbiology*, *98*(2), 357–369. https://doi.org/10.1111/mmi.13129.
Lalaouna, D., Prevost, K., Eyraud, A., & Masse, E. (2017). Identification of unknown RNA partners using MAPS. *Methods*, *117*, 28–34. https://doi.org/10.1016/j.ymeth.2016.11.011.
Lalaouna, D., Prevost, K., Laliberte, G., Houe, V., & Masse, E. (2018). Contrasting silencing mechanisms of the same target mRNA by two regulatory RNAs in Escherichia coli. *Nucleic Acids Research*, *46*(5), 2600–2612. https://doi.org/10.1093/nar/gkx1287.
Lalaouna, D., Simoneau-Roy, M., Lafontaine, D., & Masse, E. (2013). Regulatory RNAs and target mRNA decay in prokaryotes. *Biochimica et Biophysica Acta*, *1829*(6–7), 742–747. https://doi.org/10.1016/j.bbagrm.2013.02.013.
Langmead, B., & Salzberg, S. L. (2012). Fast gapped-read alignment with Bowtie 2. *Nature Methods*, *9*(4), 357–359. https://doi.org/10.1038/nmeth.1923.
Le Pabic, H., Germain-Amiot, N., Bordeau, V., & Felden, B. (2015). A bacterial regulatory RNA attenuates virulence, spread and human host cell phagocytosis. *Nucleic Acids Research*, *43*(19), 9232–9248. https://doi.org/10.1093/nar/gkv783.
Li, H., Handsaker, B., Wysoker, A., Fennell, T., Ruan, J., Homer, N., et al. (2009). The sequence alignment/map format and SAMtools. *Bioinformatics*, *25*(16), 2078–2079. https://doi.org/10.1093/bioinformatics/btp352.
Lioliou, E., Sharma, C. M., Caldelari, I., Helfer, A. C., Fechter, P., Vandenesch, F., et al. (2012). Global regulatory functions of the Staphylococcus aureus endoribonuclease III in gene expression. *PLoS Genetics*, *8*(6), e1002782. https://doi.org/10.1371/journal.pgen.1002782.
Liu, Y., Wu, N., Dong, J., Gao, Y., Zhang, X., Mu, C., et al. (2010). Hfq is a global regulator that controls the pathogenicity of Staphylococcus aureus. *PLoS One*, *5*(9). https://doi.org/10.1371/journal.pone.0013069.

Lutz, R., & Bujard, H. (1997). Independent and tight regulation of transcriptional units in Escherichia coli via the LacR/O, the TetR/O and AraC/I1-I2 regulatory elements. *Nucleic Acids Research*, *25*(6), 1203–1210.

Macias, S., Bragulat, M., Tardiff, D. F., & Vilardell, J. (2008). L30 binds the nascent RPL30 transcript to repress U2 snRNP recruitment. *Molecular Cell*, *30*(6), 732–742. https://doi.org/10.1016/j.molcel.2008.05.002.

Manna, A. C., Kim, S., Cengher, L., Corvaglia, A., Leo, S., Francois, P., et al. (2018). Small RNA teg49 is derived from a sarA transcript and regulates virulence genes independent of SarA in Staphylococcus aureus. *Infection and Immunity*, *86*(2). https://doi.org/10.1128/iai.00635-17.

Melamed, S., Peer, A., Faigenbaum-Romm, R., Gatt, Y. E., Reiss, N., Bar, A., et al. (2016). Global mapping of small RNA-target interactions in Bacteria. *Molecular Cell*, *63*(5), 884–897. https://doi.org/10.1016/j.molcel.2016.07.026.

Mohanty, B. K., & Kushner, S. R. (2016). Regulation of mRNA decay in bacteria. *Annual Review of Microbiology*, *70*, 25–44. https://doi.org/10.1146/annurev-micro-091014-104515.

Monk, I. R., Shah, I. M., Xu, M., Tan, M. W., & Foster, T. J. (2012). Transforming the untransformable: Application of direct transformation to manipulate genetically Staphylococcus aureus and Staphylococcus epidermidis. *MBio*, *3*(2), 1–11. https://doi.org/10.1128/mBio.00277-11.

Moon, K., & Gottesman, S. (2011). Competition among Hfq-binding small RNAs in Escherichia coli. *Molecular Microbiology*, *82*(6), 1545–1562. https://doi.org/10.1111/j.1365-2958.2011.07907.x.

Nitzan, M., Fechter, P., Peer, A., Altuvia, Y., Bronesky, D., Vandenesch, F., et al. (2015). A defense-offense multi-layered regulatory switch in a pathogenic bacterium. *Nucleic Acids Research*, *43*(3), 1357–1369. https://doi.org/10.1093/nar/gkv001.

Novick, R. P., & Geisinger, E. (2008). Quorum sensing in staphylococci. *Annual Review of Genetics*, *42*, 541–564. https://doi.org/10.1146/annurev.genet.42.110807.091640.

Novick, R. P., Ross, H. F., Projan, S. J., Kornblum, J., Kreiswirth, B., & Moghazeh, S. (1993). Synthesis of staphylococcal virulence factors is controlled by a regulatory RNA molecule. *The EMBO Journal*, *12*(10), 3967–3975.

Olejniczak, M., & Storz, G. (2017). ProQ/FinO-domain proteins: Another ubiquitous family of RNA matchmakers? *Molecular Microbiology*, *104*(6), 905–915. https://doi.org/10.1111/mmi.13679.

Oliveira, D., Borges, A., & Simoes, M. (2018). Staphylococcus aureus toxins and their molecular activity in infectious diseases. *Toxins (Basel)*, *10*(6). https://doi.org/10.3390/toxins10060252.

Penades, J. R., Chen, J., Quiles-Puchalt, N., Carpena, N., & Novick, R. P. (2015). Bacteriophage-mediated spread of bacterial virulence genes. *Current Opinion in Microbiology*, *23*, 171–178. https://doi.org/10.1016/j.mib.2014.11.019.

Pichon, C., & Felden, B. (2005). Small RNA genes expressed from Staphylococcus aureus genomic and pathogenicity islands with specific expression among pathogenic strains. *Proceedings of the National Academy of Sciences of the United States of America*, *102*(40), 14249–14254. https://doi.org/10.1073/pnas.0503838102.

Qin, L., McCausland, J. W., Cheung, G. Y., & Otto, M. (2016). PSM-Mec-A virulence determinant that connects transcriptional regulation, virulence, and antibiotic resistance in staphylococci. *Frontiers in Microbiology*, *7*, 1293. https://doi.org/10.3389/fmicb.2016.01293.

Quereda, J. J., & Cossart, P. (2017). Regulating bacterial virulence with RNA. *Annual Review of Microbiology*, *71*, 263–280. https://doi.org/10.1146/annurev-micro-030117-020335.

Romilly, C., Lays, C., Tomasini, A., Caldelari, I., Benito, Y., Hammann, P., et al. (2014). A non-coding RNA promotes bacterial persistence and decreases virulence by regulating a regulator in Staphylococcus aureus. *PLoS Pathogens*, *10*(3), e1003979. https://doi.org/10.1371/journal.ppat.1003979.

Said, N., Rieder, R., Hurwitz, R., Deckert, J., Urlaub, H., & Vogel, J. (2009). In vivo expression and purification of aptamer-tagged small RNA regulators. *Nucleic Acids Research*, *37*(20), e133. https://doi.org/10.1093/nar/gkp719.

Saliba, A. E., C Santos, S., & Vogel, J. (2017). New RNA-seq approaches for the study of bacterial pathogens. *Current Opinion in Microbiology*, *35*, 78–87. https://doi.org/10.1016/j.mib.2017.01.001.

Santiago-Frangos, A., & Woodson, S. A. (2018). Hfq chaperone brings speed dating to bacterial sRNA. *Wiley Interdisciplinary Reviews. Ribonucleic Acid*, *9*(4), e1475. https://doi.org/10.1002/wrna.1475.

Schu, D. J., Zhang, A., Gottesman, S., & Storz, G. (2015). Alternative Hfq-sRNA interaction modes dictate alternative mRNA recognition. *The EMBO Journal*, *34*(20), 2557–2573. https://doi.org/10.15252/embj.201591569.

Smirnov, A., Schneider, C., Hor, J., & Vogel, J. (2017). Discovery of new RNA classes and global RNA-binding proteins. *Current Opinion in Microbiology*, *39*, 152–160. https://doi.org/10.1016/j.mib.2017.11.016.

Song, T., Mika, F., Lindmark, B., Liu, Z., Schild, S., Bishop, A., et al. (2008). A new Vibrio cholerae sRNA modulates colonization and affects release of outer membrane vesicles. *Molecular Microbiology*, *70*(1), 100–111. https://doi.org/10.1111/j.1365-2958.2008.06392.x.

Tomasini, A., Francois, P., Howden, B. P., Fechter, P., Romby, P., & Caldelari, I. (2014). The importance of regulatory RNAs in Staphylococcus aureus. *Infection, Genetics and Evolution*, *21*, 616–626. https://doi.org/10.1016/j.meegid.2013.11.016.

Tomasini, A., Moreau, K., Chicher, J., Geissmann, T., Vandenesch, F., Romby, P., et al. (2017). The RNA targetome of Staphylococcus aureus non-coding RNA RsaA: Impact on cell surface properties and defense mechanisms. *Nucleic Acids Research*, *45*(11), 6746–6760. https://doi.org/10.1093/nar/gkx219.

Vanderpool, C. K., Balasubramanian, D., & Lloyd, C. R. (2011). Dual-function RNA regulators in bacteria. *Biochimie*, *93*(11), 1943–1949. https://doi.org/10.1016/j.biochi.2011.07.016.

Varet, H., Brillet-Gueguen, L., Coppee, J. Y., & Dillies, M. A. (2016). SARTools: A DESeq2- and EdgeR-based R pipeline for comprehensive differential analysis of RNA-Seq data. *PLoS One*, *11*(6), e0157022. https://doi.org/10.1371/journal.pone.0157022.

Wagner, E. G., & Romby, P. (2015). Small RNAs in bacteria and archaea: Who they are, what they do, and how they do it. *Advances in Genetics*, *90*, 133–208. https://doi.org/10.1016/bs.adgen.2015.05.001.

Wright, P. R., Georg, J., Mann, M., Sorescu, D. A., Richter, A. S., Lott, S., et al. (2014). CopraRNA and IntaRNA: Predicting small RNA targets, networks and interaction domains. *Nucleic Acids Research*, *42*(Web Server issue), W119–W123. https://doi.org/10.1093/nar/gku359.

CHAPTER SEVENTEEN

Noncoding RNAs in Archaea: Genome-Wide Identification and Functional Classification

Anne Buddeweg*, Michael Daume†, Lennart Randau†, Ruth A. Schmitz*,[1]

*Institut für allgemeine Mikrobiologie, Christian-Albrechts-Universität zu Kiel, Kiel, Germany
†Max-Planck-Institut für terrestrische Mikrobiologie, Marburg, Germany
[1]Corresponding author: e-mail address: rschmitz@ifam.uni-kiel.de

Contents

Abstract

Noncoding RNAs (ncRNAs) fulfill essential functions in eukaryotes and bacteria, but also in the third domain of life, the Archaea. Many archaeal organisms live in hostile environments that provide unique challenges for their transcriptional and translational regulatory pathways. Computational analyses and RNA-sequencing methodologies allowed for the genome-wide detection of ncRNA molecules in archaea. Several new classes of ncRNAs have been discovered and are expected to enable life in these extreme habitats. Here, we provide an overview of the current knowledge on archaeal ncRNAs and their deduced or biochemically verified functions. In addition, details of applying RNA-seq methodology for the detection of ncRNAs in *Sulfolobus acidocaldarius* are provided. Identified ncRNAs include small RNAs (sRNAs) that regulate gene expression and C/D box sRNAs that guide 2′-*O* methylation of target RNAs.

Methods in Enzymology, Volume 612
ISSN 0076-6879
https://doi.org/10.1016/bs.mie.2018.08.003

1. INTRODUCTION

Archaea, representing the third domain of life, are known to occupy extreme environments. Some members of the archaeal domain inhabit the most hostile biotopes on earth, including record holders for growth at the highest temperature (121°C, e.g., *Methanopyrus kandleri*), lowest pH (pH ~0, *Picrophilus*), or highest salt concentrations (~5.2 *M* NaCl, e.g., *Halobiforma haloterrestris*) (Fütterer et al., 2004; Grant, Kamekura, McGenity, & Ventosa, 2001; Kashefi & Lovley, 2003). In recent years, improvements in 16S rRNA-sequencing techniques allowed for culture-independent estimation of the microbial and particularly also archaeal diversity of many habitats. Archaea were detected not only in extreme environments but also in moderate habitats, such as ocean waters, freshwater sediments, soil, and even on skin or in the intestines of humans and apes (Chaban, Ng, & Jarrell, 2006; Koskinen et al., 2017; Moissl-Eichinger et al., 2018; Raymann, Moeller, Goodman, & Ochman, 2017). The best-studied archaeal phyla are crenarchaeota and euryarchaeota. Members of the crenarchaeota are mostly extremophiles representing thermophiles and hyperthermophiles as well as thermoacidophiles like the order *Sulfolobales*. In contrast, organisms belonging to the euryarchaeota grow at a broader range of temperatures. They occupy mesophilic, thermophilic, and hyperthermophilic habitats and often cover anaerobic niches (e.g., methanogenic archaea) or halophilic environments (Chaban et al., 2006).

From a phylogenetic point of view, archaea are more related to eukaryotes than bacteria, but they share several features with only one of the two other domains (Woese, Kandler, & Wheelis, 1990). Both, archaea and bacteria, employ a single DNA-dependent RNA polymerase (RNAP), and do not rely on three polymerases with different functional roles like eukaryotes. However, the archaeal RNAP composition exhibits a complexity similar to its eukaryotic counterpart and archaeal transcription is initiated by promoter elements (e.g., TATA-box) and transcription initiation proteins TFB and TBP (Transcription Factor B and TATA-box Binding Protein) that are also found in eukaryotes (Dennis, 1997; Gehring, Walker, & Santangelo, 2016; Olsen & Woese, 1997; Reeve, 1992). Archaeal ribosomes are larger than bacterial and smaller than eukaryotic ribosomes. They comprise rRNAs that are of similar size to bacterial rRNAs but contain additional eukaryotic-like ribosomal proteins. Similar to bacteria, translation is frequently initiated via Shine–Dalgarno ribosome binding sites (RBSs) (Malys & McCarthy, 2011) and regulatory noncoding RNAs (sRNA) that target RBSs were identified (Buddeweg, Sharma, Urlaub, & Schmitz, 2018; Jäger et al., 2012; Prasse, Förstner, Jäger, Backofen, & Schmitz, 2017). Processing and modification of archaeal rRNAs show similarities to the eukaryotic pathways as they share C/D box guide sRNAs, homologs of the eukaryotic small nucleolar RNAs (snoRNAs) (Malys & McCarthy, 2011).

RNA-seq methodologies allowed for the detection of large ncRNA networks and it became apparent that archaeal ncRNAs exhibit a variety of characteristics and functions shared with bacterial and/or eukaryotic counterparts (Babski et al., 2014; Wagner & Romby, 2015). In this review we will summarize the known classes of archaeal ncRNAs and describe different methods which led to their identification. As an example, the comparative RNA-seq approach to facilitate the identification of C/D box sRNAs and other RNA molecules bound by the ribosomal protein L7Ae in *Sulfolobus acidocaldarius* is described in detail in Section 4.

2. SYSTEMS BIOLOGY VIEW OF ARCHAEAL NONCODING RNA FAMILIES

Different methods and approaches led to the discovery of a variety of ncRNA families in archaea. Some of the first experimental approaches led to the identification of eukaryotic sno-like guide RNAs. Homologs of these snoRNAs were first described in *Sulfolobus* and *Pyrococcus* species in 2000 (Gaspin, Cavaille, Erauso, & Bachellerie, 2000; Omer et al., 2000) and were shown to guide the 2′-*O*-methylation (C/D box sRNAs) and pseudouridylation (H/ACA box sRNAs) of rRNAs and tRNAs. Both RNA families exhibit highly conserved sequences that form structural motifs, termed kink-turns (k-turns). The abundant ribosomal RNA-binding protein (RBP) L7Ae facilitates k-turn formation. Each C/D box sRNA is integrated into a ribonucleoprotein complex containing two copies of L7Ae, NOP56 and archaeal fibrillarin. Two guide sequences between the k-turn structures are used to identify complementary target RNA molecules, mostly rRNAs. Subsequently, the methyltransferase fibrillarin methylates the fifth nucleotide upstream of the box D sequences. Coordinated binding of two targets has been proposed to allow for an RNA chaperone role of C/D box sRNPs, assisting rRNA structuring. It is evident that archaeal life at elevated temperatures coincides with an increased number of RNA modifications and, consequently, record numbers of C/D box sRNA genes. In contrast, mesophilic archaea often contain only few of these genes (Dennis, Tripp, Lui, Lowe, & Randau, 2015; Omer et al., 2000).

The analysis of ncRNAs in *Archaeoglobus fulgidus* revealed a different RNA family, small RNA (sRNA) molecules derived from short tandem repeats, which were initially named SRSRs (short regularly spaced repeats) (Tang et al., 2002). Today these molecules are known as CRISPR RNAs (crRNAs), i.e., the guide RNAs of the prokaryotic adaptive immune system CRISPR–Cas that identifies foreign (viral) nucleic acids (Hille et al., 2018; Koonin, Makarova, & Zhang, 2017; Marchfelder et al., 2012; Nickel et al., 2013). Additionally *A. fulgidus* ncRNAs were identified and divided into groups based on their genomic locations: (i) RNAs of intergenic regions (sRNAs), (ii) RNAs with antisense orientation to an open reading frame (ORF) (asRNAs), (iii) RNAs overlapping ORFs, and (iv) RNAs within sense orientation of ORFs. In subsequent work, RNAs were investigated in additional archaeal species and it became evident that, as known for bacteria, archaeal sRNAs and asRNAs generally function in fast response stress regulation by masking the RBS or by stabilizing the transcript (Buddeweg et al., 2018; Jäger, Förstner, Sharma, Santangelo, & Reeve, 2014; Jäger et al., 2012; Jaschinski et al., 2014; Märtens, Manoharadas, Hasenöhrl, Manica, & Bläsi, 2013; Prasse et al., 2017) (Fig. 1). However, only in a few examples has the physiological role been elucidated using biochemical and genetic approaches (Buddeweg et al., 2018; Jäger et al., 2012; Prasse et al., 2017). Phenotypic screening of 27 sRNAs in *Haloferax volcanii* showed evidence that sRNAs are mostly related to long-term metabolic changes (Table 1). Unique gain-of-function phenotypes for several sRNA deletion mutants support this hypothesis (Jaschinski et al., 2014).

With the development of direct RNA sequencing (RNA-seq) methods, further classes of ncRNAs were observed. For example, sRNAs which also exhibit small ORFs (spRNAs) encoding 20–50 amino acids were first described for archaea in *Methanosarcina mazei.* Similar spRNAs were previously discovered (Sharma et al., 2010; Storz, Wolf, & Ramamurthi, 2014), and in part characterized as dual-function sRNAs in bacteria (Gimpel & Brantl, 2017; Lloyd, Park, Fei, & Vanderpool, 2017; Vanderpool, Balasubramanian, & Lloyd, 2011). Recently, it was shown for *M. mazei* that at least some of the small proteins encoded by spRNAs are translated in vivo (Prasse et al., 2015). However, their biological function in archaea remains to be unraveled. Other examples of ncRNAs which were identified by RNA-seq are Transcription Start Site-associated RNAs (TSSaRNAs) in *Halobacterium salinarum* (Zaramela, Vencio, ten-Caten, Baliga, & Koide, 2014) and

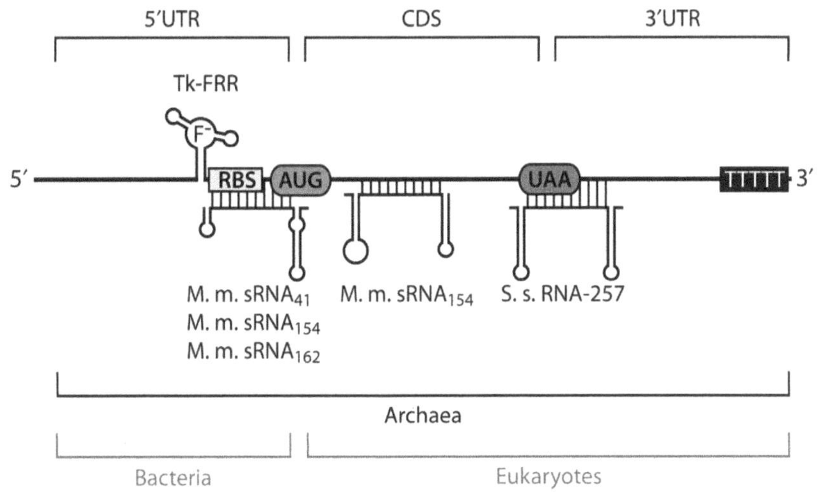

Fig. 1 Schematic of interaction mechanisms of archaeal ncRNAs. The first identified archaeal riboswitch and the only sRNAs with characterized function are shown at their respective interaction sites on the mRNA. Tk-FRR is a fluoride responsive riboswitch that was described in *T. kodakarensis*. $sRNA_{41}$, $sRNA_{154}$, and $sRNA_{162}$ are characterized in *M. mazei* (M. m.) and RNA-257 originates from *S. solfataricus* (S. s.). mRNA regions are depicted above the scheme, regions of ncRNA functions known from all three domains of life are depicted later. *AUG/UAA*, start/stop codon; *CDS*, coding sequence; *RBS*, ribosome binding site; multiple T represents transcription termination; *UTR*, untranslated region.

circular RNAs (cRNAs) (Danan, Schwartz, Edelheit, & Sorek, 2011; Richter, Mohr, & Randau, 2013; Starostina et al., 2004; Su, Tripp, & Randau, 2013). TSSaRNAs are derived from TSS of coding genes and seem to be at least in part independently expressed from their adjacent genes. cRNAs are to date solely identified among the hyperthermophilic archaea where they mostly act as C/D box sRNAs. The circular form is speculated to stabilize these RNA molecules, since formation of double stranded regions seems to be not efficient enough considering the high environmental temperatures (Danan et al., 2011; Richter et al., 2013; Starostina et al., 2004; Su et al., 2013).

Special cis-acting regulatory ncRNA elements, called *riboswitches*, have been described in the 5′UTR of bacteria. Riboswitches are structured ncRNA elements which form as the mRNA emerges from RNAP. Therefore, they are essential for the regulation of protein expression either by permitting continued transcription elongation or by mediating intrinsic transcription termination (Gong, Wang, Wang, & Zhang, 2017).

Table 1 Identified Archaeal sRNAs From Different Species With Known Characteristics

sRNA Designation	Organism	Length [nt]	Expression Regulation	Biological Mechanisms and Relevance: Phenotype	Biological Mechanisms and Relevance: Target Identification → (Potential?) Function	Identification Methods	Reference
$sRNA_{41}$	*M. mazei*	120	↑ **N-sufficiency**	No detectable growth phenotype	Masking of several RBSs in polycistronic *ACDS*-mRNAs leads to reduced protein expression → Link between N- and C-metabolism	RNA-seq	Buddeweg et al. (2018) and Jäger et al. (2009)
$sRNA_{154}$	*M. mazei*	134	↑ **N-limitation**	Expression is NrpR regulated $\Delta sRNA_{154}$ compromised growth under N-limitation	Masking of RBSs of $glnA_2$ leads to reduced protein expression Transcript stabilization of *nifH*, *nrpA*, $glnK_1$, and $glnA_1$ leads to enhanced protein expression → Central role in N-metabolism	RNA-seq	Jäger et al. (2009) and Prasse et al. (2017)
$sRNA_{162}$	*M. mazei*	191	Growth phase dependent	$sRNA_{162}$ OE elongated lag-phase on MeOH Much faster adaption after switch from MeOH to TMA	Masking of RBS of *MM2441*-mRNA (potential transcriptional regulator) in *trans* binds 5′UTR of *MM2442*-mRNA in *cis* → Adaption to TMA as C- and energy source	RNA-seq	Jäger et al. (2009) and Jäger et al. (2012)

RNA257$_{1–4}$	*S. solfataricus*	83–190	↓ Phosphate limitation	4 Paralogous sRNAs	Binding of 3′UTR of *Sso1183*-mRNA (phosphate transporter) leads to mRNA degradation → Adaption to phosphate limitation	RNA-seq	Märtens et al. (2013)
sRNA Hts$_4$	*H. volcanii*	110, 75, 65	↑ **Low salt conditions**	ΔsRNA Hts$_4$ slight growth defect on Glc and Ac Reduced growth yield at high salt	→ Adaption to osmotic changes?	RNA-seq	Heyer et al. (2012)
sRNA Hts$_{10}$	*H. volcanii*	110	↑ Stationary growth phase	ΔsRNA Hts$_{10}$ reduced growth yield on Ac and high salt Higher growth yield under low salt	→ Adaption to osmotic changes?	Bioinformatics predictions RNA-seq	Heyer et al. (2012)
sRNA Hts$_{14}$	*H. volcanii*	149	Expressed only at low salt conditions in stationary growth phase	ΔsRNA Hts$_{14}$ extended doubling times but higher growth yield on xylose Shortened lag-phase but extended doubling time and higher growth yield on acetate Higher growth yield on Glc Extended doubling time but much higher growth yield after salt stress Reduced growth yield after cold stress	→ Adaption to osmotic changes?	RNA-seq	Heyer et al. (2012)

Continued

Table 1 Identified Archaeal sRNAs From Different Species With Known Characteristics—cont'd

sRNA Designation	Organism	Length [nt]	Expression Regulation	Biological Mechanisms and Relevance: Phenotype	Biological Mechanisms and Relevance: Target Identification → (Potential?) Function	Identification Methods	Reference
sRNA Hts_{21}	*H. volcanii*	120	Expressed only at low salt conditions in exponential growth phase	ΔsRNA Hts_{21} shortened lag-phase on Ac Higher growth yield on Glc and Ac and after salt stress	→ Adaption to osmotic changes?	RNA-seq	Heyer et al. (2012)
$sRNA_{30}$	*H. volcanii*	131	↑ High temperature (48.5°C) in exponential growth phase	$\Delta sRNA_{30}$ to severely reduced growth yield at high temperature	→ Adaption to high temperatures?	Experimental RNomics	Straub et al. (2009)
$sRNA_{63}$	*H. volcanii*	178	Not reported	$\Delta sRNA_{63}$ heavily compromised growth under low salt conditions (15%) Elongated lag-phase on Cas Shortened lag-phase on xylose Higher swarming activity	→ Adaption to low salt conditions? → Regulation in motility?	Experimental RNomics	Jaschinski et al. (2014) and Straub et al. (2009)
$sRNA_{132}$	*H. volcanii*	138	↑ High temperature and low salt in exponential growth phase	$\Delta sRNA_{132}$ slightly reduced growth after oxidative stress but higher growth yield	—	Experimental RNomics RNA-seq	Jaschinski et al. (2014) and Straub et al. (2009)

$sRNA_{168}$	*H. volcanii*	145	Not reported	$\Delta sRNA_{168}$ fast growth on low Glc (1.2 *M*) Reduced swarming activity	—	Experimental RNomics	Jaschinski et al. (2014) and Straub et al. (2009)
$sRNA_{194}$	*H. volcanii*	117	↓ **Low salt conditions** ↓ Stationary growth phase	$\Delta sRNA_{194}$ shortened lag-phase on diverse Glc concentrations Extended doubling times at high (4 *M*) and low Glc Reduced growth yield at high Glc	—	Experimental RNomics RNA-seq	Jaschinski et al. (2014) and Straub et al. (2009)
$sRNA_{235}$	*H. volcanii*	200	Not reported	$\Delta sRNA_{235}$ elongated lag-phase at low and normal (2.1 *M*) Glc Reduced growth yield at high Glc	—	Experimental RNomics RNA-seq	Jaschinski et al. (2014) and Straub et al. (2009)
$sRNA_{288}$	*H. volcanii*	141	Not reported	$\Delta sRNA_{288}$ reduced growth yield on high Glc	—	Experimental RNomics	Jaschinski et al. (2014) and Straub et al. (2009)
$sRNA_{308}$	*H. volcanii*	151	Not reported	$\Delta sRNA_{308}$ higher growth yield after oxidative stress Reduced swarming activity	—	Experimental RNomics RNA-seq	Jaschinski et al. (2014) and Straub et al. (2009)
$sRNA_{362}$	*H. volcanii*	143	Not reported	$\Delta sRNA_{362}$ elongated lag-phase on Cas, xylose, Ac, normal and high Glc, shorter doubling time on Cas, much longer doubling times on Ac	—	Experimental RNomics	Jaschinski et al. (2014) and Straub et al. (2009)

Continued

Table 1 Identified Archaeal sRNAs From Different Species With Known Characteristics—cont'd

				Biological Mechanisms and Relevance			
sRNA Designation	**Organism**	**Length [nt]**	**Expression Regulation**	**Phenotype**	**Target Identification → (Potential?) Function**	**Identification Methods**	**Reference**
$sRNA_{450}$	*H. volcanii*	160	Not reported	Δ$sRNA_{450}$ elongated lag-phase on Ac, extended doubling times but higher growth yield on low Glc	—	Experimental RNomics	Jaschinski et al. (2014) and Straub et al. (2009)
$sRNA_{479}$	*H. volcanii*	166	Not reported	Δ$sRNA_{479}$ elongated lag-phase on Ac but shorter doubling time Extended doubling time on low Glc but slightly higher growth yield	—	Experimental RNomics	Jaschinski et al. (2014) and Straub et al. (2009)
$sRNA_{500}$	*H. volcanii*	155	Not reported	Δ$sRNA_{500}$ higher growth yield after oxidative stress Reduced swarming activity	—	Experimental RNomics Bioinformatics prediction RNA-seq	Jaschinski et al. (2014) and Straub et al. (2009)
$sRNA_{529}$	*H. volcanii*	149	Not reported	Δ$sRNA_{529}$ shorter doubling time but reduced growth yield on high Glc	—	Experimental RNomics RNA-seq	Jaschinski et al. (2014) and Straub et al. (2009)

$sRNA_{H225.2R}$	*H. volcanii*	96	Not reported	$\Delta sRNA_{H225.2R}$ elongated lag-phase on Ac, low and normal Glc Much longer doubling time on Ac Shorter doubling time on low Glc but reduced growth yield on low and normal Glc Higher growth yields after salt and cold stress	—	Bioinformatics predictions	Babski et al. (2011) and Jaschinski et al. (2014)
$sRNA_{htsf182}$	*H. volcanii*	17	Not reported	$\Delta sRNA_{htsf182}$ reduced swarming activity	→ Regulation of motility?	RNA-seq	Heyer et al. (2012) and Jaschinski et al. (2014)
$sRNA_{htsf242}$	*H. volcanii*	30	Not reported	$\Delta sRNA_{htsf242}$ elongated lag-phase on normal Glc	—	Bioinformatics predictions RNA-seq	Heyer et al. (2012) and Jaschinski et al. (2014)
$sRNA_{htsf339}$	*H. volcanii*	17	Not reported	$\Delta sRNA_{htsf339}$ elongated lag-phase on Ac	—	RNA-seq	Heyer et al. (2012) and Jaschinski et al. (2014)
$sRNA_{htsf416}$	*H. volcanii*	17	Not reported	$\Delta sRNA_{htsf416}$ elongated lag-phases on Ac and normal Glc but shorter doubling time on Glc	—	Bioinformatics predictions RNA-seq	Heyer et al. (2012) and Jaschinski et al. (2014)

Continued

Table 1 Identified Archaeal sRNAs From Different Species With Known Characteristics—cont'd

sRNA Designation	Organism	Length [nt]	Expression Regulation	Biological Mechanisms and Relevance: Phenotype	Biological Mechanisms and Relevance: Target Identification → (Potential?) Function	Identification Methods	Reference
$sRNA_{htsf468}$	*H. volcanii*	193	Not reported	$\Delta sRNA_{htsf468}$ elongated lag-phase on high Glc Highly reduced doubling time on xylose Shorter doubling time on normal Glc and high Glc, but reduced growth yield	—	RNA-seq	Heyer et al. (2012) and Jaschinski et al. (2014)
$sRNA_{htsf494}$	*H. volcanii*	104	Not reported	$\Delta sRNA_{htsf494}$ extended doubling times on Ac	—	RNA-seq	Heyer et al. (2012) and Jaschinski et al. (2014)
$sRNA_{htsf574}$	*H. volcanii*	21	Not reported	$\Delta sRNA_{htsf574}$ elongated lag-phase but shorter doubling time on normal Glc	—	RNA-seq	Heyer et al. (2012) and Jaschinski et al. (2014)

Vertical arrows represent up- or downregulation of sRNA expression (*bold*, high regulation), ΔsRNA represent deletion mutants of the respective sRNA genes, sRNA OE represent plasmid-based sRNA overexpression. *Ac*, acetate; *C*, carbon; *Cas*, casamino acids; *Glc*, glucose; *MeOH*, methanol; *N*, nitrogen; *TMA*, trimethylamine.

Riboswitches can also promote or limit the access of ribosomes to translation initiation sites (Barrick & Breaker, 2007; Breaker, 2018). Archaeal riboswitches have initially only been predicted based on conserved sequences and secondary structures (Barrick & Breaker, 2007; Phok et al., 2011; Weinberg et al., 2010). Rother and colleagues showed that in principle a bacterial riboswitch was functional in the methanogenic archaeon *Methanosarcina acetivorans* (Demolli et al., 2014). Only very recently, the first functional archaeal riboswitch was described to respond to changing fluoride concentrations in the hyperthermophilic organism *Thermococcus kodakarensis* (Tk-FRR). A combination of comparative genomics and experimental approaches demonstrates for the first time the functional role of a riboswitch in archaea (Speed, Burkhart, Picking, & Santangelo, 2018) (Fig. 1).

3. APPROACHES FOR IDENTIFICATION OF ARCHAEAL NONCODING RNAs

Successful identification of ncRNAs was performed in several archaeal model organisms. Initially, ncRNA was investigated by a combination of experimental approaches and bioinformatics predictions that relied on the presence of sequenced archaeal species. With the improvement of different RNA-seq methods, genome-wide detection of different ncRNAs classes became feasible. In the following sections, we will summarize the development of identification methods in regard to archaeal classes of ncRNAs.

3.1 First Bioinformatics Predictions and Experimental ncRNA Identification

Different ncRNAs were first identified in *Sulfolobus* and *Pyrococcus* species using bioinformatics predictions and biochemical experiments (Gaspin et al., 2000; Omer et al., 2000). Coimmunoprecipitation of the archaeal proteins fibrillarin and NOP56 with total RNA led to the identification of sno-like C/D box RNAs in *S. acidocaldarius*. Computational searches for conserved sequence and structure motifs led to the identification of two sno-like RNAs in other *Sulfolobus* species and several putative sno-like RNAs in three different *Pyrococcus* species (Gaspin et al., 2000; Omer et al., 2000). A subsequent study combining computational predictions and experiments in *P. abyssi* identified additional ncRNAs candidates including crRNAs and riboswitches (Phok et al., 2011).

Furthermore, ncRNA gene prediction has been attempted by bioinformatics approaches. The initial screening for conserved promoter and terminator sequences in intergenic regions (IGRs) of several methanoarchaea resulted in a large number of predicted ncRNAs, which at the time were not further analyzed (W.R. Hess, H. Liesegang, R.A. Schmitz, unpublished). The increasing number of genome sequences is beneficial for computational searches, but the lack of ORFs and statistical bias in nucleotide composition makes it hard to identify ncRNA genes with classical gene finder programs. One way around this is to determine the GC content of structured RNAs in hyperthermophilic archaea. These archaea often exhibit AT-rich genomes and GC-rich islands can be used as indicators for genes that encode for structure RNA molecules. This approach was used to identify known ncRNAs (e.g., rRNA, tRNA) and nine novel ncRNA candidates in *Methanocaldococcus jannaschii* (Klein, Misulovin, & Eddy, 2002). However, genetic and biochemical studies are needed to unravel their biological function.

In parallel, experimental approaches, first described for *A. fulgidus*, allow for the identification of different ncRNA classes (Tang et al., 2002). After extraction of total RNA and size separation by denaturing PAGE, RNA with a length of 50–500 nt was reverse transcribed to generate cDNA libraries. After Sanger sequencing of the libraries and exclusion of rRNA-, tRNA-, and mRNA-derived fragments, ncRNA could be identified representing only about 1% of the overall identified RNA sequences. In vivo expression of those ncRNAs was generally confirmed with Northern blots. Using experimental RNomics, ncRNAs were additionally identified in *Sulfolobus solfataricus* (Tang et al., 2005) and *H. volcanii* (Straub et al., 2009).

In *H. volcanii*, widely conserved ncRNA genes were also identified by bioinformatics predictions using comparative genomics of IGRs (Babski et al., 2011). Due to the lack of consensus promoter and terminator sequences in *H. volcanii*, tools for sRNA prediction in bacteria cannot be used. Hence, IGRs were compared with four other haloarchaeal species as well with *P. abyssi* and the halophilic bacterium *Salinibacter ruber.* The identified ncRNA candidates were experimentally validated using microarrays.

It is evident that a combination of complementary bioinformatics prediction and experimental RNomics approaches are needed to provide a comprehensive view of the ncRNA potential in archaea.

3.2 RNA Sequencing

In recent years, the number of sequenced archaeal genomes has increased drastically. In addition, sequencing techniques have developed rapidly and several high-throughput RNA-sequencing (HTS) techniques were invented. In general, RNA molecules are isolated from archaeal species and converted into cDNA libraries subjected to HTS. Sequence reads can be mapped to the whole genome sequence allowing for the genome-wide detection of transcripts and possible transcription start sites (TSS), which facilitates detection of IGR transcripts. Furthermore, primary transcripts can be enriched by Terminator 5′-Phosphate-dependent Exonuclease (TEX) treatment, which degrades processed transcripts with a 5′ mono-phosphate. This method enables the differentiation between primary transcripts and processed RNA allowing the identification of ncRNAs and mRNA degradation products and has been established by Sharma et al. and termed differential RNA-seq (dRNA-seq) (Sharma et al., 2010) (see chapter "Characterizing the role of exoribonucleases in the control of microbial gene expression: Differential RNA-seq" by Pobre and Arraiano).

RNA-seq offers a variety of options depending on the study design. The first study used for ncRNA identification in an archaeon was performed in *M. mazei* (Jäger et al., 2009). A dRNA-seq approach under varying nitrogen availabilities was used for the identification of ncRNAs. RNA-seq simplifies the execution of comparative studies to identify sRNAs related to specific regulatory pathways by the application of certain stress conditions. In *M. mazei* 248 ncRNAs were discovered using the 454 pyrosequencing technique, yielding a maximum of ~185,000 cDNA reads. The majority of those ncRNAs seem to function as potential regulators of gene expression rather than in rRNA or tRNA maturation (C/D box or H/ACA box RNAs); a few of them were confirmed in later studies and their physiological function determined (Buddeweg et al., 2018; Jäger et al., 2012; Prasse et al., 2017). Furthermore, 40 of the identified ncRNAs contain sORFs giving the first hint to the existence of small proteins in archaea. In a subsequent study, two different sequencing approaches, whole transcriptome and 5′ end sequencing, were combined for *S. solfataricus* ncRNAs. This combination enabled the detection of new transcripts that were previously not annotated in *S. solfatoricus*, including new ncRNAs (Wurtzel et al., 2010). Moreover, in this study, Illumina sequencing technology was used resulting in more than 10-fold higher read coverage compared to the earlier 454 pyrosequencing study allowing for the detection of transcripts of low abundance. Additionally, known sRNAs and snoRNAs (Omer et al., 2000; Tang et al., 2005) were confirmed.

A transcriptome of *P. abyssii* was obtained which enabled the identification of numerous ncRNAs using a protocol that clusters RNA-seq reads in a strand-specific fashion. Novel asRNAs, as well as trans-encoded sRNAs and long 5′UTRs, were identified (Toffano-Nioche et al., 2013). Only one-third and approximately one-half of the sRNAs and 5′UTRs, respectively, were identified or at least predicted before, showing the great potential of ncRNA identification using RNA-seq. Utilization of this potential also reveals a compelling number of ncRNA candidates in *T. kodakarensis* and over 1000 asRNA transcripts were detected; however, until today no targets or in vivo functions have been identified so far. Besides, the expression of C/D box and H/ACA box sRNAs, crRNAs and conserved sRNAs were confirmed, whereas also previously unrecognized ncRNAs were detected (Jäger et al., 2014).

Modification of the RNA-seq protocol with a 5′ independent ligation strategy of cDNA preserves the strand orientation of the transcripts, captures the 5′ and 3′ ends, and generates no bias for molecule selection based on 5′ phosphorylation state (Bernick, Dennis, Hochsmann, & Lowe, 2012). Using this approach, broad ranges of ncRNAs as well as highly conserved novel sRNAs in four *Pyrobaculum* species were identified. In a second study, whole transcriptome analyses were coupled with a sRNA transcriptome sequencing approach. This led to the discovery of new H/ACA box sRNAs in *Pyrobaculum* (Bernick, Dennis, Lui, & Lowe, 2012), which differ in a number of structural features compared to other archaeal species and thereby were not discovered by computational methods. In subsequent studies for identifying ncRNAs in *H. volcanii* (Straub et al., 2009), an RNA-seq approach was used to compare different growth phases and conditions (Heyer et al., 2012). In addition to validating known ncRNAs, a new family of archaeal ncRNAs, so-called tRNA-derived fragments (tRFs), was discovered.

Protocol modifications, e.g., different RNA enrichment strategies, offer possibilities to focus on specific ncRNA features in archaea. The discovery of cRNAs in a transcriptome-wide manner in *S. solfataricus* resulted from RNase R treatment prior to sequencing to selectively remove linear RNA molecules (Danan et al., 2011). Additionally, careful alignment of permuted reads was necessary to identify circular C/D box and H/ACA box sRNAs as well as some asRNAs. Circular C/D box sRNAs were also identified in *M. kandleri* and *Pyrococcus furiosus* (Starostina et al., 2004; Su et al., 2013). The sRNome of *Nanoarchaeum equitans*, which has a minimal fragmented genome of 480 kb, identified several classes of sRNAs including circular and linear C/D box sRNAs and crRNAs (Richter et al., 2013).

In addition to ncRNAs, deep sequencing of archaeal transcriptomes revealed a plethora of other RNA-based regulatory mechanisms. For example,

RNA-seq showed that *M. mazei* exhibits some unusually long 5′UTRs of up to 500 nt, pointing to a potential regulatory function, e.g., by encoding for potential riboswitches. In contrast, *S. solfataricus* generally lacks 5′UTRs, but instead exhibits 3′UTRs of significant size (Wurtzel et al., 2010). The UTR length can be used to deduce potential interaction and regulation mechanisms of sRNAs. As known from bacteria, sRNA interactions often occur in the 5′UTR by masking the ribosome binding site or by melting secondary structures. This is also true for at least three sRNAs in *M. mazei*, as shown in three independent target identification studies (Fig. 1; Table 1) (Buddeweg et al., 2018; Jäger et al., 2012; Prasse et al., 2017). Moreover, $sRNA_{162}$ from *M. mazei* was characterized as a simultaneously *cis*- and *trans*-acting sRNA, binding to two different RBSs and thereby challenging the old paradigm of the differences of *cis*- and *trans*-acting sRNA functions (Jäger et al., 2012). In the case of *S. solfataricus*, which lacks long 5′UTRs, targeting 3′UTRs has been shown for $RNA257_{1-4}$, which binds the 3′UTR of a phosphate transporter gene and thereby reduces stability and consequently protein expression (Fig. 1; Table 1) (Märtens et al., 2013). However, there is evidence in other archaeal species sRNA that interactions generally occur in the 3′UTR or coding region, similar to the mechanism known from eukaryotic miRNA (Babski et al., 2014; Bernick, Dennis, Lui, et al., 2012; Dar, Prasse, Schmitz, & Sorek, 2016; Märtens et al., 2013; Straub et al., 2009). Finally, multiple RNAP pausing sites along gene sequences were discovered in *Halobacterium salinarium* (Zaramela et al., 2014).

3.3 Sequencing of RNA 3′ Termini (Term-Seq)

A recently developed specific method of RNA-seq called Term-seq was first applied in *Bacillus subtilis* (Dar, Shamir, et al., 2016). This approach enables a transcriptome-wide mapping of 3′ ends by sequencing from the end of transcripts instead of the beginning. Using this tool, terminator consensus sequences were also detected for *M. mazei* and *S. acidocaldarius* coding genes and ncRNAs, showing uridine enrichment directly before the termination site. Using this information can improve bioinformatics prediction of ncRNA termini. Furthermore, this approach provides insights into an, until then, undiscovered regulatory potential of 3′UTRs in archaea. For *S. acidocaldarius*, a significant potential antisense overlap by gene pairs representing convergent orientation was observed, which supports that this phenomenon can be exploited for gene regulation in archaeal cells. One-third of the analyzed protein-coding genes and one-quarter of the ncRNAs of both, *M. mazei* and *S. acidocaldarius,* exhibit multiple termination sites, arguing for potential gene expression control by including or excluding regulatory regions (e.g., sRNA-binding sites) as it was shown in eukaryotes. Identification of computationally predicted binding sites of sRNAs in 3′UTRs supports this hypothesis which remains to be experimentally evaluated (Dar, Prasse, et al., 2016).

3.4 RNA-Seq of RNA-Binding Proteins

Many ncRNA species will have extensive interactions with RBPs and the elucidation of global RNA–protein interaction maps will be highly beneficial for the functional classification of ncRNAs. For bacteria, it is known that most *trans*-encoded sRNAs require RBPs (e.g., Hfq or ProQ) for target interactions. The role of such RBPs is not fully established for most archaeal ncRNA families. In this regard, RNA-seq approaches can be very useful for identifying novel RNA interaction partners and for mapping the interaction network of individual proteins. In general, two main approaches were developed: RNA immunoprecipitation followed by sequencing (RIP-seq) and cross-linked immunoprecipitation followed by sequencing (CLIP-Seq). Both methods enable the identification of RBP-associated RNAs and allow for the additional identification of conserved RNA-binding motifs for RBPs and the possible elucidation of their interaction mechanism (Holmqvist et al., 2016). In the following section, we will highlight how RIP-seq methodology can be applied for an archaeal species. The ribosomal protein L7Ae from *S. acidocaldarius* was genomically tagged and interaction partners were identified. As previously described, L7Ae binds to k-turn structures in C/D box sRNA, rRNA, and other structured ncRNAs. Therefore, L7Ae RIP-seq constitutes an approach to identify the genome-wide occurrence of k-turns in archaeal RNAs. It was discovered that 5′UTRs of mRNAs are able to form k-turns which allows for L7Ae-mediated translational regulation (Daume, Uhl, Backofen, & Randau, 2017).

4. DETAILED METHOD FOR IDENTIFICATION OF L7Ae INTERACTOME

In this section a detailed description of the method for a comparative RNA-seq approach to facilitate the identification of C/D box sRNAs and other RNA molecules bound by the ribosomal protein L7Ae in *S. acidocaldarius* is provided.

4.1 Generation of Genomic Tags and Mutations in *S. acidocaldarius*

Genomic Flag-HA/His tagging of the l7ae gene was performed using the pop-in/pop-out method as previously described (Wagner et al., 2012). *S. acidocaldarius* MW001 cells were aerobically grown at 180 rpm, 75°C, and pH 3.5 in Brock media (Brock, Brock, Belly, & Weiss, 1972). The medium was supplemented with 0.1% (w/v) tryptone or NZ-Amine, as well as 0.2% (w/v) sucrose and 20 μg/mL uracil was added for strains without plasmid. *S. acidocaldarius* MW001 cells were transformed with the plasmid pSVA406 containing the C-terminally Flag-HA-tagged l7ae gene (saci_1520) sequence with 1000 bp upstream and downstream regions from the tag sequence for homologous recombination. This fragment was cloned into the restriction sites *Sac*II and *Sal*I (Daume et al., 2017; Wagner et al., 2012). Transformants were grown to the logarithmic phase in 50 mL Brock media supplemented with 0.1% (w/v) tryptone and 0.2% (w/v) sucrose. 10 μL of the cell cultures were diluted in 100 μL ddH_2O and plated on second selection plates bearing 5′-FOA and uracil. The 5'-FOA induced the loss of the plasmid, yielding a 50% chance of obtaining the desired mutation. The plates were incubated for 5 days and colonies were screened for correct mutations by colony PCR and sequencing.

4.2 Isolation of *S. acidocaldarius* Total and sRNAs

To obtain an overview of the sRNAs that can be detected in parallel to the L7Ae enrichment strategy, an sRNome approach was performed. Total and sRNAs (<200 nt) of *S. acidocaldarius* were isolated using the mirVana™ miRNA isolation kit according to the manufacturer's instructions. Briefly, 500 ng of *S. acidocaldarius* cells were resuspended in 1 mL lysis/binding buffer and lysed by a homogenizer for 3 min on ice. 100 μL miRNA Homogenate Additive™ was added to the cell lysate and incubated for 10 min on ice. Afterward, the RNA was isolated by acid-phenol/chloroform extraction. sRNAs with a size of less than 200 nt were separated from total RNAs by adjusting the ethanol concentration in the aqueous phase to 25%. This approach immobilized larger RNAs on a glass-fiber filter tube, while sRNAs were collected in the filtrate. Afterward, the EtOH concentration of the filtrate was increased to 55%, and the sRNAs were immobilized on a second glass-fiber filter tube. The two immobilized RNA fractions (sRNA-depleted total RNA and sRNAs) were washed and subsequently eluted in 100 μL nuclease-free water (Fig. 2A).

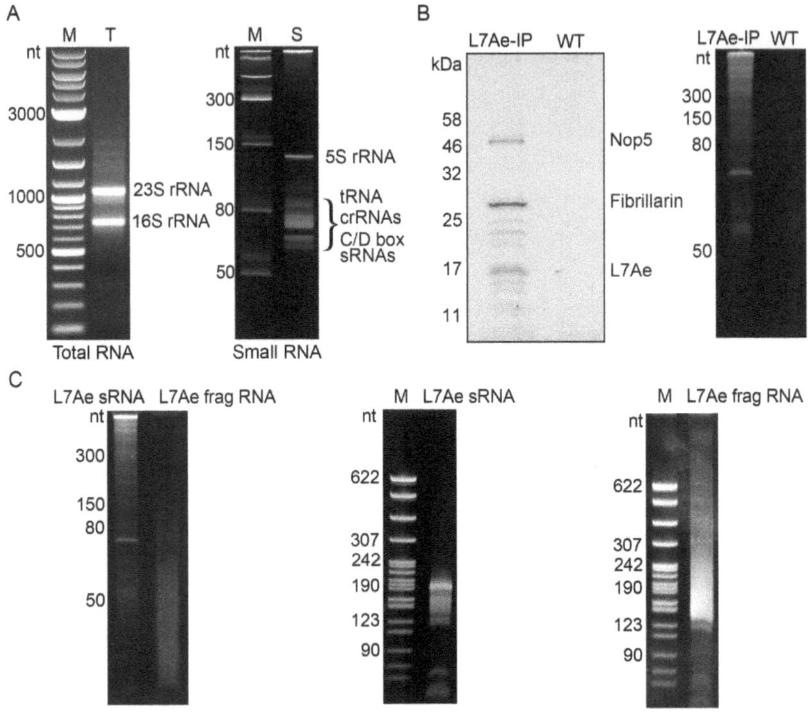

Fig. 2 RNA isolation and cDNA library preparation. (A) Native agarose gel separation of isolated total RNA (T) and denaturing PAGE of small RNA (S) of *S. acidocaldarius*. (B) SDS-PAGE analysis illustrates the final purification fractions of L7Ae (16 kDa) after Flag- and HA-IP and shows co-IP of Nop5 (47 kDa), fibrillarin (25 kDa), and other proteins. The mock purification (WT) displays no protein bands. The *right panel* shows corresponding co-IP of RNAs separated via denaturing PAGE. (C) Denaturing PAGE of L7Ae co-IP RNA (L7Ae sRNA) and fragmentation of the sample (L7Ae frag RNA) after $ZnCl_2$ treatment. PCR amplified cDNA libraries of L7Ae sRNA (middle) and fragmented RNA (*right*) were size selected between 120 and 270 bp, which corresponds to RNA insert sizes of 0–150 nt.

4.3 Coimmunoprecipitation of RNAs That Interact With L7Ae

The RNAs that interact with L7Ae were isolated using the Flag® HA Tandem Affinity Purification kit, which involves two subsequent immunoprecipitation steps. For this, 2–2.5 g *S. acidocaldarius* L7Ae-CFHA or MW001 (control) cells were resuspended in purification buffer (100 m*M* MOPS pH 6.5, 300 m*M* NaCl, 10% (v/v) glycerol, 5 mL/g cells) and lysed by French Pressure (three times at 25,000 psi). After centrifugation of the cell lysate (30,000 × *g*, 20 min, 4°C), 100 μL of washed 50% Flag-Ab slurry were added to 10 mL of the filtered supernatant (Millex® syringe filter, pore size 20 μm). The mixture was incubated for 1 h at 4°C on a rotary shaker.

After centrifugation (3000 × *g*, 2 min, 4°C), most of the supernatant was carefully removed without disturbing the pelleted resin. The remaining supernatant (about 1 mL) was then used to suspend the pellet and to transfer the resin to a SigmaPrep™ spin column. The resin was washed three times (3000 × *g*, 2 min, 4°C) with 500 μL RIPA + PI buffer (50 m*M* Tris pH 8.0, 150 m*M* NaCl, 1% (v/v) Nonidet P-40, 0.5% (w/v) sodium deoxycholate, 0.1% (w/v) SDS, 1% (w/v) protease inhibitor cocktail powder (Sigma-Aldrich). Afterward, the L7Ae-CFHA protein was eluted from the resin by two subsequent elutions with 250 μL 3 × Flag peptide in TBS (0.015% (w/v) 3 × Flag peptide, 50 m*M* Tris–HCl pH 7.6, 150 m*M* NaCl) and 15 min incubation at 4°C on a rotary shaker. 75 μL of washed 50% HA slurry was added to the pooled eluates, and the mixture was incubated for 30 min at 4°C on a rotary shaker. After three washing steps (3000 × *g*, 2 min, 4°C) with 500 μL RIPA + PI buffer, the L7Ae protein was eluted from the HA resin by the addition of 75 μL 8 *M* urea and incubation for 15 min at RT (shaking). This eluate contained coimmunoprecipitated RNAs of L7Ae. All steps were performed on ice or at 4°C to avoid degradation of the RNA. The column was furthermore eluted with 36 μL SDS-loading buffer (75 m*M* Tris–HCl pH 6.8, 0.6% (w/v) SDS, 0.001% (w/v) bromophenol blue, 15% (v/v) glycerol, 1 *M* β-mercaptoethanol) for 15 min at RT (shaking) to elute the L7Ae protein and interacting proteins. The L7Ae-interacting RNAs of the urea eluate were afterward separated by denaturing PAGE and purified via gel extraction (Fig. 2B). The SDS eluate was analyzed by SDS-PAGE to detect L7Ae protein interaction partners (Fig. 2B).

4.4 RNA Fragmentation by $ZnCl_2$ Treatment and Curation of RNA Termini

Aliquots of L7Ae coimmunoprecipitated RNAs were fragmented by $ZnCl_2$ treatment as previously described to allow for Illumina sequencing of longer RNAs (>100 nt) (Dominissini, Moshitch-Moshkovitz, Salmon-Divon, Amariglio, & Rechavi, 2013). Therefore, 16 μL of the L7Ae immunoprecipitation urea eluates were mixed with 1.8 μL of 10 × $ZnCl_2$ fragmentation buffer (100 m*M* Tris–HCl pH 6.8, 100 m*M* $ZnCl_2$), incubated for 150 s at 94°C, and the fragmentation was immediately stopped on ice by the addition of 2 μL 500 m*M* sodium EDTA pH 8.0. The fragmented RNA was subsequently separated by denaturing PAGE and gel extracted. Bivalent metal ion fragmentation is presumed to result in 2′,3′-cyclic phosphate and 5'-OH RNA termini (Forconi & Herschlag, 2009). Therefore, the termini were cured by a dephosphorylation (2′,3′-cyclic phosphate curation) and phosphorylation (5'-OH curation) procedure using T4 PNK. To achieve this, 15 µL of the fragmented RNA were mixed with 6 µL of 5 × dephosphorylation buffer (500 m*M* Tris–HCl pH 6.5, 500 MgAc, 25 m*M* β-mercaptoethanol) and 1 µL of T4 PNK (Ambion) in a total volume of 30 µL and incubated for 6 h at 37°C. Subsequently, 1 m*M* ATP and 1 µL T4 PNK (Ambion) were added and incubated for 1 h at 37°C. The RNA was afterward purified by EtOH precipitation using 1% (v/v) glycogen and separated by denaturing PAGE (Fig. 2C).

4.5 Preparation of cDNA Libraries for Illumina Sequencing

The isolated sRNAs of *S. acidocaldarius* (sRNA-Seq) and the coimmunoprecipitated RNAs of L7Ae proteins (RIP-Seq) were used to prepare cDNA libraries for Illumina sequencing. The library preparation was performed using the NEBNext® Multiplex Small RNA Library Prep Set for Illumina according to the instructions of the manufacturer. Briefly, 5′ and 3' RNA adapters were ligated to the isolated RNA (100 ng RNA input material or the maximum volume of 6.5 µL for low-yield RNA preparations, e.g., mock purifications), and cDNA was produced by a reverse transcriptase, initiated from a primer, which hybridized to the ligated 3′' RNA adapter. Next, the produced cDNA was amplified by PCR (15 cycles and 20 s elongation) utilizing barcoded primers, which allowed distinguishing the different libraries after sequencing. The amplified cDNA libraries were afterward separated by native PAGE and size selected by gel extraction (excision from 120 to 270 and 170 to 270 bp for RIP-Seq/narrow-range sRNome libraries and wide-range sRNome libraries, respectively) (Fig. 2C). After gel elution, the concentration of the different libraries was measured fluorometrically, and the libraries were pooled using a concentration of 2 n*M* cDNA in 10 m*M* Tris pH 8.5 and 0.1% Tween 20. The cDNA libraries were sequenced by Illumina HiSeq2500 methodology.

4.6 Processing and Mapping of Illumina Sequencing Data

The CLC Genomics Workbench 9.5.3 software (Qiagen, Hilden) was used for processing and mapping of the Illumina sequencing data sets. The data were processed by (i) removal of sequences of low quality (quality score limit: 0.05; max. Ambiguities: 2), (ii) trimming of adapter sequences, and (iii) filtering by length (15 nt cutoff). The trimmed sequences were mapped to the *S. acidocaldarius* reference genome (accession no.: CP000077) by using default settings (Fig. 3). The sRNome and L7Ae RIP-Seq data sets are

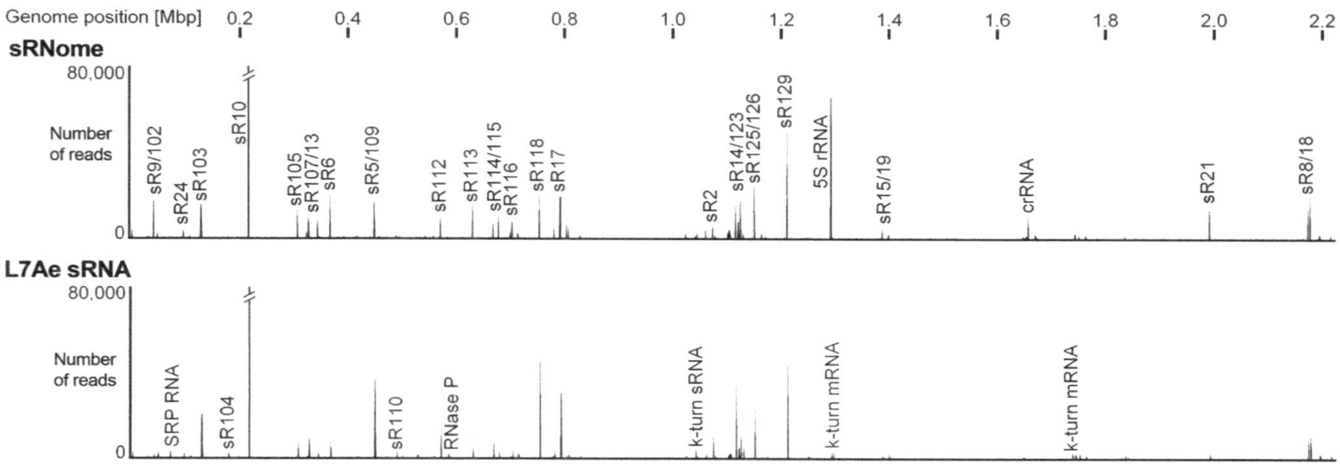

Fig. 3 Global overview of the L7Ae-RNA interactome of *S. acidocaldarius*. The coverage plots of the sequenced sRNome and L7Ae-IP cDNA libraries are illustrated. Peaks representing C/D box sRNAs are labeled by sR#. For clearness, only the peaks which are unique for the L7Ae plot were labeled. Each RNA profile contains 1 million mapped reads. The most abundant RNA is the C/D box sRNA sR10. L7Ae interactions with k-turns structures were observed for C/D box sRNAs, RNase P RNA, SRP RNA, and several 5′UTRs of mRNAs.

available at Gene Expression Omnibus (GSE94748). To identify L7Ae-interacting RNAs a custom peak calling pipeline was applied to the mapped reads of the L7Ae and LSm RIP-Seq data sets as described earlier (Holmqvist et al., 2016). First, the pipeline defined potential binding regions from the read data by utilizing the Blockbuster tool (Langenberger et al., 2009). Next, the statistical significance of these regions in terms of their differential levels of abundance was assessed using the DESeq2 tool by comparing two replicates of each L7Ae RIP-Seq data set with a control data set of untagged L7Ae (one WT control data set for each analyzed protein) (Love, Huber, & Anders, 2014). The input to DESeq2 was composed of count tables that contained the number of reads counted for each identified binding region in the experiments as well as in the control libraries. On the basis of these counts, DESeq2 determined size factors for each library that were then used to calculate normalized read counts by dividing each read count by the corresponding library size factor. The determined normalized read counts were then used for comparisons between libraries, specifically for differential expression testing, and to calculate fold change and false discovery rate (q) values as described (Holmqvist et al., 2016). Identified L7Ae-interacting RNAs with a q-value of <0.1 were further filtered by their normalized read count (at least 3000 reads) to remove false-positive hits due to low-abundance RNAs.

5. PERSPECTIVES AND DISCUSSION

Many ncRNA families were discovered in archaea so far and RNA-seq approaches offer great opportunities for ncRNA identification. RNA-seq was used to detect ncRNAs in archaea of different genera, covering euryarchaeota like *M. mazei, M. kandleri*, and *H. volcanii*; various *Sulfolobus* and *Pyrococcus* species, which belong to the crenarchaeota, and even one member of the Nanoarchaeota, *N. equitans*. The Lowe lab provides databases for archaeal genomes and snoRNAs as well as webserver for snoRNA prediction in archaea (http://lowelab.ucsc.edu/databases.html). Nevertheless, although much is known about ncRNAs in archaea, RNA-seq is limited to cultivatable organisms. Hence, a lot of ncRNAs remain possibly undiscovered, even those which might be responsible for life at extreme conditions.

The development of different RNA-seq protocols enables the identification of primary transcripts as well as detection of RNAs associated with RBPs to unravel the whole ncRNA landscape in cultivable archaea. These modified protocols have been rarely used for archaea to date. For example, dRNA-seq (Sharma et al., 2010), based on the enrichment of primary transcripts, was only used for sRNA identification in *M. mazei* and *H. salinarium* (Jäger et al., 2009; Zaramela et al., 2014). dRNA-seq offers opportunities for quantitative expression profiling and the concomitant TSS mapping for all transcribed genes under the tested condition. With the help of determined TSS, other transcriptome features can be unraveled such as operon structures, 5′UTR length or RNA maturation pathways (Sharma & Vogel, 2014). Although mostly leaderless transcripts were assumed for archaea, dRNA-seq in *M. mazei* showed that 5′UTRs can exhibit impressive lengths of up to 500 nt. This knowledge can be used for identification of sRNA-target interaction mechanisms or structured 5′UTR RNA elements such as riboswitches or RNA thermometer.

Considering all identified ncRNA families to date, no family is unique to the domain of Archaea. They reveal similarities to both, bacterial and eukaryotic RNA networks. Identified regulatory mechanisms of ncRNA underline this fact. C/D box sRNAs show similar characteristics and modification mechanisms as known from eukaryotes. Small regulatory RNAs in archaea are related to those of bacteria comparing structural and mechanistic features. One exception is RNA-257 from *S. solfataricus* which functions as previously described for eukaryotic miRNAs or siRNAs by binding to the 3′UTR (Fig. 1). However, functional classification of ncRNAs in archaea is still lagging behind.

Global RNA-seq approaches are needed for identification of RNA-protein networks. In general two approaches have been developed, RNA immunoprecipitation followed by sequencing (RIP-seq) and cross-linked immunoprecipitation followed by sequencing (CLIP-seq). Both methods enable the identification of RBP-associated RNAs and allow the additional identification of RNA-binding motifs for RBP and possibly their interaction mechanism (Holmqvist et al., 2016). Although RNA-seq offers great opportunities to investigate the ncRNA landscape of archaea, it provides only limited insights into the regulatory mechanisms of ncRNAs. Thus, additional RNA-seq modifications for detection of RNA–protein interaction were developed in bacteria and eukaryotes, but have not been tested in archaea so far. Digestion-optimized RIP-seq (DO-RIP-seq) provides a quantitative score of whole transcripts targeting by RBP and determines RNA-binding site strengths in vivo (Nicholson, Friedersdorf, & Keene, 2017). These methods can be particularly useful to map the modifications mediated by C/D box sRNAs.

Only four archaeal sRNAs target mRNAs were identified to date using a combination of different experiments, including computational target prediction using IntaRNA or CopraRNA (Busch, Richter, & Backofen, 2008; Wright et al., 2014). Both tools give reliable predictions for bacterial as well as archaeal sRNA interactions. In bacteria, sRNA–mRNA interactions were also detected using RNA interaction by ligation followed by sequencing (RIL-seq) which was not used for archaea so far. In this approach sRNA-target pairs bound by Hfq were ligated prior to sequencing. The result is an enrichment of chimeric cDNA reads which can be subsequent aligned to sRNAs and mRNAs (Melamed et al., 2016), helping to identify novel sRNA–mRNA pairs. But RIL-seq is based on RNA–protein interactions and until now it is not clear, whether archaeal RBPs such as Sm-Like proteins (Lsm) are required for sRNA–mRNA interactions. This might be unraveled using gradient profiling sequencing (Grad-seq). Grad-seq relies on sedimentation of whole RNA bound to proteins in a glycerol gradient and thereby sorting complexes by size and shape (Smirnov et al., 2016). The subsequent RNA-seq of defined fractions detects all RNAs associated to different RBPs, whereby mRNA–ribosome complexes can be distinguished from sRNA–RBP complexes. Detected sRNAs can be used for pull-downs with cell lysates which enable the identification of novel RBP.

RNA-seq is a very valuable tool which has developed rapidly in the last decade. Protocol modifications helped with the identification of novel ncRNA classes and possible interaction mechanisms or partner, e.g., RBP, in all three domains of life. Future applications of these RNA-seq approaches and complementary bioinformatics studies will allow us to obtain a more complete picture of the functional roles of the discovered ncRNAs and their impact on regulatory networks in organisms that often experience extreme growth conditions.

REFERENCES

Babski, J., Maier, L.-K., Heyer, R., Jaschinski, K., Prasse, D., Jäger, D., et al. (2014). Small regulatory RNAs in archaea. *RNA Biology*, *11*(5), 484–493.

Babski, J., Tjaden, B., Voss, B., Jellen-Ritter, A., Marchfelder, A., Hess, W. R., et al. (2011). Bioinformatic prediction and experimental verification of sRNAs in the haloarchaeon *Haloferax volcanii*. *RNA Biology*, *8*(5), 806–816.

Barrick, J. E., & Breaker, R. R. (2007). The distributions, mechanisms, and structures of metabolite-binding riboswitches. *Genome Biology*, *8*(11), R239.

Bernick, D. L., Dennis, P. P., Hochsmann, M., & Lowe, T. M. (2012). Discovery of *Pyrobaculum* small RNA families with atypical pseudouridine guide RNA features. *RNA*, *18*(3), 402–411.

Bernick, D. L., Dennis, P. P., Lui, L. M., & Lowe, T. M. (2012). Diversity of antisense and other non-coding RNAs in archaea revealed by comparative small RNA sequencing in four *Pyrobaculum* species. *Frontiers in Microbiology, 3*, 231.

Breaker, R. R. (2018). Riboswitches and translation control. *Cold Spring Harbor Perspectives in Biology*. https://doi.org/10.1101/cshperspect.a032797.

Brock, T. D., Brock, K. M., Belly, R. T., & Weiss, R. L. (1972). *Sulfolobus*: A new genus of sulfur-oxidizing bacteria living at low pH and high temperature. *Archiv für Mikrobiologie, 84*(1), 54–68.

Buddeweg, A., Sharma, K., Urlaub, H., & Schmitz, R. A. (2018). $sRNA_{41}$ affects ribosome binding sites within polycistronic mRNAs in *Methanosarcina mazei* Gö1. *Molecular Microbiology, 107*(5), 595–609.

Busch, A., Richter, A. S., & Backofen, R. (2008). IntaRNA: Efficient prediction of bacterial sRNA targets incorporating target site accessibility and seed regions. *Bioinformatics, 24*(24), 2849–2856.

Chaban, B., Ng, S. Y., & Jarrell, K. F. (2006). Archaeal habitats-from the extreme to the ordinary. *Canadian Journal of Microbiology, 52*(2), 73–116.

Danan, M., Schwartz, S., Edelheit, S., & Sorek, R. (2011). Transcriptome-wide discovery of circular RNAs in archaea. *Nucleic Acids Research, 40*(7), 3131–3142.

Dar, D., Prasse, D., Schmitz, R. A., & Sorek, R. (2016). Widespread formation of alternative 3′ UTR isoforms via transcription termination in archaea. *Nature Microbiology, 1*, 16143.

Dar, D., Shamir, M., Mellin, J. R., Koutero, M., Stern-Ginossar, N., Cossart, P., et al. (2016). Term-seq reveals abundant ribo-regulation of antibiotics resistance in bacteria. *Science, 352*(6282), aad9822.

Daume, M., Uhl, M., Backofen, R., & Randau, L. (2017). RIP-Seq suggests translational regulation by L7Ae in archaea. *MBio, 8*(4), e00730-17.

Demolli, S., Geist, M. M., Weigand, J. E., Matschiavelli, N., Suess, B., & Rother, M. (2014). Development of beta-lactamase as a tool for monitoring conditional gene expression by a tetracycline-riboswitch in *Methanosarcina acetivorans*. *Archaea, 2014*, 725610.

Dennis, P. P. (1997). Ancient ciphers: Translation in archaea. *Cell, 89*(7), 1007–1010.

Dennis, P. P., Tripp, V., Lui, L., Lowe, T., & Randau, L. (2015). C/D box sRNA-guided 2'-O-methylation patterns of archaeal rRNA molecules. *BMC Genomics, 16*, 632.

Dominissini, D., Moshitch-Moshkovitz, S., Salmon-Divon, M., Amariglio, N., & Rechavi, G. (2013). Transcriptome-wide mapping of N(6)-methyladenosine by m(6) A-seq based on immunocapturing and massively parallel sequencing. *Nature Protocols, 8*(1), 176–189.

Forconi, M., & Herschlag, D. (2009). Metal ion-based RNA cleavage as a structural probe. *Methods in Enzymology, 468*, 91–106.

Fütterer, O., Angelov, A., Liesegang, H., Gottschalk, G., Schleper, C., Schepers, B., et al. (2004). Genome sequence of *Picrophilus torridus* and its implications for life around pH 0. *Proceedings of the National Academy of Sciences of the United States of America, 101*(24), 9091–9096.

Gaspin, C., Cavaille, J., Erauso, G., & Bachellerie, J. P. (2000). Archaeal homologs of eukaryotic methylation guide small nucleolar RNAs: Lessons from the *Pyrococcus* genomes. *Journal of Molecular Biology, 297*(4), 895–906.

Gehring, A. M., Walker, J. E., & Santangelo, T. J. (2016). Transcription regulation in archaea. *Journal of Bacteriology, 198*(14), 1906–1917.

Gimpel, M., & Brantl, S. (2017). Dual-function small regulatory RNAs in bacteria. *Molecular Microbiology, 103*(3), 387–397.

Gong, S., Wang, Y., Wang, Z., & Zhang, W. (2017). Co-transcriptional folding and regulation mechanisms of riboswitches. *Molecules, 22*(7), 1169.

Grant, W. D., Kamekura, M., McGenity, T. J., & Ventosa, A. (2001). Family I. Halobacteriaceae. In J. A. Bell, T. G. Lilburn, & G. M. Garrity (Eds.), *Bergey's manual of systematic bacteriology (2nd ed): Vol. 1* (pp. 299–301). New York: Springer-Verlag.

Heyer, R., Dorr, M., Jellen-Ritter, A., Späth, B., Babski, J., Jaschinski, K., et al. (2012). High throughput sequencing reveals a plethora of small RNAs including tRNA derived fragments in *Haloferax volcanii*. *RNA Biology*, *9*(7), 1011–1018.

Hille, F., Richter, H., Wong, S. P., Bratovic, M., Ressel, S., & Charpentier, E. (2018). The biology of CRISPR-Cas: Backward and forward. *Cell*, *172*(6), 1239–1259.

Holmqvist, E., Wright, P. R., Li, L., Bischler, T., Barquist, L., Reinhardt, R., et al. (2016). Global RNA recognition patterns of post-transcriptional regulators Hfq and CsrA revealed by UV crosslinking in vivo. *The EMBO Journal*, *35*(9), 991–1011.

Jäger, D., Förstner, K. U., Sharma, C. M., Santangelo, T. J., & Reeve, J. N. (2014). Primary transcriptome map of the hyperthermophilic archaeon *Thermococcus kodakarensis*. *BMC Genomics*, *15*(1), 684.

Jäger, D., Pernitzsch, S. R., Richter, A. S., Backofen, R., Sharma, C. M., & Schmitz, R. A. (2012). An archaeal sRNA targeting cis- and trans-encoded mRNAs via two distinct domains. *Nucleic Acids Research*, *40*, 10964–10979.

Jäger, D., Sharma, C. M., Thomsen, J., Ehlers, C., Vogel, J., & Schmitz, R. A. (2009). Deep sequencing analysis of the *Methanosarcina mazei* Gö1 transcriptome in response to nitrogen availability. *Proceedings of the National Academy of Sciences of the United States of America*, *106*(51), 21878–21882.

Jaschinski, K., Babski, J., Lehr, M., Burmester, A., Benz, J., Heyer, R., et al. (2014). Generation and phenotyping of a collection of sRNA gene deletion mutants of the haloarchaeon *Haloferax volcanii*. *PLoS One*, *9*(3), e90763.

Kashefi, K., & Lovley, D. R. (2003). Extending the upper temperature limit for life. *Science*, *301*(5635), 934.

Klein, R. J., Misulovin, Z., & Eddy, S. R. (2002). Noncoding RNA genes identified in AT-rich hyperthermophiles. *Proceedings of the National Academy of Sciences of the United States of America*, *99*(11), 7542–7547.

Koonin, E. V., Makarova, K. S., & Zhang, F. (2017). Diversity, classification and evolution of CRISPR-Cas systems. *Current Opinion in Microbiology*, *37*, 67–78.

Koskinen, K., Pausan, M. R., Perras, A. K., Beck, M., Bang, C., Mora, M., et al. (2017). First insights into the diverse human archaeome: Specific detection of archaea in the gastrointestinal tract, lung, and nose and on skin. *MBio*, *8*(6), e00824–17.

Langenberger, D., Bermudez-Santana, C., Hertel, J., Hoffmann, S., Khaitovich, P., & Stadler, P. F. (2009). Evidence for human microRNA-offset RNAs in small RNA sequencing data. *Bioinformatics*, *25*(18), 2298–2301.

Lloyd, C. R., Park, S., Fei, J., & Vanderpool, C. K. (2017). The small protein SgrT controls transport activity of the glucose-specific phosphotransferase system. *Journal of Bacteriology*, *199*(11), e00869-16.

Love, M. I., Huber, W., & Anders, S. (2014). Moderated estimation of fold change and dispersion for RNA-seq data with DESeq2. *Genome Biology*, *15*(12), 550.

Malys, N., & McCarthy, J. E. (2011). Translation initiation: Variations in the mechanism can be anticipated. *Cellular and Molecular Life Sciences*, *68*(6), 991–1003.

Marchfelder, A., Fischer, S., Brendel, J., Stoll, B., Maier, L. K., Jäger, D., et al. (2012). Small RNAs for defence and regulation in archaea. *Extremophiles*, *16*(5), 685–696.

Märtens, B., Manoharadas, S., Hasenöhrl, D., Manica, A., & Bläsi, U. (2013). Antisense regulation by transposon-derived RNAs in the hyperthermophilic archaeon *Sulfolobus solfataricus*. *EMBO Reports*, *14*(6), 527–533.

Melamed, S., Peer, A., Faigenbaum-Romm, R., Gatt, Y. E., Reiss, N., Bar, A., et al. (2016). Global mapping of small RNA-target interactions in bacteria. *Molecular Cell*, *63*(5), 884–897.

Moissl-Eichinger, C., Pausan, M., Taffner, J., Berg, G., Bang, C., & Schmitz, R. A. (2018). Archaea are interactive components of complex microbiomes. *Trends in Microbiology, 26*(1), 70–85.

Nicholson, C. O., Friedersdorf, M., & Keene, J. D. (2017). Quantifying RNA binding sites transcriptome-wide using DO-RIP-seq. *RNA, 23*(1), 32–46.

Nickel, L., Weidenbach, K., Jager, D., Backofen, R., Lange, S. J., Heidrich, N., et al. (2013). Two CRISPR-Cas systems in *Methanosarcina mazei* strain Go1 display common processing features despite belonging to different types I and III. *RNA Biology, 10*(5), 779–791.

Olsen, G. J., & Woese, C. R. (1997). Archaeal genomics: An overview. *Cell, 89*(7), 991–994.

Omer, A. D., Lowe, T. M., Russell, A. G., Ebhardt, H., Eddy, S. R., & Dennis, P. P. (2000). Homologs of small nucleolar RNAs in archaea. *Science, 288*(5465), 517–522.

Phok, K., Moisan, A., Rinaldi, D., Brucato, N., Carpousis, A. J., Gaspin, C., et al. (2011). Identification of CRISPR and riboswitch related RNAs among novel noncoding RNAs of the euryarchaeon *Pyrococcus abyssi. BMC Genomics, 12*, 312.

Prasse, D., Förstner, K. U., Jäger, D., Backofen, R., & Schmitz, R. A. (2017). sRNA154 a newly identified regulator of nitrogen fixation in *Methanosarcina mazei* strain Gö1. *RNA Biology, 14*(11), 1544–1558.

Prasse, D., Thomsen, J., De Santis, R., Muntel, J., Becher, D., & Schmitz, R. A. (2015). First description of small proteins encoded by spRNAs in *Methanosarcina mazei* strain Go1. *Biochimie, 117*, 138–148.

Raymann, K., Moeller, A. H., Goodman, A. L., & Ochman, H. (2017). Unexplored archaeal diversity in the great ape gut microbiome. *mSphere*, 2(1), e00026–17.

Reeve, J. N. (1992). Molecular biology of methanogens. *Annual Review of Microbiology, 46*, 165–191.

Richter, H., Mohr, S., & Randau, L. (2013). C/D box sRNA, CRISPR RNA and tRNA processing in an archaeon with a minimal fragmented genome. *Biochemical Society Transactions, 41*(1), 411–415.

Sharma, C. M., Hoffmann, S., Darfeuille, F., Reignier, J., Findeiss, S., Sittka, A., et al. (2010). The primary transcriptome of the major human pathogen *Helicobacter pylori. Nature, 464*(7286), 250–255.

Sharma, C. M., & Vogel, J. (2014). Differential RNA-seq: The approach behind and the biological insight gained. *Current Opinion in Microbiology, 19*, 97–105.

Smirnov, A., Förstner, K. U., Holmqvist, E., Otto, A., Günster, R., Becher, D., et al. (2016). Grad-seq guides the discovery of ProQ as a major small RNA-binding protein. *Proceedings of the National Academy of Sciences of the United States of America, 113*(41), 11591–11596.

Speed, M. C., Burkhart, B. W., Picking, J. W., & Santangelo, T. J. (2018). An archaeal, fluoride-responsive riboswitch provides an inducible expression system for hyperthermophiles. *Applied and Environmental Microbiology, 84*, e02306–17.

Starostina, N. G., Marshburn, S., Johnson, L. S., Eddy, S. R., Terns, R. M., & Terns, M. P. (2004). Circular box C/D RNAs in *Pyrococcus furiosus. Proceedings of the National Academy of Sciences of the United States of America, 101*(39), 14097–14101.

Storz, G., Wolf, Y. I., & Ramamurthi, K. S. (2014). Small proteins can no longer be ignored. *Annual Review of Biochemistry, 83*, 753.

Straub, J., Brenneis, M., Jellen-Ritter, A., Heyer, R., Soppa, J., & Marchfelder, A. (2009). Small RNAs in haloarchaea: Identification, differential expression and biological function. *RNA Biology, 6*(3), 281–292.

Su, A. A., Tripp, V., & Randau, L. (2013). RNA-Seq analyses reveal the order of tRNA processing events and the maturation of C/D box and CRISPR RNAs in the hyperthermophile *Methanopyrus kandleri. Nucleic Acids Research, 41*, 6250–6258.

Tang, T. H., Bachellerie, J. P., Rozhdestvensky, T., Bortolin, M. L., Huber, H., Drungowski, M., et al. (2002). Identification of 86 candidates for small non-messenger RNAs from the archaeon *Archaeoglobus fulgidus*. *Proceedings of the National Academy of Sciences of the United States of America*, *99*(11), 7536–7541.

Tang, T. H., Polacek, N., Zywicki, M., Huber, H., Brugger, K., Garrett, R., et al. (2005). Identification of novel non-coding RNAs as potential antisense regulators in the archaeon *Sulfolobus solfataricus*. *Molecular Microbiology*, *55*(2), 469–481.

Toffano-Nioche, C., Ott, A., Crozat, E., Nguyen, A. N., Zytnicki, M., Leclerc, F., et al. (2013). RNA at 92°C: The non-coding transcriptome of the hyperthermophilic archaeon *Pyrococcus abyssi*. *RNA Biology*, *10*(7), 1211–1220.

Vanderpool, C. K., Balasubramanian, D., & Lloyd, C. R. (2011). Dual-function RNA regulators in bacteria. *Biochimie*, *93*(11), 1943–1949.

Wagner, & Romby, P. (2015). Small RNAs in bacteria and archaea: Who they are, what they do, and how they do it? *Advances in Genetics*, *90*, 133–208.

Wagner, van Wolferen, M., Wagner, A., Lassak, K., Meyer, B. H., Reimann, J., et al. (2012). Versatile genetic tool box for the crenarchaeote Sulfolobus acidocaldarius. *Frontiers in Microbiology*, *3*, 214.

Weinberg, Z., Wang, J. X., Bogue, J., Yang, J., Corbino, K., Moy, R. H., et al. (2010). Comparative genomics reveals 104 candidate structured RNAs from bacteria, archaea, and their metagenomes. *Genome Biology*, *11*(3), R31.

Woese, C. R., Kandler, O., & Wheelis, M. L. (1990). Towards a natural system of organisms: Proposal for the domains archaea, bacteria, and eucarya. *Proceedings of the National Academy of Sciences of the United States of America*, *87*(12), 4576–4579.

Wright, P. R., Georg, J., Mann, M., Sorescu, D. A., Richter, A. S., Lott, S., et al. (2014). CopraRNA and IntaRNA: Predicting small RNA targets, networks and interaction domains. *Nucleic Acids Research*, *42*(web server issue), W119–W123.

Wurtzel, O., Sapra, R., Chen, F., Zhu, Y., Simmons, B. A., & Sorek, R. (2010). A single-base resolution map of an archaeal transcriptome. *Genome Research*, *20*(1), 133–141.

Zaramela, L. S., Vencio, R. Z., ten-Caten, F., Baliga, N. S., & Koide, T. (2014). Transcription start site associated RNAs (TSSaRNAs) are ubiquitous in all domains of life. *PLoS One*, *9*(9), e107680.

CHAPTER EIGHTEEN

Methodology for Ribosome Profiling of Key Stages of the *Caulobacter crescentus* Cell Cycle

James R. Aretakis, Nadra Al-Husini, Jared M. Schrader[1]
Department of Biological Sciences, Wayne State University, Detroit, MI, United States
[1]Corresponding author: e-mail address: schrader@wayne.edu

Contents

Abstract

Bacterial cell division is the result of a productive round of the cell cycle to yield two daughter cells. The cell cycle is highly coordinated in *Caulobacter crescentus* where it is driven by a cell cycle gene-regulatory network that coordinates gene expression with the major cell cycle events such as chromosome replication and cell division. Recent ribosomes profiling data showed that 484 genes undergo changes in translation efficiency during the cell cycle, suggesting a broad role for translational control in cell cycle regulation. In this chapter, we focus on how to perform ribosome profiling to measure the translation efficiency across cellular mRNAs at key stages in the *Caulobacter* cell cycle. This methodology relies on the high-yield ludox gradient synchronization of *Caulobacter* cells followed by ribosome profiling to measure ribosome density and total RNA-seq to measure mRNA levels.

Methods in Enzymology, Volume 612
ISSN 0076-6879
https://doi.org/10.1016/bs.mie.2018.07.008

1. INTRODUCTION

In bacteria, cell cycle-regulatory networks control progression of the cell cycle by integrating information about key cell cycle events into their gene expression programs. *Caulobacter crescentus* has proven to be a powerful model of the bacterial cell cycle due to the high-yield and rapid ability to perform cell synchronization (Evinger & Agabian, 1977; Schrader & Shapiro, 2015). In *Caulobacter*, the well-defined cell cycle-regulatory network is controlled by a set of master regulators that are arranged in a transcriptional regulatory circuit (Collier, 2016; Lasker, Mann, & Shapiro, 2016; McAdams & Shapiro, 2011). While transcriptional control, DNA methylation, phosphosignaling, protein localization, and regulated proteolysis have well-established roles in the cell cycle-regulatory circuit, translational control was recently found to play a significant role where 12.4% of its genes have cell cycle-regulated translation (Schrader et al., 2016). Ribosome profiling now allows genome-wide interrogation of mRNA translation, making it a powerful tool toward understanding the gene-regulatory networks controlling complex cellular events (Brar et al., 2012; Ingolia, Ghaemmaghami, Newman, & Weissman, 2009; Stumpf, Moreno, Olshen, Taylor, & Ruggero, 2013). The methodology described here integrates the high-yield of *Caulobacter* cell synchronization with the power of ribosome profiling to measure translation efficiency at key stages of the cell cycle. This method allows unbiased interrogation of translational control in bacterial cell cycles with genome-wide coverage.

2. *CAULOBACTER* CELL CYCLE RIBOSOME PROFILING

Ribosome profiling described here is performed in multiple phases including Sections 2.1—2.6 (Fig. 1). This protocol takes several days to perform (Table 1) so it is important to include control samples throughout the process. Typically, ribosome profiling in bacteria uses large volumes of cultures but this protocol is adapted to small samples containing only 30 mL of log-phase cells retrieved upon ludox gradient synchronization. To measure translation efficiency total RNA-seq is compared to ribosome footprints to calculate the translation efficiency, which is the metric measured across the cell cycle.

2.1 Cell Growth, Synchronization, and Harvesting

Caulobacter synchronization by ludox gradient synchronization yields a large amount of swarmer cells that can be followed during the cell cycle

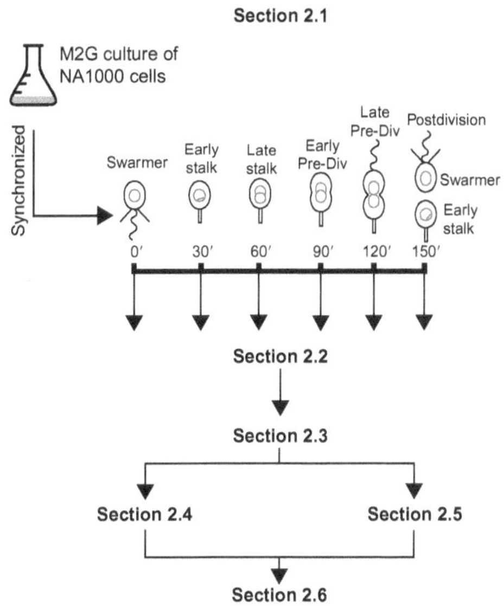

Fig. 1 Schematic of *Caulobacter* ribosome profiling procedure. Section 2.1: Grow 1 L of NA1000 cells, synchronize, and then grow to one of the cell cycle time points. Section 2.2: To each cell population, harvest by chloramphenicol pretreatment (2 min at 28°C) followed by rapid cell collection on ice. Section 2.3: Lysis is performed under liquid nitrogen, and a small portion is saved as a frozen aliquot for total RNA-seq. Section 2.4: Footprint mRNA fragments, select for size, and recover digested mRNA fragments. Section 2.5: Use the lysate aliquot from Section 2.3 and prepare for total RNA-seq. Section 2.6: Sequencing libraries are generated in the same way from the products of both ribosome footprinting and total RNA-seq.

Table 1 Expected Completion Time for Each Section

Section	Approximate Work Time
2.1	3 days
2.2	1 h
2.3	1.5 h
2.4	10 h
2.5	13 h
2.6	3 days

(Evinger & Agabian, 1977). Briefly, 1 L of mid-log phase NA1000 cells are arrested in growth by carbon starvation, then swarmer cells are isolated by ludox gradient centrifugation. This procedure has been covered in depth recently at both large and small scales (Schrader & Shapiro, 2015). For each ribosome profiling sample, 1 L of cells grown in M2G should be synchronized, and the 30 mL of resulting swarmer cells at $OD_{600} = 0.5$ should be grown to the desired cell cycle stage and rapidly harvested for ribosome profiling and total RNA-seq.

2.1.1 Equipment

- Standard Shaker Incubator
- Spectrophotometer
- Optical Cuvettes
- Corex Tubes
- JA-20 Fixed-Angle Aluminum Rotor (Beckman Coulter)

2.1.2 Buffers and Reagents

- *C. crescentus* Strain NA1000 (Synchronizable Variant).
- PYE Media: 2 g of Bactopeptone, 1 g of Yeast Extract, 2 mL of 0.5 *M* $MgSO_4$, 5 mL of 0.1 *M* $CaCl_2$, fill up to 1 L with Milli-Q filtered water (autoclaved solution).
- 20 × M2 Salts: 17.4 g of Na_2HPO_4, 10.6 g of KH_2PO_4, 10 g of NH_4Cl, Resuspend in 1 L of Milli-Q filtered water (autoclaved solution).
- M2G Media: 50 mL of 20 × M2 Salts, 1 mL of 0.5 *M* $MgSO_4$, 10 mL of 20% Glucose, 1 mL of Ferrous Sulfate Chelate Solution (Sigma-Aldrich), fill to 995 mL with Milli-Q filtered water, and then add 5 mL of 0.1 *M* $CaCl_2$ (add last after dilution to avoid precipitation). Sterilize with a 0.22-μm filter.
- Ludox AS-40 Solution (Sigma).

2.1.3 Procedure

- In a shaker incubator at 28°C and 250 rpm, grow an overnight culture of *Caulobacter* strain NA1000 in 5 mL of PYE media.
- Transfer 200 μL of saturated culture to 25 mL of M2G media and grow overnight in a shaker incubator at 28°C and 250 rpm.
- Read the OD_{600} of the 25 mL culture using spectrophotometer.
- Using a doubling time of 2.33 h, calculate how much of the 25 mL culture needs to be added to 1 L of culture to give a mid-log ($OD_{600} = 0.5$) 1 L culture in the morning.

- Verify the presence of swarmer cells using a phase-contrast microscope and a wet mount of the cells.
- A detailed video protocol of the synchronization procedure can be found in Schrader and Shapiro (2015).
- Harvest cells by centrifugation and suspend in 180 mL cold M2 salts.
- 60 mL of cold ludox solution should be mixed, and the resulting suspension should be pipetted into eight 30 mL corex tubes and spun for 30 min at 6400 × *g* in a JA-20 rotor.
- The lower band is the swarmer band. The top of the solution can be aspirated off and discarded. The lower swarmer cell band should be collected from each tube and pooled in a new 50-mL conical tube.
- Check again by phase microscopy that the cells are highly enriched for swarmer cells (90%–95% motile swarmer cells).
- The swarmer cells should be pelleted and washed twice in 30 mL of cold M2 each. Finally, the cells should be suspended in 30 mL of 28°C M2G media at an $OD_{600} = 0.5$ and placed into a sterile 250-mL flask and shaken at 250 rpm until the desired cell cycle stage.

2.1.4 Notes

- After synchronization, an entire 1 L cell synchrony allows for the isolation of one cell cycle ribosome profiling sample.
- It is important to examine and confirm that the cells are in the proper phase of the cell cycle by imaging with a wet mount on a phase-contrast microscope. Swarmer cells should be small and motile, stalked cells should be small and immotile, and predivisional cells should be larger with a dumbbell shape morphology.

2.2 Harvesting Cells

Caulobacter cells can only be harvested by centrifugation as their cells rapidly clog 0.22-μm filters used for ribosome profiling of *Bacillus subtilis* or *Escherichia coli* cells (Oh et al., 2011). Harvesting cells by centrifugation is a process that must be performed in a timely matter to minimize artifacts arising from the chloramphenicol addition. After removal from the shaker incubator, cell temperature needs to be rapidly lowered to 4°C and maintained there until frozen in liquid nitrogen. A quick harvesting time and rapidly lowering cell temperatures by addition of ice cubes are necessary for the efficient harvesting of ribosomes trapped mRNA fragments.

2.2.1 Equipment

- Sorvall Legend X1R Centrifuge (Thermo Scientific)
- Fiberlite F15-8 × 50cy Fixed-Angle Rotor (Thermo Scientific)
- 50-mL Conical Tube
- Standard Ice Cube Tray
- Hammer
- Zip-Top Bag
- Standard Syringe Needle

2.2.2 Buffers and Reagents

- 50 mg/mL Chloramphenicol solution in ethanol.
- Ice cubes of M2G minimal media with 100 μg/mL of Chloramphenicol, 15 mL each, stored at −80°C.
- Resuspension Buffer: 20 m*M* Tris–HCl pH 8.0, 10 m*M* $MgCl_2$, 100 m*M* NH_4Cl, 1 m*M* chloramphenicol.
- Lysis Buffer: 20 m*M* Tris pH 8.0, 10 m*M* $MgCl_2$, 100 m*M* NH_4Cl, 5 m*M* $CaCl_2$, 0.4% Triton X-100, 0.1% NP-40, 1 m*M* chloramphenicol, 100 U/mL RNase-free DNase I (Roche).
- Liquid Nitrogen.

2.2.3 Procedure

- Before harvesting the cells, take an ice cube out of the −80°C freezer, place ice cube in zip-top bag, crush using a hammer and place each crushed cube into a 50-mL conical tube, and store the tube on ice. Each tube is sufficient for harvesting 30 mL of cells.
- For each cell cycle time point, 30 mL of cells at $OD_{600} = 0.5$ are treated with chloramphenicol (addition of 60 μL chloramphenicol to a final concentration of 100 μg/mL) and incubate in 28°C shaker for 2 min at 250 rpm.
- Immediately add the cell culture to a 50-mL prechilled conical tube (4°C) containing 15 mL of crushed ice (M2G broth with 100 μg/mL of chloramphenicol), cap the tube and invert twice.
- Place tubes in the prechilled, −10°C F15-8 rotor and spin at 14,000 rpm (22,789 × *g*) for 1.5 min to pellet cells.
- Discard supernatant by decanting and suspend cell pellet in 20 mL of prechilled Resuspension Buffer.
- Pellet cells again at 14,000 rpm (22,789 × *g*) for 1.5 min to pellet cells in −10°C F15-8 rotor.

- Discard supernatant by decanting and suspend cell pellet in 300 μL of Lysis Buffer.
- Slowly drip the cell pellet/lysis buffer suspension into a liquid nitrogen filled 50-mL conical tube.
- Poke holes in 50-mL conical lid using a hot needle and pour out remaining liquid nitrogen in the tube. Store the cell pellet in a −80°C freezer.

2.2.4 Notes

- Cell collection should be performed as quickly as possible to avoid artifacts resulting from changes in translation occurring in response to the cell collection procedure.
- Cells should be cooled down rapidly by addition to crushed ice (M2G broth with 100 μg/mL of chloramphenicol) and maintained chilled throughout the cell collection procedure.
- After the initial spin, the supernatant should still contain small pieces of ice.
- Frozen cell pellets can be stored at −80°C for at least a year.

2.3 Cell Lysis

Cell lysis is performed in a mixer mill with samples frozen in liquid nitrogen to minimize changes in ribosome position occurring in a liquid lysate (Oh et al., 2011). Save a small frozen aliquot for RNA-seq. Most of the lysate is used for ribosome footprinting.

2.3.1 Equipment

- Mixer Mill MM 400 (Roche)
- Grinding Jar 10 mL (Roche)
- 12-mm Grinding Ball (Roche)
- 1.5-mL RNase-Free Microcentrifuge Tube

2.3.2 Procedure

- Chill assembled jar and grinding ball in liquid nitrogen.
- Open jar and add frozen cell pellet/lysis buffer suspension to chilled jar with grinding ball and close lid.
- Mill at 15 Hz for 3 min. Chill the sealed jar in liquid nitrogen between each run. Five runs total.

- A small scoop of the frozen pulverized cells should be scraped out of the jars using a prechilled scoopula and stored in a prechilled 1.5-mL microcentrifuge tube at −80°C for total RNA-seq.
- To thaw the remaining lysate, place each half, chamber side up, in warm 30°C water.

2.3.3 Notes

- Use cryogenic gloves when working with liquid nitrogen.

2.4 Footprinting and Footprint Extraction

The thawed lysates are then footprinted by addition of micrococcal nuclease (MNase) which generates heterogeneous mRNA footprint sizes unlike the 28 nt footprints generated by eukaryotic ribosomes (Ingolia et al., 2009; O'Connor, Li, Weissman, Atkins, & Baranov, 2013; Oh et al., 2011). MNase reactions can be quenched by addition of EGTA to chelate Ca^{2+} ions needed for nuclease activity. After footprinting with MNase, the 70S ribosomes are purified on a sucrose gradient. It is important to also run an undigested control on the sucrose gradient to ensure that polysomes were preserved upon cell harvesting and lysis (Fig. 2).

2.4.1 Equipment

- Standard Microcentrifuge
- Nanodrop Spectrophotometer (Thermo Fisher)
- Open-Top Polyclear Tubes (7030 Seton)
- TH-641 Swinging Bucket Rotor Package (Thermo Scientific) (SW 41 Ti equivalent)

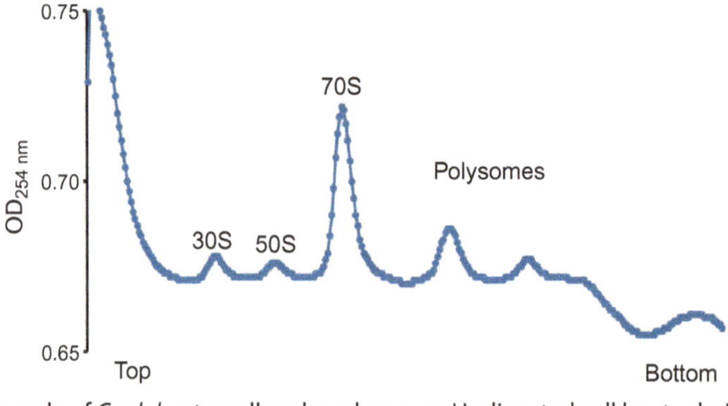

Fig. 2 Example of *Caulobacter* cell cycle polysomes. Undigested cell lysate derived from a single cell cycle time point was separated on a 10%–55% sucrose gradient.

- Sorvall WX+ Ultracentrifuge (Thermo Scientific)
- 2-mL Siliconized RNase-Free Microcentrifuge Tubes
- Standard Thermomixer
- Gradient Station (BIO-COMP)

2.4.2 Buffers and Reagents

- 10 m*M* Tris–HCl pH 7.0, from 1 *M* stock (Ambion)
- Lysis Buffer: 20 m*M* Tris–HCl pH 8.0, 10 m*M* $MgCl_2$, 100 m*M* NH_4Cl, 5 m*M* $CaCl_2$, 0.4% Triton X-100, 0.1% NP-40, 1 m*M* chloramphenicol, 100 U/mL RNase-free DNase I (Roche)
- MNase Enzyme 1500 U/μL (Roche)
- SUPERase·In RNase Inhibitor Enzyme 20 U/μL (Ambion)
- 0.5 *M* EGTA pH 8.0
- 10% Sucrose solution: 10% sucrose (w/v), 20 m*M* Tris–HCl pH 8.0, 10 m*M* $MgCl_2$, 100 m*M* NH_4Cl, 1 m*M* Chloramphenicol
- 50% Sucrose solution: 50% sucrose (w/v), 20 m*M* Tris–HCl pH 8.0, 10 m*M* $MgCl_2$, 100 m*M* NH_4Cl, 1 m*M* Chloramphenicol
- Gradient Balance Buffer: 20 m*M* Tris–HCl pH 8.0, 10 m*M* $MgCl_2$, 100 m*M* NH_4Cl, 1 m*M* Chloramphenicol
- 20% SDS
- Acid: Phenol Chloroform (Ambion)
- Chloroform
- 3 *M* NaOAc pH 5.5 (Ambion)
- 100% Isopropanol
- 80% Ethanol from 100% stock
- Novex 2 × TBE-Urea Sample Buffer (Invitrogen)

2.4.3 Procedure

- Continue from Section 2.3.2 by recovering the pulverized cells from the thawing chamber and placing them into a prechilled 1.5-mL microcentrifuge tube for footprinting.
- Incubate the lysate for 20 min on ice.
- Pellet cell debris at 20,000 × *g* for 10 min in a 4°C microcentrifuge.
- Transfer supernatant to a prechilled siliconized microcentrifuge tube without disturbing pelleted cells.
- Dilute 1 μL of lysate with 99 μL of 10 m*M* Tris–HCl pH 7.0. Measure A_{260} by Nanodrop spectrophotometer (blank with 1 μL lysis buffer diluted into 99 μL of 10 m*M* Tris–HCl pH 7.0).

- Digest 500 mg of lysate RNA with 1.3 µL of MNase and 6 µL of SUPERase·In. Add lysis buffer for a 200 µL reaction. Treat undigested polysome control similarly, except without the addition of MNase.
- Incubate at 25°C for 1 h while shaking in a thermomixer at 1400 rpm.
- Quench reaction with 2 µL of EGTA and immediately return to ice.
- Chill TH-641 swinging rotors and buckets to 4°C.
- Fill open-top polyclear tube with 55% sucrose solution until you reach the line on the maker block metal stand (~6 mL).
- Carefully layer 10% sucrose solution on top.
- Cap open-top polyclear tubes.
- From gradient using Gradient Master parameters for a SW 41 Ti rotor, short capped tubes, and 7%–47% weight by volume sucrose gradient.
- Remove cap and carefully layer nuclease digested cell lysate and undigested polysome control cell lysate into separate tubes.
- Use gradient buffer to balance tubes.
- Spin at 35,000 rpm (209,627 × *g*) for 2.5 h at 4°C.
- Fractionate using Gradient Station fractionator with a piston speed of 0.2 mm/s.
- Manually collect fractionated monosomes, cap 2.0-mL siliconized microcentrifuge tube, flash-freeze in liquid nitrogen, and stored at −80°C indefinitely.
- To extract total RNA from footprinted monosomes, add 40 µL of 20% SDS to 0.7 mL of thawed monosomes fraction solution. Monosomes may need to be split into two tubes of ~0.7 mL each.
- Add 0.7 mL of acid phenol prewarmed to 65°C to the tube and mix by vortexing.
- Incubate at 65°C for 5 min while shaking in a thermomixer.
- Chill on ice for 5 min.
- Spin at 20,000 × *g* for 2 min in a microcentrifuge and transfer supernatant to fresh tube.
- Add 0.7 mL of room temperature acid phenol, mix by vortexing, and then incubate at room temperature for 5 min.
- Spin at 20,000 × *g* for 2 min in a microcentrifuge and transfer aqueous layer to fresh tube.
- Add 0.6 mL of chloroform and mix by vortexing.
- Immediately spin at 20,000 × *g* for 1 min in a microcentrifuge and transfer aqueous layer to fresh tube.
- Add 75 µL of 3 *M* NaOAc pH 5.5 and 0.8 mL of 100% isopropanol. Mix by vortexing.

- Chill at −80°C for 30 min.
- Pellet RNA at 20,000 × *g* for 1 h in a 4°C microcentrifuge.
- Wash pellet with 0.8 mL of −20°C 80% ethanol.
- Air-dry pellet for 5 min and suspend in 20 μL of 10 m*M* Tris–HCl pH 7.0.
- Dilute 1 μL of precipitated RNA with 9 μL of 10 m*M* Tris pH 7.0 and then quantify with Nanodrop spectrophotometer.
- Dilute 50 μg of RNA to 20 μL with 10 m*M* Tris–HCl pH 7.0 and then add 20 μL of 2 × TBE-Urea sample buffer.
- Save samples at − 20°C until Section 2.5.3 and run size selection for both samples at once on 1 × TBE 7 *M* UREA PAGE gel.

2.4.4 Notes

- When layering 10% sucrose on top of 55% sucrose in an open-top polyclear tube care must be taken, especially in the beginning, to ensure that there is no mixing at the boundary between the two layers.
- Carefully transport the sucrose gradient tubes to not disturb the boundary layer. Use smooth, even movements, especially when attaching and removing the buckets from the rotor.
- Surplus lysis buffer can be used to balance rotor buckets instead of gradient buffer.
- All six rotor buckets should be filled with sucrose and balanced for the spin even if all buckets are not loaded with samples.
- Fractionated monosomes typically have a final volume of 1.2 mL.
- Use siliconized microcentrifuge tubes to avoid transfer loss.

2.5 Total RNA-Seq Sample Preparation and PAGE Size Selection

To measure translation efficiency, mRNA levels are required to normalize the ribosome footprint density. To get high-quality total RNA-seq data, rRNA, which is abundant, should be removed from the sample. rRNA-depleted total RNA should then be hydrolyzed under denaturing conditions to yield a size distribution that is similar to ribosome footprint fragment length.

2.5.1 Equipment

- Standard Thermomixer
- Standard Microcentrifuge
- Standard 1-D Vertical Gel Electrophoresis Equipment

- 2-mL Siliconized Microcentrifuge Tube
- Corning Costar Spin-X Centrifuge Tube Filters (Fisher Scientific)
- 2100 Bioanalyzer (Agilent)

2.5.2 Buffers and Reagents

- 20% SDS
- Acid: Phenol Chloroform (Ambion)
- Dry Ice
- 100% Ethanol
- Chloroform
- 3 *M* NaOAc pH 5.5 (Ambion)
- 0.5 *M* EDTA
- 100% Isopropanol
- 80% Ethanol from 100% Stock
- 10 m*M* Tris–HCl pH 7.0, from 1 *M* stock (Ambion)
- Novex 2 × TBE-Urea Sample Buffer (Invitrogen)
- 10 Base Pair Ladder (Invitrogen)
- Novex TBE-Urea Gels, 15%, 10-well (Invitrogen)
- 1 × TBE from 10 × stock (Ambion)
- SYBR Gold (Invitrogen)
- Elution Buffer: 10 m*M* Tris–HCl pH 7.0, 1 m*M* EDTA, 0.5 *M* NaOAc pH 5.5
- Glycogen (20 mg/mL)
- Bioanalyzer RNA 6000 Nano Kit (Agilent)
- Ribo-Zero rRNA Removal Kit: Gram-Negative Bacteria (Illumina)
- 2 × Alkaline hydrolysis buffer: 1 μL of 0.5 *M* EDTA, 30 μL of 0.1 *M* Na_2CO_3, 220 μL of 0.1 *M* $NaHCO_3$
- o199-p RNA oligo (IDT)

2.5.3 Procedure

- To frozen pulverized cell powder aliquot from Section 2.3.2, add 0.6 mL of RNA lysis buffer including 40 μL of 20% SDS.
- Immediately add 0.7 mL of acid phenol prewarmed to 65°C to the tube and mix by vortexing.
- Incubate at 65°C for 5 min while shaking in a thermomixer.
- Chill in a dry ice—ethanol bath for 5 min (or until completely frozen).
- Spin at 20,000 × *g* for 2 min in a microcentrifuge and transfer supernatant to a fresh tube.
- Add 0.7 mL of room temperature acid phenol, mix by vortexing, and then incubate at room temperature for 5 min.

- Spin at 20,000 × *g* for 2 min in a microcentrifuge and transfer aqueous layer to fresh tube.
- Add 0.6 mL of chloroform and mix by vortexing.
- Immediately spin at 20,000 × *g* for 1 min in a microcentrifuge and transfer aqueous layer to fresh tube.
- Add 65 μL of 3 *M* NaOAc pH 5.5 (or 1/9 volume) and 0.7 mL of 100% isopropanol (or 1.1 volume), and mix by vortexing.
- Chill at −80°C for 30 min.
- Pellet RNA at 20,000 × *g* for 1 h in a 4°C microcentrifuge.
- Quickly wash pellet with 0.8 mL of −20°C 80% ethanol.
- Suspend in 50 μL of 10 m*M* Tris–HCl pH 7.0.
- Assess RNA quality before and after ribosomal removal (Fig. 3) using Bioanalyzer and RNA 6000 Nano Kit, following manufactures instructions.

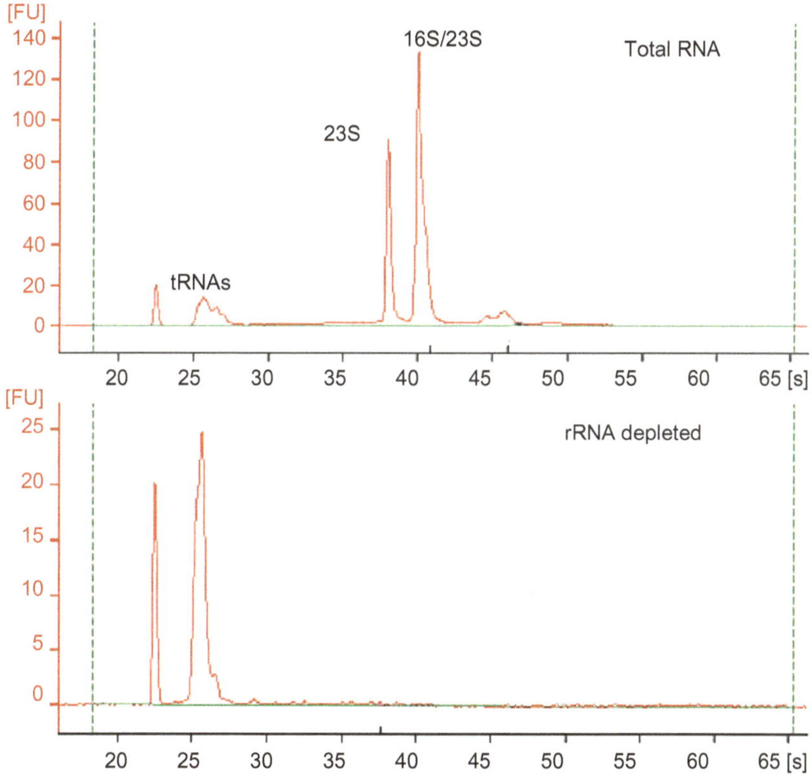

Fig. 3 rRNA depletion of total RNA samples for total RNA-seq. Bioanalyzer traces before (*top*) and after rRNA removal (*bottom*). Stable RNA peaks are labeled. *Caulobacter* 23S rRNA is processed into two fragments, one runs at the same size as the 16S rRNA and the other runs at a shorter length. *Y*-axis scales are different for the two graphs.

- 5 μg of RNA in a volume of no more than 26 μL can used for the subsequent steps.
- Remove 16S and 23S ribosomal RNAs by subtractive hybridization using Ribo-Zero rRNA Removal Kit (Gram-Negative Bacteria). Follow manufactures instructions.
- The protocol requires washing the magnetic beads, treating samples with rRNA removal solution, removing RNA, and purifying rRNA-depleted samples.
- Prepare 2 × alkaline hydrolysis buffer and add 12.5 μL of 2 × alkaline hydrolysis buffer to 12.5 μL of enriched mRNA.
- Incubate at 95°C for 23 min and immediately return to ice.
- Add 560 μL of stop buffer to end hydrolysis.
- Add 650 μL of 100% isopropanol. Mix by vortexing
- Chill at −80°C for 30 min.
- Pellet hydrolyzed RNA at 20,000 × *g* for 1 h in a 4°C microcentrifuge.
- Wash pellet with 0.8 mL of −20°C 80% ethanol.
- Dry and suspend pellet in 10 μL of 10 m*M* Tris–HCl pH 7.0.
- Add 10 μL of 2 × TBE-Urea sample buffer.
- Prepare 10 base pair ladder in parallel with a solution containing: 1 μL of 10 base pair ladder, 9 μL of 10 m*M* Tris–HCl pH 7.0, and 10 μL of 2 × TBE-Urea sample buffer.
- Prepare o199-P control oligo in parallel with a solution containing: 2 μL of 10 μ*M* o199-P, 8 μL of 10 m*M* Tris–HCl pH 7.0, and 10 μL of 2 × TBE-Urea sample buffer.
- Thaw extracted ribosome footprints sample from Section 2.4.3.
- Denature ladder and RNA samples at 80°C for 1 min and return to ice.
- Set up TBE-Urea Gel 15% in 1 × TBE and prerun for 1 h at 200 V.
- Wash and then load lanes. Total RNA samples and footprinting samples from Section 2.4.3 should be loaded on the same gel. Run for 65 min at 200 V.
- Stain gel with 6 μL of SYBR Gold in 60 mL of 1 × TBE for 5 min and excise desired bands from 30 to 40 nt based on ladder and o199-p oligo.
- *RNA recovery from acrylamide gels* will be repeated in this procedure so it will be explained in detail here and only referenced subsequently.
- Pierce a needle through a 0.5-mL tube and insert gel slice in pierced tube and nest pierced tube into a 2-mL siliconized microcentrifuge tube.
- Spin at 20,000 × *g* for 2 min in a microcentrifuge, or until most of the gel has extruded through, and transfer the remaining gel pieces.
- Add 0.5 mL of elution buffer and freeze at −80°C for 20 min.

- Nutate overnight in a cold room or 4°C refrigerator.
- Incubate at 25°C for 5 min while shaking in a thermomixer at 1400 rpm.
- Transfer gel slurry to a Spin-X centrifuge tube filter using a wide bore pipette or a cut standard tip.
- Add 2 μL of glycogen and 550 μL of 100% isopropanol (1.1 equivalence volume), then mix by vortexing and chill at −20°C for 1 h.
- Pellet RNA at 20,000 × *g* for 1 h in a 4°C microcentrifuge and aspirate away the supernatant.
- Wash pellet with 0.8 mL of 4°C 80% ethanol.
- Pulse spin to aspirate away remaining ethanol and air-dry for 5 min.
- Suspend recovered RNA in 15 μL of 10 m*M* Tris–HCl pH 7.0.

2.5.4 Notes

- o199-p control and ladder help to identify appropriate RNA fragments.
- Ribosome footprints may need to be spread across up to four lanes of the gel due to high concentration to avoid overloading.

2.6 Library Generation

Library preparation is performed by 3′ end repair, 3′ linker ligation, reverse transcription, single strand cDNA circularization, and finally PCR (Ingolia, Brar, Rouskin, McGeachy, & Weissman, 2012). These steps should be performed on both the footprinted mRNA fragments and the size-selected mRNA fragments generated for mRNA-seq. Control o199-p RNA oligo can be utilized throughout the library prep to ensure each step has worked efficiently. Quality control steps are noted below to ensure final DNA libraries contain a maximal amount of mRNA fragments. Library generation is performed as described in Ingolia et al. (2012) but using the T4 RNA ligase KQ mutant (NEB) due to higher efficiency of ligation. DNA oligos utilized here have been adapted to work on the newer illumina sequencers including the HiSeq 4000 and MiniSeq, as well as older sequencers (Table 2).

2.6.1 Equipment

- Standard Thermomixer
- Standard Microcentrifuge
- Standard 1-D Vertical Gel Electrophoresis Equipment
- 2-mL Siliconized Microcentrifuge Tubes
- 0.2-mL PCR Tube
- Agilent Bioanalyzer
- Qubit Fluorimeter

Table 2 Oligos for Library Prep Generation

o199-P	5′ AUGUACACGGAGUCGACCCGCAACGCGA/3phos/3′
Linker-1 (NEB)	5′ App/CTGTAGGCACCATCAAT/3ddC 3′
o225-linker-mini-indexed	5′/5Phos/AGATCGGAAGAGCGTCGTGTAGGGAAAGAGTGT /iSp18/CACTCA/iSp18/CAAGCAGAAGACGGCATACGAGAT ATTGATGGTGCCTACAG 3'
JS-ind1	5′ AATGATACGGCGACCACCGAGATCTACACGATCGGAA GAGCACACGTCTGAACTCCAGTCACATCACGACACTCTT TCCCTACACGACG 3′
JS-ind2	5′ AATGATACGGCGACCACCGAGATCTACACGATCGGAA GAGCACACGTCTGAACTCCAGTCACCGATGTACACTCTT TCCCTACACGACG 3′
JS-ind3	5′ AATGATACGGCGACCACCGAGATCTACACGATCGGAA GAGCACACGTCTGAACTCCAGTCACTTAGGCACACTCTT TCCCTACACGACG 3′
JS-ind4	5′ AATGATACGGCGACCACCGAGATCTACACGATCGGAA GAGCACACGTCTGAACTCCAGTCACTGACCAACACTCTT TCCCTACACGACG 3′
JS-ind5	5′ AATGATACGGCGACCACCGAGATCTACACGATCGGAA GAGCACACGTCTGAACTCCAGTCACACAGTGACACTCTT TCCCTACACGACG 3′
JS-ind6	5′ AATGATACGGCGACCACCGAGATCTACACGATCGGAA GAGCACACGTCTGAACTCCAGTCACGCCAATACACTCTT TCCCTACACGACG 3′
JS-ind7	5′ AATGATACGGCGACCACCGAGATCTACACGATCGGAA GAGCACACGTCTGAACTCCAGTCACCAGATCACACTCTT TCCCTACACGACG 3′
JS-ind8	5′ AATGATACGGCGACCACCGAGATCTACACGATCGGAA GAGCACACGTCTGAACTCCAGTCACACTTGAACACTCTT TCCCTACACGACG 3′
JS-ind9	5′ AATGATACGGCGACCACCGAGATCTACACGATCGGAA GAGCACACGTCTGAACTCCAGTCACGATCAGACACTCTT TCCCTACACGACG 3′
JS-ind10	5′ AATGATACGGCGACCACCGAGATCTACACGATCGGAA GAGCACACGTCTGAACTCCAGTCACTAGCTTACACTCTT TCCCTACACGACG 3′
JS-ind11	5′ AATGATACGGCGACCACCGAGATCTACACGATCGGAA GAGCACACGTCTGAACTCCAGTCACGGCTACACACTCTT TCCCTACACGACG 3′
JS-ind12	5′ AATGATACGGCGACCACCGAGATCTACACGATCGGAA GAGCACACGTCTGAACTCCAGTCACCTTGTAACACTCTT TCCCTACACGACG 3′
oJS231	5′ CAAGCAGAAGACGGCATACGAGATATTGATGGTGCC 3′

- Corning Costar Spin-X Centrifuge Tube Filters (Fisher Scientific)
- Standard Nutator

2.6.2 Buffers and Reagents

- 10 × T4 PNK Buffer supplied with enzyme (New England Biolabs)
- T4 PNK Enzyme (New England Biolabs)
- SUPERase·In RNase Inhibitor Enzyme 20 U/μL (Ambion)
- Dephosphorylation Buffer Master Mix: 2 μL of 10 × T4 PNK buffer, 1 μL SUPERase·In (for each reaction)
- 10 m*M* Tris–HCl pH 7.0 from 1 *M* stock (Ambion)
- 0.5 *M* EDTA
- 3 *M* NaOAc pH 5.5 (Ambion)
- Precipitation Master Mix: 10 m*M* Tris pH 7.0, 3 m*M* EDTA, 0.375 *M* NaOAc pH 5.5
- Glycogen (20 mg/mL)
- 100% Isopropanol
- 100% Ethanol
- 80% Ethanol from 100% stock
- 50% Sterile filtered PEG M.W. 8000
- 10 × T4 RNA Ligase 2 Buffer, supplied with enzyme (New England Biolabs)
- 100% DMSO
- 1 μg/μL of Linker-1 (New England Biolabs)
- T4 RNA Ligase 2, Truncated K227Q Enzyme (New England Biolabs)
- Novex 2 × TBE-Urea Sample Buffer (Invitrogen)
- Ligation Master Mix: 8 μL of 50% sterile filter PEG M.W., 2 μL of 10 × T4 Ligase 2 buffer, 2 μL of 100% DMSO, 1 μL of SUPERase·In, 1 μL of 1 μg/μL of Linker-1 (for each reaction).
- 10 Base Pair Ladder (Invitrogen)
- Novex TBE-Urea Gels, 15%, 10-well (Invitrogen)
- 1 × TBE from 10 × stock (Ambion)
- SYBR Gold (Invitrogen)
- Elution Buffer: 10 m*M* Tris–HCl pH 7.0, 1 m*M* EDTA, 0.5 *M* NaOAc pH 5.5
- 10 m*M* dNTP (New England Biolabs)
- 20 μ*M* o225-linker-mini-indexed (IDT)
- 5 × FSB Buffer supplied with enzyme (Invitrogen)
- 0.1 *M* DTT supplied with enzyme (Invitrogen)
- Superscript III (Invitrogen)
- 1 N NaOH

- 10 m*M* Tris–HCl pH 8.0 from 1 *M* stock (Ambion)
- Nucleic Acid Master Mix: 1 μL of 10 m*M* dNTP, 1 μL of 25 μ*M* o225-linker-mini-indexed, 1.5 μL of Milli-Q filtered water (for each reaction)
- Reverse Transcription Buffer Master Mix: 4 μL of 5 × FSB buffer, 1 μL of 0.1 *M* DTT, 1 μL of SUPERase·In (for each reaction)
- CircLigase Enzyme Kit (Epicentre)
- Circularization Master Mix: 2 μL of 10 × CircLigase buffer, 1 μL of 1 m*M* ATP, 1 μL of 50 m*M* $MnCl_2$ (all components supplied with enzyme)
- High-Fidelity Phusion Enzyme (New England Biolabs)
- 10 m*M* dNTP (New England Biolabs)
- 100 μ*M* DNA oligos (from Table 2) (IDT)
- Amplification Master Mix: 16.7 μL of 5 × High-Fidelity buffer (supplied with enzyme), 1.7 μL of dNTP, 0.4 μL of 100 μ*M* JS-ind (use separate oligos #1–12 for each biological sample), 0.4 μL of 100 μ*M* oJS231, 59.2 μL of Milli-Q filtered water, 0.8 μL of High-Fidelity Phusion Enzyme (for each sample)
- 6 × DNA loading dye
- 0.05% Tween 20
- EBT: 10 m*M* Tris–HCl pH 8.0, 0.05% Tween 20
- High Sensitivity DNA kit (Agilent)

2.6.3 Procedure

- Place the size-selected mRNA footprints, size-selected total mRNA fragments, and o199-P control sample on ice.
- To dephosphorylate the samples, prepare dephosphorylation buffer master mix and add to 15 μL of the sample.
- Add 2 μL of T4 PNK and incubate reaction at 37°C for 1 h.
- Prepare precipitation master mix and add 90 μL to each tube.
- Heat kill enzyme at 75°C for 10 min. Acid:Phenol chloroform extraction will work as well.
- Add 0.3 mL of 100% ethanol, 2 μL glycogen and mix by vortexing.
- Chill at −20°C for 1 h.
- Pellet dephosphorylated RNA at 20,000 × *g* for 1 h in a 4°C microcentrifuge.
- Remove supernatant, wash pellet with 0.8 mL of 4°C 80% ethanol, pellet by spinning at 20,000 × *g* for 15 min in a 4°C microcentrifuge. Repeat washes two additional times.
- Air-dry pellet for 5 min.

- Suspend pellet in 5 μL of 10 m*M* Tris pH 7.0.
- Prepare the ligation master mix. The PEG will crash out of solution upon addition of DMSO. Warm the tube in your hands to solubilize.
- Add 2 μL of T4 RNA Ligase 2, Truncated K227Q recording lot number, mix by vigorously pipetting and incubate reaction at 25°C for 2.5 h.
- Add 20 μL of 2 × TBE-Urea sample buffer to the samples.
- Prepare 10 bp ladder in parallel: 1 μL of 10 bp ladder, 5 μL of 10 m*M* Tris–HCl pH 7.0, 6 μL of 2 × TBE-Urea sample buffer.
- Denature samples at 70°C for 1 min and return to ice.
- Set up 10% TBE-Urea gel in 1 × TBE and prerun for 1 h at 200 V. It is useful that o199-P negative control and ligated 0199-P positive control are included as sizing controls and to assess efficiency of ligation.
- Wash wells and load the gel. Every other lane should be skipped as to not cross contaminate samples.
- Run for 50 min at 200 V.
- Stain gel with 6 μL of SYBR gold in 60 mL of 1 × TBE for 5 min.
- Excise the desired band and recover ligated products.
- To recover RNA, follow the procedure detailed at the end of Section 2.5.3.
- Resuspend in 10 μL of 10 m*M* Tris–HCl pH 7.0.
- For reverse transcription, prepare nucleic acid master mix and add 3.5–10 μL of ligated RNA.
- Denature at 60°C for 5 min and return to ice.
- Prepare reverse transcription buffer master mix, add 6 μL to the sample, and then add 1 μL of Superscript III.
- Incubate at 50°C for 30 min and then add 2.3 μL of 1 N NaOH.
- Incubate at 95°C for 15 min.
- To ethanol precipitate add 23 μL of −20°C 100% ethanol (3 volumes), 2.3 μL of sodium acetate (1/10 volume), and 2 μL of glycogen then mix by vortexing and chill at −20°C for 1 h.
- Pellet cDNA at 20,000 × *g* for 1 h in a 4°C microcentrifuge and then aspirate away the supernatant. Wash pellet with 1 mL of −20°C 80% ethanol. Repeat wash two more times.
- Pulse spin to aspirate away remaining ethanol and air-dry for 5 min.
- After the pellet is dry, suspend the sample in 5 μL of 1 × Novex TBE-Urea Sample Buffer.
- Prepare 10-bp ladder in parallel: 1 μL of 10 bp ladder, 9 μL of 10 m*M* Tris–HCl pH 8.0, 10 μL of 2 × TBE-Urea sample buffer.
- Denature all samples at 80°C for 2 min and return to ice.

- Set up 9% TBE-Urea gel in 1 × TBE. Prerun for 1 h at 200 V.
- Wash lanes and load in samples. Run for 70 min at 200 V checking predicted size vs dye.
- Stain gel with 6 μL of SYBR gold in 60 mL of 1 × TBE for 5 min.
- Excise desired bands and recover the reverse transcription products.
- To recover DNA pierce a needle through a 0.5-mL tube, insert gel slice in pierced tube, and nest pierced tube into a 2-mL siliconized microcentrifuge tube.
- Spin at 20,000 × *g* for 2 min in a microcentrifuge or until most of the gel has extruded through. Transfer the remaining gel pieces.
- Add 0.5 mL of elution buffer (1/9 volume) and freeze at −80°C for 20 min.
- Incubate at 70°C for 10 min. Shake in thermomixer at 1400 rpm.
- Transfer gel slurry to a Spin-X centrifuge tube filter using a wide bore pipette or a cut standard tip.
- Add 2 μL of glycogen and 550 μL of 100% isopropanol (1.1 volume), then mix by vortexing and chill at −20°C for 1 h.
- Pellet DNA at 20,000 × *g* for 1 h in a 4°C microcentrifuge and aspirate away the supernatant.
- Wash pellet with 0.8 mL of 4°C 80% ethanol.
- Pulse spin to aspirate away remaining ethanol and air-dry for 5 min.
- After DNA has been recovered, resuspend in 15 μL of 10 m*M* Tris–HCl pH 8.0.
- Add 4 μL of circularization master mix and 1 μL of CircLigase enzyme and incubate at 60°C for 1 h. Include o225-linker-mini-indexed alone as negative control.
- Heat kill enzyme at 80°C for 10 min and then place on ice.
- For PCR amplification, prepare amplification master mix and add 79.2–5 μL of circularized ssDNA from proceeding steps. *PCR primers containing DNA barcodes to allow for multiplexed sequencing on illumina sequencers should be used (Table 2) with a different index oligo for each sample.*
- Aliquot 17 μL of PCR mix into four separate PCR strip tubes and perform the following PCR reaction:

Thermal Cycling Conditions

Initial denaturing	98°C	30 s
Denaturing	98°C	10 s
Annealing	60°C	10 s
Extension	72°C	5 s

- Remove PCR tubes after cycles 6, 8, 10, and 12. *The sample with the fewest number of cycles with visible PCR product will be gel extracted.* As a control, include the o225-linker-mini-indexed to mark the size of no-insert products.
- Set up 8% TB polyacrylamide gel in 1 × TBE.
- Directly load samples and run at 180 V for 40 min.
- Stain gel with 6 μL of SYBR gold in 60 mL of 1 × TBE for 5 min.
- Excise insert DNA band and recover double-stranded DNA products by piercing a needle through a 0.5-mL tube.
- Insert gel slice in pierced tube and nest pierced tube into a 2-mL siliconized microcentrifuge tube.
- Spin at 20,000 × *g* for 2 min in a microcentrifuge, or until most of the gel has extruded through. Transfer the remaining gel pieces.
- Add 0.5 mL of elution buffer (1/9 volume) and freeze at −80°C for 20 min.
- Incubate at 70°C for 10 min in thermomixer at 1400 rpm.
- Transfer gel slurry to a Spin-X centrifuge tube filter using a wide bore pipette or a cut standard tip.
- Add 2 μL of glycogen and 550 μL of 100% isopropanol (1.1 volume), mix by vortexing, and then chill at −20°C for 1 h.
- Pellet DNA at 20,000 × *g* for 1 h in a 4°C microcentrifuge and aspirate away the supernatant.
- Wash pellet with 0.8 mL of 4°C 80% ethanol.
- Pulse spin to aspirate away remaining ethanol and air-dry for 5 min.
- Suspend amplified DNA in 10 μL of EBT.
- For quantification of amplified DNA, dilute 1 μL of amplified DNA with 9 μL of 10 m*M* Tris–HCl pH 8.0.
- Assess DNA quality using the bioanalyzer by the steps outlined by Agilent for the High Sensitivity DNA Kit. Quantify DNA concentration using the Qubit fluorimeter.
- Samples should then be pooled in equal amounts for Illumina sequencing (50 bp single end) on an Illumina HiSeq instrument.

2.6.4 Notes

- Bring the o199 control oligo through each step-in library prep to assess problematic reactions.
- Ligation efficiency markedly decreases if PEG is old so it is recommended that PEG be made within a month of use.
- T4 RNA Ligase 2, Truncated can be used for ligation instead of T4 RNA Ligase 2, Truncated K227Q mutant, but higher yields have been obtained with the K227Q mutant.

- Addition of 1 N NaOH and heating at 95°C during the reverse transcription steps quenches the reaction by hydrolyzing RNA template.
- Loading the gel with small volumes avoids formation of "bunny ears" on gels.

3. SUMMARY AND CONCLUSION

If followed correctly, this procedure allows the genome-wide measurement of translation at the key stages of the *Caulobacter* cell cycle. For each ORF, the ratio of ribosome footprint density to total RNA-seq density can be used to calculate translation efficiency, which can be assessed across the cell cycle. Currently, *Caulobacter* is the only organism with a high-yield drug-free method of cell cycle synchronization making it a powerful organism for the study of the bacterial cell cycle. Innovative new methodologies in cell synchronization, such as "baby machines" (Bates et al., 2005), should allow for the expansion of ribosome profiling studies into the cell cycles of other bacterial species.

ACKNOWLEDGMENTS

Research reported in this publication was supported by NIGMS of the National Institutes of Health under award number R35GM124733. The authors thank Wayne State University startup funds to J.M.S.

COMPETING INTERESTS

The authors declare no competing interests exist.

REFERENCES

Bates, D., Epstein, J., Boye, E., Fahrner, K., Berg, H., & Kleckner, N. (2005). The Escherichia coli baby cell column: A novel cell synchronization method provides new insight into the bacterial cell cycle. *Molecular Microbiology*, *57*(2), 380–391. https://doi.org/10.1111/j.1365-2958.2005.04693.x.

Brar, G. A., Yassour, M., Friedman, N., Regev, A., Ingolia, N. T., & Weissman, J. S. (2012). High-resolution view of the yeast meiotic program revealed by ribosome profiling. *Science*, *335*(6068), 552–557. https://doi.org/10.1126/science.1215110.

Collier, J. (2016). Cell cycle control in Alphaproteobacteria. *Current Opinion in Microbiology*, *30*, 107–113. https://doi.org/10.1016/j.mib.2016.01.010.

Evinger, M., & Agabian, N. (1977). Envelope-associated nucleoid from *Caulobacter crescentus* stalked and swarmer cells. *Journal of Bacteriology*, *132*(1), 294–301.

Ingolia, N. T., Brar, G. A., Rouskin, S., McGeachy, A. M., & Weissman, J. S. (2012). The ribosome profiling strategy for monitoring translation in vivo by deep sequencing of ribosome-protected mRNA fragments. *Nature Protocols*, 7(8), 1534–1550. https://doi.org/10.1038/nprot.2012.086.

Ingolia, N. T., Ghaemmaghami, S., Newman, J. R. S., & Weissman, J. S. (2009). Genome-wide analysis in vivo of translation with nucleotide resolution using ribosome profiling. *Science, 324*(5924), 218–223. https://doi.org/10.1126/science.1168978.

Lasker, K., Mann, T. H., & Shapiro, L. (2016). An intracellular compass spatially coordinates cell cycle modules in Caulobacter crescentus. *Current Opinion in Microbiology, 33*, 131–139. https://doi.org/10.1016/j.mib.2016.06.007.

McAdams, H. H., & Shapiro, L. (2011). The architecture and conservation pattern of whole-cell control circuitry. *Journal of Molecular Biology, 409*(1), 28–35. https://doi.org/10.1016/j.jmb.2011.02.041.

O'Connor, P. B., Li, G. W., Weissman, J. S., Atkins, J. F., & Baranov, P. V. (2013). rRNA: mRNA pairing alters the length and the symmetry of mRNA-protected fragments in ribosome profiling experiments. *Bioinformatics, 29*(12), 1488–1491. https://doi.org/10.1093/bioinformatics/btt184.

Oh, E., Becker, A. H., Sandikci, A., Huber, D., Chaba, R., Gloge, F., et al. (2011). Selective ribosome profiling reveals the cotranslational chaperone action of trigger factor in vivo. *Cell, 147*(6), 1295–1308. https://doi.org/10.1016/j.cell.2011.10.044.

Schrader, J. M., Li, G. W., Childers, W. S., Perez, A. M., Weissman, J. S., Shapiro, L., et al. (2016). Dynamic translation regulation in Caulobacter cell cycle control. *Proceedings of the National Academy of Sciences of the United States of America, 113*(44), E6859–E6867. https://doi.org/10.1073/pnas.1614795113.

Schrader, J. M., & Shapiro, L. (2015). Synchronization of Caulobacter crescentus for investigation of the bacterial cell cycle. *Journal of Visualized Experiments, 98,* e52633. https://doi.org/10.3791/52633.

Stumpf, C. R., Moreno, M. V., Olshen, A. B., Taylor, B. S., & Ruggero, D. (2013). The translational landscape of the mammalian cell cycle. *Molecular Cell, 52*(4), 574–582. https://doi.org/10.1016/j.molcel.2013.09.018.

CHAPTER NINETEEN

Transcriptome-Wide Analysis of Protein–RNA and RNA–RNA Interactions in Pathogenic Bacteria

Jai J. Tree*[,1], Kenn Gerdes[†], David Tollervey[‡,1]

*School of Biotechnology and Biomolecular Sciences, University of New South Wales, Sydney, NSW, Australia

[†]Department of Biology, Centre for Bacterial Stress Response and Persistence, University of Copenhagen, Copenhagen, Denmark

[‡]Wellcome Centre for Cell Biology, University of Edinburgh, Edinburgh, United Kingdom

[1]Corresponding authors: e-mail address: j.tree@unsw.edu.au; D.Tollervey@ed.ac.uk

Contents

Methods in Enzymology, Volume 612
ISSN 0076-6879
https://doi.org/10.1016/bs.mie.2018.08.009

Abstract

Regulatory RNAs and RNA-binding proteins (RBPs) play critical roles in virulence gene expression in pathogenic bacteria. A wealth of regulatory RNAs have been identified in bacterial pathogens using RNA-seq and recent technical advances are uncovering their mRNA targets. UV-crosslinking is a powerful tool for identifying protein binding sites throughout the transcriptome providing base-pair resolution of sites in vivo. With minor modifications to the protocol, RNA–RNA interactions can also be captured by proximity-dependent ligation of RNA pairs on the protein. Here, we described a high-stringency UV-crosslinking method for recovery of both protein–RNA interactions (CRAC) and RNA–RNA interactions occurring on the bait protein (CLASH). These analyses provide complementary data that provide insights into RBP, and regulatory RNA function.

1. INTRODUCTION

Free-living, unicellular organisms require continual signal perception to sense and respond to changing environmental cues. The ability to sense and exploit new environments is exemplified by many pathogenic bacteria. Transitions from environmental reservoir, to reservoir/vector host, and humans can often require distinct virulence factors that promote survival in each environment. A complex, interleaved network of transcriptional and posttranscriptional control is emerging from recent studies (Holmqvist et al., 2010; Melamed et al., 2016; Papenfort, Espinosa, Casadesús, & Vogel, 2015; Sedlyarova et al., 2016; Waters et al., 2017). One- and two-component sensor-regulators play critical roles in shaping the transcriptional landscape in response to sensory cues. Additional regulatory inputs are post-transcriptionally layered onto this landscape, to enhance accurate signal perception, and integrate responses with broader regulatory networks.

RNA-binding proteins (RBPs) and regulatory RNAs play prominent roles in regulating virulence genes in pathogenic bacteria. The most thoroughly characterized to date remains the Hfq-dependent regulatory small RNA (sRNA) network (Chao & Vogel, 2010; Feliciano, Grilo, Guerreiro, Sousa, & Leitão, 2016). However, additional global RBPs play important roles in pathogenicity, including ProQ, CsrA (RsmA), CspA, and CspE (Holmqvist, Li, Bischler, Barquist, & Vogel, 2018; Michaux et al., 2017; Vakulskas, Potts, Babitzke, Ahmer, & Romeo, 2015). High-throughput techniques for profiling the RNA-binding proteome (RBPome) are poised to expand the RBP repertoire in bacteria even further. Pathogenic bacteria exploit posttranscriptional regulation for host-sensing and CsrA is required for contact-dependent translation of effector proteins that are injected into host cells by pathogenic *Escherichia coli* (Katsowich et al., 2017). Novel regulatory RNAs are also being uncovered that regulate gene expression in response to host cell invasion (exemplified by PinT) (Westermann et al., 2016), and many regulatory RNAs are known to be key regulators of virulence factors (Bardill & Hammer, 2012; Bronesky et al., 2016; Srikumar, Colgan, & Hinton, 2012).

The RNA binding sites for RBPs can be identified by immunoprecipitation techniques (RIP-seq), which can be enhanced by using chemical or UV-crosslinking (Lee & Ule, 2018). Treatment with formaldehyde chemically crosslinks RNA, DNA, and proteins and can lead to recovery of indirect RNA targets associated with RBP complexes. These effects can be used to stabilize larger complexes for recovery of transient RNA–RNA interactions (FLASH, fCLIP) (Bharathavikru et al., 2017; Kim, Jeong, & Kim, 2017). However, the most widely used crosslinker for identifying protein–RNA interaction sites is UV, which provides zero-length, covalent crosslinking of direct RNA–protein contacts, while the comparatively sparse crosslinking within cells prevents indirect recovery via large complex formation. Variations on the methodology confer different advantages: Increased sensitivity by incorporating photoactivated nucleotides (PAR-CLIP) (Webb, Hector, Kudla, & Granneman, 2014), improved recovery of crosslinking sites by circularization of reverse transcribed RNAs (iCLIP, eCLIP) (König et al., 2010; Van Nostrand et al., 2016) or high-stringency recovery of protein–RNA interaction sites using denaturing purification (CRAC) (Granneman, Kudla, Petfalski, & Tollervey, 2009; Tree, Granneman, McAteer, Tollervey, & Gally, 2014). The sequencing datasets recovered by these techniques contain a wealth of data.

RNA–RNA interactions, both inter- and intramolecular, can be recovered via proximity-dependent ligation; in techniques termed hi-CLIP, RIL-seq, or CLASH (Helwak, Kudla, Dudnakova, & Tollervey, 2013; Melamed et al., 2016; Sugimoto et al., 2015; Waters et al., 2017). These interactions occur in vivo and are maintained in the purified complex with the "bait" RBP, stabilized by RNA–protein crosslinking. Notably, in our own studies, we have found that not all bait proteins recover RNA–RNA interaction data equivalently. This it likely due to differing stabilities of interacting RNAs on the bait protein, as only one RNA strand is UV-crosslinked to the scaffold. Alternatively, RNA–RNA interactions can be specifically enriched and identified in the absence of a protein partner using psoralen derivatives (Aw et al., 2016; Lu et al., 2016; Nguyen et al., 2016; Sharma, Sterne-Weiler, O'Hanlon, & Blencowe, 2016). These results allow profiling of global RNA folding patterns (the "RNA structurome") and the RNA interactome, providing complementary data that can be collated with RNA–protein interactions.

In our own studies, we have analyzed the sRNA chaperone Hfq and the endoribonucleases RNase E and YbeY in enterohaemorrhagic *E. coli*, the (p)ppGpp synthetase RelA in commensal *E. coli*, and the VapC-family PIN-domain endoribonucleases from *Mycobacterium tuberculosis* (McAteer et al., 2018; Tree et al., 2014; Waters et al., 2017; Winther, Roghanian, & Gerdes, 2018; Winther, Tree, Tollervey, & Gerdes, 2016). We identified the RNA interaction partners of these proteins using UV-crosslinking and analysis of cDNA (CRAC). When compared to CLIP-Seq, this includes an additional denaturing purification of RNA–protein complexes, which provides higher stringency, resulting in lower background and higher confidence in RNA–protein interaction sites. In these experiments, up to ~2% of reads recovered represent in vivo RNA–RNA interactions that are recovered in association with the bait protein. Small modifications to the protocol, including preparation of RNA 5′ and 3′ ends before linker ligation, improve recovery of RNA–RNA interactions and are described here.

The methodology requires that the bait protein is engineered to contain a tandem affinity tag, typically consisting of His_6 and 3xFLAG separated by a TEV protease cleavage site. Protein–RNA complexes are UV-crosslinked in vivo and purified using M2 anti-FLAG affinity resin (Fig. 1). Following elution by tobacco etch virus (TEV) protease cleavage, RNAs are fragmented by partial RNase A + T1 digestion. This generates RNAs with 5′-OH and 3′-P termini, which prevent RNA ligation. RNase activity is halted by addition of 4 *M* guanidinium, which denatures proteins but not RNA structure. Protein–RNA complexes are then bound to a Ni-NTA column, still under denaturing conditions. The immobilized complexes are enzymatically treated to prepare the RNA 5′ and 3′ ends, with easy wash steps to exchange constituents between reactions. The order of reactions can be manipulated to favor linker ligation and recovery of single reads (as in CRAC analysis), or initially converted to ligation-compatible, 5′P and 3′OH termini to promote intermolecular ligation (as in CLASH analysis). Once linker sequences have been appended to the RNA ends, protein–RNA complexes can be eluted, size separated by SDS-PAGE electrophoresis, and the RNAs recovered following proteinase digestion.

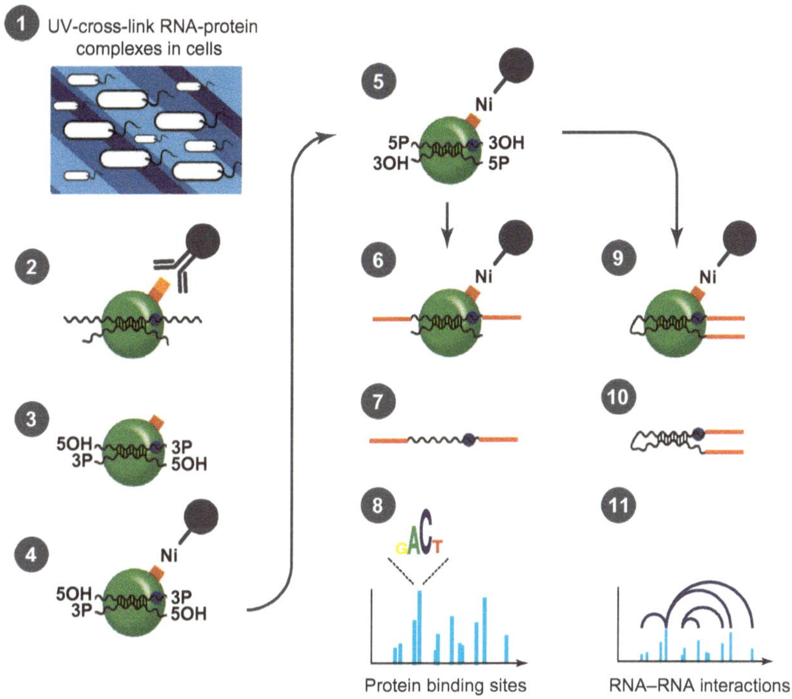

Fig. 1 Overview of UV-crosslinking and analysis of cDNA (CRAC) and UV-crosslinking and analysis of hybrids (CLASH) methodologies for identification and protein–RNA interaction sites and RNA–RNA interactions.

1. Cells expressing a dual affinity-tagged RNA-binding proteins are irradiated with ultraviolet light (UV-C, 254 nm) to covalently crosslink RNA–protein complexes.
2. The dual affinity tag contains both FLAG (*orange*) and His (*red*) epitopes, separated by a TEV cleavage site. The cells are lysed and the crosslinked RNA–protein complexes are initially purified using M2 anti-FLAG resin.
3. The complexes are eluted from the resin by TEV protease digestion to release the FLAG tag. The crosslinked RNAs are trimmed to yield a short fragment that surround the protein binding site. The RNases used for digestion leave products with 5′ hydroxyl and 3′ monophosphate ends.
4. The trimmed RNA–protein complex is immobilized on a Ni-NTA column that allows stringent washing of the complex.
5. The RNA is radiolabeled with ^{32}P and 5′ and 3′ ends of the immobilized RNAs are converted to 5′ monophosphate and 3′ OH termini that are substrates for T4 RNA ligase.
6. RNA linkers are ligated to 5′ and 3′ ends of crosslinked RNA fragments.
7. The linker ligated RNAs are released from the RNA–protein complex by proteinase digestion.
8. The RNA fragments are converted to cDNA, PCR amplified, and sequenced using an Illumina platform. Protein binding sites throughout the transcriptome are represented as read peaks.

(Continued)

Libraries are prepared by reverse transcription and PCR amplification of often low-abundance cDNA libraries. The libraries can be sequenced using single-end or paired-end protocols, depending on the downstream application of the data.

2. PRIMARY PURIFICATION AND TRIMMING OF RNA–PROTEIN COMPLEXES

Construction of affinity-tagged RBPs: A dual affinity purification tag is incorporated into the C-terminus of the RBP. We used a 12.6 kDa dual 3xFLAG and His_6 tag (GenBank accession KX714724) where the affinity tags are separated by a TEV protease cleavage site. Where possible it is desirable to incorporate the tag into the chromosomal copy of the gene to retain native expression levels and avoid off target interactions from artificially high or nonphysiological expression patterns. However, in some cases expression in a heterologous host is required and plasmid-based expression systems are the only viable approach. Once constructed, the activity of the fusion protein should be confirmed to ensure the affinity tag has not impaired protein function. A potential concern in analyses of endoribonucleases is the complete degradation of the covalently attached RNA during purification. This could be countered by the use of catalytically inactive variants. However, this was not necessary during our analyses of RNase E, YbeY, or VapC-family toxins (McAteer et al., 2018; Waters et al., 2017; Winther et al., 2016).

UV-crosslinking RNA–protein complexes: The optimal UV dose can be determined empirically; however, we routinely use 1800 mJ cm^{-2} of UV-C (254 nm) to crosslink RNA–protein complexes in 800–2000 mL of ~0.8 OD_{600} culture. UV transmittance can be limited for some media

Fig. 1—cont'd

9. A small fraction of RNAs UV-crosslinked to the bait protein are base-paired with RNA interaction partners. The 5′ and 3′ ends of these RNAs are in close proximity after RNase digestion and may be ligated together by T4 RNA ligase during RNA linker ligation reactions. RNA linkers are ligated to the free 5′ and 3′ ends forming an RNA hybrid.
10. The RNA hybrids are released from the RNA–protein complex by proteinase digestion.
11. The ligated RNA hybrids are converted to cDNA, PCR amplified, and sequenced. This yields chimeric reads that can be identified by discontiguous mapping of read halves and represent RNA–RNA interactions occurring on the bait protein in vivo.

and dense cultures, so the path-length through the culture should be minimized where possible. To improve the capture of transient interactions we used a modified water sterilization unit (W5, UVO3) and irradiated the culture for 100 s. The recently developed Vari-X-linker (UVO3) can provide this UV dose in a few seconds, facilitating time-resolved analysis of RNA–protein interactions (Van Nues et al., 2017).

2.1 Equipment

- W5 small diameter UV-crosslinker (LT 40-205, van Remmen UV Techniek, distributed by UVO3) or comparable UV-crosslinker.
- Liquid nitrogen
- Refrigerated centrifuge with 1 L centrifuge bottles
- Refrigerated microcentrifuge
- Refrigerated centrifuge for 50 and 15 mL tubes
- Orbital shaker or nutator at 4°C
- 0.1 mm zirconia beads
- Vortex (Raytek VM1 or comparable)

2.2 Buffers and Reagents

- Phosphate-buffered saline (PBS)
- LYSIS buffer (50 m*M* Tris–HCl pH 7.8, 150 m*M* NaCl, 1.5 m*M* $MgCl_2$, 0.1% NP-40, 5 m*M* β-mercaptoethanol, 1 cOmplete EDTA-free protease inhibitor tablet [Roche] per 50 mL final volume)
- M2 anti-FLAG resin (Sigma)
- TMN150 (50 m*M* Tris–HCl pH 7.8, 150 m*M* NaCl, 0.1% NP-40, 5 m*M* β-mercaptoethanol)
- TMN1000 (50 m*M* Tris–HCl pH 7.8, 1 *M* NaCl, 0.1% NP-40, 5 m*M* β-mercaptoethanol)
- TEV-GST protease (prepared in-house)
- RNace IT (Agilent)
- Ni-NTA agarose (Invitrogen)
- Guanidine–HCl
- 5 *M* NaCl
- 2.5 *M* Imidazole pH 8.0

2.3 Procedure

Note: All steps are performed on ice unless otherwise indicated.

2.3.1 Preparation of UV-Crosslinked RNA–Protein Complexes

1. Depending on the UV-crosslinking equipment used, 800 mL–2 L of the wild-type parental culture, and an isogenic culture expressing the affinity tagged RBP are grown and UV-crosslinked by exposure to 1800 mJ cm^{-2} of UV-C (254 nm). Decant the culture into centrifuge bottles and centrifuge for 20 min at 4000 × *g* and 4°C.
2. Aspirate the culture supernatant and suspend the pellet in 50 mL of ice cold PBS. Transfer the culture to a 50-mL centrifuge tube and pellet at 4000 × *g* for an additional 20 min at 4°C. Aspirate the supernatant.
3. Determine the wet weight of the pellet and suspend the pellet in ice cold PBS. Divide the culture into aliquots that contain 1 g of culture and centrifuge at 4000 × *g* and 4°C for 10 min. Aspirate and snap freeze the pellet in liquid nitrogen.

2.3.2 Cell Lysis

1. Defrost the cell pellet on ice and add 1 vol. of ice cold LYSIS buffer (~1–1.5 mL) and 3 vol. of 0.1 mm zirconia beads. Vortex the cells 5 × for 1 min with 1 min on ice between vortexing to prevent overheating.
2. Add an additional 3 vol. of LYSIS buffer and vortex briefly to suspend the lysate. Centrifuge the lysate for 20 min at 4000 × *g* and 4°C to collect the cell debris. Transfer to supernatant to 1.5-mL microcentrifuge tubes and centrifuge for an additional 20 min at 14,000 × *g* and 4°C to remove residual cell debris.

2.3.3 Primary Purification of RNA–Protein Complexes Using M2 Anti-FLAG Resin

1. In a 15-mL centrifuge tube, wash an aliquot (200 μL) of 50% M2 anti-FLAG resin slurry twice with 5 mL of ice cold LYSIS buffer and aspirate the wash supernatant.
2. Transfer the lysate supernatant to the M2 anti-FLAG resin and incubate on an orbital shaker or nutator at 4°C for 2 h.
3. Collect the resin by pulsing the centrifuge tube to 1000 × *g* and aspirate the supernatant. Wash the resin twice with 10 mL of ice cold TNM1000 buffer and twice with 10 mL of ice cold TNM150 buffer by pulsing the centrifuge tube to 1000 × *g* and aspirating the wash supernatant.
4. Resuspend the resin in 600 μL of ice cold TNM150 and transfer to a 1.5-mL microcentrifuge tube.
5. Add 20–30 U of TEV-GST protease to the M2 anti-FLAG resin and incubate at 18°C for 2 h with agitation to cleave the RNA–protein complex from the M2 anti-FLAG resin.

6. Transfer the resin slurry to an empty microcentrifuge spin collection column and briefly zip spin to elute the RNA–protein complexes into a clean 1.5-mL microcentrifuge tube.

2.3.4 Trimming of Crosslinked RNA Fragments

To footprint the RBP on the crosslinked RNA fragments, the crosslinked RNA is partially degraded with a cocktail of RNase T1 and RNase A. The concentration of RNase is determined empirically and will affect the length of RNA fragment recovered in the sequencing library. For analysis of RNA–protein binding sites we use 0.15 U of RNace IT (Agilent) and incubate at 37°C for 5 min. For analysis of RNA–RNA interactions, we predict that only a single RNA strand is UV-crosslinked to the RBP and avoid incubating the RNA–protein complex at elevated temperatures until after the base paired RNAs are ligated, to avoid melting and dissociation of the RNA duplex. For these analyses we incubate with 0.15 U of RNace IT for 7 min at 20°C.

1. Transfer 100 μL of Ni-NTA resin slurry to a 1.5-mL microcentrifuge tube and wash twice with 750 μL of ice cold WASH buffer I. Aspirate the wash supernatant and store the resin on ice until required.
2. Prepare 1.5-mL microcentrifuge tubes containing 0.4 g of guanidine–HCl and set aside until required.
3. To ~500 μL of M2 anti-FLAG resin eluate, add RNace IT and incubate to trim crosslinked RNAs (time and temperature as earlier).
4. Add the eluate to 0.4 g of guanidine–HCl from step 2 and vortex to dissolve the guanidine–HCl and inactivate the RNase IT.
5. To each tube (containing ~700 μL) add 27 μL of 5 *M* NaCl (final concentration ~300 m*M*) and 3 μL of 2.5 *M* imidazole pH 8.0 (final concentration ~10 m*M*).
6. Transfer the RNase treated eluate to the tube containing washed Ni-NTA resin from step 1 and incubate at 4°C with agitation for 2–16 h.

Break. The RNA–protein complexes will be stable in guanidine–HCl at 4°C overnight.

3. PREPARATION OF RNA 5′ AND 3′ ENDS FOR LINKER LIGATION AND INTERMOLECULAR LIGATION OF RNA–RNA INTERACTIONS

3.1 Equipment

- Dry block incubator (capable of cooling to 16°C)
- Microcentrifuge spin collection columns (Pierce)

3.2 Buffers and Reagents

- PNK wash buffer (50 m*M* Tris–HCl pH 7.8, 10 m*M* $MgCl_2$, 0.5% NP-40, and 5 m*M* β-mercaptoethanol)
- WASH buffer I (6 *M* guanidine–HCl, 50 m*M* Tris–HCl pH 7.8, 300 m*M* NaCl, 0.1% NP-40, and 5 m*M* β-mercaptoethanol)
- WASH buffer II (50 m*M* Tris–HCl pH 7.8, 50 m*M* NaCl, 10 m*M* imidazole, 0.1% NP-40, 5 m*M* β-mercaptoethanol)
- PNK reaction buffer (50 m*M* Tris–HCl pH 7.8, 10 m*M* $MgCl_2$, and 10 m*M* β-mercaptoethanol)
- Thermosensitive alkaline phosphatase (TSAP, Promega)
- RNasIN (Promega)
- Milli Q dH_2O
- T4 polynucleotide kinase (PNK, Sigma)
- ^{32}P-γ-ATP (10 mCi/mL, Perkin Elmer)
- ATP (100 m*M*)
- T4 RNA ligase (New England Biolabs)
- 3′ RNA linker (Table 1)
- 5′ RNA linker (L5Aa-d, Table 1)

3.3 Procedure

RNA cleavage by either RNase T1 or RNase A produces 5′-OH and a 2′,3′-cyclic monophosphate intermediate that spontaneously resolves into a 3′ phosphate group. Neither of these RNA ends are a substrate for RNA ligase. Prior to ligation they must be phosphorylated to generate 5′ phosphate or dephosphorylated to a 3′ hydroxyl group, respectively. For analysis of RNA–protein binding sites, concatemerization and circularization of RNA fragments can be blocked by initial dephosphorylation followed by 3′ linker ligation, prior to phosphorylation of the 5′ terminus. For analysis of RNA–RNA interactions, the RNA 5′ and 3′ ends must be prepared for intermolecular ligation before ligation of the linkers. The 3′ dephosphorylation and 5′ phosphorylation steps are therefore performed before linker ligation. In practice, the amount of concatemerization that occurs during linker ligation using RNA ligase I in the absence of ATP (see below) is minimal and >98% of reads recovered represent a single RNA fragment. In the detailed protocol, we describe the steps used for preparation of libraries for analysis of RNA–RNA interactions on RNase E. However, these libraries were also used to map RNA–protein interactions.

Table 1 Oligonucleotides Used to Prepare CRAC and CLASH Libraries

Oligonucleotide Name	Oligonucleotide Sequence
5′ RNA linker (L5Aa)[a]	5′-invddT-ACACrGrArCrGrCrUrCrUrUrCrCrGrA rUrCrU*rNrNrN*rUrArArGrCrN-3′
5′ RNA linker (L5Ab)[a]	5′-invddT-ACACrGrArCrGrCrUrCrUrUrCrCrGrA rUrCrU*rNrNrN*rArUrUrArGrCrN-3′
5′ RNA linker (L5Ac)[a]	5′-invddT-ACACrGrArCrGrCrUrCrUrUrCrCrGrA rUrCrU*rNrNrN*rGrCrGrCrArGrCrN-3′
5′ RNA linker (L5Ad)[a]	5′-invddT-ACACrGrArCrGrCrUrCrUrUrCrCrGrA rUrCrU*rNrNrN*rCrGrCrUrUrArGrCrN-3′
3′ RNA linker	5′AppNAGATCGGAAGAGCACACGTCTG-ddC-3′
RT oligo	5′-CAGACGTGTGCTCTTCCGATCT-3′
P5	5′-AATGATACGGCGACCACCGAGATCTACACT CTTTCCCTACACGACGCTCTTCCGATCT-3′
P3 (BC1)[b]	5′-CAAGCAGAAGACGGCATACGAGATCGTGAT GTGACTGGAGTTCAGACGTGTGCTCTTCCG ATCT-3′
P3 (BC2)[b]	5′-CAAGCAGAAGACGGCATACGAGATACATCG GTGACTGGAGTTCAGACGTGTGCTCTTCCG ATCT-3′
P3 (BC3)[b]	5′-CAAGCAGAAGACGGCATACGAGATGCCTAA GTGACTGGAGTTCAGACGTGTGCTCTTCCG ATCT-3′

[a]Barcode sequences used to multiplex libraries are underlined. Random nucleotide used to identify PCR duplicates are italicized.
[b]Barcode sequences used to multiplex libraries are underlined.

The reaction temperatures used for RNA–RNA recovery in the CLASH protocol are generally lower and incubations longer than in our CRAC protocol (Sy et al., 2018) to favor RNA duplex recovery.

3.3.1 Preparation of RNA 3′ Ends

1. Pulse the microcentrifuge tubes containing Ni-NTA resin to 1000 × *g* and aspirate the resin. Wash the resin twice with 750 μL ice-cold WASH buffer I and 3 × with 750 μL of ice-cold PNK wash buffer. Transfer the Ni-NTA resin to an empty microcentrifuge spin collection column.

2. Briefly spin the collection column to remove the residual PNK wash buffer and cap the bottom of the column. Add 80 μL of TSAP reaction buffer containing: 16 μL 5 × PNK reaction buffer, 8 μL of thermosenstive alkaline phosphatase (TSAP, Promega), 2 μL RNasIN (Promega), and 54 μL of milli Q dH_2O. Incubate the reaction for 45 min at 20°C.
3. Wash the resin with 400 μL of ice-cold WASH buffer I by gravity filtration (to inactivate TSAP).
4. Wash the resin 3 × with 400 μL of PNK wash buffer to remove traces of guanidine–HCl.

3.3.2 Preparation of RNA 5′ Ends

1. Briefly spin the collection column containing Ni-NTA resin to remove residual wash buffer and recap the column. Add 76 μL of reaction buffer containing: 16 μL of PNK reaction buffer, 4 μL of T4 PNK (Sigma), 56 μL of milli Q dH_2O.
2. Add 4 μL of ^{32}P-γ-ATP (10 μCi μL^{-1}) and incubate the reaction for 100 min at 20°C to label the RNA fragments.
3. Add 1 μL of 100 m*M* ATP and incubate at 20°C for a further 50 min to ensure that all available 5′ ends are phosphorylated.
4. Wash the resin with 400 μL of ice-cold WASH buffer I by gravity filtration (to inactivate PNK).
5. Wash the resin 3 × with 400 μL of PNK wash buffer to remove traces of guanidine–HCl.

3.3.3 Ligation of 3′ and 5′ Linkers

For both linker ligation reactions we use RNA ligase 1. This ligase is more promiscuous than RNA ligase 2 when using preadenylated linkers and is capable of adenylation and may ligate RNA ends in the absence of ATP addition. We do not include a dedicated intermolecular ligation step in our protocol as we (and others Grosswendt et al., 2014) have found that these ligation events occur during the linker ligation steps (without a dedicated ligation step). In theory, the RNA–RNA interaction ends will potentially be ligated during both the 3′ and 5′ linker ligation steps.

3.3.4 Ligation of 3′ RNA Linker

1. Briefly spin the collection column to remove the residue wash buffer. Recap the base of the column and add 68 μL of the following reaction

buffer: 16 µL of PNK reaction buffer, 2 µL of RNasIN, 50 µL of milli Q dH_2O.

2. Add 8 µL of 3′ RNA linker (10 µ*M*), and 4 µL of T4 RNA ligase 1. Incubate the ligation reaction at 16°C for 16 h (or overnight).

Break. The ligation reactions are incubated overnight.

3. Wash the resin with 400 µL of ice-cold WASH buffer I.
4. Wash the resin 3 × with 400 µL of PNK wash buffer to remove traces of guanidine–HCl.

3.3.5 Ligation of 5′ RNA Linker

1. Briefly spin the collection column to remove the residue wash buffer. Recap the base of the column and add 76 µL of the following reaction buffer: 16 µL of PNK reaction buffer, 8 µL of 10 m*M* ATP, 2 µL of RNasIN, 49 µL of milli Q dH_2O.
2. Add 1 µL of 5′ RNA linker (100 µ*M*) and 4 µL of T4 RNA ligase 1. Incubate the ligation reaction at 16°C for 16 h (or overnight).

Break. The ligation reactions are incubated overnight.

3. Wash the resin 3 × with 400 µL of WASH buffer II (does not contain guanidine–HCL).

4. SIZE SEPARATION OF RNA–PROTEIN COMPLEXES

4.1 Equipment

- Geiger counter
- SDS-PAGE equipment (Invitrogen Mini Gel Tank or comparable)
- Western blot apparatus (BioRad Mini Trans-Blot or comparable)
- Scalpel

4.2 Buffer and Reagents

- ELUTION buffer (50 m*M* Tris–HCl pH 7.8, 50 m*M* NaCl, 150 m*M* imidazole, 0.1% NP-40, 5 m*M* β-mercaptoethanol)
- DIGEST buffer (50 m*M* Tris–HCl pH 7.8, 50 m*M* NaCl, 0.1% NP-40, 5 m*M* β-mercaptoethanol, 1% SDS, 5 m*M* EDTA)
- Glycogen (coprecipitant)
- Proteinase K
- Trichloroacetic acid (100% (w/v), TCA)
- Acetone
- 3 *M* NaOAc pH 5.2

- Phenol:chloroform:isoamylalcohol pH 8.0 (25:24:1, PCI)
- Ethanol
- NuPAGE LDS sample buffer (Invitrogen)
- NuPAGE MOPS SDS running buffer (20 ×, Invitrogen)
- NuPAGE 4%–12% Bis–Tris precast polyacrylamide gel (Invitrogen)
- NuPAGE Transfer buffer (20 ×, Invitrogen)
- Nitrocellulose membrane
- Whatman paper
- Chemiluminescent labels for autoradiograms or comparable
- Biomax MS autoradiography film (Kodak)

4.3 Procedure

The RNA–protein complexes are size separated by SDS-PAGE and visualized by autoradiography of the ^{32}P-labeled RNA ends. This is often the first indication that the RNA–protein complex has been successfully purified and is intact. Because the crosslinked RNA is often from a heterogeneous pool of RNAs and has been randomly trimmed with RNase T1 and RNase A, the size of the RNA complex will be heterogeneous. This results in a characteristic smear from slightly higher than the expected molecular weight for the RBP to high molecular weights. The smear contains the longer RNA fragments, presumably with a higher proportion of mappable RNA–RNA interactions. These are transferred to a nitrocellulose membrane and excised for cDNA library preparation.

1. Briefly spin the collection column to remove the residual wash buffer and recap the column end.
2. Add 200 μL of ELUTION buffer and incubate on ice for 5 min. Allow the eluate to gravity filter into a clean microcentrifuge tube.
3. Repeat step 2 with a second volume of ELUTION buffer. Briefly spin the collection column to recover the residual bed volume in the collection column.
4. Add 40 μg of glycogen coprecipitant to the eluate and 100 μL of trichloroacetic acid (TCA, 100% (w/v)). Vortex to mix and incubate on ice for 1 h.
5. Centrifuge for 30 min at 16,000 × *g* and 4°C to collect the precipitated RNA–protein complexes. Aspirate the pellet. The pellet should be radiolabeled and a Geiger counter can be used to ensure that RNA–protein complexes are not decanted at this stage.
6. Add 800 μL of ice-cold acetone.
7. Centrifuge for an additional 20 min at 16,000 × *g* and 4°C. Aspirate the pellet. Leave the pellet to air dry.

8. Resuspend the pellet in 30 μL of 1 × NuPAGE loading buffer and load onto a 4%–12% gradient, Bis–Tris NuPAGE gel (Invitrogen). Load a prestained protein molecular weight marker and electrophorese the RNA–protein complexes to obtain good separation.
9. Remove the PAGE gel from the plastic gel castor and transfer the RNA–protein complexes to a nitrocellulose membrane using a Mini-Trans Blot system (BioRad) with NuPAGE transfer buffer (Invitrogen) at 100 V for 1.5 h.
10. Remove the nitrocellulose membrane from the transfer sandwich and wrap in cling film (saran wrap). Tape the membrane into an autoradiography cassette with an intensifier screen. Add a luminescent marker to allow alignment of the autoradiography film and membrane after development. Place a sheet of autoradiography film into the cassette and expose at −80°C until the smeared RNA–protein complex can be detected on the autoradiogram. The exposure time will depend on the abundance of RNA–protein complexes and the efficiency of radiolabeling (typically 2–24 h).
11. Realign the developed autoradiogram with the nitrocellulose membrane and tape in place. Using the autoradiogram to identify the region of the nitrocellulose membrane that contains the smeared RNA–protein complex, cut through the film and membrane to excise a piece of membrane containing the smeared complexes.The lowest molecular weight complexes contain short RNA fragments crosslinked to the protein and may be difficult to map unambiguously to the transcriptome. The higher molecular weight smeared complexes are usually less abundant, but are more likely to contain longer chimeric RNA fragments that can be unambiguously mapped to the genome.
12. Transfer the excised membrane pieces to a clean 1.5-mL microcentrifuge tube.

Break. The nitrocellulose membrane can be stored at −80°C overnight.

13. Incubate the membrane pieces in 400 μL of DIGEST buffer with 100 μg of Proteinase K at 55°C for 2 h to release the labeled RNA fragments from the nitrocellulose-bound protein.
14. Add 50 μL of 3 *M* NaOAc pH 5.2 and 500 μL of PCI. Vortex the membrane pieces and centrifuge at room temperature for 5 min at 16,000 × *g*.
15. Add 1 mL of ice-cold ethanol and 20 μg of glycogen coprecipitant. Incubate the RNA fragments at −80°C for 30 min to precipitate. Centrifuge the samples for 20 min at 16,000 × *g* and 4°C.

16. Aspirate the supernatant and wash the pellet with ice-cold 70% ethanol. Centrifuge the samples for 20 min at 16,000 × *g* and 4°C. Aspirate the pellets and air dry the purified, linker-ligated RNA fragments.

5. PREPARATION OF SEQUENCING LIBRARIES FROM PURIFIED, LINKER-LIGATED RNA FRAGMENTS

5.1 Equipment

- Thermocycler
- Agarose gel electrophoresis equipment

5.2 Buffer and Reagents

- Superscript III First Strand Reverse Transcription kit (Invitrogen)
- RT oligo (10 μ*M*, Table 1)
- dNTPs (5 m*M*)
- RNasIN (Promega)
- RNase H (New England Biolabs)
- LA Taq (Takara)
- 10 × LA Taq buffer (Takara)
- P5 primer (10 μ*M*, Table 1)
- P3 primer (10 μ*M*, BC1-3, Table 1)
- NaOAc pH 5.2 (3 *M*)
- Ethanol
- TBE buffer
- Metaphor high resolution agarose gel or comparable
- DNA gel stain
- Overhead transparency
- Scalpel
- MinElute gel extraction kit (Qiagen)

5.3 Procedure

The purified RNA fragments have ligated 5′ and 3′ linkers that are used to reverse transcribe the RNA and append sequences for hybridization to Illumina platform sequencer flow cells by PCR. The number of PCR cycles should be as few as needed. An initial test run may be required to estimate how much amplification is required to generate sufficient amplicons for sequencing. Importantly, the 5′ RNA linker contains three random nucleotides that are designed to "barcode" each independent RNA ligation event. This allows PCR duplicates within the sequencing library to be identified and distinguished from individual RNA fragments that fortuitously start and end at the same position.

1. Suspend the RNA pellet in 13 μL of Reverse Transcriptase reaction buffer I, 2 μL of RT oligo (10 μ*M*), 2 μL of 5 m*M* dNTPs, 9 μL of milli Q dH_2O. Incubate the reaction at 80°C for 3 min and then snap chill on ice for 5 min.
2. To each reaction add 6 μL of Reverse Transcriptase reaction buffer II containing: 4 μL of 5 × First Strand reaction buffer (Superscript III, Invitrogen), 1 μL of 100 m*M* DTT, 1 μL of RNasIN (Promega). Incubate the reaction at 50°C for 3 min and then add 1 μL of Superscript III reverse transcriptase. Incubate the reaction for a further 1 h at 50°C.
3. Inactivate the reverse transcription reaction by incubating at 65°C for 15 min. Add 2 μL of RNase H (NEB) and incubate at 37°C for 30 min to remove the RNA fragments.
4. The number of PCR reactions and cycles should be determined from an initial amplification experiment. We routinely start with 3 × 50 μL PCR reactions using 21 cycles and will increase the number of PCR reactions (up to ~10 reactions) and PCR cycles (up to ~24 cycles) to obtain sufficient material for a sequencing run. Increasing the number of independent PCR reactions has the advantage of limiting PCR duplication of sequencing reads. For each PCR reaction add: 5 μL of 10 × LA Taq buffer (Lonza), 1 μL of 10 μ*M* P5 primer, 1 μL of 10 μ*M* P3 primer, 2.5 μL of 5 m*M* dNTPs, 0.5 μL of LA Taq (Lonza), 2 μL of cDNA library, and 38 μL of milli Q dH_2O.
5. Cycle the reactions for: 1 cycle of 95°C for 2 min; 21 cycles of 98°C for 20 s, 52°C for 20 s, 68°C for 20 s; 1 cycle of 72°C for 5 min.
6. Pool the PCR reactions and add 0.1 V of NaOAc pH 5.2 and 2.5 V of ice-cold ethanol. Incubate at −80°C for 30 min, and centrifuge for 30 min at 16,000 × *g* and 4°C to precipitate the PCR amplified library. Aspirate the pellet and air dry briefly.
7. Resuspend the library in 20 μL of milli Q dH_2O.

5.3.1 Size Selection of the PCR Amplified cDNA Library

The libraries are submitted to a second round of size selection to remove primer dimers introduced during the amplification step, and to bias the library toward larger amplicons that will contain a higher proportion of mappable RNA fragments.

1. Prepare a 2% high-resolution Metaphor agarose gel in TBE buffer. Prestain the gel with a DNA gel stain.
2. Add 4 μL of 6 × DNA loading buffer to the precipitated library and electrophorese at 120 V to get good separation of fragments ~100–300 nucleotides in length.
3. Image the agarose gel and print the gel image at 1:1 scaling. Using a plastic sheet (overhead transparency) to separate the image and the gel, lay the agarose gel onto the image and excise the smeared library from the region above the primer dimers, which can be identified in the control library.
4. Transfer the excised gel fragment to a clean 1.5-mL microcentrifuge tube. The DNA is purified using a MinElute gel extraction kit.
5. Add 750 μL of buffer QG and incubate at 42°C for 30 min (incubate for longer if needed to dissolve the gel). Add the dissolved gel fragment to a MinElute column and centrifuge for 1 min at 16,000 × *g*.
6. Add 750 μL of buffer PE and incubate the column at room temperature for 10 min. Centrifuge for 1 min at 16,000 × *g* and remove the flow through. Centrifuge for an additional 1 min at 16,000 × *g*.
7. Add 20 μL of milli Q dH_2O and incubate for 1 min at room temperature. Centrifuge for 1 min at 16,000 × *g* to elute the library.
8. Quantify the size selected library. If the library concentration is too low for Illumina sequencing, the library can be reamplified using additional PCR cycles and reactions.

6. IN SILICO ANALYSIS OF RNA–PROTEIN INTERACTION SITES AND RNA–RNA INTERACTIONS

We routinely use the *pyCRAC* software package (Webb et al., 2014) to identify protein binding sites throughout the transcriptome (recently described in Sy et al., 2018). Many groups have now demonstrated that RNA–RNA interaction data can be captured with UV-crosslinking datasets (Grosswendt et al., 2014; Helwak et al., 2013; Melamed et al., 2016; Sugimoto et al., 2015; Waters et al., 2017). In these datasets, RNA ends that were in close proximity on the bait protein are ligated together and linkers appended to the opposite ends. In the sequence dataset these appear as "hybrid" reads, in which the 5′ read half maps to a different transcriptome position than the 3′ read half. To identify these reads we use the *hyb* software package (Travis, Moody, Helwak, Tollervey, & Kudla, 2014).

For paired-end sequencing data, the reads should be merged into a single read representing an RNA fragment. We use *flash* (Magoč & Salzberg, 2011) to merge the two read halves together where possible.

```
flash input_R1.fastq input_R2.fastq -t 8 -M 70
```

The *hyb* pipeline is then used to identify hybrid reads within the dataset.

```
hyb demultiplex code=barcode_list.txt preprocess
qc=fastx check detect db=database anti=1 in=merge.fastq
```

Where barcode_list.txt is a tab delimited file of 5′ linker barcodes, database is a database constructed with *make_hyb_db*, and merge.fastq is the output from *flash* (or fastq file of single-end reads).

The *hyb* pipeline is designed to annotate the hyb file with features extracted from an ensembl transcriptome database file. However, these do not exist for most bacterial pathogens. The output .hyb file contains genomic coordinates for the start and end of each hybrid read half and these can be used to annotate the hybrids with overlapping genomic features. As read halves can map across feature boundaries (particularly intergenic regions) we use an iterative approach to map read halves to classes of genomic feature so that each is annotated with a single feature (e.g., genomic features > predicted UTRs > antisense features > intergenic regions). The in-house scripts used to annotate hybrids and determine statistical significance for RNA interactions are deposited at https://bitbucket.org/jaitree/hyb_stats. Importantly, `annotate_hyb_2.py` will assign the hybrid half to the first annotation that it overlaps.

```
annotate_hyb_2.py replicate_1.hyb -g genome_annot.gtf >
input_annot.hyb
```

To calculate the probability that an RNA–RNA interaction is observed by chance we use the abundance of RNA fragments in the pool of single mapping (nonhybrid) reads to model the likelihood of randomly ligating two RNA fragments together (we review different methods for calculating probability for RNA interactome datasets in Wong, Pang, Wilkins, and Tree (2018). We model the likelihood of randomly ligating two RNA fragments as two draws from a multinomial distribution. To determine the abundance of single mapping reads we use the *pyCRAC* software package as previously described (Sy et al., 2018). The .sgr format files generated by *pyCRAC* and the .hyb file generated by *hyb* are used as input for the statistical analysis.

The positive and negative strand .sgr files are combined into a single .pws file.

```
sgr2pws.py -p positive_strand.sgr -n negative_strand.
sgr>replicate_1.pws
```

The FDR corrected *P*-values for each unique RNA–RNA interaction can then be determined using the shell script hyb_p_values.sh.

```
hyb_p_value.sh datasets.txt
```

where datasets.txt is a space delimited table with three columns containing the file names for the .hyb file, the .pws file, and the total number of mapped reads determined by *pyCRAC*. The total number of mapped reads (cDNAs) can be taken from the `file_statistics_cDNAs.txt` file produced by *pyCRAC*. If there are multiple datasets, they can be entered as additional rows in the datasets.txt file and the shell script will identify overlapping hybrids in independent datasets and multiply the *p*-values together. For example, the datasets.txt file might look like:

```
replicate_1.hyb replicate_1.pws 1234567
replicate_2.hyb replicate_2.pws 1443584
replicate_3.hyb replicate_3.pws 135463
```

The results are written to a table-delimited table in the file cumulative_hyb_p_values.txt.

REFERENCES

Aw, J. G. A., Shen, Y., Wilm, A., Sun, M., Lim, X. N., Boon, K. L., et al. (2016). In vivo mapping of eukaryotic RNA interactomes reveals principles of higher-order organization and regulation. *Molecular Cell*, *62*, 603–617.

Bardill, J. P., & Hammer, B. K. (2012). Non-coding RNAs regulate virulence in the bacterial pathogen *Vibrio cholera*. *RNA Biology*, *9*, 392–401.

Bharathavikru, R., Dudnakova, T., Aitken, S., Slight, J., Artibani, M., Hohenstein, P., et al. (2017). Transcription factor Wilms' tumor 1 regulates developmental RNAs through 3′ UTR interaction. *Genes & Development*, *31*, 347–352.

Bronesky, D., Wu, Z., Marzi, S., Walter, P., Geissmann, T., Moreau, K., et al. (2016). *Staphylococcus aureus* RNAIII and its regulon link quorum sensing, stress responses, metabolic adaptation, and regulation of virulence gene expression. *Annual Review of Microbiology*, *70*, 299–316.

Chao, Y., & Vogel, J. (2010). The role of Hfq in bacterial pathogens. *Current Opinion in Microbiology*, *13*, 24–33.

Feliciano, J. R., Grilo, A. M., Guerreiro, S. I., Sousa, S. A., & Leitão, J. H. (2016). Hfq: A multifaceted RNA chaperone involved in virulence. *Future Microbiology*, *11*, 137–151.

Granneman, S., Kudla, G., Petfalski, E., & Tollervey, D. (2009). Identification of protein binding sites on U3 snoRNA and pre-rRNA by UV cross-linking and high-throughput analysis of cDNAs. *Proceedings of the National Academy of Sciences of the United States of America*, *106*(24), 9613–9618.

Grosswendt, S., Filipchyk, A., Manzano, M., Klironomos, F., Schilling, M., Herzog, M., et al. (2014). Unambiguous identification of miRNA: Target site interactions by different types of ligation reactions. *Molecular Cell*, *54*(6), 1042–1054.

Helwak, A., Kudla, G., Dudnakova, T., & Tollervey, D. (2013). Mapping the human miRNA interactome by CLASH reveals frequent noncanonical binding. *Cell*, *153*(3), 654–665.

Holmqvist, E., Li, L., Bischler, T., Barquist, L., & Vogel, J. (2018). Global maps of ProQ binding *in vivo* reveal target recognition via RNA structure and stability control at mRNA 3′ ends. *Molecular Cell*, *70*(5), 971–982.

Holmqvist, E., Reimegård, J., Sterk, M., Grantcharova, N., Römling, U., & Wagner, E. G. H. (2010). Two antisense RNAs target the transcriptional regulator CsgD to inhibit curli synthesis. *EMBO Journal*, *29*(11), 1840–1850.

Katsowich, N., Elbaz, N., Pal, R. R., Mills, E., Kobi, S., Kahan, T., et al. (2017). Host cell attachment elicits posttranscriptional regulation in infecting enteropathogenic bacteria. *Science*, *355*(6326), 735–739.

Kim, B., Jeong, K., & Kim, V. N. (2017). Genome-wide mapping of DROSHA cleavage sites on primary microRNAs and noncanonical substrates. *Molecular Cell*, *66*(2), 258–269.

König, J., Zarnack, K., Rot, G., Curk, T., Kayikci, M., Zupan, B., et al. (2010). iCLIP reveals the function of hnRNP particles in splicing at individual nucleotide resolution. *Nature Structural and Molecular Biology*, *17*(7), 909–915.

Lee, F. C. Y., & Ule, J. (2018). Advances in CLIP technologies for studies of protein-RNA interactions. *Molecular Cell*, *69*(3), 354–369.

Lu, Z., Zhang, Q. C., Lee, B., Flynn, R. A., Smith, M. A., Robinson, J. T., et al. (2016). RNA duplex map in living cells reveals higher-order transcriptome structure. *Cell*, *165*(5), 1267–1279.

Magoč, T., & Salzberg, S. L. (2011). FLASH: Fast length adjustment of short reads to improve genome assemblies. *Bioinformatics*, *27*(21), 2957–2963.

McAteer, S. P., Sy, B. M., Wong, J. L., Tollervey, D., Gally, D. L., & Tree, J. J. (2018). Ribosome maturation by the endoribonuclease YbeY stabilises a type III secretion system transcript required for virulence of enterohemorrhagic *Escherichia coli*. *The Journal of Biological Chemistry*, *293*(23), 9006–9016.

Melamed, S., Peer, A., Faigenbaum-Romm, R., Gatt, Y. E., Reiss, N., Bar, A., et al. (2016). Global mapping of small RNA-target interactions in bacteria. *Molecular Cell*, *63*(5), 884–897.

Michaux, C., Holmqvist, E., Vasicek, E., Sharan, M., Barquist, L., Westermann, A. J., et al. (2017). RNA target profiles direct the discovery of virulence functions for the cold-shock proteins CspC and CspE. *Proceedings of the National Academy of Sciences, 114*(26), 201620772.

Nguyen, T. C., Cao, X., Yu, P., Xiao, S., Lu, J., Biase, F. H., et al. (2016). Mapping RNA-RNA interactome and RNA structure *in vivo* by MARIO. *Nature Communications, 7*, 12023.

Papenfort, K., Espinosa, E., Casadesús, J., & Vogel, J. (2015). Small RNA-based feedforward loop with AND-gate logic regulates extrachromosomal DNA transfer in Salmonella. *Proceedings of the National Academy of Sciences of the United States of America, 112*(34), E4772–E4781.

Sedlyarova, N., Shamovsky, I., Bharati, B. K., Epshtein, V., Chen, J., Gottesman, S., et al. (2016). sRNA-mediated control of transcription termination in *E. coli*. *Cell, 167*(1), 111–121.

Sharma, E., Sterne-Weiler, T., O'Hanlon, D., & Blencowe, B. J. (2016). Global mapping of human RNA-RNA interactions. *Molecular Cell, 62*(4), 618–626.

Srikumar, S., Colgan, A., & Hinton, J. C. D. (2012). sRNA and the virulence of *Salmonella enterica* serovar Typhimiurium. *RNA Biology, 9*, 437–445.

Sugimoto, Y., Vigilante, A., Darbo, E., Zirra, A., Militti, C., D'Ambrogio, A., et al. (2015). hiCLIP reveals the *in vivo* atlas of mRNA secondary structures recognized by Staufen 1. *Nature, 519*(7544), 491–494.

Sy, B., Wong, J., Granneman, S., Tollervey, D., Gally, D., & Tree, J. J. (2018). High-resolution, high-throughput analysis of Hfq-binding sites using UV-crosslinking and analysis of cDNA (CRAC). *Methods in Molecular Biology, 1737*, 251–272.

Travis, A. J., Moody, J., Helwak, A., Tollervey, D., & Kudla, G. (2014). Hyb: A bioinformatics pipeline for the analysis of CLASH (crosslinking, ligation and sequencing of hybrids) data. *Methods, 65*(3), 263–273.

Tree, J. J., Granneman, S., McAteer, S. P., Tollervey, D., & Gally, D. L. (2014). Identification of bacteriophage-encoded anti-sRNAs in pathogenic *Escherichia coli*. *Molecular Cell, 55*(2), 199–213.

Vakulskas, C. A., Potts, A. H., Babitzke, P., Ahmer, B. M. M., & Romeo, T. (2015). Regulation of bacterial virulence by Csr (Rsm) systems. *Microbiology and Molecular Biology Reviews, 79*(2), 193–224.

Van Nostrand, E. L., Pratt, G. A., Shishkin, A. A., Gelboin-Burkhart, C., Fang, M. Y., Sundararaman, B., et al. (2016). Robust transcriptome-wide discovery of RNA-binding protein binding sites with enhanced CLIP (eCLIP). *Nature Methods, 13*(6), 508–514.

Van Nues, R., Schweikert, G., De Leau, E., Selega, A., Langford, A., Franklin, R., et al. (2017). Kinetic CRAC uncovers a role for Nab3 in determining gene expression profiles during stress. *Nature Communications, 8*(1), 12.

Waters, S. A., McAteer, S. P., Kudla, G., Pang, I., Deshpande, N. P., Amos, T. G., et al. (2017). Small RNA interactome of pathogenic *E. coli* revealed through crosslinking of RNase E. *The EMBO Journal, 36*(3), 374–387.

Webb, S., Hector, R. D., Kudla, G., & Granneman, S. (2014). PAR-CLIP data indicate that Nrd1-Nab3-dependent transcription termination regulates expression of hundreds of protein coding genes in yeast. *Genome Biology, 15*(1), R8.

Westermann, A. J., Förstner, K. U., Amman, F., Barquist, L., Chao, Y., Schulte, L. N., et al. (2016). Dual RNA-seq unveils noncoding RNA functions in host-pathogen interactions. *Nature, 529*(7587), 496–501.

Winther, K. S., Roghanian, M., & Gerdes, K. (2018). Activation of the stringent response by loading of RelA-tRNA complexes at the ribosomal A-site. *Molecular Cell, 70*(1), 95–105.

Winther, K., Tree, J. J., Tollervey, D., & Gerdes, K. (2016). VapCs of *Mycobacterium tuberculosis* cleave RNAs essential for translation. *Nucleic Acids Research, 44*(20), 9860–9871.

Wong, J., Pang, I., Wilkins, M., & Tree, J. J. (2018). Systems level analysis of bacterial regulatory small RNA networks. In N. Rajewsky, S. Jurga, & J. Barciszewski (Eds.), *RNA technologies: Systems biology*: Springer [in press].

CHAPTER TWENTY

Profiling RNA Polymerase II Phosphorylation Genome-Wide in Fission Yeast

Tea Kecman, Dong-Hyuk Heo, Lidia Vasiljeva[1]
Department of Biochemistry, University of Oxford, Oxford, United Kingdom
[1]Corresponding author: e-mail address: lidia.vasilieva@bioch.ox.ac.uk

Contents

Abstract

The RNA polymerase II carboxyl-terminal domain (CTD) consists of tandem repeats of consensus sequence Tyr^1-Ser^2-Pro^3-Thr^4-Ser^5-Pro^6-Ser^7. Dynamic posttranslational modifications of the CTD generate a CTD code crucial for the cotranscriptional recruitment of factors that control transcription, chromatin modification, and RNA processing.

Analysis of CTD phosphorylation by ChIP (Chromatin ImmunoPrecipitation) coupled with high-throughput DNA sequencing (ChIP-seq) is a powerful tool to investigate the changes in CTD phosphorylation during the transcription cycle. In this chapter, we describe a ChIP-seq protocol to profile the different CTD phospho-marks in fission yeast. Using this protocol, we have found that Tyr1P, Ser2P, and Thr4P signals are highest at gene 3′ ends, whereas Ser5P is enriched across the gene bodies.

1. INTRODUCTION

In eukaryotes, RNA polymerase II (Pol II) is responsible for the transcription of all protein-coding genes as well as many noncoding RNA genes. The carboxyl-terminal domain (CTD) of Rpb1, the largest subunit of Pol II,

Methods in Enzymology, Volume 612
ISSN 0076-6879
https://doi.org/10.1016/bs.mie.2018.07.009

coordinates the entire transcription cycle. The CTD is composed of tandemly repeated heptapeptides of consensus sequence Tyr^1-Ser^2-Pro^3-Thr^4-Ser^5-Pro^6-Ser^7. The number of repeats varies from 26 to 52 depending on the species (Corden, 2013; Eick & Geyer, 2013).

The CTD is subject to extensive posttranslational modifications, with the serine (Ser), tyrosine (Tyr), and threonine (Thr) being phosphorylated or glycosylated in vivo, and the two prolines existing in either *cis*- or *trans*-configuration. The combination of differential CTD posttranslational modifications generates a CTD code that orchestrates transcription and RNA processing (Eick & Geyer, 2013).

Among the CTD modifications, reversible phosphorylation is critical for the activity of Pol II and is regulated by the combined action of specific kinases and phosphatases that determine specific phosphorylation patterns during the different stages of transcription. Analysis of CTD phosphorylation by ChIP (Chromatin ImmunoPrecipitation) has been a powerful tool to study transcription. ChIP coupled to quantitative PCR (qPCR) was used initially to analyze the pattern of transcription factors on a few genes at a time (Keogh & Buratowski, 2004). The development of ChIP coupled to DNA microarray (ChIP-chip) first, and to next-generation sequencing (ChIP-seq) later, provided a tool to study the DNA-binding patterns genome-wide. ChIP of Pol II CTD phosphoisoforms relies on monoclonal antibodies that recognize specific phosphorylated residues (Tyr1P, Ser2P, Thr4P, Ser5P, Ser7P). ChIP-seq has shown that the phosphorylation of the Ser2 and Ser5 in the CTD is similar in budding yeast and mammals (Bataille et al., 2012; Hintermair et al., 2012; Kim et al., 2010; Mayer et al., 2012, 2010; Milligan et al., 2016; Tietjen et al., 2010). Ser5P is enriched close to the transcription start sites (TSS), where it is important to recruit the mRNA capping enzyme during transcription initiation (Komarnitsky, Cho, & Buratowski, 2000; Schroeder, Schwer, Shuman, & Bentley, 2000). Conversely, Ser2P peaks at the 3′ end of genes, where it plays an important role for the recruitment of the termination and 3′ processing factors at the 3′ end of protein-coding genes (Ahn, Kim, & Buratowski, 2004).

While the patterns of serine phosphorylation are conserved between different species, the distributions of Tyr1P and Thr4P are not. While Tyr1P is enriched at the beginning and end of mammalian protein-coding genes (Descostes et al., 2014; Hsin, Li, Hoque, Tian, & Manley, 2014; Schlackow et al., 2017), in budding yeast Tyr1P is enriched in gene bodies and depleted at the 5′ and 3′ ends of genes (Mayer et al., 2012). Thr4P is highest at the end of genes in mammals (Hintermair et al., 2012; Schlackow et al., 2017), while in *S. cerevisiae* Thr4P is present within the gene body and 3′ ends (Harlen et al., 2016; Mayer et al., 2012).

Despite progress in many model systems, the role of the CTD in gene expression in *Schizosaccharomyces pombe* is poorly understood. Fission yeast is a useful experimental model system since it can be easily genetically manipulated and shares many similarities with humans compared to the budding yeast. The *S. pombe* CTD is composed of 29 heptapeptides and so far, only the phosphorylation patterns of Ser2 and Ser5 have been characterized genome-wide by ChIP-chip (Coudreuse et al., 2010). We decided to use ChIP-seq to map the distribution of Pol II phospho-marks in fission yeast. The following protocol describes *S. pombe* ChIP-seq, from ChIP to sequence data analysis.

2. METHODS

2.1 Chromatin Immunoprecipitation

2.1.1 Equipment

- Incubator shaker
- Table-top centrifuge
- Centrifuge for 15/50-mL Falcon tubes, for example, Multifuge 3 SR Heraeus
- Beckman centrifuge for JLA-16.250 rotor or equivalent
- Beckman ultracentrifuge, for example, Beckman Coulter Optima L-100 XP, and Type 70.1 Ti rotor or equivalent
- Ultracentrifuge tubes, for example, Beckman 10.4-mL polycarbonate tubes
- 14-mL Round-bottom tubes (Falcon, #352059)
- Glass beads, acid washed (Sigma, #G-8772)
- Qiagen adapters
- 22-Gauge ½-in. needle
- Wooden sticks (S Murray Applicator Wood, Fisher Scientific)
- 15-mL Polystyrene Falcon tubes (Corning, #352095)
- Sonicator with microprobe tip, for example, Diagenode Bioruptor UCD-200T-EX
- Magnetic stand for 1.5-mL tubes
- Thermoblock
- Qubit fluorometer
- Bioanalyzer

2.1.2 Buffers and Reagents

- YES medium: 5 g/L yeast extract, 30 g/L glucose, 225 mg/L adenine hemisulfate, 225 mg/L histidine, 225 mg/L leucine, 225 mg/L uracil, 225 mg/L lysine hydrochloride. Autoclave
- 37% Formaldehyde (HCHO)
- Cross-linking solution: 0.143 *M* NaCl, 1.43 m*M* EDTA, 71.43 m*M* HEPES–KOH. Adjust pH to 7.5 with KOH
- Quenching solution: 3 *M* glycine, 20 m*M* Tris. Do not adjust pH. Dissolve the glycine by microwave. With time the glycine may precipitate; in this case dissolve by microwave
- TBS buffer: 20 m*M* Tris–HCl pH 7.5, 150 m*M* NaCl
- Lysis buffer: 50 m*M* HEPES–KOH pH 7.5, 150 m*M* NaCl, 1 m*M* EDTA, 1% Triton X-100, 0.1% Sodium Deoxycholate, 0.1% SDS
- 100 × PMSF: 100 m*M* phenylmethylsulfonyl fluoride in 100% ethanol. Store at −20°C
- 10,000 × PIC, protease inhibitor cocktail IV, fungal (Fisher Scientific)
- TE buffer: 10 m*M* Tris–HCl pH 8.0, 1 m*M* EDTA
- IP wash buffer: 10 m*M* Tris–HCl pH 8.0, 0.25 *M* LiCl, 1 m*M* EDTA, 0.5% NP-40, 0.5% Sodium Deoxycholate
- Elution buffer: 50 m*M* Tris–HCl pH 7.5, 10 m*M* EDTA, 1% SDS
- Dynabeads Protein G, 30 mg/mL (Invitrogen)
- 20 mg/mL Pronase (Boehringer). Store at −20°C
- 10 mg/mL RNase A (Sigma). Store at −20°C
- Antibodies: α-Pol II (8WG16, Millipore), α-Tyr1P Pol II (3D12, ActiveMotif), α-Ser2P Pol II (3E10, Millipore), α-Thr4P Pol II (6D7, ActiveMotif), α-Ser5P Pol II (3E8, Millipore)
- ChIP DNA Clean & Concentrator Kit (ZymoResearch)
- Bioanalyzer High Sensitivity DNA Kit (Agilent)

2.1.3 Procedure

2.1.3.1 Cell Culture

1. Grow cells in YES medium overnight to late log phase/early stationary phase at 30°C (preculture).
2. Next morning, dilute the preculture to an $OD_{600} \sim 0.15$ in 200 mL YES (see Note 1).
3. Grow cells to exponential mid-log phase ($OD_{600} = 0.6$–0.8).

2.1.3.2 Cross-linking

1. To 200 mL culture, add 20 mL of 11% formaldehyde (freshly made by mixing 6 mL of 37% HCHO and 14 mL of cross-linking solution).

Incubate 20 min at room temperature, swirling every 2 min. This allows cross-linking of protein–DNA complexes.

2. Stop the cross-linking reaction by adding 30 mL of quenching solution. Incubate for another 5 min (keep swirling).
3. Collect cells by centrifugation in a Beckman JLA-16.250 rotor (2500 × *g*/4000 rpm, 4 min at 4°C). All subsequent steps are performed on ice unless otherwise stated. Wash pellet with 100 mL TBS. Spin down and resuspend cells in 8 mL lysis buffer/1 m*M* PMSF. Transfer to a 14-mL round-bottom Falcon tube. Spin down cells for 4 min. At this point pellets can be frozen at −80°C for long-term storage.

2.1.3.3 Chromatin Isolation and Sonication

1. Resuspend cells in 1 mL lysis buffer with 0.5% SDS and 1 × PIC. Add 1 mL of glass beads and vortex 30 × with a cycle of 30 s of vortexing at maximum speed followed by 30 s on ice. Make sure the beads can move freely while vortexing.
2. Add 6.5 mL of lysis buffer to the tube and puncture its bottom with a 22-gauge ½-in. needle. Insert tube into a Qiagen adapter, and place into a 50-mL Falcon tube (Fig. 1A). Collect lysate by centrifugation at 1000 × *g* for 1 min. Caution: The lysate is now in the 50-mL Falcon tube (see Fig. 1A). Discard the round-bottom tubes containing the glass beads.
3. Transfer the lysate (both cell pellet and supernatant) to an ultracentrifuge tube. Spin at 170,000 × *g* (44,000 rpm) at 4°C for 20 min.
4. Decant the supernatant. The presence of a translucid top layer on the pellet is an indication that cross-linking occurred (Fig. 1B). Resuspend the pellet in 1 mL of lysis buffer with the help of a wooden stick (Fig. 1C, see Note 2). Then add 7 mL of lysis buffer.
5. Centrifuge at 170,000 × *g*, 4°C for 20 min.
6. Decant the supernatant and resuspend the pellet in 1.5 mL lysis buffer with a wooden stick. Transfer to a 15-mL polystyrene Falcon tube (see Note 3) and insert the sonicator microprobe tip.
7. Sonicate 15 s on, 45 s off on high power for 80 min to shear the chromatin. Keep the sample tubes at 4°C throughout sonication (see Note 4). Chromatin should be broken into fragments with a median size of 150–200 bp (Fig. 1D, see Note 4).
8. Transfer the sonicated lysate to a 2-mL Eppendorf tube. Centrifuge at 25,000 × *g* for 20 min at 4°C. The sheared chromatin is now in the supernatant.

9. Transfer the supernatant to a 15-mL Falcon tube. Bring up the volume to 3 mL by adding 1.5 mL lysis buffer. Adjust the NaCl concentration if necessary (see Note 5).
10. Save 50 μL of chromatin prior to immunoprecipitation to generate input DNA for Illumina sequencing.
11. Divide the rest equally (~1 mL) in three 1.5-mL Eppendorf tubes. At this point chromatin can be frozen at −80°C for long-term storage. All three aliquots are used for the next steps but are kept separate until elution (Section 2.1.3.5, step 4).

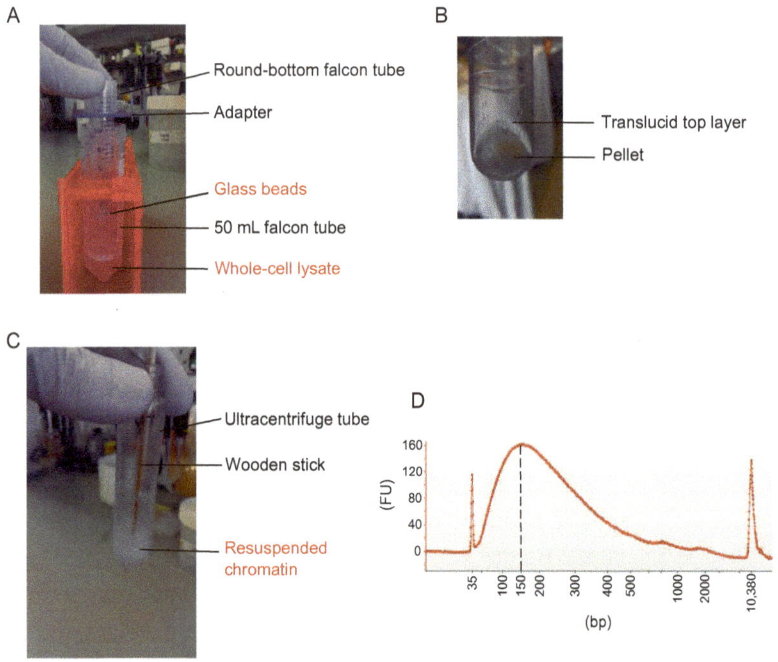

Fig. 1 (A) Recovery of whole-cell lysate after disruption with glass beads and centrifugation. (B) Pellet after ultracentrifugation. (C) Mechanical resuspension of chromatin with a wooden stick. (D) Bioanalyzer trace (High Sensitivity DNA kit) of chromatin input DNA after sonication and reverse cross-linking. 150–200 bp long DNA fragments are enriched. *FU*, arbitrary fluorescence unit.

2.1.3.4 Immunoprecipitation

1. Equilibrate 60 μL protein G beads with 1 mL of TE buffer at room temperature for 4 min on a rotator.
2. Place the tube on a magnetic stand until the solution appears clear. Remove and discard the supernatant.

3. Remove the tube from the magnetic stand. Resuspend beads in 700 μL of TE buffer and add 15 μL of antibody (see Note 6). Incubate at 4°C for at least 2 h on a rotator.
4. Collect beads by placing the tube on a magnetic stand. Discard the supernatant taking care not to disturb the beads.
5. Add 1 mL of TE buffer and place on a rotator for 4 min at room temperature.
6. Place tube on a magnetic stand. When the solution is clear, carefully discard the supernatant.
7. Remove the tube from the magnetic stand. Resuspend beads in 60 μL TE buffer.
8. Add 20 μL of beads conjugated with antibody in each 1 mL chromatin aliquot from Section 2.1.3.3, step 11. Immunoprecipitate overnight (18–20 h) on a rotator at 4°C.
9. Collect beads by placing tubes on a magnetic stand. Wash the immunoprecipitated samples once with 1 mL of appropriate buffer. At each washing step: collect beads by placing the tubes on a magnetic stand; discard the supernatant and add the next washing buffer; place on a rotator for 4 min at room temperature. Wash with:
   (a) Lysis buffer with 275 m*M* NaCl, if possible (see Note 7)
   (b) Lysis buffer with 500 m*M* NaCl, if possible (see Note 7)
   (c) IP wash buffer
   (d) TE buffer
10. Add 200 μL of elution buffer to the washed beads and elute by shaking for 20 min at 65°C at 800 rpm.
11. Collect beads by placing tubes in a magnetic stand. Transfer the supernatant to a new 1.5-mL tube. (Caution: The immunoprecipitated DNA is now in the supernatant; do not discard the supernatant.)
12. Wash beads with 200 μL TE buffer. Collect beads by placing them on a magnetic stand. Wait until the solution is clear and pool this supernatant with previous eluate from step 11. This sample is the IP. Discard beads.

2.1.3.5 Protein Digestion and Reverse Cross-linking

1. To reverse protein–DNA cross-links, add 20 μL of 20 mg/mL pronase. Incubate for 1 h at 42°C, then overnight at 65°C.
2. To prepare the input sample, add 150 μL lysis buffer and 200 μL of TE buffer to 50 μL of input DNA from Section 2.1.3.3, step 10. Add 20 μL of 20 mg/mL pronase and incubate for 1 h at 42°C,

then overnight at 65°C. For the following steps, the input DNA is processed in parallel with the IP DNA.

3. Add 2 μL of 10 mg/mL RNase A and incubate at 37°C for 1 h.
4. Purify DNA using ChIP DNA Clean & Concentrator Kit following manufacturer's instructions. Elution is done in 20 μL elution buffer per column. The three IP samples coming from the same chromatin (Section 2.1.3.3, step 11) are now pooled together (total volume of 60 μL). Samples can be stored at −20°C.
5. Quantify the amounts of DNA by Qubit (see Note 8).

Notes

1. 200 mL of yeast culture gives enough ChIP DNA for sequencing. The exception is Thr4P-IP DNA, where yield is lower. For this antibody, start with 600 mL of yeast culture to get enough DNA for sequencing.
2. It is not possible to completely resuspend the chromatin pellet; disrupt as much as possible.
3. To transfer the crushed pellet, use a 200-μL cut tip.
4. The conditions described in this protocol are optimized for the Bioruptor UCD-200T-EX. This sonicator is equipped with a water bath. To ensure the best results, keep the water in the water bath at 4°C. During the experiment setup, it is important to control that the sheared chromatin is of the correct size. To assess the sonication efficiency, load 1 ng of input DNA on a Bioanalyzer High Sensitivity DNA chip. You should observe an enrichment of 150–200 bp long DNA fragments.
5. To reduce nonspecific binding to the antibody-conjugated beads, we recommend adjusting the final NaCl concentration to 275 m*M*. Higher salt concentration does not interfere with antibody binding to Protein G, with the exception of Thr4P antibody. When immunoprecipitating with Thr4P antibody, keep all buffers with NaCl concentration at 150 m*M*.
6. A negative control with beads only (no-Ab IP) is recommended to control for nonspecific binding to the beads. This sample is processed in parallel with the others, i.e., same starting culture volume, same IP conditions, etc.
7. High-salt washes help reducing nonspecific binding. Do not increase NaCl concentration above 150 m*M* when immunoprecipitating with Thr4P antibody.
8. Typical yields for ChIP DNA are between 5 and 100 ng. Yields for input DNA are around 150–500 ng.

2.2 Library Preparation for Illumina Sequencing

Bioinformatic simulations based on the *S. cerevisiae* genome showed that only 260,000 uniquely mapped reads are required to identify at least 95% of the transcription factor binding sites, given a five-fold enrichment of these binding in the ChIP sample (Lefrançois et al., 2009). *S. pombe* genome is similar in size to that of *S. cerevisiae*. A single NextSeq 500 Illumina run generates about 400 million mapped reads; hence, up to 48 yeast ChIP-Seq samples can be sequenced simultaneously using multiplex Illumina sequencing.

2.2.1 Equipment

- 1.5-mL Nuclease-free tubes
- PCR machine
- PCR tubes
- Magnetic stand for 1.5-mL tubes
- Bioanalyzer
- qPCR machine

2.2.2 Reagents

- NEBNext® Ultra™ DNA Library Prep Kit for Illumina
- NEBNext® Multiplex Oligos for Illumina (set1, set2, set3, or set4)
- Agencourt Ampure XP beads (Beckman Coulter)
- Bioanalyzer High Sensitivity DNA chip
- Kapa Library Quantification DNA standards for Illumina (Kapa Biosystems)
- 2 × SensiMix SyBR No-Rox (Bioline)
- qPCR primers (5 μ*M* stock):
  Primer 1: 5′-AAT GAT ACG GCG ACC GA-3′
  Primer 2: 5′-CAA GCA GAA GAC GGC ATA CGA-3′
- qPCR mix (12 μL): 7.5 μL SensiMix SyBR, 1 μL Primer 1, 1 μL Primer 2, 2.5 μL ddH_2O

2.2.3 Procedure

2.2.3.1 Library Preparation

Prepare the library as described in NEBNext® Ultra™ DNA Library Prep Kit for Illumina. Reagents are supplied with the kit. This kit allows library preparation from small amounts of DNA (5 ng to 1 μg). For ChIP DNA, use 5–30 ng as starting material for library preparation (see Note 9). For input DNA use 25 ng.

Detailed protocols for library preparation are available from Illumina's web site. The protocol consists of five steps:

1. *End repair.* In this step, DNA with overhangs is converted to blunt-ended DNA.
2. *Adaptor ligation.* This step ligates adapters to the ends of DNA fragments.
3. *Bead clean-up of adaptor-ligated DNA without size-selection.* This step removes any excess adapters to prevent them from being amplified during PCR. NEBNext® Ultra™ DNA Library Prep Manual gives the option at this point to clean up the adaptor-ligated DNA with or without size-selection. Since the sheared chromatin is already enriched of 150–200 bp DNA fragments (see Fig. 1D), there is no need to size-select. Thus, follow the protocol for clean up without size-selection.
4. *PCR amplification.* PCR is carried out to selectively amplify the DNA fragments that have adapters bound to them and to introduce a specific barcode/index to the PCR products. There are 48 different primer indexes (barcodes) across four sets, available from NEB (set1, set2, set3, or set4). To avoid PCR duplicates in the sequencing, perform the minimum number of PCR cycles possible (see Note 10).
5. *Double bead clean-up.* This step removes excess primer and primer dimers. Two rounds of bead clean-up are suggested as there is still some primer contamination after the first clean-up.

2.2.3.2 Quality Control

It is important to check the size and quality of DNA libraries by using the Bioanalyzer High DNA Sensitivity chip. Dilute the library three-folds and load 1 μL on Bioanalyzer. If the size of sonicated DNA is 150–200 bp, you should observe an enrichment of 250–300 bp long DNA fragments after library preparation, due to adapter ligation (58 bp).

2.2.3.3 Library Quantification and Pooling for Sequencing

To get consistent sequencing depth across different samples, it is important to accurately quantify DNA libraries. We use qPCR to assess the library concentration.

1. Make 1:1000 and 1:10,000 dilutions of each library in water.
2. Mix 3 μL of diluted library with 12 μL of qPCR mix.

3. Mix 3 μL of Kapa standards with 12 μL of qPCR mix.
4. Set up the qPCR cycles as follows:

95°C, 10 min	
95°C, 10 s	50 ×
60°C, 15 s	
72°C, 15 s	
65–95°C (melt)	

5. Analyze the qPCR data and calculate the library concentration as described in the Technical Data Sheet for the Kapa kit.
6. Dilute the concentration of each library to 4 n*M* with water. Pool samples at equimolar concentration (see Note 11).

Notes

9. In our hands, 1 ng of DNA can still be used for library preparation and give good quality sequencing data.
10. For 25 ng of starting material, do 5–6 cycles. For 5 or 10 ng, perform 8 cycles.
11. If more sequence depth is needed for the IP DNA, we suggest pooling IP and input samples at 2:1 *M* ratio.

2.3 High-Throughput Sequencing

Set up a paired-end run on the Illumina NextSeq 500 (or equivalent). The process of demultiplexing (dividing the sequenced reads into separate files for each barcode/sample) and generating the FASTQ files is carried out on BaseSpace (Illumina's cloud-based resource).

2.4 Data Analysis

2.4.1 Software

- Trimmomatic 0.36 (Bolger, Lohse, & Usadel, 2014)
- Bowtie2 (Langmead & Salzberg, 2012)
- SAMtools (Li et al., 2009)
- deepTools (Ramírez et al., 2016)

2.4.2 Procedure

1. Scan reads with Trimmomatic with a four-base wide sliding window, trimming when the average quality per base drops below 20. Reads less than 20 bp after trimmings were discarded.

2. Align to *S. pombe* genome (version ASM294v2.28) with Bowtie2. Mixed mode and search for discordant alignment are disabled (–no-mixed –no-discordant). After mapping, we keep reads that align uniquely to the reference genome and we remove reads of quality less than Q5 using SAMtools.
3. Remove PCR duplicates using SAMtools.
4. Generated BAM files are converted to Genome Browser visualization files (e.g., bigwig) using "bamCoverage" from deepTools. To normalize for differences in sequencing depth among ChIP-seq samples, the sequence reads are "down-sampled" to the lowest number of the uniquely mapped reads using the "scaleFactor" parameter from "bamCoverage." Fig. 2 shows representative profiles from a successful ChIP-seq experiment. This example shows ChIP-seq profiles of the *rps002* gene. The CTD-IP profiles are compared to the input and to the no-Ab IP. For Ser5P, higher reads are detected at the 5′ end of the gene, whereas Tyr1P, Ser2P, and Thr4P signals are highest downstream the poly(A) site.

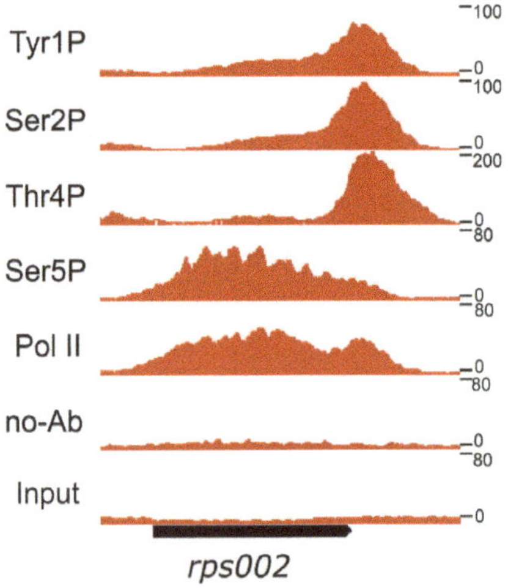

Fig. 2 Example of ChIP-seq profiles for the 40S ribosomal protein S0B (*rps002*) gene. Five different Pol II antibodies, α-Pol II (8WG16), α-Tyr1P Pol II, α-Ser2P Pol II, α-Thr4P Pol II, and α-Ser5P Pol II, were used for the ChIP-seq. Input and the no-Ab IP tracks show the specificity of the phospho-CTD IP signal.

5. Prepare data for plotting with "computeMatrix" in deepTools. Create metagene ("plotProfile") and heatmap ("plotHeatmap") plots using deepTools for a preliminary analysis of the data. Metagene and heatmap plots are useful to identify where the CTD phospho-marks are enriched on Pol II-transcribed genes and whether this enrichment is similar among genes. Fig. 3A and B shows plots for Ser5P-IP. Genes are scaled

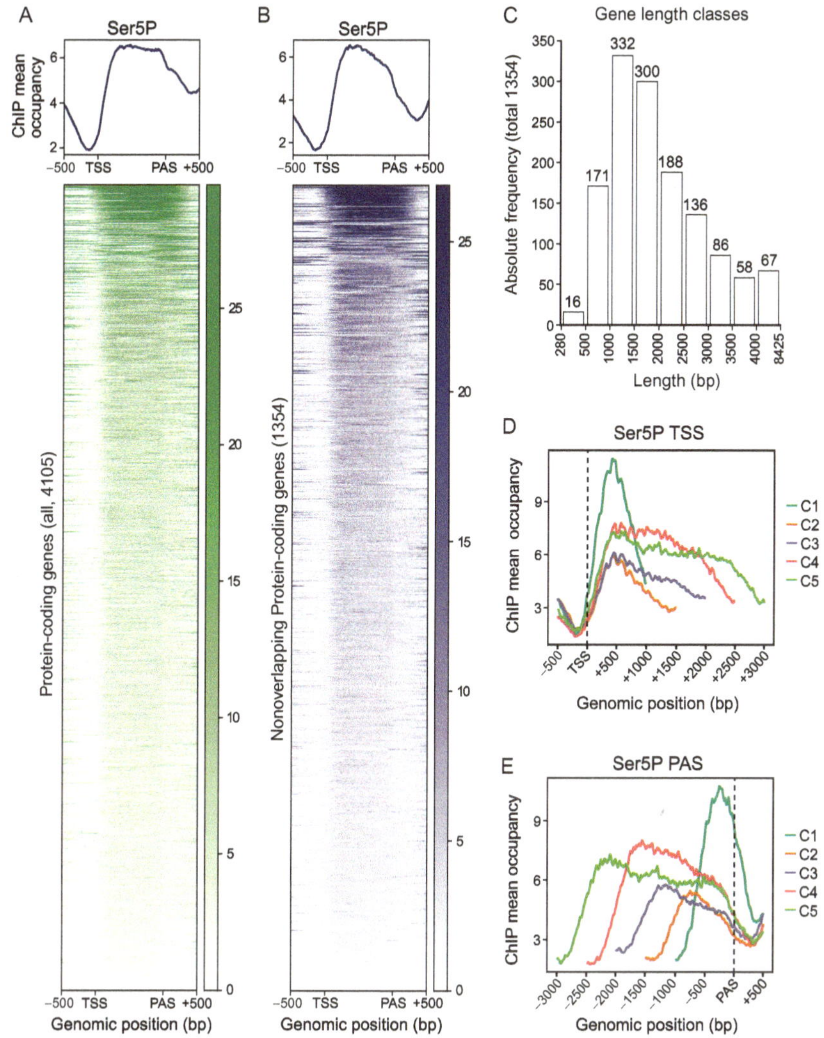

Fig. 3 Ser5P-IP analysis. (A) Averaged occupancy profiles of Ser5P from ChIP-seq on all 4105 yeast protein-coding genes (*upper panel*). Color-coded heatmap illustrating Ser5P profiles across single protein-coding genes (*lower panel*). Profiles are aligned to the TSS and PAS as indicated. (B) Averaged profile and heatmap for Ser5P on nonoverlapping protein-coding genes that have a distance greater than 250 bp from flanking genes (1354 ORFs). (C) Division of genes from (B) into length classes: from 280 to 500 bp, from 500 to 1000 bp (C1), from 1000 to 1500 bp (C2), from 1500 to 2000 bp (C3), from 2000 to 2500 bp (C4), from 2500 to 3000 bp (C5), from 3000 to 3500 bp, from 3500 to 4000 bp, from 4000 to 8425 bp. (D and E) Averaged occupancy profiles of Ser5P on C1–C5 gene length classes aligned at TSS (D) and PAS (E).

to the same length using "computeMatrix/scale-regions" and aligned at both the TSS (transcription start site) and PAS (poly(A) site). This is very useful to determine where a phospho-mark peaks relative to gene 5′ and 3′ ends. Fig. 3A shows plots for Ser5P-IP on all protein-coding genes (4105 ORFs). Ser5 phospho-mark is placed early during the transcription cycle, is enriched on gene bodies, and then drops at gene 3′ ends. Because yeast genome is very compact and ChIP-seq is not strand specific, plotting all genes may give mixed signals coming from overlapping genes and thus mask—at least partially—the real pattern of CTD phosphorylation. Thus, Fig. 3B shows metagene and heatmap plots generated considering only protein-coding genes that do not overlap with any other transcriptional unit and have a distance greater than 250 bp from flanking genes (1354 ORFs). Although the general pattern is unchanged, the drop of Ser5P at the 3′ end of genes becomes more obvious when considering only nonoverlapping genes. Thus, filtering out signals coming from overlapping genes gives an unbiased idea of the relative enrichment of Ser5P.

6. In Fig. 3A and B, genes are scaled to the same length. Although useful to determine the position of a peak relative to both TSS and PAS, scaling genes to the same length can hide some information. Indeed, gene length may affect phosphorylation patterns. To assess whether this is the case for Ser5P, the 1354 nonoverlapping genes are divided into different length classes (Fig. 3C). Ser5P-IP is then plotted for the five most abundant classes (C1–C5, as defined in Fig. 3C). Fig. 3D and E shows metagene profiles for the different groups. Genes are aligned at TSS (Fig. 3D) or PAS (Fig. 3E) using "computeMatrix/reference-point." These plots show that Ser5P peak is highest 500 bp downstream the TSS, independent of gene length. Interestingly, the signal then drops quickly for short genes, while it persists on long genes, suggesting that dephosphorylation of Ser5 might be differently regulated depending on gene length. Ser5P enrichment then drops at the PAS for all gene classes.

3. CONCLUSIONS

We find ChIP-seq a powerful and robust method to characterize genome-wide Pol II CTD phosphorylation patterns. Given the small size of *S. pombe* genome and the possibility of multiplex sequencing, ChIP-seq also emerges as a high-throughput but cost-effective technique.

REFERENCES

Ahn, S. H., Kim, M., & Buratowski, S. (2004). Phosphorylation of serine 2 within the RNA polymerase II C-terminal domain couples transcription and 3J end processing. *Molecular Cell*, *13*, 67–76.

Bataille, A. R., Jeronimo, C. L., Jacques, P.-É. T., Laramé, L., Fortin, M.-V., Forest, A., et al. (2012). A universal RNA polymerase II CTD cycle is orchestrated by complex interplays between kinase, phosphatase, and isomerase enzymes along genes. *Molecular Cell*, *45*(2), 158–170. https://doi.org/10.1016/j.molcel.2011.11.024.

Bolger, A. M., Lohse, M., & Usadel, B. (2014). Trimmomatic: A flexible trimmer for Illumina sequence data. *Bioinformatics*, *30*(15), 2114–2120. https://doi.org/10.1093/bioinformatics/btu170.

Corden, J. L. (2013). RNA polymerase II C-terminal domain: Tethering transcription to transcript and template. *Chemical Reviews*, *113*(11), 8423–8455. https://doi.org/10.1021/cr400158h.

Coudreuse, D., Van Bakel, H., Dewez, M., Soutourina, J., Parnell, T., Vandenhaute, J., et al. (2010). A gene-specific requirement of RNA polymerase II CTD phosphorylation for sexual differentiation in S. pombe. *Current Biology*, *20*(12), 1053–1064. https://doi.org/10.1016/j.cub.2010.04.054.

Descostes, N., Heidemann, M., Spinelli, L., Schüller, R., Maqbool, M. A., Fenouil, R., et al. (2014). Tyrosine phosphorylation of RNA polymerase II CTD is associated with antisense promoter transcription and active enhancers in mammalian cells. *eLife*, *2014*(3), 1–19. https://doi.org/10.7554/eLife.02105.

Eick, D., & Geyer, M. (2013). The RNA polymerase II carboxy-terminal domain (CTD) code. *Chemical Reviews*, *113*(11), 8456–8490. https://doi.org/10.1021/cr400071f.

Harlen, K. M., Trotta, K. L., Smith, E. E., Mosaheb, M. M., Fuchs, S. M., & Churchman, L. S. (2016). Comprehensive RNA polymerase II interactomes reveal distinct and varied roles for each phospho-CTD residue. *Cell Reports*, *15*(10), 2147–2158. https://doi.org/10.1016/j.celrep.2016.05.010.

Hintermair, C., Heidemann, M., Koch, F., Descostes, N., Gut, M., Gut, I., et al. (2012). Threonine-4 of mammalian RNA polymerase II CTD is targeted by Polo-like kinase 3 and required for transcriptional elongation. *The EMBO Journal*, *31*(12), 2784–2797. https://doi.org/10.1038/emboj.2012.123.

Hsin, J. P., Li, W., Hoque, M., Tian, B., & Manley, J. L. (2014). RNAP II CTD tyrosine 1 performs diverse functions in vertebrate cells. *eLife*, *2014*(3), 1–10. https://doi.org/10.7554/eLife.02112.

Keogh, M.-C., & Buratowski, S. (2004). Using chromatin immunoprecipitation to map cotranscriptional mRNA processing in Saccharomyces cerevisiae. In *mRNA processing and metabolism* (pp. 1–16). New Jersey: Humana Press. https://doi.org/10.1385/1-59259-750-5:001.

Kim, H., Erickson, B., Luo, W., Seward, D., Graber, J. H., Pollock, D. D., et al. (2010). Gene-specific RNA polymerase II phosphorylation and the CTD code. *Nature Structural & Molecular Biology*, *17*(10), 1279–1286. https://doi.org/10.1038/nsmb.1913.

Komarnitsky, P., Cho, E. J., & Buratowski, S. (2000). Different phosphorylated forms of RNA polymerase II and associated mRNA processing factors during transcription. *Genes & Development*, *14*(19), 2452–2460.

Langmead, B., & Salzberg, S. L. (2012). Fast gapped-read alignment with Bowtie 2. *Nature Methods*, *9*(4), 357–359. https://doi.org/10.1038/nmeth.1923.

Lefrançois, P., Euskirchen, G. M., Auerbach, R. K., Rozowsky, J., Gibson, T., Yellman, C. M., et al. (2009). Efficient yeast ChIP-Seq using multiplex short-read DNA sequencing. *BMC Genomics*, *10*, 37. https://doi.org/10.1186/1471-2164-10-37.

Li, H., Handsaker, B., Wysoker, A., Fennell, T., Ruan, J., Homer, N., et al. (2009). The sequence alignment/map format and SAMtools. *Bioinformatics*, *25*(16), 2078–2079. https://doi.org/10.1093/bioinformatics/btp352.

Mayer, A., Heidemann, M., Lidschreiber, M., Schreieck, A., Sun, M., Hintermair, C., et al. (2012). CTD tyrosine phosphorylation impairs termination factor recruitment to RNA polymerase II. *Science*, *336*(6089), 1723–1725. https://doi.org/10.1126/science.1219651.

Mayer, A., Lidschreiber, M., Siebert, M., Leike, K., Söding, J., & Cramer, P. (2010). Uniform transitions of the general RNA polymerase II transcription complex. *Nature Structural & Molecular Biology*, *17*(10), 1272–1278. https://doi.org/10.1038/nsmb.1903.

Milligan, L., Huynh-Thu, V. A., Delan-Forino, C., Tuck, A., Petfalski, E., Lombraña, R., et al. (2016). Strand-specific, high-resolution mapping of modified RNA polymerase II. *Molecular Systems Biology*, *12*(6), 874. https://doi.org/10.15252/MSB.20166869.

Ramírez, F., Ryan, D. P., Grüning, B., Bhardwaj, V., Kilpert, F., Richter, A. S., et al. (2016). deepTools2: A next generation web server for deep-sequencing data analysis. *Nucleic Acids Research*, *44*(W1), W160–W165. https://doi.org/10.1093/nar/gkw257.

Schlackow, M., Nojima, T., Gomes, T., Dhir, A., Carmo-Fonseca, M., & Proudfoot, N. J. (2017). Distinctive patterns of transcription and RNA processing for human lincRNAs. *Molecular Cell*, *65*(1), 25–38. https://doi.org/10.1016/j.molcel.2016.11.029.

Schroeder, S. C., Schwer, B., Shuman, S., & Bentley, D. (2000). Dynamic association of capping enzymes with transcribing RNA polymerase II. *Genes & Development*, *14*(19), 2435–2440.

Tietjen, J. R., Zhang, D. W., Rodríguez-Molina, J. B., White, B. E., Akhtar, M. S., Heidemann, M., et al. (2010). Chemical-genomic dissection of the CTD code. *Nature Structural & Molecular Biology*, *17*(9), 1154–1161. https://doi.org/10.1038/nsmb.1900.

CHAPTER TWENTY-ONE

Toward Cell Type-Specific In Vivo Dual RNA-Seq

Lutz Frönicke*, Denise N. Bronner†, Mariana X. Byndloss†, Bridget McLaughlin‡, Andreas J. Bäumler†, Alexander J. Westermann§,¶,1

*University of California Davis Genome Center, Davis, CA, United States
†Department of Medical Microbiology and Immunology, School of Medicine, University of California Davis, Davis, CA, United States
‡Comprehensive Cancer Center Flow Cytometry Shared Resource, University of California Davis, Davis, CA, United States
§Institute of Molecular Infection Biology (IMIB), University of Würzburg, Würzburg, Germany
¶Helmholtz Institute for RNA-Based Infection Research (HIRI), Würzburg, Germany
[1]Corresponding author: e-mail address: alexander.westermann@uni-wuerzburg.de

Contents

Abstract

Dual RNA-seq has emerged as a genome-wide expression profiling technique, simultaneously measuring RNA transcript levels in a given host and its pathogen during an infection. Recently, the method was transferred from cell culture to in vivo models of bacterial infections; however, specific host cell-type resolution has not yet been achieved. Here we present a detailed protocol that describes the application of Dual RNA-seq to murine colonocytes isolated from mice infected with the enteric pathogen *Salmonella* Typhimurium. At day 5 after oral infection, the mice were humanely euthanized, their colons extracted, and colonocytes isolated and fixed. Upon antibody staining of cell type-specific surface markers, the fraction of *Salmonella*-invaded

Methods in Enzymology, Volume 612
ISSN 0076-6879
https://doi.org/10.1016/bs.mie.2018.08.013

colonocytes was collected by fluorescence-activated cell sorting based on a fluorescent signal emitted by the internalized bacteria. Total RNA was extracted from cells enriched by this method, and ribosomal transcripts from host and microbial cells were removed prior to cDNA synthesis and library generation. We compared different protocols for library preparation and discuss their respective advantages and caveats when applied to minute RNA amounts that constitute an inherent challenge for in vivo transcriptomics. Our results introduce an ultralow input protocol that holds promise for cell type-specific in vivo Dual RNA-seq for charting gene expression of a bacterial pathogen within its respective in vivo niche, along with the consequent host response.

1. INTRODUCTION

The term "Dual RNA-seq" refers to the simultaneous quantification of host and pathogen gene expression during their interaction. The concept of jointly extracting and sequencing RNA from both a pathogen and its host organism and separating the resulting sequencing reads computationally was first put forward in a thought experiment in 2012 (Westermann, Gorski, & Vogel, 2012). Since, the practicability of this technique was demonstrated for the interaction of mammalian cells with various bacterial (reviewed in Westermann, Barquist, & Vogel, 2017) and viral pathogens (e.g., Lisnic et al., 2013; Park et al., 2015; Wesolowska-Andersen et al., 2017; Xu et al., 2014) as well as with pathogenic fungi (reviewed in Wolf, Kammer, Brunke, & Linde, 2018) and eukaryotic parasites (e.g., Choi, Aliota, Mayhew, Erickson, & Christensen, 2014; Dillon et al., 2015; Ehret, Spork, Dieterich, Lucius, & Heitlinger, 2017; Pittman, Aliota, & Knoll, 2014). In most cases, however, the technique was applied to cell culture-based host models challenged with the respective infectious agent in vitro. For example, we previously published a detailed protocol for the performance of Dual RNA-seq for *Salmonella*-infected epithelial cells in tissue culture (Westermann & Vogel, 2018).

Only recently, Dual RNA-seq is being transferred to in vivo models of bacterial infections (Nuss et al., 2017; Thanert, Goldmann, Beineke, & Medina, 2017). While these pioneering studies provided new insights into the pathogenesis of the extracellular pathogens *Yersinia pseudotuberculosis* and *Staphylococcus aureus*, respectively, they analyzed host gene expression in bulk tissues, namely, Peyer's patches or kidneys, composed of numerous individual cell types. As a direct consequence, differential abundances of host transcripts before and upon infection could be due to gene expression changes, an altered cellular composition of the targeted host organ, or both.

To truly detect host gene expression changes in vivo, and differentiate between the host response mounted by invaded or attached cells versus uninfected bystanders, transcriptome studies are therefore required that are based on specific host cell subpopulations. In other words, global host–pathogen transcriptome profiling must be made compatible with protocols to dissociate whole tissues into individual cells, antibody staining, and the sort-enrichment of specific populations of host cells.

Here, we lay the groundwork for Dual RNA-seq analysis of the enteric, facultative intracellular pathogen *Salmonella enterica* serovar Typhimurium within one of its in vivo niches, namely, inside colonic epithelial cells (colonocytes) (Fig. 1A). To this end, C57BL/6J mice are infected with fluorescent *Salmonella* by gavage and euthanized at day 5 of the infection. The murine colon is prepared, colonocytes isolated and further dissociated into single-cell suspensions compatible with cell sorting. The colonocytes are then fixed and stained with fluorochrome-conjugated antibodies against specific cell surface markers. Using fluorescence-activated cell sorting (FACS), *Salmonella*-containing colonocytes are collected, their total RNA content is harvested, ribosomal RNA (rRNA) of either origin depleted, and the resulting rRNA-free sample used for reverse transcription and sequencing library generation. Due to the low amount of starting material (~8000 sorted cells per animal, corresponding to ~3 ng of total RNA), rRNA depletion and cDNA library preparation are technically challenging, and we therefore compare the performance of different low input protocols (Fig. 1B) and discuss benefits and pitfalls of individual steps.

2. INFECTION OF C57BL/6J MICE WITH *S.* TYPHIMURIUM AND COLONOCYTE ISOLATION

2.1 The C57BL/6J Model of *Salmonella* Infection

C57BL/6J mice are devoid of a functional version of the ferrous iron transporter NRAMP (Natural Resistance-Associated Macrophage Protein) that is normally produced by dendritic cells and macrophages and, as a result, are genetically susceptible to lethal *Salmonella* infections (Blackwell et al., 2001). When delivered orally to these mice, *Salmonella* traverse the Peyer's patches within a few hours, colonize the caecum and colon, and after 2–3 days disseminate to systemic sites including the spleen and liver. At day 5 of the infection, the mice show severe signs of acute intestinal inflammation and are humanely euthanized. By this time, a small fraction of *Salmonella* reside inside colonocytes, probably boosting intestinal inflammation, as the arising

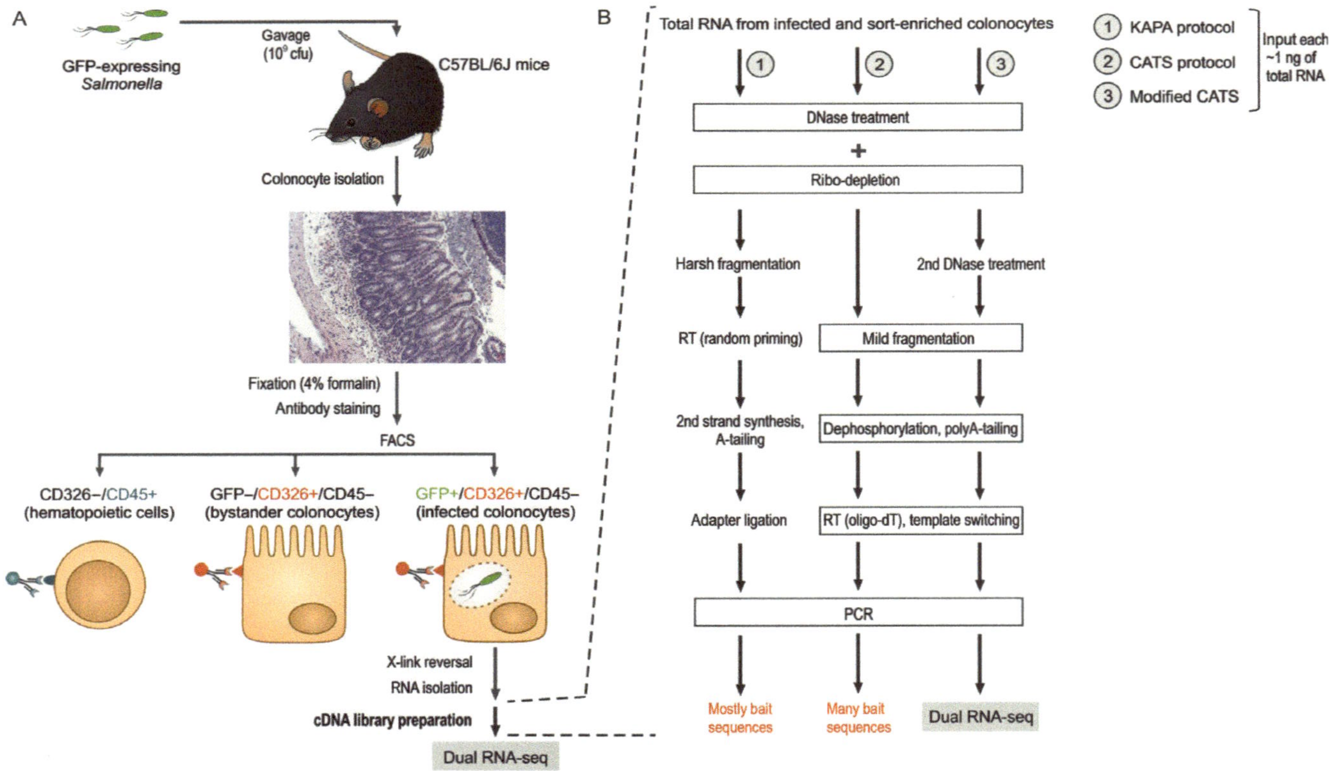

Fig. 1 Cell type-specific in vivo Dual RNA-seq. (A) Schematic representation of the experimental pipeline described here. The histopathology image depicts *Salmonella*-infected mouse colonocytes. (B) Overview of the workflows of the three cDNA library preparation protocols tested.

changes in the host's metabolism allow luminal *Salmonella* to outcompete the resident microbiota (Baumler & Sperandio, 2016). Here, in a first step toward the establishment of cell type-specific in vivo Dual RNA-seq, invaded colonocytes are collected and their RNA content is sequenced. This becomes feasible with the help of a *S.* Typhimurium strain (JVS-3858) (Papenfort et al., 2009) that stably expresses the green fluorescent protein (GFP), enabling the enrichment of invaded (GFP-positive) colonocytes by FACS.

2.1.1 Equipment

- Petri dishes
- Incubator for agar plates (at 37°C and 5% CO_2)
- Glass test tubes and Erlenmeyer flasks
- Shaker (at 37°C)
- Plastic cuvettes
- Spectrophotometer
- Plastic tubes (15 mL; 50 mL)
- Centrifuge (cooled to 4°C)
- Gavage needles (sterilized in 70% ethanol)
- Syringes (1 mL; 10 mL)
- CO_2 chamber
- Ice bucket
- Tweezers and scissors (sterilized in 70% ethanol)
- 18-gauge blunt needles
- Cell strainer (70 μm; 40 μm)
- FACS tubes

2.1.2 Buffers and Reagents

- Luria broth (LB) agar
- LB liquid media
- Chloramphenicol (Cm; 0.03 mg/mL)
- 70% ethanol (v/v)
- Phosphate-buffered saline (PBS)
- PBS/10% fetal bovine serum (FBS)
- Dissociation reagent #1: 30 m*M* ethylenediaminetetraacetic acid (EDTA), 1.5 m*M* dithiothreitol, PBS
- Dissociation reagent #2: 30 m*M* EDTA, PBS
- TrypLE express (Gibco)
- Ammonium/chloride/potassium (ACK) buffer: 0.15 *M* NH_4Cl, 1 m*M* $KHCO_3$, 0.1 m*M* Na_2EDTA (pH 7.2)

2.1.3 Procedure

2.1.3.1. All experiments in this study were approved by the Institutional Animal Care and Use Committee at the University of California at Davis. Female C57BL/6J mice, aged 8 weeks, were obtained from The Jackson Laboratory. Routinely, mice are housed in ventilated cage racks on corn bedding, providing water and mouse chow *ad libitum*, monitored twice daily, and cage bedding is exchanged every 2 weeks.

2.1.3.2. An overnight LB-Cm culture of *S.* Typhimurium strain JVS-3858 is used for infection. After determining the optical density at 600 nm (OD_{600}), an equivalent volume of 82.5 OD is pelleted for 20 min at 4000 rpm and 4°C, and suspended in 8.25 mL of fresh LB. 100 μL of this suspension correspond to 10^9 colony forming units (c.f.u.) and are administered by gavage using a sterile 1-mL syringe. For the mock control, 100 μL of pure LB are applied to the mouse in the same way.

2.1.3.3. To confirm the correct infectious dose, a serial dilution of the input suspension is set up in PBS. The 10^{-6} and 10^{-7} dilutions are plated in technical duplicates on LB-Cm plates and incubated overnight at 37°C, before counting the colonies on the next day.

2.1.3.4. The weight of the infected mice is measured each day during the course of infection. At day 5 of infection, the mice are euthanized and their proximal colon is collected and flushed with ice-cold PBS using a 10-mL syringe fitted with an 18-gauge blunt needle to remove remnant feces. The colon is opened lengthwise, cut into ~1 cm pieces, which are collected in 10 mL of ice-cold dissociation reagent #1, and incubated therein for 20 min on ice.

2.1.3.5. The intestinal pieces are transferred from dissociation reagent #1 into 10 mL of dissociation reagent #2 prewarmed to 37°C and incubated therein for 10 min at 37°C.

2.1.3.6. The tube is shaken manually for 30 s (1–2 times per second) to release the epithelium from the basement membrane. The solution turns cloudy. The remnant intestinal tissue is removed using forceps and the solution (containing the colonic epithelial fraction) is pelleted for 5 min at 800 × *g*, 4°C.

2.1.3.7. The colonocytes are washed once with 10 mL ice-cold PBS/10% FBS, pelleted as above, suspended in 5 mL of TrypLE express, and incubated for 20 min at 37°C with vigorous shaking every 5 min.

2.1.3.8. 5 mL of PBS/10% FBS are added, and the suspension is mixed by gently pipetting up and down.

2.1.3.9. The solution is passed through a 70-μm filter and then through a 40-μm filter to exclude larger clumps, collected in a 50-mL plastic tube, pelleted (5 min at 800 × *g*, 4°C), and suspended in 1 mL of ACK buffer to lyse red blood cells.

2.1.3.10. After an incubation for 5 min at room temperature, 10 mL of PBS/10% FBS are added, the sample is centrifuged (5 min at 800 × *g*, 4°C), and the pellet suspended in 2 mL of PBS and transferred into a FACS tube.

2.1.4 Notes

The majority of cells of the initial sample will inevitably be lost during the single-cell suspension protocol described above. After the filtering steps and treatment with ACK buffer, the cell pellet may hardly be visible with the naked eye. While the present approach (particularly the cDNA library preparation; see step 4.2) is compatible with such small cell numbers, for other applications it might be necessary to pool colonocytes from multiple animals at this step of the protocol. Either way, caution needs to be taken during the downstream antibody staining procedure in order not to lose the residual cell pellet during the multiple rounds of centrifugation and resuspension that will follow.

3. ANTIBODY STAINING, FIXATION, AND CELL SORTING

3.1 Cell Staining With Antibodies Against Specific Surface Markers

The above isolation protocol yields cell samples mostly consisting of colonocytes (~67%–84%; see Fig. 2C). However, to further increase purity, the cell mixture is stained with antibodies specifically binding to cells of epithelial origin (CD326) or hematopoietic cells (CD45). In the course of antibody staining, the cells will be fixed using 4% formalin that sterilizes the sample (Fig. 2A) and thus minimizes changes of the infected cells' transcriptomes during the lengthy sorting procedure. Formalin, however, induces unspecific cross-links between cellular RNA and other macromolecules, which need to be reversed prior to the RNA extraction (Evers, Fowler, Cunningham, Mason, & O'Leary, 2011; Hamatani et al., 2006; Kuramochi et al., 2006) (Fig. 2B).

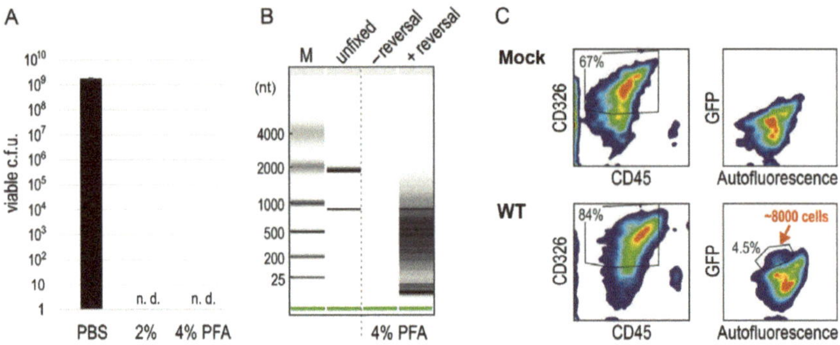

Fig. 2 Fixation and sorting of infected colonocytes. (A) Paraformaldehyde (PFA) sterilizes *Salmonella* samples. 1 mL of an overnight culture of *S.* Typhimurium strain JVS-3858 was washed once in PBS and serially diluted in PBS prior to plating of the 10^{-8} and 10^{-9} dilutions in technical triplicates on LB agar (PBS). Alternatively, the same amount of bacteria were incubated for 20 min in either 2% or 4% PFA on ice, washed with PBS, and 100 μL of the undiluted samples plated on LB agar. Individual colonies—if present—were counted after incubation over night at 37°C. *n. d.*, not detected. (B) Heat reverses the PFA-induced cross-link between cellular RNA and other macromolecules. Total RNA was extracted from unfixed HeLa-S3 cells infected with *S.* Typhimurium, or from the same cells after their fixation with 4% PFA with or without reversal of the cross-links via a 15 min heating step at 70°C. RNA samples were then loaded on an Agilent RNA Nano chip and analyzed. (C) FACS plots of colonocytes from *Salmonella*- or mock-infected C57BL/6J mice. *Left*: CD326 served as an epithelial and CD45 as a hematopoietic marker. $CD326^{+}/CD45^{-}$ cells were gated and are replotted on the right. *Right*: GFP was constitutively expressed by *Salmonella*, and thus a GFP-positive colonocyte population was exclusively detected in the infected sample (*red arrow*). Panel B: *Modified from Westermann, A. J., Forstner, K. U., Amman, F., Barquist, L., Chao, Y., Schulte, L. N., et al. (2016). Dual RNA-seq unveils noncoding RNA functions in host-pathogen interactions.* Nature, 529, *496–501.*

3.1.1 Equipment

- Ice bucket
- Centrifuge (cooled to 4°C)
- Fume hood

3.1.2 Buffers and Reagents

- LIVE/DEAD™ Fixable Aqua Dead Cell Stain kit (Thermo Fisher Scientific)
- FACS buffer: PBS containing 1% bovine serum albumin
- Fc Receptor Binding Inhibitor Polyclonal Antibody, eBioscience™ (Thermo Fisher Scientific)
- Antimouse CD326 antibodies coupled to APC (BioLegend)
- Antimouse CD45 antibodies coupled to PE (BioLegend)
- 4% formalin

3.1.3 *Procedure*

3.1.3.1 All steps of the surface staining protocol should be performed in the dark and on ice. All reagents should be cooled to 4°C. Formalin is handled under the fume hood.

3.1.3.2 7.5 μL of Aqua Dead Cell Stain are added to the cell sample (from step 2.1.3.10). The suspension is incubated for 20 min on ice. The cells are pelleted (5 min, 2000 rpm, 4°C) and the supernatant is carefully removed. Then, 50 μL of FACS buffer and 2 μL of Fc Receptor Binding Inhibitor are added. The suspension is incubated for 15 min on ice.

3.1.3.3 In the meantime, the monoclonal antibody cocktail is prepared; in this case a 1:1 mixture of antimouse CD326 and antimouse CD45.

3.1.3.4 4 μL of the antibody cocktail are added to each tube, the cells are incubated for 20 min on ice, pelleted (5 min at 2000 rpm, 4°C), the supernatant taken off, and the cell pellet fixed by suspending it in 200 μL of 4% formalin, and an incubation step for at least 20 min on ice.

3.1.3.5 To prepare the samples for sorting, the cells are centrifuged (5 min at 2000 rpm, 4°C), the formalin is removed, and the pellet suspended in 300 μL of FACS buffer.

3.1.4 *Notes*

It is recommended to initially run a pilot experiment, including unstained and single-stained cell samples, to determine the appropriate gating strategy. For the actual Dual RNA-seq experiment we excluded these controls to save material.

3.2 FACS Enrichment of Infected Colonocytes and Cross-Link Reversal

As the invaded colonocytes will usually emit a rather weak GFP signal, as they contain only a few internalized bacteria at this stage of infection, strict gating is essential (Fig. 2C). The mock-infected sample should be analyzed first, in order to define the cluster of uninfected, GFP-negative cells. Based on the GFP-minus population, the fraction of GFP-positive cells can subsequently be identified. Using the above infection and colonocyte isolation protocols, ~8000 infected (GFP-positive) colonocytes were obtained per mouse.

3.2.1 *Equipment*

- MoFlo Astrios EQ cell sorter (Beckman Coulter)
- 2-mL reaction tubes
- Heat block

3.2.2 Buffers and Reagents

- FACS Flow and Clean buffers
- Buffer RL (NorGen)

3.2.3 Procedure

3.2.3.1 The gating is adjusted at 405/526 nm (Aqua Dead Cell Stain), 633/660 nm (CD326-APC), 488/575 nm (CD45-PE), and 488/509 nm (GFP-expressing *Salmonella*) excitation/emission wavelengths as illustrated in Fig. 2C.

3.2.3.2 The cell samples are sorted using the 100-μm nozzle at medium flow rate and under continuous cooling of both the sample input and collection chambers to 4°C. The population of $CD326^+/CD45^-$ cells is collected in a 2-mL plastic tube preloaded with 350 μL of Buffer RL (NorGen).

3.2.3.3 The formalin-induced cross-links are reversed by heating the cell lysates for 10 min to 70°C in a heat block. The samples are then frozen until further processing.

3.2.4 Notes

When sorting the infected and stained colonocytes, it is important to minimize cell loss. Therefore, the cells are directly sorted into the lysis buffer used for the downstream RNA isolation procedure. This is possible, due to the low cell number (and thus small FACS Flow buffer volume) that will not markedly dilute the lysis buffer. The benefit of this strategy, compared to collecting the sorted cells in a tube prefilled with FACS buffer or PBS, is twofold. First, by collecting cells in the lysis buffer an otherwise obligate centrifugation step to reduce the liquid volume prior to cell lysis can be avoided that would result in additional cell loss. Second, due to the inhibition of RNA degradation in the lysis buffer, the transcriptomes of the cells are immediately stabilized after sorting.

4. RNA EXTRACTION, DNase I TREATMENT, RIBO-DEPLETION, cDNA SYNTHESIS, AND SEQUENCING

4.1 Total RNA Extraction and gDNA Removal

For RNA extraction, purification columns are used that allow the isolation of both bacterial and eukaryotic RNA (Fig. 3A). As judged from capillary electrophoresis, RNAs from a broad size range can be extracted using these

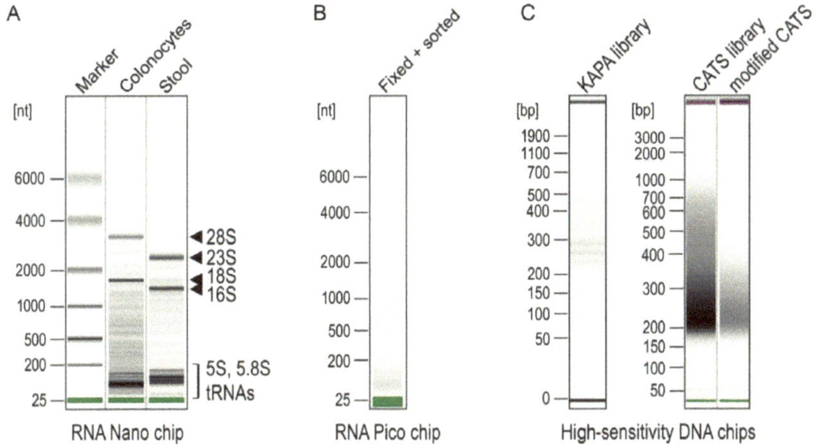

Fig. 3 RNA samples and cDNA libraries at different stages of the protocol. (A) The RNA isolation columns (RNA Extraction kits from NorGen) can be used to efficiently extract total RNA from bacterial and mammalian samples. RNA was extracted from mouse colonocytes or a murine fecal pellet and analyzed on an Agilent Bioanalyzer. The positions of prominent transcript species are denoted to the *right*. (B) RNA profile after fixation and cell sorting. Note that the heat-based cross-link reversal and mechanical stress applied during FACS lead to partial RNA fragmentation. Due to the ultralow input, the RNA traces are barely visible even when using the sensitive Pico chip. (C) Final cDNA libraries obtained from the three protocols (see Fig. 1B).

columns: transfer RNAs (tRNAs; ~75 nt) and 5S rRNA (120 nt) were similarly efficiently isolated than 23S (2900 nt) or 28S rRNAs (4700 nt) (Fig. 3A). Here, using *S.* Typhimurium as a bacterial pathogen model, a mild lysis buffer, as is usually used for pure mammalian RNA extractions, is sufficient to crack the cell walls of the internalized bacteria. This might however be different for other—particularly Gram-positive—bacteria; efficient lysis of which may depend on mechanical or enzymatic pretreatment (Raynaud, Le Pabic, & Felden, 2018). Either way, the harsh conditions associated with formalin fixation and reversal of the resulting cross-links between RNA and other cellular macromolecules will result in partial RNA fragmentation (Fig. 2B). To account for this, only a mild fragmentation step is implemented in the optimized cDNA library generation protocol.

4.1.1 Equipment

- Benchtop centrifuge
- Spin and collection columns provided with the Total RNA Extraction Plus kit (NorGen)

- 1.5-mL reaction tubes
- Qubit Fluorometer (Thermo Fisher Scientific) or similar device
- Agilent Bioanalyzer 2100 and corresponding RNA chips (Nano, Pico)

4.1.2 Buffers and Reagents

- Buffers provided with the Total RNA Extraction Plus kit (NorGen)
- RNase-free water
- TURBO DNA-*free*™ kit (Thermo Fisher Scientific)
- 30:1 ethanol/3 *M* sodium acetate (pH 5.4) (v/v)
- 100% ethanol
- 70% ethanol (v/v)

4.1.3 Procedure

4.1.3.1 The cell lysates from step 3.2.3.3 are vortexed for 15 s. Then, 210 μL of 100% ethanol are added and the lysate is mixed by vortexing, transferred to an RNA Purification Column, and spun for 1 min at 6000 rpm, room temperature.

4.1.3.2 The column is washed three times with each 400 μL of Wash Solution A and centrifugation for each 1 min at 14,000 rpm, room temperature.

4.1.3.3 After the third wash, the column is spun for another 2 min to dry the resin. RNA is eluted by the addition of 44 μL of RNase-free water and centrifugation for 2 min at 2000 rpm, room temperature, followed by an additional centrifugation step for another minute at 14,000 rpm.

4.1.3.4 The purified RNA is directly applied to DNase treatment using the TURBO DNA-*free* kit to remove genomic DNA contaminations. To this end, each 5 μL of 10 × DNase I buffer and 1 μL of DNase I are added to the 44 μL RNA sample from step 4.1.3.3. The resulting reaction mix is incubated for 30 min at 37°C.

4.1.3.5 5 μL of DNase inactivation reagent are added to the reaction mix, the solution is mixed briefly by vortexing and incubated for 2 min at room temperature with brief vortexing steps every 30 s.

4.1.3.6 The samples are centrifuged (3 min at 14,000 rpm, room temperature) to pellet the inactivation reagent. The gDNA-free RNA sample in the supernatant is transferred to a fresh 1.5-mL reaction tube.

4.1.3.7 The RNA concentration in the final samples may be determined using a Qubit instrument (or similar) and checked on a Bioanalyzer (Fig. 3B). The samples should be frozen or directly submitted to rRNA depletion and cDNA library generation.

4.1.4 Notes

The Total RNA Extraction Plus kit (NorGen) offers an optional on-column gDNA removal step. However, we skipped this step here as we found that for our samples it did not efficiently deplete contaminations with genomic DNA. Instead, we performed a dedicated DNase I treatment (TURBO DNA-*free*) after RNA isolation.

4.2 Evaluation of Different cDNA Library Generation Protocols

In the following, we tested two distinct protocols, and further optimized one of them, to obtain high-quality cDNA libraries from the sorted colonocytes (Fig. 1B). To this end, the total RNA samples from step 4.1.3.7 were first split into three parts, each comprising of ~1 ng of total RNA and subjected to rRNA depletion and cDNA library preparation. The Ribo-Zero (epidemiology) kit depletes rRNA of both bacterial and eukaryotic origin. The KAPA RNA HyperPrep protocol (#1 in Fig. 1B) relies on random-primed reverse transcription followed by second-strand synthesis, adapter ligation, and PCR amplification. On the other hand, the CATS (Capture and Amplification by Tailing and Switching) protocol (#2 in Fig. 1B) is a ligation-free method that relies on polyA-tailing followed by oligo-dT-primed reverse transcription and template switching prior to PCR amplification. Due to the minimization of individual enzymatic treatment steps and a reduction of reaction purifications, both approaches lend themselves for the generation of high-quality cDNA libraries from ultralow inputs—according to the respective instruction manuals down to 1 ng (KAPA) or 100 pg (CATS) of RNA. Similarly important for Dual RNA-seq, both protocols are capable of detecting diverse transcript classes independent of their initial polyA status or size range.

4.2.1 Equipment

- Agilent Bioanalyzer 2100 and corresponding DNA chips
- Heat block
- PCR cycler
- HiSeq 4000 (Illumina)

4.2.2 Buffers and reagents

- Ribo-Zero Gold (epidemiology) kit (Illumina)
- RNase-free water

- AMPure XP Magnetic Beads (Agencourt)
- DNase I (NEB)
- 10 × DNase I Reaction Buffer (NEB)
- Exonuclease VII (NEB)
- 0.5 *M* EDTA
- PureSpin RNA Clean-up kit (OmniGenX)
- KAPA RNA HyperPrep kit (Kapa Biosystems)
- CATS RNA-seq kit v2 (Diagenode)

4.2.3 Procedure

4.2.3.1 Ribo-Zero treatment is as per the manufacturer's instructions, except that only 0.125 × of the reaction volumes are used to account for the low RNA inputs and avoid probe carryover. The elution is in 20 μL of RNase-free water.

4.2.3.2 Ribo-depleted RNA samples are cleaned up with each 1.8 × of AMPure RNA Clean XP beads according to the manufacturer's instructions. Then the purified samples are either subjected to the KAPA (step 4.2.3.3) or CATS protocol (step 4.2.3.4).

4.2.3.3 For the KAPA RNA HyperPrep protocol, RNA samples are first chemically fragmented for 6 min at 85°C. First- and second-strand cDNA synthesis as well as adapter ligation are performed as described in the instruction manual.

4.2.3.4 Alternatively, the RNA samples from step 4.2.3.2 are subjected to the CATS RNA-seq protocol. An optional DNase step is performed by treating the 20 μL of RNA sample with 0.28 U of DNase I and 1.4 U of exonuclease VII in 1 × DNase I buffer (final concentration) and a total reaction volume of 50 μL for 20 min at 37°C. The enzymes are inactivated by the addition of 0.5 μL of 0.5 *M* EDTA per reaction and an incubation for 10 min at 75°C. The samples are purified over PureSpin RNA Clean-up columns and eluted in 20 μL of RNase-free water. Irrespective of whether this optional DNase step is included (#3 in Fig. 1B) or not (#2), RNA samples are then chemically fragmented for 30 s at 94°C and subjected to the CATS protocol as described in the instruction manual.

4.2.3.5 Individual cDNA libraries from step 4.2.3.3 or step 4.2.3.4 (Fig. 3C) are pooled in equimolar ratios and sequenced on a HiSeq 4000 instrument (Illumina).

4.2.4 Notes

Since analysis of rRNA, which makes up ~90% of the total cellular RNA in both bacteria and eukaryotes, provides little information about the physiological state of the cells, these molecules are typically removed for RNA-seq studies. The Ribo-Zero protocol, that pulls down ribosomal transcripts with the use of rRNA capture oligonucleotides coupled to magnetic beads, proved suitable for infection models based on *S.* Typhimurium and mammalian cells (Afonso-Grunz et al., 2015; Bhagwat, Ying, & Smith, 2014). During the Ribo-Zero treatment, however, the bait DNA oligonucleotides may occasionally detach from the magnetic beads. While this is a rare event and can be neglected for standard applications (using ≥500 ng of input RNA, as suggested in the instruction manual), we observed that the relatively few detached bait DNA molecules exceeded the RNA transcripts that remained after rRNA depletion. This was problematic as the reverse transcriptases typically used for cDNA synthesis have partial DNA-dependent polymerase activity (Levin et al., 1988) and consequently, bait reads dominated our initial sequencing data (Fig. 4A). Therefore, the reagents used for

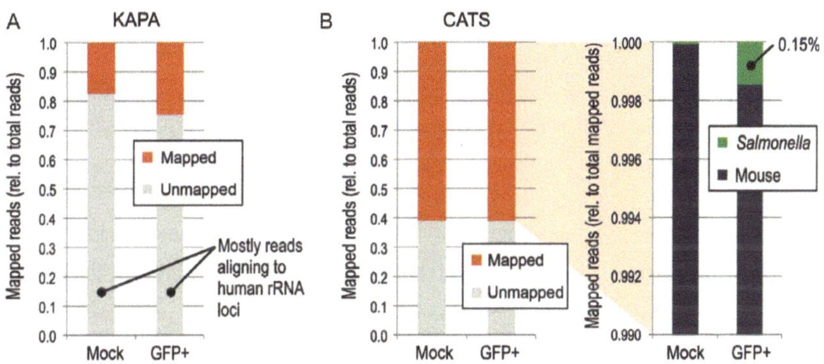

Fig. 4 Dual RNA-seq read mapping overview. (A) Read distribution in the uninfected or *Salmonella*-infected (GFP+) colonocyte samples when processed according to the KAPA protocol. Quality-filtered sequencing reads were aligned in parallel to the *S.* Typhimurium (NC_016810.1, NC_017718.1, NC_017719.1, NC_017720.1) or mouse (GENCODE M2, GRCm38.p2) reference sequences. The majority of sequencing reads, however, did not align to either one of these genomes. Reinspection of the fraction of unmapped reads revealed that the majority of them mapped against the human genome, particularly to spacer regions within human rRNA genes, and are likely derived from probe carryover during ribo-depletion. (B) Read distribution in the uninfected or *Salmonella*-infected colonocyte samples when processed according to the modified CATS protocol. In this case, the majority of reads could be mapped against the murine and *Salmonella* references, with *Salmonella* reads being almost exclusively detected in the library derived from the infected sample (GFP+).

ribo-depletion were substantially scaled down as compared to the manufacturer's default instructions. Additionally, in our optimized protocol (#3 in Fig. 1B) a second DNase treatment step using DNase I and exonuclease VII, after ribo-depletion, was included to remove any residual bait oligonucleotides prior to cDNA synthesis.

5. SUMMARY AND CONCLUSION

Above we present a detailed protocol for the sampling, purification, and transcriptome detection of *Salmonella*-infected murine colonocytes. The colon of infected mice was harvested, dissociated into single-cell suspensions from which colonocytes were specifically enriched after labelling them by antibody staining. In the course of this procedure, the cells needed to be formalin-fixed, which in turn required reversal of the induced covalent modifications of nucleic acid bases after cell sorting and prior to RNA extraction. From the minute RNA yields obtained, rRNA was depleted and cDNA libraries were generated. We compared the performance of two commercial library preparation protocols and identified one pipeline that—upon further optimization for minute host–pathogen RNA mixtures—appears highly suitable for Dual RNA-seq analyses of such low-input samples. Despite the bacterial read coverage in the final dataset being rather low (~0.15% of the total mapped reads; Fig. 4B), it is comparable to that obtained from similar in vivo RNA-seq (Avican et al., 2015; Szafranska et al., 2014) and in vivo Dual RNA-seq studies on bulk tissue samples (Nuss et al., 2017; Thanert et al., 2017). Therefore, our protocol—once coupled with sufficient sequencing depth and inclusion of replicates—should allow for quantitative gene expression analyses of specific, infected host cell types and of the respective bacterial pathogen in the future.

ACKNOWLEDGMENTS

The authors thank Briana Young, Jonathan Van Dyke, and Emily Kumimoto for technical support and Sandy R. Pernitzsch (www.scigraphix.com) for graphical support with Fig. 1. RNA-seq data were processed by the Bioinformatics Core Facility at the UC Davis Genome Center. A.J.W. acknowledges Jörg Vogel and the University of Würzburg for financial support.

REFERENCES

Afonso-Grunz, F., Hoffmeier, K., Muller, S., Westermann, A. J., Rotter, B., Vogel, J., et al. (2015). Dual 3' Seq using deepSuperSAGE uncovers transcriptomes of interacting Salmonella enterica Typhimurium and human host cells. *BMC Genomics*, *16*, 323.

Avican, K., Fahlgren, A., Huss, M., Heroven, A. K., Beckstette, M., Dersch, P., et al. (2015). Reprogramming of Yersinia from virulent to persistent mode revealed by complex in vivo RNA-seq analysis. *PLoS Pathogens*, *11*, e1004600.

Baumler, A. J., & Sperandio, V. (2016). Interactions between the microbiota and pathogenic bacteria in the gut. *Nature*, *535*, 85–93.

Bhagwat, A. A., Ying, Z. I., & Smith, A. (2014). Evaluation of ribosomal RNA removal protocols for Salmonella RNA-Seq projects. *Advances in Microbiology*, *4*, 8.

Blackwell, J. M., Goswami, T., Evans, C. A., Sibthorpe, D., Papo, N., White, J. K., et al. (2001). SLC11A1 (formerly NRAMP1) and disease resistance. *Cellular Microbiology*, *3*, 773–784.

Choi, Y. J., Aliota, M. T., Mayhew, G. F., Erickson, S. M., & Christensen, B. M. (2014). Dual RNA-seq of parasite and host reveals gene expression dynamics during filarial worm-mosquito interactions. *PLoS Neglected Tropical Diseases*, *8*, e2905.

Dillon, L. A., Suresh, R., Okrah, K., Corrada Bravo, H., Mosser, D. M., & El-Sayed, N. M. (2015). Simultaneous transcriptional profiling of Leishmania major and its murine macrophage host cell reveals insights into host-pathogen interactions. *BMC Genomics*, *16*, 1108.

Ehret, T., Spork, S., Dieterich, C., Lucius, R., & Heitlinger, E. (2017). Dual RNA-seq reveals no plastic transcriptional response of the coccidian parasite Eimeria falciformis to host immune defenses. *BMC Genomics*, *18*, 686.

Evers, D. L., Fowler, C. B., Cunningham, B. R., Mason, J. T., & O'Leary, T. J. (2011). The effect of formaldehyde fixation on RNA: Optimization of formaldehyde adduct removal. *The Journal of Molecular Diagnostics*, *13*, 282–288.

Hamatani, K., Eguchi, H., Takahashi, K., Koyama, K., Mukai, M., Ito, R., et al. (2006). Improved RT-PCR amplification for molecular analyses with long-term preserved formalin-fixed, paraffin-embedded tissue specimens. *The Journal of Histochemistry and Cytochemistry*, *54*, 773–780.

Kuramochi, H., Hayashi, K., Uchida, K., Miyakura, S., Shimizu, D., Vallbohmer, D., et al. (2006). Vascular endothelial growth factor messenger RNA expression level is preserved in liver metastases compared with corresponding primary colorectal cancer. *Clinical Cancer Research*, *12*, 29–33.

Levin, J. G., Crouch, R. J., Post, K., Hu, S. C., McKelvin, D., Zweig, M., et al. (1988). Functional organization of the murine leukemia virus reverse transcriptase: Characterization of a bacterially expressed AKR DNA polymerase deficient in RNase H activity. *Journal of Virology*, *62*, 4376–4380.

Lisnic, V. J., Cac, M. B., Lisnic, B., Trsan, T., Mefferd, A., Das Mukhopadhyay, C., et al. (2013). Dual analysis of the murine cytomegalovirus and host cell transcriptomes reveal new aspects of the virus-host cell interface. *PLoS Pathogens*, *9*, e1003611.

Nuss, A. M., Beckstette, M., Pimenova, M., Schmuhl, C., Opitz, W., Pisano, F., et al. (2017). Tissue dual RNA-seq allows fast discovery of infection-specific functions and riboregulators shaping host-pathogen transcriptomes. *Proceedings of the National Academy of Sciences of the United States of America*, *114*, E791–E800.

Papenfort, K., Said, N., Welsink, T., Lucchini, S., Hinton, J. C., & Vogel, J. (2009). Specific and pleiotropic patterns of mRNA regulation by ArcZ, a conserved, Hfq-dependent small RNA. *Molecular Microbiology*, *74*, 139–158.

Park, S. J., Kumar, M., Kwon, H. I., Seong, R. K., Han, K., Song, J. M., et al. (2015). Dynamic changes in host gene expression associated with H5N8 avian influenza virus infection in mice. *Scientific Reports*, *5*, 16512.

Pittman, K. J., Aliota, M. T., & Knoll, L. J. (2014). Dual transcriptional profiling of mice and Toxoplasma gondii during acute and chronic infection. *BMC Genomics*, *15*, 806.

Raynaud, S., Le Pabic, H., & Felden, B. (2018). Selective recovery of RNAs from bacterial pathogens after their internalization by human host cells. *Methods*, *143*, 4–11.

Szafranska, A. K., Oxley, A. P., Chaves-Moreno, D., Horst, S. A., Rosslenbroich, S., Peters, G., et al. (2014). High-resolution transcriptomic analysis of the adaptive response of Staphylococcus aureus during acute and chronic phases of osteomyelitis. *MBio*, *5*, e01775.

Thanert, R., Goldmann, O., Beineke, A., & Medina, E. (2017). Host-inherent variability influences the transcriptional response of Staphylococcus aureus during in vivo infection. *Nature Communications*, *8*, 14268.

Wesolowska-Andersen, A., Everman, J. L., Davidson, R., Rios, C., Herrin, R., Eng, C., et al. (2017). Dual RNA-seq reveals viral infections in asthmatic children without respiratory illness which are associated with changes in the airway transcriptome. *Genome Biology*, *18*, 12.

Westermann, A. J., Barquist, L., & Vogel, J. (2017). Resolving host-pathogen interactions by dual RNA-seq. *PLoS Pathogens*, *13*, e1006033.

Westermann, A. J., Gorski, S. A., & Vogel, J. (2012). Dual RNA-seq of pathogen and host. *Nature Reviews. Microbiology*, *10*, 618–630.

Westermann, A. J., & Vogel, J. (2018). Host-pathogen transcriptomics by dual RNA-seq. *Methods in Molecular Biology*, *1737*, 59–75.

Wolf, T., Kammer, P., Brunke, S., & Linde, J. (2018). Two's company: Studying interspecies relationships with dual RNA-seq. *Current Opinion in Microbiology*, *42*, 7–12.

Xu, G., Strong, M. J., Lacey, M. R., Baribault, C., Flemington, E. K., & Taylor, C. M. (2014). RNA CoMPASS: A dual approach for pathogen and host transcriptome analysis of RNA-seq datasets. *PLoS One*, *9*, e89445.

CHAPTER TWENTY-TWO

Genomic Analysis of DNA Double-Strand Break Repair in *Escherichia coli*

A.M. Mahedi Hasan[2], Benura Azeroglu[2], David R.F. Leach[1]
Institute of Cell Biology, School of Biological Sciences, University of Edinburgh, King's Buildings, Edinburgh, United Kingdom
[1]Corresponding author: e-mail address: d.leach@ed.ac.uk

Contents

Abstract

Counting DNA whole genome sequencing reads is providing new insight into DNA double-strand break repair (DSBR) in the model organism *Escherichia coli*. We describe the application of RecA chromatin immunoprecipitation coupled to genomic DNA sequencing (RecA-ChIP-seq) and marker frequency analysis (MFA) to analyze the genomic consequences of DSBR. We provide detailed procedures for the preparation of DNA and the analysis of data. We compare different ways of visualizing ChIP data and show that alternative protocols for the preparation of DNA for MFA differentially affect the recovery of branched DNA molecules containing Holliday junctions.

ABBREVIATIONS

ChIP chromatin immunoprecipitation
DSB DNA double-strand break
DSBR DNA double-strand break repair

[2] These authors have contributed equally to this work.

Methods in Enzymology, Volume 612
ISSN 0076-6879
https://doi.org/10.1016/bs.mie.2018.09.001

gDNA genomic DNA
IGB integrated genome browser
IGV integrative genomics viewer
IN input
IP immunoprecipitated
MFA marker frequency analysis
PCR polymerase chain reaction
RNase ribonuclease

1. INTRODUCTION

Being able to visualize the consequences of DNA double-strand break repair (DSBR) by genomic sequencing is powerful at two scales of analysis. First, it can depict in unprecedented detail the events occurring over distances of hundreds of base pairs to hundreds of kilobase pairs at a specified chromosomal locus at which damage is generated. Second, it can reveal the occurrence of unexpected secondary consequences of DSBR, aberrant DSBR, or the absence of DSBR. These may occur preferentially at specific genomic locations such as the chromosome terminus or highly transcribed genes. In this review, we provide details of the methods of RecA chromatin immunoprecipitation analyzed by genomic DNA sequencing (RecA-ChIP-seq) and marker frequency analysis (MFA) that we have developed to analyze DSBR in *Escherichia coli*.

1.1 RecA-ChIP-seq

ChIP involves the use of antibodies to a specific DNA-binding protein to precipitate protein–DNA complexes from which the DNA is subsequently released and identified. The mapping of specific proteins to DNA-binding sites was initiated in UV cross-linking studies of RNA polymerase in bacteria (Gilmour & Lis, 1984). The technique was improved by using formaldehyde cross-linking (Solomon, Larsen, & Varshavsky, 1988) and applied to genome wide investigations using microarrays (ChIP-on-chip) (Blat & Kleckner, 1999). High-throughput genomic DNA sequencing technologies have superseded microarrays for the identification of DNA sequences identified by ChIP. Early examples of ChIP-seq include mapping histone modifications (Barski et al., 2007) and the transcription factor STAT1 (Robertson et al., 2007) to their targets in human cells.

The workflow for the ChIP-seq data analysis involves uniquely mapping of the sequence reads to an appropriate reference genome and subsequent conversion of the aligned (`.sam`) file to a `.bedgraph` file for the visualization of DNA enrichment at a specific region in the genome using a stand-alone genome viewer (like integrative genomics viewer (IGV) or integrated genome browser (IGB)) (Nicol, Helt, Blanchard, Raja, & Loraine, 2009; Robinson et al., 2011). The mapped reads are then used to generate count data using HTSeq (Anders, Pyl, & Huber, 2015). The count data can then be used to compare DNA enrichment between two strains or conditions. We describe two methods to do this. The first is a manual method (step 7a in Section 2.1.3.3) and the second is to use the Bioconductor package "DESeq" (step 7b in Section 2.1.3.3) (Anders & Huber, 2010). At least three biological replicates of each strain or condition are needed for a statistically significant fold-change measurement of DNA enrichment at a defined region in the genome, otherwise, the same method can be used for exploratory data analysis without any statistical significance. In the latter case, a minimum of two independent biological repeats are recommended to confirm qualitative reproducibility.

We have developed the technique of RecA-ChIP-seq to analyze the in vivo activity of RecBCD enzyme, which loads RecA protein during DSBR in *E. coli* (Cockram, Filatenkova, Danos, El Karoui, & Leach, 2015). RecA-ChIP-seq provides both information on location and orientation of broken ends engaged in DSBR. This is so, not only at sites of induced DNA double-strand breaks (DSBs), but also at sites of secondary DSBs that would be silent in most assays because of their efficient repair in recombination proficient *E. coli*. RecBCD enzyme is a helicase–nuclease which unwinds DNA from the site of a DSB to the site of a Chi sequence (5′-GCTGGTGG-3′) (Dillingham & Kowalczykowski, 2008; Smith, 2012). At Chi, the enzyme cleaves the DNA and continues unwinding to generate a 3′ single-strand upon which it facilitates the loading of RecA protein (Anderson & Kowalczykowski, 1997). RecA is the protein mediating homology recognition and strand exchange during recombination (Bell & Kowalczykowski, 2016). The loading of RecA by RecBCD generates a characteristic RecA-ChIP-seq signature in which the binding of RecA rises rapidly at the Chi site (nearest to a DSB) and then declines with an approximately exponential profile (Cockram et al., 2015). This has been used to model the in vivo activity of RecBCD enzyme (Cockram et al., 2015); to show that an induced DSB can stimulate another break, with ends facing the *dif* site required for dimer resolution, in the terminus region of the chromosome (Cockram et al., 2015); and to show that outward oriented DNA ends are generated at the termination sites TerA and TerB in the terminus of a *recG* mutant (Azeroglu et al., 2016) and at the site of predicted replication–transcription collision in a *recG* mutant (Azeroglu et al., 2016).

RecA-ChIP-seq has also been used to show that the DNA breaks detected in the terminus of the chromosome of a *recB* mutant in slow growth conditions (in a minimal medium supplemented with glucose) are not formed in the presence of RecBCD enzyme, demonstrating that this class of breaks occurs specifically in the absence of RecBCD (Sinha et al., 2017).

1.2 Marker Frequency Analysis

A decade after the discovery of the structure of DNA (Watson & Crick, 1953), a theoretical basis for MFA was developed to describe chromosomal DNA replication of *Bacillus subtilis* (Sueoka & Yoshikawa, 1965; Yoshikawa & Sueoka, 1963). Later, MFA was used to detect the origin of replication and establish the bidirectionality of the DNA replication in *E. coli* (Bird, Louarn, Martuscelli, & Caro, 1972; Jonasson, 1973; Masters & Broda, 1971). In these early experiments, MFA was performed by measuring the dosage of chromosomal regions based on DNA–DNA hybridization of specific DNA sequences or by the frequency of gene transfer by generalized transduction. These methods were superseded by DNA–DNA hybridization using microarrays (Khan, Mahaseth, Kouzminova, Cronan, & Kuzminov, 2016; Kouzminova & Kuzminov, 2008, 2012; Kuong & Kuzminov, 2012). Finally, the advent of high-throughput DNA sequencing has enabled the construction of MFA maps with a resolution of approximately 1 kb depicting the replication (timing) profile of actively replicating unsynchronized cells (Skovgaard, Bak, Lobner-Olesen, & Tommerup, 2011). This has been used to investigate the replication profiles of *E. coli* mutants affecting recombination (Azeroglu et al., 2016; Dimude et al., 2015; Midgley-Smith et al., 2018; Rudolph, Upton, Stockum, Nieduszynski, & Lloyd, 2013; Sinha et al., 2017; Wendel, Cole, Courcelle, & Courcelle, 2018; Wendel, Courcelle, & Courcelle, 2014; White, Azeroglu, Lopez-Vernaza, Hasan, & Leach, 2018). The most striking observations made have been the overreplication of the terminus region of the *E. coli* chromosome in *recG* and in multiple exonuclease mutants (Azeroglu et al., 2016; Rudolph et al., 2013; Wendel et al., 2018) and the loss of DNA sequences in the chromosomal terminus of a *recB* mutant (Courcelle, Wendel, Livingstone & Courcelle, 2015; Dimude, Midgley-Smith & Rudolph, 2018; Rudolph et al., 2013; Sinha et al., 2017, 2018; Wendel et al., 2014). MFA has also been used to visualize DNA sequence loss associated with DSBs (Azeroglu et al., 2016; Badrinarayanan, Le, Spille, Cisse, & Laub, 2017; Sivaramakrishnan et al., 2017).

MFA of the *E. coli* genome by high-throughput DNA sequencing is currently performed by normalizing the uniquely mapped sequence reads derived from an exponentially growing cell culture to their total read counts (to correct for differences in the depth of sequencing or the library size) and then to the normalized reads (calculated in the same way) of a corresponding nonreplicating stationary-phase cell culture (to correct for differences in sequence-based recovery across the genome). As the normalized data points are noisy, it is customary to apply a smoothing approach to the raw data. This may include the binning of data in nonoverlapping bins (e.g., 1 or 10 kb) across the genome and/or the application of a nonparametric smoothing method (LOESS) (Azeroglu et al., 2016; Maduike, Tehranchi, Wang, & Kreuzer, 2014). Other smoothing approaches include the exclusion of reads below a defined depth/count. These include the elimination of reads with fewer than one quarter of the expected number in a bin (Muller et al., 2014), or an empirically determined threshold of 250 reads (Muller & Nieduszynski, 2012), and/or applying Fourier convolution smoothing (Muller & Nieduszynski, 2012; Raghuraman et al., 2001; Rudolph et al., 2013). We have concerns that smoothing methods that exclude regions of the genome with low read counts have the potential to obscure experimentally meaningful chromosomal DNA loss during DSBR and so we have restricted our processing of the data to binning and nonparametric smoothing.

2. EXPERIMENTAL METHODS

Here, we will describe our examination of recombination proficient (wild-type) and deficient (Δ*ruvAB*) *E. coli* strains after an hour induction of a single site-specific chromosomal DSB per replication cycle. For this we use the palindrome/SbcCD system which was described in (Eykelenboom, Blackwood, Okely, & Leach, 2008). This system utilizes the presence of an inducible structure-specific endonuclease, SbcCD, to cleave a DNA hairpin that forms on the lagging-strand template following replication of a 246-bp interrupted palindrome, which has been inserted into the chromosomal *lacZ* gene. In order to control induction of the DSB, the *sbcDC* operon was placed under the control of an inducible P_{araBAD} promoter. The promoter can be induced by the presence of arabinose in the medium and can be further repressed by the presence of glucose. For better quantitative expression from the P_{araBAD} promoter, the *E. coli* K12 BW27784 strain background was used in this study. An isogenic strain without a palindrome was used as a negative control.

The bacterial cultures were obtained for examination as follows. Overnight cultures were grown in 5 mL of LB medium supplemented with 0.5% glucose. The following day, cultures were diluted to an OD_{600nm} of 0.02 and grown shaking at 37°C to an OD_{600nm} of 0.2. Cultures were then rediluted to an OD_{600nm} of 0.02 and grown shaking at 37°C to an OD_{600nm} of 0.2. Expression from the P_{araBAD}-*sbcDC* construct was induced by the addition of 0.2% arabinose to the culture medium. Cultures were then incubated at 37°C for 1 h before samples were isolated for analysis.

2.1 RecA-ChIP-seq

2.1.1 Experimental Protocol

2.1.1.1 Equipment

- Temperature controlled shaking water bath
- Sonicator: Diagenode Bioruptor
- Magnetic rack
- Tube rotator
- Thermal cycler
- Gel Doc™ XR+ gel documentation system (Bio-Rad)

2.1.1.2 Reagents and Buffers

- 2.5 *M* glycine
- Formaldehyde (Sigma-Aldrich, Cat# 252549)
- 1 × PBS pH 7.4 (137 m*M* NaCl, 2.7 m*M* KCl, 10 m*M* Na_2HPO_4, 2 m*M* KH_2PO_4)
- ChIP buffer (200 m*M* Tris–HCl pH 8.0, 600 m*M* NaCl, 4% Triton X-100)
- cOmplete™, mini, EDTA-free Protease Inhibitor Cocktail (Roche, Cat# 4693159001)
- Anti-RecA antibody (Abcam, Cat# ab63797)
- Protein G Dynabeads® (ThermoFisher, Cat# 10003D)
- RNase A
- Tween-20
- TE buffer pH 8.0 (10 m*M* Tris–HCl, 1 m*M* EDTA)
- 10% SDS

- QIAquick polymerase chain reaction (PCR) Purification kit (Qiagen, Cat# 28104)
- NEBNext® ChIP-Seq Library Prep Master Mix Set for Illumina (New England Biolabs, Cat# E6240)
- NEBNext® Multiplex Oligos for Illumina (Index Primers Sets 1) (New England Biolabs, Cat# E7335)
- MinElute PCR Purification kit (Qiagen, Cat# 28004)
- Loading buffer (50 m*M* Tris–HCl pH 8, 40 m*M* EDTA, 40% w/v sucrose)
- Agarose (Melford, Cat# MB1200)
- Nucleic acid stain, SafeView (NBS Biologicals, Cat# NBS-SV5)

2.1.1.3 ChIP

1. Add 37% (w/v) formaldehyde to the medium containing exponentially grown *E. coli* cells to obtain a final concentration of 1%. Incubate for 10 min at 22.5°C in a shaking water bath.
2. Quench the formaldehyde by adding 2.5 *M* glycine to obtain a final concentration of 0.5 *M*. Incubate for 10 min at 22.5°C in a shaking water bath.
3. Transfer the media into a 50-mL Falcon tube and place on ice immediately. Collect cells by centrifugation (2000 × *g* for 7 min at 4°C), then resuspend the pellet in 1 mL ice-cold 1 × PBS. Transfer the cells into a 1.5-mL Eppendorf tube. Wash cells three times in 1 mL ice-cold 1 × PBS at 4°C.

   Pause point: Cell pellets can be stored at − 80°C for at least 1 year.
4. Resuspend cell pellets in 250 μL of ChIP buffer containing protease inhibitor cocktail by pipetting up and down several times in a 1.5-mL Eppendorf tube (*Note*: The cOmplete™ protease inhibitor cocktail is supplied in the form of tablets. One tablet of cOmplete™ protease inhibitor cocktail was dissolved in 10 mL of ChIP buffer at 4°C. It is recommended to prepare ChIP buffer containing protease inhibitor cocktail fresh each time).
5. To shear the chromatin, sonicate the washed pellet resuspended in 250 mL of ChIP buffer containing protease inhibitors using a Diagenode Bioruptor at 30 s intervals for 10 min at high amplitude, keep samples at 5°C at all times using a temperature-controlled water bath. (These parameters are optimized for the bespoken model of sonicator and volume of ChIP buffer-suspended cells. The

parameters for a particular sonicator and conditions must be empirically determined.)

6. Add fresh 350 μL of ChIP buffer containing protease inhibitors to the sonicated sample and mix well by pipetting up and down several times.
7. Transfer an aliquot of sheared chromatin (100 μL) to a new microcentrifuge tube and store at −20°C. This aliquot will be used for isolation of total DNA, to determine shearing efficiency, and to use as input (IN) DNA.
8. Add 5 μL of anti-RecA antibody to the remaining sheared DNA (500 μL) and incubate the samples overnight at 4°C on a rotating platform (30–40 rotations per minute).
9. Following morning, resuspend Dynabeads either by vortexing for 30 s or by rotating for 5 min. Once resuspended, aliquot 100 μL Dynabeads into a 1.5-mL Eppendorf tube. Place the tube on the magnetic rack to separate the beads from the solution and remove the supernatant.
10. Remove the tubes containing Dynabeads from the rack and add immunoprecipitated (IP) samples from step 8 to the tubes and incubate with rotation for 2 h at room temperature.
11. Place the tube containing the slurry on the magnetic rack to separate the beads from the solution.
12. Remove supernatant and add 1 mL of 1 × PBS with 0.02% Tween-20 to the tubes. Gently mix the beads with the solution. Once fully mixed, place the tube back on the magnetic rack. Repeat this washing step three times. (*Note*: do not let the beads dry out between washes. This would affect the efficiency of immunoprecipitation.)
13. After the last wash, resuspend the beads with 200 μL of TE buffer with 1% SDS and 20 mg/mL RNase A solution.
14. Add 100 μL of TE buffer with 1% SDS and 20 mg/mL RNase A solution into tubes containing the IN DNA (from step 7) and transfer them into PCR tubes.
15. Incubate all the samples at 65°C for 10 h in a thermal cycler to reverse the cross-linking.
16. After the reverse cross-linking step, transfer the IP samples into 1.5 mL Eppendorf tubes and place the tubes onto the magnetic rack to remove the beads from solution.
17. Take the supernatant from the IP samples, which now contains the DNA.
18. Carry out PCR purification using Qiagen PCR Purification kit following the manufacturer's instructions to isolate DNA from the supernatant.

19. In parallel, carry out PCR purification of IN samples.
20. Elute the DNA in 50 μL of MilliQ water.

2.1.1.4 Library Preparation

The last step of sample preparation for the ChIP-seq protocol is to construct the libraries. There are various commercially available kits to do this. Here, we will describe how to prepare a single library prep using NEBNext® ChIP-Seq Library Preparation Master Mix Set for Illumina. Briefly, the library preparation consists of five steps: end repair of sheared DNA, dA-tailing of end repaired DNA, adaptor ligation of dA-tailed DNA, PCR enrichment of adaptor-ligated DNA, and gel purification of the library. The amplified library is purified on an agarose gel to select a specific size range for cluster generation, as well as to remove potential adapter dimers.

21. Pipette the following mix into a PCR tube and keep on ice. Mix well and incubate in a thermal cycler for 30 min at 20°C.

ChIP sample (5–50 ng)	40 μL
NEBNext® End Repair Reaction Buffer (10 ×)	5 μL
NEBNext® End Repair Enzyme Mix	1 μL
MilliQ H_2O	4 μL

22. Once the reaction is done, clean up the samples using Qiagen PCR Purification kit following the manufacturer's instructions.
23. Elute the DNA in 45 μL of MilliQ H_2O.
24. Pipette the following mix into a PCR tube and keep on ice. Mix well and incubate in a thermal cycler for 30 min at 37°C.

End repaired DNA	44 μL
NEBNext® dA-Tailing Reaction Buffer (10 ×)	5 μL
Klenow fragment	1 μL

25. Once the reaction is done, clean up the samples using MinElute PCR Purification kit following the manufacturer's instructions.
26. Elute the DNA in 20 μL of MilliQ H_2O.

27. Pipette the following mix into a PCR tube and keep on ice. Mix well and incubate in a thermal cycler for 15 min at 20°C.

dA-tailed DNA	19 μL
Quick ligation reaction buffer (5×)	6 μL
NEBNext® adaptor (1.5 μ*M*)	1 μL
Quick T4 DNA ligase	4 μL

28. Add 3 μL USER Enzyme Mix and mix well by pipetting up and down couple of times.
29. Incubate at 37°C for 15 min.
30. Once the reaction is done, clean up the samples using MinElute PCR Purification kit following the manufacturer's instructions.
31. Elute the DNA in 21 μL of MilliQ H_2O.
32. Pipette the following mix into a PCR tube and keep on ice.

Adaptor-ligated DNA fragments	20 μL
Index (i7) primer	2.5 μL
Universal (i5) primer	2.5 μL
NEBNext® Q5 Hot Start HiFi PCR master mix	25 μL

33. Mix the content of tube well and place into a thermal cycle. Run following program:

Cycle Step	Temp	Time	Number of Cycles
Initial denaturation	98°C	30 s	1
Denaturation Annealing/extension	98°C 65°C	10 s 75 s	15
Final extension	65°C	5 min	1
Hold	4°C	∞	

34. Once the reaction is done, clean up the samples using MinElute PCR Purification kit following the manufacturer's instructions.
35. Elute the DNA in 10 μL of MilliQ H_2O.

36. Prepare 2% (w/v) agarose in 1 × TAE containing 0.5 μg/mL of SafeView and cast a gel.
37. Add 3 μL of loading buffer to each sample and mix well.
38. Load 4 μL of 100 bp DNA ladder into first and last lane of gel. Leave one lane empty and load the sample into the next well. You can load more samples into the same gel but make sure to leave one lane empty between the samples.
39. Run the gel at 5 V/cm for 90 min.
40. Visualize the gel using a gel imagining system (*Note*: it is important not to expose the DNA to UV light directly. For this purpose, the Gel Doc™ XR+ gel documentation system with UV/blue conversion screen can be used). Excise the gel region containing DNA fragments between 200 and 500 bp using a sterile scalpel. Carry out isolation of DNA from gel with MinElute Gel Extraction Kit following the manufactures instructions. However, make sure not to warm gel slice above 37°C at any stage.
41. Elute the DNA in 10 μL EB with 0.1% Tween-20 (elution solution may change depending on sequencing protocol and machine used. We advise checking this with the sequencing facility).
42. Determine the concentration of library using a fluorometric-based method and average size of fragments using a Bioanalyzer, which then can be used to determine the molarity of the sample (the requirement for concentration and molarity may change depending on the sequencing protocol and machine used).
43. If the libraries are going to be multiplexed, then an equal molarity of each sample should be pooled unless different samples require different depths of sequencing.

2.1.2 DNA Sequencing

All of the sequencing that has been done for this study was carried out by Edinburgh Genomics (Edinburgh, UK) using the Illumina sequencing platform. The principle of the sequencing reaction is based on a sequencing-by-synthesis (SBS) method in which fluorescent labelled, reversible terminated nucleotides are incorporated in a cyclic manner using a special DNA polymerase (Ansorge, 2009). Briefly, DNA libraries are hybridized to a primer lawn on a flow cell and extended by the polymerase. The double-stranded DNA is denatured, and the original template is washed away. The newly synthesized single-stranded DNA flips over and forms a bridge by hybridizing to an adjacent, complementary primer on the flow cell. The hybridized primer is extended by the polymerases, and a double-stranded bridge is formed. This initiates a so-called bridge amplification cycle which is repeated creating approximately one million copies of each template in tight physical clusters on the flow cell surface. dsDNA bridges are then denatured, and the reverse strands are cleaved and washed away. Free 3′ ends are blocked to prevent any unwanted DNA priming. A sequencing primer is introduced to the flow cell and hybridized to the adapter sequence annealing site. SBS is performed in a cyclic manner. Each cycle consists of (1) addition of polymerase and fluorescently labelled reversibly 3′ terminated nucleotides, (2) incorporation of the complementary reversibly terminated nucleotide, (3) washing off excess reagents, (4) imaging of the flow cell to determine the incorporated nucleotide, and finally (5) removing fluorescent dye and blocking group from the 3′ end of the base (Ansorge, 2009; Mardis, 2008). At the end, a camera is used to capture a temporal series of color images, which can be computationally converted into sequence reads. The length of the reads depends on the particular instrument and type of flow cell that is used. In this study, a NovaSeq 6000 instrument was used with an S1 flow cell. The reads were 50 bp paired-end reads.

2.1.3 Data Analysis

Here, we describe how to analyze and compare ChIP-seq data on a command-line environment using a Linux-based operating system and statistical environment R (https://www.r-project.org/). The command line codes and file formats are given in "`courier new`" font. An R script is available at https://github.com/AMMHasan/DEAChIP.

2.1.3.1 Software

- Bowtie 2 (version 2.4; Langmead & Salzberg, 2012)
- BEDTools (version 2.27.0; Quinlan & Hall, 2010)
- FastQC (version 0.11.2, http://www.bioinformatics.babraham.ac.uk/projects/fastqc/)
- HTSeq (version 0.9.1; Anders et al., 2015)
- Picard (version 1.95, http://broadinstitute.github.io/picard)
- Rstudio (version 1.1.442, https://www.rstudio.com/)
- SAMtools (version 1.3.1; Li, 2011)
- DESeq (R/Bioconducter package, version 1.30.0; Anders & Huber, 2010)
- seqinR (R/Bioconducter package, version 3.4-5; Charif & Lobry, 2007)
- ggplot2 (R/Bioconducter package, version 3.0.0; Wickham, 2009)

2.1.3.2 Data

The data files used for RecA-ChIP-seq analysis have been deposited in the NCBI GEO database under accession number GSE117541.

Data type: 50 bp paired-end Illumina reads in `.fastq` format.

2.1.3.3 Workflow

1. Check the quality of demultiplexed raw Illumina reads for the following features prior to further analysis:
   - **a.** Per base sequence quality
   - **b.** Per sequence quality score
   - **c.** Per base sequence content
   - **d.** Per base GC content
   - **e.** Sequence length distribution
   - **f.** Sequence duplication

   The following command can be used for quality control check on individual raw read files:

   ```
   fastqc reads.fq
   ```
2. The reference ggenome (in `.fasta` format) must be FM-indexed, only once, before proceeding to the mapping step (step 3) using the following command:

   ```
   bowtie2-build reference_genome.fasta base_name
   ```

Explanation of parameters:

`base_name`	the base-name of the index files, which will be used in the mapping step (step 3)

3. Map the paired-end reads to an FM-indexed reference genome (done in the previous step) using Bowtie 2:

   ```
   bowtie2 -p 10 -x ref_genome_base_name -1 X_1.fq -2 X_2.fq -S X.sam
   ```

Explanation of parameters:

`-p`	number of threads
`-x`	base-name of the indexed reference genome
`-1`	individual or comma-separated list of files containing mate 1 s (filename usually includes _1), e.g., `-1 X_1.fq, Y_1.fq.`
`-2`	individual or comma-separated list of files containing mate 2 s (filename usually includes _2), e.g., `-2 X_2.fq, Y_2.fq.`
`-S`	file to write the SAM alignments to

4. Convert the SAM file to its binary version BAM, using SAMtools:

```
samtools view -b -o X.bam -T ref_genome.fasta X.sam -@ 10
```

Explanation of parameters:

-b	output in .bam format
-o	output to the file mentioned (X.bam)
-T	reference .fasta file that has been indexed by "samtools faidx" command already. If not, it will be indexed automatically at the beginning of this step
-@	number of threads

However, for batch conversion, use a bash command like the following:

```
for i in *.sam; do j=${i%.sam}; samtools view -b -o $j.bam -T ref_genome.fasta $i -@ 10; done.
```

5. To get rid of duplicates, first, sort read alignments BAM file by coordinates using Picard software:

```
picard-tools SortSam INPUT=./X.bam OUTPUT=./coordinate-sorted_X.bam SORT_ORDER=coordinate
```

Explanation of parameters:

INPUT (File)	The BAM or SAM file to sort
OUTPUT (File)	The sorted BAM or SAM output file
SORT_ORDER	Sort order of output file. Possible values: unsorted, queryname, coordinate, duplicate, unknown

Then, remove the duplicate (or deduplicate) reads using the following commands:

```
picard-tools MarkDuplicates INPUT=./coordinate-sorted_X.bam OUTPUT=./deduped_X.bam METRICS_FILE=dedup_mat_X.txt REMOVE_DUPLICATES=TRUE
```

Explanation of parameters:

INPUT (String)	One or more input SAM or BAM files to analyze. Must be coordinate sorted
OUTPUT (File)	The output file to write marked records to
METRICS_FILE (File)	File to write duplication metrics to
REMOVE_DUPLICATES (Boolean)	Possible values: true, false

6. At this point, visualize the DNA enrichment on a stand-alone genome viewer (for example, IGV or IGB) using `.bedgraph` files. To generate `BedGraphs` from respective deduplicated BAM files, use the following command:

```
bedtools genomecov -bg -ibam ./deduped_X.bam > X.bedgraph
```

Explanation of parameters:

`-bg`	Report depth in BedGraph format
`-ibam`	BAM file as input for coverage
`-g`	indexed reference genome (can be done using "`samtools faidx`" command) with the extension "`.fai`"

7. Comparative ChIP-seq data analysis can be performed manually (a) or using R/Bioconductor package DESeq from count data (b).
   a. Manual comparison of ChIP-seq data requires normalization of the IP data for their library sizes. Further normalization to IN data may be undertaken if desired. For the library size normalization, we recommend a median-based method (as reported by Anders & Huber, 2010). The method minimizes the effect of extreme values of the count data. An R script for such normalization is provided on GitHub: https://github.com/AMMHasan/DEAChIP. This generates a profile of ratios. Alternatively, normalization can be carried out by subtraction if appropriate.
   b. For comparative ChIP-seq data analysis using R/Bioconductor package DESeq,
      i. Extract the count data for any predefined feature (here, 1 kb nonoverlapping bins). The list of features is provided in `surrogate_gff.gff` file. This step is crucial as the width of the bin relative to the read length affects the calculation of read counts. Any prior normalization of data would result in a spurious outcome by DESeq.

```
htseq-count -f bam -r pos -s no -m intersection-strict
--type Bin --idattr ID ./deduped_X.bam ./surrogate_gff.gff >
./count_data_X.txt
```

Explanation of parameters:

-f	Format of the input file
-r	Mode of ordering aligned paired-end reads. For paired-end data, the alignment is sorted either by read name [name] or by alignment position [pos]. The default is by name. If sorted by name, htseq-count expects all the alignments for the reads of a given read pair to appear in adjacent records in the input data. For sorting by position (or coordinate), read alignments whose mate alignments have not yet been seen are kept in a buffer in memory until the mate is found. For larger alignment files the buffer is likely to overflow, hence after deduplication, further sorting by name is advisable
-s	Whether the data are from a strand-specific assay
-m	Mode to handle reads overlapping more than one feature
--type	Feature type: the third column in the .gff file
--idattr	GFF attribute to be used as feature ID
surrogate_gff.gff	A surrogate annotation file, with a nonoverlapping bin of 1 kb covering the total length of the genome. A few lines of the surrogate_gff file are given in Table 1

ii. Trim the content of each count data file, so that only the bin identifier and corresponding count data are captured.

iii. Combine all the count data to compare, in a single `count_data.csv` file. An example of the content is given in Table 2

iv. Analyze the differential DNA enrichment using DESeq Bioconductor package. R commands and codes for visualization are provided in an R script on GitHub: https://github.com/AMMHasan/DEAChIP.

Table 1 Example Content of the `surrogate_gff.gff` File

Fasta_identifier_ ref_genome	NA	Bin	1	1000	.	+	.	ID=Bin_0001
Fasta_identifier_ ref_genome	NA	Bin	1001	2000		+		ID=Bin_0002
Fasta_identifier_ ref_genome	NA	Bin	2001	3000		+		ID=Bin_0003
...	...	...	...	...	...	...	...	...
Fasta_identifier_ ref_genome	NA	Bin	4629001	4630000		+		ID=Bin_4630

Table 2 Example Content of the `count_data.csv` File

Bin	Wt_DSB_A_IP	Wt_DSB_B_IP	Wt_A_IP	Wt_B_IP
Bin_0001	1450	1381	1589	2203
Bin_0002	622	671	884	1157
Bin_0003	770	830	1118	1524
...	...	...	...	...
Bin_4630	1163	1249	1568	2072

2.1.4 Lessons From Data

RecA-ChIP-seq is a powerful method to determine the location and orientation of DNA double-strand ends that have engaged with RecA protein during repair by homologous recombination (Azeroglu et al., 2016; Cockram et al., 2015). Both the shape of the RecA-binding profile and its location relative to a correctly oriented Chi site, implicated in the RecBCD-RecA reaction, define unambiguously the orientation of the DNA end involved. The location of the break can be further refined using a *recD* mutant where the binding of RecA rises directly from the site of the break (White et al., 2018).

Both the shape and the location of the RecA-binding profile are clearly visible from immunoprecipitated reads (IP) alone (Fig. 1A), and the baseline of input DNA is sufficiently flat over short regions of the genome (e.g., 100 kb) to allow comparisons to be made between areas under peaks without normalization to input DNA (Azeroglu et al., 2016; Cockram et al., 2015; White et al., 2018). IP reads from an isogenic control strain, not undergoing DSBR, provide an understanding of the background (Fig. 1B), and a ratio of reads from the strain undergoing DSBR to reads from the control strain normalizes for background effects (Fig. 1C). When visualizing effects over genomic distances, binning (e.g., 1 kb bins) is advised (Fig. 1A left), but fine features of the read distribution in regions of less than 100 kb may be most clearly visible in the raw data (Fig. 1A right). It is worth noting that methods of normalizing the DSB^+ reads to control reads require the binning of data in order to avoid the problem of dividing by zeros. Binning tends to blur features of the distribution (Fig. 1C). Furthermore, the experimental conditions may alter the background in ways that require understanding for interpretation. For example, the highly transcribed gentamicin resistance gene generates a RecA-ChIP signal "artifact" in the control strain but may not do so in the experimental strain as the ratio of reads is dramatically lowered at this position in the ratio (Fig. 1C right, marked with *), and a peak above the RecA-ChIP signal is not seen in the raw data from

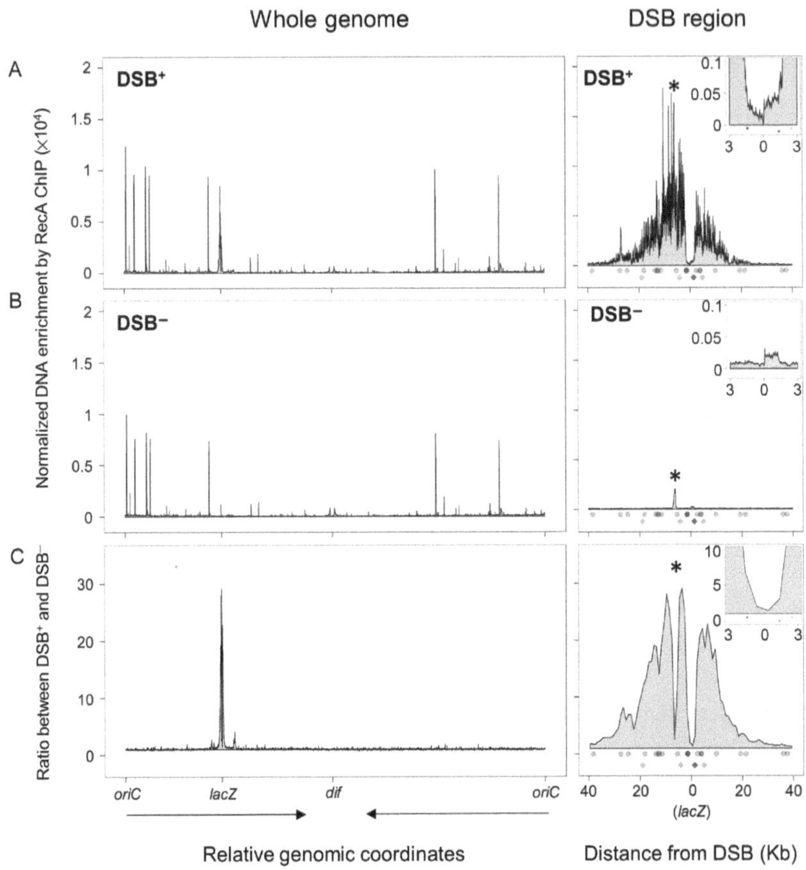

Fig. 1 RecA-ChIP-seq profiles for DSB$^+$ and DSB$^-$ cells and their ratio. RecA-ChIP-seq profiles for DSB$^+$ and DSB$^-$ cells, along the whole chromosome of *E. coli*, are shown in (A) *left* and (B) *left*, respectively. The data have been normalized for library size using the median-based method (step 7a in Section 2.1.3.3) and then binned in 1 kb non-overlapping bins. 80 kb regions containing the DSB locus (*lacZ*) are shown in (A) *right* and (B) *right*. The relative RecA-ChIP-seq enrichment in DSB$^+$ cells compared to DSB$^-$ is shown in (C) *left*, and an 80 kb region containing the DSB locus (*lacZ*) is shown in (C) *right*. Further zoomed in views of the 3 kb region containing the DSB locus (*lacZ*) are shown in the *insets* (A *right*, B *right*, and C *right*). The genomic positions of the origin of replication (*oriC*), the DSB locus (*lacZ*), and the *dif* sequence are shown on the *x*-axis. The highly expressed gentamicin gene is denoted with a *star* *. *Circles* (forward) and *diamonds* (reverse) depict the positions and orientations of Chi sequences. *Arrows* under the *x*-axis depict the direction of replication. Data are averages of two biological replicates.

experimental strain (Fig. 1A right, marked with *). On the other hand, the small signal origin–distal to the palindrome, visible in the control IP reads (Fig. 1B right, inset), does contribute to the DSB^+ reads (Fig. 1A right) and is removed by normalization (Fig. 1C right). However, this is at the expense of precise mapping of the position of RecA binding in relation to the Chi sites 1.5 kb from the DSB site. Alternatively, IP reads may be further normalized to input DNA reads (Fig. 2A and B). This normalizes for the level of background reads caused by the replication profile of the cell but also requires binning of data to avoid the problem of dividing by zeros and results in blurring of the signal (Fig. 2A right). Further normalization of the input-normalized IP for DSB^+ to input-normalized IP for a control strain provides a theoretically optimized outcome (Fig. 2C). However, this retains the problem of blurring because of binning and provides little advantage over the simpler normalizations presented in Figs. 1C and 2A. Furthermore, the DESeq comparison (Fig. 2D) does not provide any advantage over these simpler normalizations either.

2.2 Marker Frequency Analysis

2.2.1 Experimental Protocol

2.2.1.1 Equipment

- Shaker
- Benchtop refrigerated centrifuge
- PCR machine
- Gel Doc™ XR+ gel documentation system (Bio-Rad)
- Qubit fluorometer

2.2.1.2 Reagents and Buffers

- TE pH 8.0 (10 m*M* Tris–HCl, 1 m*M* EDTA)
- Tris-EDTA-sucrose solution pH 8.0 (50 m*M* Tris–HCl pH 8.0, 40 m*M* EDTA, 25% w/v sucrose)
- Lysozyme 20 mg/mL
- Triton lysis buffer (10% Triton X-100, 5 m*M* Tris–HCl pH 8.0, 62.5 m*M* EDTA)
- Proteinase K 50 mg/mL
- RNase A 10 mg/mL
- SDS 10% w/v
- Phenol:chloroform:isoamylalchol 25:24:1 (Acros Organics, Cat# 136112000)

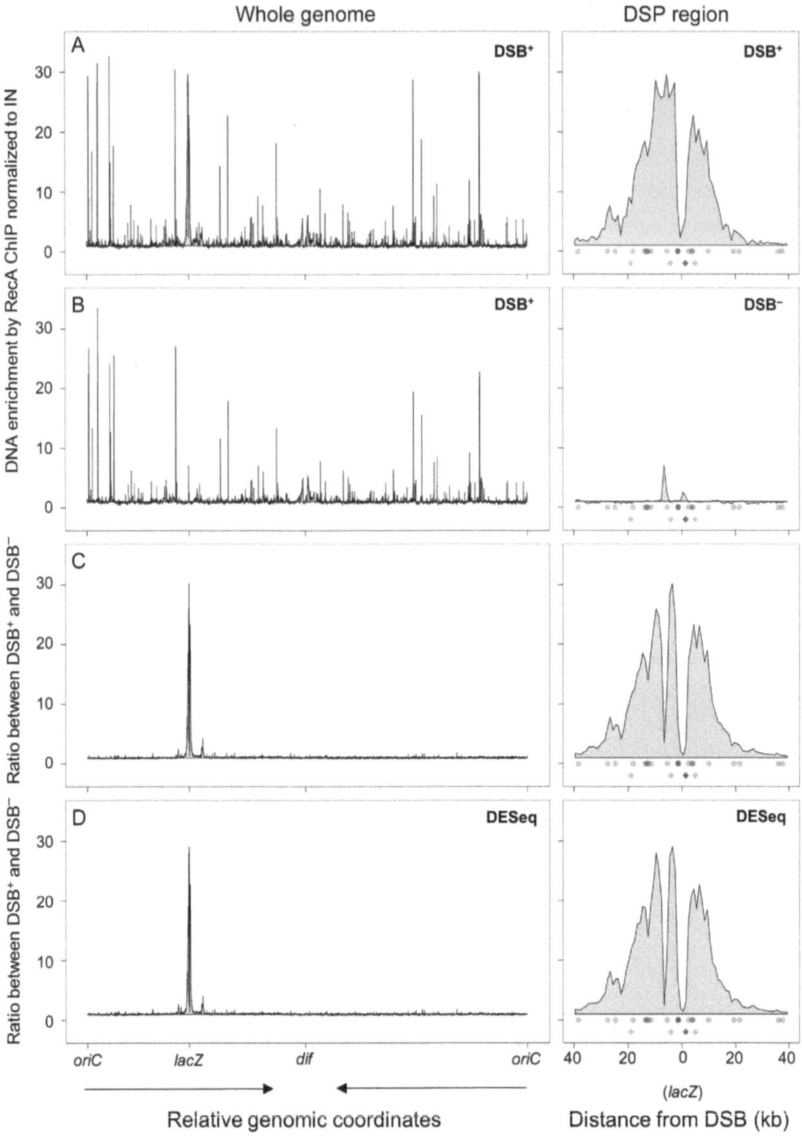

Fig. 2 Input (IN) normalized RecA-ChIP-seq profiles for DSB$^+$ and DSB$^-$ cells and their ratio compared to DESeq normalized ratio. Input normalized RecA-ChIP-seq profiles for DSB$^+$ and DSB$^-$ cells, along the whole chromosome of *E. coli*, are shown in (A) *left* and (B) *left*, respectively. 1 kb binned RecA-ChIP-seq count data were obtained using HTSeq. They were then normalized for library size using the median-based method (step 7a in Section 2.1.3.3) and further normalized by their respective input (IN) data using the same method. 80 kb regions containing the DSB locus (*lacZ*) are shown in (A) *right* and (B) *right*. The ratio of RecA-ChIP-seq count data from DSB$^+$ to DSB$^-$ cells obtained by the method in step 7a in Section 2.1.3.3 is shown in (C). The ratio of RecA-ChIP-seq IP count data from DSB$^+$ to DSB$^-$ cells, calculated using DESeq, is shown in (D). *Note*: the DESeq method does not make use of input data. The genomic positions of the origin of replication (*oriC*), the DSB locus (*lacZ*), and the *dif* sequence are shown on the *x*-axis. *Circles* (forward) and *diamonds* (reverse) depict the positions and orientations of Chi sequences. *Arrows* under the *x*-axis depict the direction of replication. Data are from two biological replicates.

- Phase separator, MaXtract High Density (Scientific lab supplies, Cat# 129073)
- Sodium acetate 3 *M* pH 5.2
- Ethanol, Laboratory grade
- RiboShredder RNase blend (Epicentre, Cat# RS12100)
- Nextera XT DNA Library Preparation Kit (Illumina, Cat# FC-131-1096)
- Nextera XT index Kit (Illumina, Cat# FC-131-1002)
- Qubit dsDNA HS assay (ThermoFisher, Cat# Q32854)
- 100 bp DNA ladder (New England Biolabs, Cat# N3231)
- Loading buffer (50 mM Tris–HCl pH 8, 40 mM EDTA, 40% w/v sucrose)
- MinElute PCR Purification kit (Qiagen, Cat# 28004)
- Agarose (Melford, Cat# MB1200)
- Nucleic acid stain, SafeView (NBS Biologicals, Cat# NBS-SV5)

2.2.1.3 gDNA Isolation

1. Transfer 25 mL of media containing exponentially grown *E. coli* cells into a 50-mL Falcon tube and place on ice immediately. In parallel, transfer 1 mL of stationary-phase culture into a 1.5-mL Eppendorf tube. Collect cells by centrifugation (2000 × *g* for 7 min at 4°C), then resuspend the pellet in 1 mL ice-cold TE. Transfer the cells into a 2-mL Eppendorf tube.
2. Centrifuge cells at 4000 × *g* for 5 min at 4°C and remove the supernatant.
3. Resuspend the pellet in 700 μL of Tris-EDTA-sucrose solution.
4. Add 100 μL of 20 mg/mL lysozyme and incubate at room temperature for 7 min.
5. Add 200 μL of 0.5 *M* EDTA and incubate at room temperature for 5 min.
6. Add following solutions in the following order: 700 μL of triton lysis buffer, 15 μL of 50 mg/mL proteinase K, 5 μL of 10 mg/mL RNase and 135 μL of 10% SDS.
7. Incubate the tube overnight at room temperate with gentle shaking.
8. In the morning, carry out a phenol-chloroform extraction. To do so, first transfer the sample from 2 mL Eppendorf tube in to a 15-mL Falcon tube and add equal volume (1.9 mL) of phenol:chloroform: isoamylalchol 25:24:1. To make the following steps easier, some phase separator can be added to the tube as well.

9. Vortex the tube briefly to mix the phases and centrifuge at 2000 × *g* for 20 min at room temperature.
10. Carefully collect the aqueous phase avoiding the organic phase and transfer it into a new 15-mL Eppendorf tube.
11. Repeat steps 8–10.
12. Aliquot the aqueous phase into four 1.5 mL Eppendorf tubes (400 μL into each). Add 40 μL (1/10 volume) of 3 *M* sodium acetate (pH 5.2) and 880 μL (2 volumes) of 100% ethanol.
13. Incubate the tubes at −20°C for an hour and then centrifuge at 4000 × *g* at 4°C for 15 min.
14. Discard the supernatant and resuspend the pellet with 500 μL of ice-cold 70% ethanol. Wash the pellet by inverting the tube 3–4 times.
15. Centrifuge the tube at 4000 × *g* at room temperature for 5 min.
16. Remove all ethanol and air dry the pellet.
17. Once all the residual ethanol is evaporated add 100 μL of MilliQ H_2O and rehydrate the DNA at 4°C overnight.
18. Collect the DNA from all four Eppendorf tubes.
19. Add 3 μL of RiboShredder RNase blend into the phenol extracted DNA and incubate at 37°C for 3 h.
20. Remove RiboShredder by phenol-chloroform extraction adjusting the starting volume to 400 μL (steps 8–17).

2.2.1.4 Library Preparation

The last step of sample preparation for MFA analysis is preparing the libraries. Here, we describe how to prepare a single library prep using Nextera XT DNA Library Preparation Kit using the Nextera XT index Kit. The Nextera technology allows the DNA to be simultaneously fragmented and tagged with the sequencing adapters in a single enzymatic reaction. As this reaction, called tagmentation, relies on the activity of modified transposase, the initial amount and concentration of the gDNA are very important. Briefly, the workflow consists of five steps: tagmentation of the gDNA, amplification of libraries, cleaning up the amplified libraries, normalizing the libraries, and finally pooling the libraries. The normalization step of this protocol generates libraries containing single-stranded DNA, which is not an ideal material for quality control. To overcome this problem, we describe a modified version of this protocol where the amplified library is purified on an agarose gel to select a specific size range of double-stranded DNA for cluster generation. This step also removes any adapter dimers.

21. Quantify the concentration of isolated DNA using a fluorometric method (for example, Qubit dsDNA HS assay following the manufacturer's instructions). Adjust the concentration of the starting material to 0.2 ng/μL in either MilliQ H_2O or 10 m*M* Tris–HCl (pH 7.5–8.5).
22. In a PCR tube add 10 μL of TD (Tagment DNA) buffer and 5 μL of normalized gDNA and mix well by pipetting.
23. Add 5 μL of ATM (Amplicon Tagment Mix) and mix well by pipetting. Centrifuge at 280 × *g* at room temperature for a minute.
24. Place the tube in a thermal cycler and run the following program:
    Preheat the lid to 55°C
    55°C for 5 min
    Hold at 10°C

Note: during this reaction random tagmentation of the DNA is going to occur and maintaining the correct temperature is very important.

25. Once the thermal cycler reaches 10°C, immediately remove the sample from the machine and add 5 μL of NT (Neutralize Tagment Buffer). Pipette to mix and Centrifuge at 280 × *g* at the room temperature for a minute.
26. Incubate at room temperature for 5 min to neutralize further tagmentation.
27. To amplify the libraries, add 5 μL of Index 1 (i7) adapter and 5 μL of Index 2 (i5) adapter into the PCR tube.
28. Add 15 μL of NPM (Nextera PCR Master Mix) and mix well by pipetting and centrifuge at 280 × *g* at room temperature for 1 min.
29. Place the tube in a thermal cycler and run the following program:

Preheat the lid

Temp	Time	Number of Cycles
72°C	3 min	1
95°C	30 s	1
95°C 55°C 72°C	10 s 30 s 30 s	12
72°C	5 min	1
10°C	∞	

30. Once the reaction is done, clean up the samples using MinElute PCR Purification kit (Qiagen) following the manufacturer's instructions.

31. Elute the DNA in 10 μL of MilliQ H_2O.
32. Prepare 2% (w/v) agarose in 1× TAE containing 0.5 μg/mL of SafeView and cast a gel.
33. Add 3 μL of loading buffer to each sample and mix well.
34. Load 4 μL of 100 bp DNA ladder into first and the last lane of the gel. Leave one lane empty and load the sample into the next well. You can load more samples into the same gel but make sure to leave one lane empty between the samples.
35. Run the gel at 5 V/cm for 90 min.
36. Visualize the gel using Gel Doc™ XR+ gel documentation system with UV/blue conversion screen. Excise the gel region containing DNA fragments between 200 and 500 bp using a sterile scalpel. Carry out isolation of DNA from gel with MinElute Gel Extraction Kit following the manufactures instructions. However, make sure not to warm up slice of the gel above 37°C at any stage.
37. Elute the DNA in 20 μL EB with 0.01% Tween-20.

2.2.1.5 An Alternative Sonication-Based Method to Isolate DNA and Prepare Libraries

Alternatively, gDNA isolation can be carried out by sonication of cells after the induction of DSBs. To achieve this, gDNA isolation and library preparation can be done using the ChIP-seq protocol described in Section 2.1.2. Briefly, harvest by pelleting cells either collected directly from exponentially grown cultures or after cross-linking as described in Section 2.1.1.3. Then follow the steps of ChIP protocol until step 6. Following sonication add fresh 250 μL of ChIP buffer containing protease inhibitors to the sonicated sample and mix well by pipetting up and down several times. Then proceed directly to step 13 and add 500 μL of TE buffer with 1% SDS and 20 mg/mL RNase A solution and transfer into PCR tubes. Continue the sample preparation from step 14, following the protocol until the end of library preparation.

2.2.2 DNA Sequencing

The sequencing of MFA samples was carried out by Edinburgh Genomics (Edinburgh, UK) using either the HiSeq 4000 (for samples prepared using phenol-chloroform method) or the NovaSeq 6000 (for the rest of the samples) instrument. The reads were 50 bp paired-ended regardless of which platform was used. See Section 2.1.2 for more details of the principle of the sequencing reaction.

2.2.3 Data Analysis

Here, we describe how to implement data analysis in a command-line environment using a Linux-based operating system and statistical environment R (https://www.r-project.org/). The command line codes and file formats are given in "`courier new`" font. R scripts are available at https://github.com/AMMHasan/MFA (data analysis) and https://github.com/AMMHasan/plot_MFA (data visualization).

2.2.3.1 Software

- BWA (version 0.7.5a-r405; Li & Durbin, 2010)
- FastQC (version 0.11.2, http://www.bioinformatics.babraham.ac.uk/projects/fastqc/)
- Picard (version 1.95, http://broadinstitute.github.io/picard)
- Rstudio (version 1.1.442, https://www.rstudio.com/)
- SAMtools (version 1.3.1; Li, 2011)
- seqinR (R package, version 3.4-5; Charif & Lobry, 2007)
- ggplot2 (Bioconducter package, version 3.0.0; Wickham, 2009)

2.2.3.2 Data

The data files used for MFA have been deposited in the NCBI GEO database under accession number is GSE117952.

Data type: 50 bp paired-end Illumina reads.

2.2.3.3 Workflow

1. Check the quality of demultiplexed raw Illumina reads as described in step 1 in Section 2.1.3.3.
2. FM-index the reference genome (only once) prior to mapping the reads using BWA:
```
bwa index ref_genome.fasta
```
3. Using the BWA-MEM algorithm, map the paired-end reads to the FM-indexed reference genome. The following command will generate an output alignment file in `.sam` format:
```
bwa mem -M -t 10 ref_genome.fasta Y_1.fastq Y_2.fastq > Y.sam
```

Explanation of parameters:

`-M`	marks shorter split hits as secondary. This parameter is needed for Picard compatibility in step 5
`-t`	number of threads

4. Convert the SAM file to its binary version BAM using SAMtools as described in step 4 in Section 2.1.3.3.
5. Remove read duplicates as described in step 5 in Section 2.1.3.3.
6. Calculate the depth of coverage at each of the 4.6 million nucleotides using SAMtools:

```
samtools depth -d 1000000 deduped_Y.bam > depth_Y.txt
```

Explanation of parameters:

`-d` or `--max-depth INT`	refers to the maximum number of reads found at a position. The default value is 8000. For our data, the maximum depth had been arbitrarily increased to 1000000

7. Calculate MFA profile from the depth of coverage (an R script is provided at https://github.com/AMMHasan/MFA) following the steps below:
   a. Normalize the depth of coverage for the exponentially growing culture by its total depth. This step normalizes the data for the library size.
   b. Do the same to the depth of coverage for the stationary-phase culture (if available).
   c. Divide the library size-normalized data from the exponentially growing culture (step 7a) by the library size-normalized data from the stationary-phase culture (step 7b). This step eliminates any sequencing bias.
   d. For visualization purposes, calculate 1 kb nonoverlapping bins of the normalized data from step 7c. Also, calculate a nonparametric smoothing function (LOESS) on the 1 kb nonoverlapping binned data.
   e. Plot and compare 1 kb binned data and the LOESS data from different strains. An R script for plotting can be found on https://github.com/AMMHasan/plot_MFA.

2.2.4 Lessons From Data

MFA is a powerful tool for the analysis of the replication profile of DNA. The slope of the replication profile (the frequency (N_x) of reads at genomic position (m_x)) can be used to calculate the time, C, taken to replicate the genome (if the generation time, τ, of the growing culture from which the DNA was prepared is known) from the equation:

$$d(\log_2 N_x)/dm_x = -C/\tau \qquad \text{(Skovgaard et al., 2011)}$$

If an event, such as DSBR occurs at a specific chromosomal location, its effect on the replication profile of the cell can be determined. Clearly

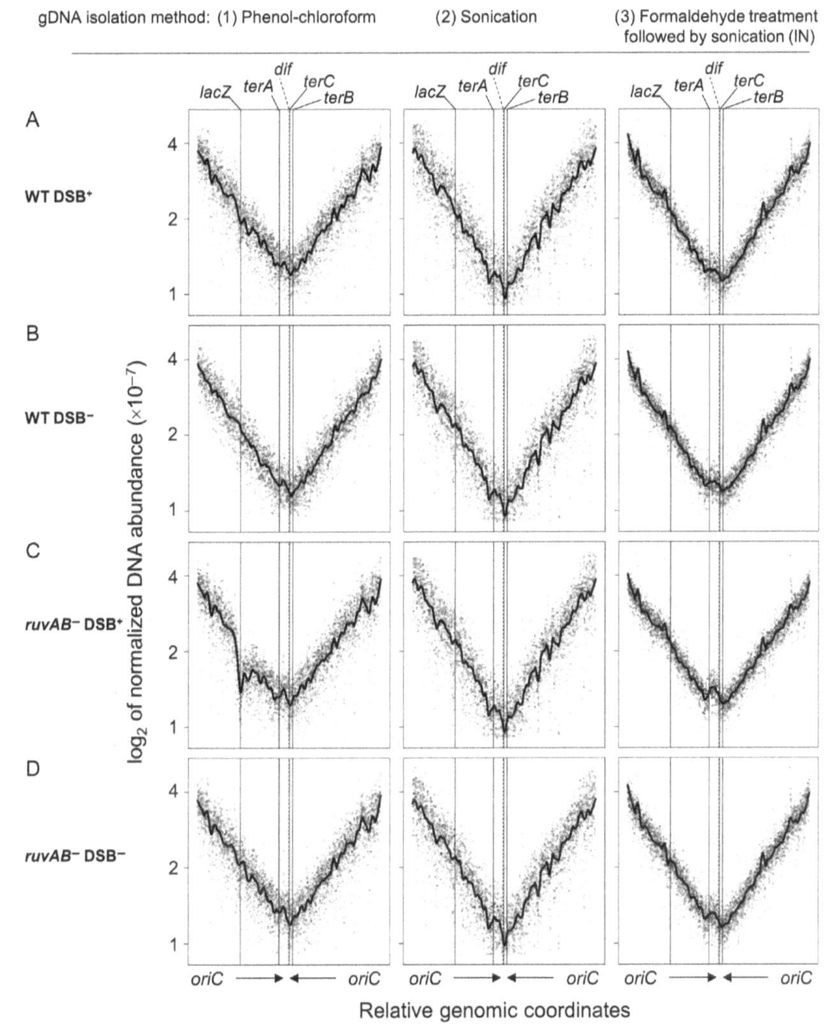

Fig. 3 Comparison of three different methods to obtain MFA profiles of wild-type and *ruvAB*$^-$ cells as a function of DSB induction. Individual depth of coverage data (wild-type DSB$^+$, A; wild-type DSB$^-$, B; *ruvAB*$^-$ DSB$^+$, C; and *ruvAB*$^-$ DSB$^-$, D) was normalized to the total depth prior to binning in 1 kb nonoverlapping bins and plotted on a $\log_2$ *y*-axis. *Solid profile lines* depict the LOESS smoothed data. The DSB locus (*lacZ*), the replication termination sequences (*terA*, *terC*, and *terB*), and the *dif* sequence are denoted by *solid* and *dashed lines*. *Arrows* under the *x*-axis depict the direction of replication. Data are averages of two biological replicates.

DSBR in a wild-type cell has little effect on the replication profile, irrespective of which of the three methods of preparing gDNA fragments for sequencing is used (Fig. 3A and B). However, this is not the case for the *ruvAB* mutant, which is defective in branch migration and resolution of

Holliday junctions (HJs) (Fig. 3C and D). Here, it can be seen that the traditional phenol-chloroform method for preparation of gDNA followed by library preparation with the Nextera XT kit results in a loss of reads in the ~500 kb region surrounding the DSB site and this loss is not observed if the gDNA is prepared by the sonication method, with or without formaldehyde treatment, followed by library preparation with the NEBNext® ChIP-Seq Library Kit (Fig. 3C and D). The observation that gDNA reads are not significantly lost from a *ruvAB* mutant strain when the DNA is prepared by either sonication method, or when assayed for by Southern hybridization (Mawer & Leach, 2014), argues that the loss observed following the phenol-chloroform method is primarily caused by the presence of unresolved HJs in the DNA.

This is confirmed by taking the ratio of gDNA sequencing reads in the presence of the palindrome (DSB^+) to the absence of the palindrome (DSB^-) where it can clearly be seen that the phenol-chloroform method results in a loss of sequencing reads from the *ruvAB* mutant (Fig. 4). This

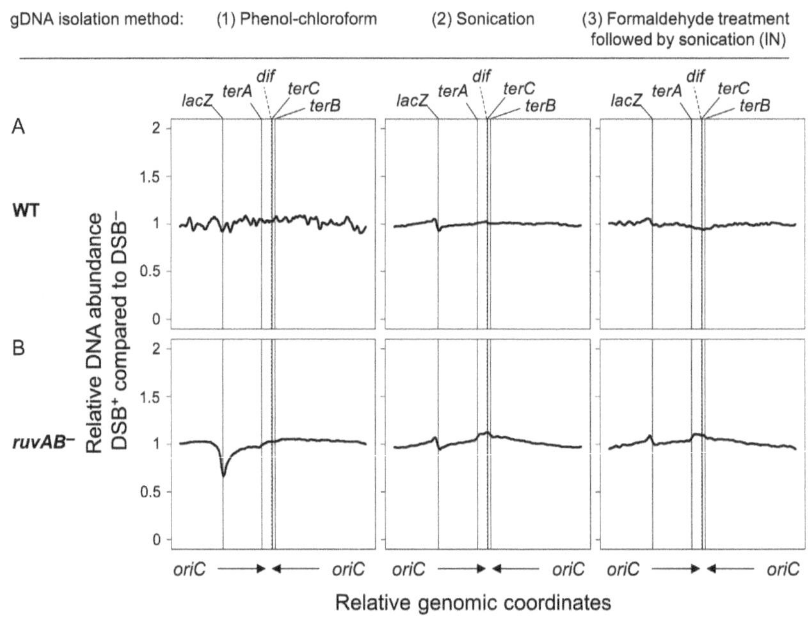

Fig. 4 Comparison of three different methods to obtain MFA profiles using ratios of wild-type and $ruvAB^-$ cells as a function of DSB induction. The data shown in Fig. 3 were used to obtain the ratios (wild type, A and $ruvAB^-$, B). The DSB locus (*lacZ*), the replication termination sequences (*terA, terC,* and *terB*), and the *dif* sequence are denoted by *solid* and *dashed lines*. *Arrows* under the *x*-axis depict the direction of replication. Data are averages of two biological replicates.

loss of sequencing reads is not observed in the wild-type strain. By contrast, the sonication method reveals a small drop in reads between the origin proximal and origin distal side of the palindrome that is independent of the *ruvAB* genotype. Interestingly, together these data argue strongly that replication forks pass through HJs with little or no impediment. Replication through HJs to generate crossover products of recombination has previously been predicted (Wardrope, Okely, & Leach, 2009).

3. CONCLUSIONS

Methods of genomic analysis are continuously being updated, from the isolation of DNA, via its sequencing to the analysis of data. This rapidly moving field means that it is critical to share expertise so that new methods can build upon existing protocols. In this chapter, we describe the methods we are currently using for RecA-ChIP and MFA. In the course of preparing this chapter, we discovered that preparation of MFA libraries by sonication using a method adapted from our ChIP input protocol avoided the loss of DNA caused by the presence of HJs in a *ruvAB* mutant. We have incorporated this alternative method for preparing MFA libraries so that future studies do not misinterpret loss of sequencing reads in the vicinity of a DSB site for evidence of resection unless they can distinguish between alternative reasons for DNA sequencing read loss. As the field moves forward, greater understanding will inform our methods of analysis and increase the power of genomic sequencing in the investigation of DSBR.

ACKNOWLEDGMENTS

We thank Edinburgh Genomics for whole genome sequencing. Edinburgh Genomics is partly supported through core grants from Natural Environment Research Council (UK) [R8/H10/56]; Medical Research Council (UK) [MR/K001744/1]; and Biotechnology and Biological Sciences Research Council (UK) [BB/J004243/1]. We thank Dr. Charlie Cockram, Dr. Elise Darmon, and Dr. Martin White for initiating these genomic studies. D.R.F.L. is supported by Grant MR/M019160/1 from the Medical Research Council (UK).

REFERENCES

Anders, S., & Huber, W. (2010). Differential expression analysis for sequence count data. *Genome Biology*, *11*(10), R106. https://doi.org/10.1186/gb-2010-11-10-r106.

Anders, S., Pyl, P. T., & Huber, W. (2015). HTSeq—A Python framework to work with high-throughput sequencing data. *Bioinformatics*, *31*(2), 166–169. https://doi.org/10.1093/bioinformatics/btu638.

Anderson, D. G., & Kowalczykowski, S. C. (1997). The translocating RecBCD enzyme stimulates recombination by directing RecA protein onto ssDNA in a chi-regulated manner. *Cell*, *90*(1), 77–86.

Ansorge, W. J. (2009). Next-generation DNA sequencing techniques. *New Biotechnology, 25*(4), 195–203. https://doi.org/10.1016/j.nbt.2008.12.009.

Azeroglu, B., Mawer, J. S., Cockram, C. A., White, M. A., Hasan, A. M., Filatenkova, M., et al. (2016). RecG directs DNA synthesis during double-strand break repair. *PLoS Genetics, 12*(2), e1005799. https://doi.org/10.1371/journal.pgen.1005799.

Badrinarayanan, A., Le, T. B. K., Spille, J. H., Cisse, I. I., & Laub, M. T. (2017). Global analysis of double-strand break processing reveals in vivo properties of the helicase-nuclease complex AddAB. *PLoS Genetics, 13*(5), e1006783. https://doi.org/10.1371/journal.pgen.1006783.

Barski, A., Cuddapah, S., Cui, K., Roh, T. Y., Schones, D. E., Wang, Z., et al. (2007). High-resolution profiling of histone methylations in the human genome. *Cell, 129*(4), 823–837. https://doi.org/10.1016/j.cell.2007.05.009.

Bell, J. C., & Kowalczykowski, S. C. (2016). RecA: Regulation and mechanism of a molecular search engine. *Trends in Biochemical Sciences, 41*(6), 491–507. https://doi.org/10.1016/j.tibs.2016.04.002.

Bird, R. E., Louarn, J., Martuscelli, J., & Caro, L. (1972). Origin and sequence of chromosome replication in *Escherichia coli*. *Journal of Molecular Biology, 70*(3), 549–566.

Blat, Y., & Kleckner, N. (1999). Cohesins bind to preferential sites along yeast chromosome III, with differential regulation along arms versus the centric region. *Cell, 98*(2), 249–259.

Charif, D., & Lobry, J. R. (2007). SeqinR 1.0-2: A contributed package to the R project for statistical computing devoted to biological sequences retrieval and analysis. In U. Bastolla, M. Porto, H. E. Roman, & M. Vendruscolo (Eds.), *Structural approaches to sequence evolution: Molecules, networks, populations* (pp. 207–232). Berlin, Heidelberg: Springer Berlin Heidelberg.

Cockram, C. A., Filatenkova, M., Danos, V., El Karoui, M., & Leach, D. R. (2015). Quantitative genomic analysis of RecA protein binding during DNA double-strand break repair reveals RecBCD action *in vivo*. *Proceedings of the National Academy of Sciences of the United States of America, 112*(34), E4735–E4742. https://doi.org/10.1073/pnas.1424269112.

Courcelle, J., Wendel, B. M., Livingstone, D. D., & Courcelle, C. T. (2015). RecBCD is required to complete chromosomal replication: Implications for double-strand break frequencies and repair mechanisms. *DNA Repair (Amst), 32*, 86–95. https://doi.org/10.1016/j.dnarep.2015.04.018.

Dillingham, M. S., & Kowalczykowski, S. C. (2008). RecBCD enzyme and the repair of double-stranded DNA breaks. *Microbiology and Molecular Biology Reviews, 72*(4), 642–671. Table of Contents. https://doi.org/10.1128/MMBR.00020-08.

Dimude, J. U., Midgley-Smith, S. L., & Rudolph, C. J. (2018). Replication-transcription conflicts trigger extensive DNA degradation in Escherichia coli cells lacking RecBCD. *DNA Repair, 70*, 37–48. https://doi.org/10.1016/j.dnarep.2018.08.002.

Dimude, J. U., Stockum, A., Midgley-Smith, S. L., Upton, A. L., Foster, H. A., Khan, A., et al. (2015). The consequences of replicating in the wrong orientation: Bacterial chromosome duplication without an active replication origin. *MBio, 6*(6), e01294-01215. https://doi.org/10.1128/mBio.01294-15.

Eykelenboom, J. K., Blackwood, J. K., Okely, E., & Leach, D. R. (2008). SbcCD causes a double-strand break at a DNA palindrome in the *Escherichia coli* chromosome. *Molecular Cell, 29*(5), 644–651. https://doi.org/10.1016/j.molcel.2007.12.020.

Gilmour, D. S., & Lis, J. T. (1984). Detecting protein-DNA interactions in vivo: Distribution of RNA polymerase on specific bacterial genes. *Proceedings of the National Academy of Sciences of the United States of America, 81*(14), 4275–4279.

Jonasson, J. (1973). Evidence for bidirectional chromosome replication in *Escherichia coli* C based on marker-frequency analysis by DNA-DNA hybridization with P2 and lambda prophages. *Molecular & General Genetics, 120*(1), 68–90.

Khan, S. R., Mahaseth, T., Kouzminova, E. A., Cronan, G. E., & Kuzminov, A. (2016). Static and dynamic factors limit chromosomal replication complexity in *Escherichia coli*, avoiding dangers of runaway overreplication. *Genetics*, *202*(3), 945–960. https://doi.org/10.1534/genetics.115.184697.

Kouzminova, E. A., & Kuzminov, A. (2008). Patterns of chromosomal fragmentation due to uracil-DNA incorporation reveal a novel mechanism of replication-dependent double-stranded breaks. *Molecular Microbiology*, *68*(1), 202–215. https://doi.org/10.1111/j.1365-2958.2008.06149.x.

Kouzminova, E. A., & Kuzminov, A. (2012). Chromosome demise in the wake of ligase-deficient replication. *Molecular Microbiology*, *84*(6), 1079–1096. https://doi.org/10.1111/j.1365-2958.2012.08076.x.

Kuong, K. J., & Kuzminov, A. (2012). Disintegration of nascent replication bubbles during thymine starvation triggers RecA- and RecBCD-dependent replication origin destruction. *The Journal of Biological Chemistry*, *287*(28), 23958–23970. https://doi.org/10.1074/jbc.M112.359687.

Langmead, B., & Salzberg, S. L. (2012). Fast gapped-read alignment with Bowtie 2. *Nature Methods*, *9*(4), 357–359. https://doi.org/10.1038/nmeth.1923.

Li, H. (2011). A statistical framework for SNP calling, mutation discovery, association mapping and population genetical parameter estimation from sequencing data. *Bioinformatics*, *27*(21), 2987–2993. https://doi.org/10.1093/bioinformatics/btr509.

Li, H., & Durbin, R. (2010). Fast and accurate long-read alignment with Burrows-Wheeler transform. *Bioinformatics*, *26*(5), 589–595. https://doi.org/10.1093/bioinformatics/btp698.

Maduike, N. Z., Tehranchi, A. K., Wang, J. D., & Kreuzer, K. N. (2014). Replication of the *Escherichia coli* chromosome in RNase HI-deficient cells: Multiple initiation regions and fork dynamics. *Molecular Microbiology*, *91*(1), 39–56. https://doi.org/10.1111/mmi.12440.

Mardis, E. R. (2008). Next-generation DNA sequencing methods. *Annual Review of Genomics and Human Genetics*, *9*, 387–402. https://doi.org/10.1146/annurev.genom.9.081307.164359.

Masters, M., & Broda, P. (1971). Evidence for the bidirectional replications of the *Escherichia coli* chromosome. *Nature: New Biology*, *232*(31), 137–140.

Mawer, J. S., & Leach, D. R. (2014). Branch migration prevents DNA loss during double-strand break repair. *PLoS Genetics*, *10*(8), e1004485. https://doi.org/10.1371/journal.pgen.1004485.

Midgley-Smith, S. L., Dimude, J. U., Taylor, T., Forrester, N. M., Upton, A. L., Lloyd, R. G., et al. (2018). Chromosomal over-replication in *Escherichia coli recG* cells is triggered by replication fork fusion and amplified if replichore symmetry is disturbed. *Nucleic Acids Research*. https://doi.org/10.1093/nar/gky566.

Muller, C. A., Hawkins, M., Retkute, R., Malla, S., Wilson, R., Blythe, M. J., et al. (2014). The dynamics of genome replication using deep sequencing. *Nucleic Acids Research*, *42*(1), e3. https://doi.org/10.1093/nar/gkt878.

Muller, C. A., & Nieduszynski, C. A. (2012). Conservation of replication timing reveals global and local regulation of replication origin activity. *Genome Research*, *22*(10), 1953–1962. https://doi.org/10.1101/gr.139477.112.

Nicol, J. W., Helt, G. A., Blanchard, S. G., Jr., Raja, A., & Loraine, A. E. (2009). The integrated genome browser: Free software for distribution and exploration of genome-scale datasets. *Bioinformatics*, *25*(20), 2730–2731. https://doi.org/10.1093/bioinformatics/btp472.

Quinlan, A. R., & Hall, I. M. (2010). BEDTools: A flexible suite of utilities for comparing genomic features. *Bioinformatics*, *26*(6), 841–842. https://doi.org/10.1093/bioinformatics/btq033.

Raghuraman, M. K., Winzeler, E. A., Collingwood, D., Hunt, S., Wodicka, L., Conway, A., et al. (2001). Replication dynamics of the yeast genome. *Science*, *294*(5540), 115–121. https://doi.org/10.1126/science.294.5540.115.

Robertson, G., Hirst, M., Bainbridge, M., Bilenky, M., Zhao, Y., Zeng, T., et al. (2007). Genome-wide profiles of STAT1 DNA association using chromatin immunoprecipitation and massively parallel sequencing. *Nature Methods*, *4*(8), 651–657. https://doi.org/10.1038/nmeth1068.

Robinson, J. T., Thorvaldsdottir, H., Winckler, W., Guttman, M., Lander, E. S., Getz, G., et al. (2011). Integrative genomics viewer. *Nature Biotechnology*, *29*(1), 24–26. https://doi.org/10.1038/nbt.1754.

Rudolph, C. J., Upton, A. L., Stockum, A., Nieduszynski, C. A., & Lloyd, R. G. (2013). Avoiding chromosome pathology when replication forks collide. *Nature*, *500*(7464), 608–611. https://doi.org/10.1038/nature12312.

Sinha, A. K., Durand, A., Desfontaines, J. M., Iurchenko, I., Auger, H., Leach, D. R. F., et al. (2017). Division-induced DNA double strand breaks in the chromosome terminus region of *Escherichia coli* lacking RecBCD DNA repair enzyme. *PLoS Genetics*, *13*(10), e1006895. https://doi.org/10.1371/journal.pgen.1006895.

Sinha, A. K., Possoz, C., Durand, A., Desfontaines, J. M., Barre, F. X., Leach, D. R. F., et al. (2018). Broken replication forks trigger heritable DNA breaks in the terminus of a circular chromosome. *PLoS Genetics*, *14*(3), e1007256. https://doi.org/10.1371/journal.pgen.1007256.

Sivaramakrishnan, P., Sepulveda, L. A., Halliday, J. A., Liu, J., Nunez, M. A. B., Golding, I., et al. (2017). The transcription fidelity factor GreA impedes DNA break repair. *Nature*, *550*(7675), 214–218. https://doi.org/10.1038/nature23907.

Skovgaard, O., Bak, M., Lobner-Olesen, A., & Tommerup, N. (2011). Genome-wide detection of chromosomal rearrangements, indels, and mutations in circular chromosomes by short read sequencing. *Genome Research*, *21*(8), 1388–1393. https://doi.org/10.1101/gr.117416.110.

Smith, G. R. (2012). How RecBCD enzyme and Chi promote DNA break repair and recombination: A molecular biologist's view. *Microbiology and Molecular Biology Reviews*, *76*(2), 217–228. https://doi.org/10.1128/MMBR.05026-11.

Solomon, M. J., Larsen, P. L., & Varshavsky, A. (1988). Mapping protein-DNA interactions *in vivo* with formaldehyde: Evidence that histone H4 is retained on a highly transcribed gene. *Cell*, *53*(6), 937–947.

Sueoka, N., & Yoshikawa, H. (1965). The chromosome of *Bacillus subtilis*. I. Theory of marker frequency analysis. *Genetics*, *52*(4), 747–757.

Wardrope, L., Okely, E., & Leach, D. (2009). Resolution of joint molecules by RuvABC and RecG following cleavage of the *Escherichia coli* chromosome by EcoKI. *PLoS One*, *4*(8), e6542. https://doi.org/10.1371/journal.pone.0006542.

Watson, J. D., & Crick, F. H. (1953). Molecular structure of nucleic acids: A structure for deoxyribose nucleic acid. *Nature*, *171*(4356), 737–738.

Wendel, B. M., Cole, J. M., Courcelle, C. T., & Courcelle, J. (2018). SbcC-SbcD and ExoI process convergent forks to complete chromosome replication. *Proceedings of the National Academy of Sciences of the United States of America*, *115*(2), 349–354. https://doi.org/10.1073/pnas.1715960114.

Wendel, B. M., Courcelle, C. T., & Courcelle, J. (2014). Completion of DNA replication in *Escherichia coli*. *Proceedings of the National Academy of Sciences of the United States of America*, *111*(46), 16454–16459. https://doi.org/10.1073/pnas.1415025111.

White, M. A., Azeroglu, B., Lopez-Vernaza, M. A., Hasan, A. M. M., & Leach, D. R. F. (2018). RecBCD coordinates repair of two ends at a DNA double-strand break, preventing aberrant chromosome amplification. *Nucleic Acids Research*, *46*(13), 6670–6682. https://doi.org/10.1093/nar/gky463.

Wickham, H. (2009). *ggplot2: Elegant graphics for data analysis*. New York, NY: Springer New York.

Yoshikawa, H., & Sueoka, N. (1963). Sequential replication of *Bacillus subtilis* chromosome. I. Comparison of marker frequencies in exponential and stationary growth phases. *Proceedings of the National Academy of Sciences of the United States of America*, *49*, 559–566.

Kolibri 360 GmbH
Ferdinand-Jühlke-Straße 7
99095 Erfurt, Deutschland
produktsicherheit@kolibri360.de